INTERNATIONAL COMPENDIUM OF COASTAL ENGINEERING

INTERNATIONAL COMPENDIUM OF COASTAL ENGINEERING

edited by

Shinji Sato
The University of Tokyo, Japan

Masahiko Isobe
Kochi University of Technology, Japan

World Scientific

NEW JERSEY • LONDON • SINGAPORE • BEIJING • SHANGHAI • HONG KONG • TAIPEI • CHENNAI

Published by

World Scientific Publishing Co. Pte. Ltd.

5 Toh Tuck Link, Singapore 596224

USA office: 27 Warren Street, Suite 401-402, Hackensack, NJ 07601

UK office: 57 Shelton Street, Covent Garden, London WC2H 9HE

British Library Cataloguing-in-Publication Data
A catalogue record for this book is available from the British Library.

INTERNATIONAL COMPENDIUM OF COASTAL ENGINEERING

ISBN 978-981-4449-42-7

Typeset by Stallion Press
Email: enquiries@stallionpress.com

Printed in Singapore by B & Jo Enterprise Pte Ltd

Preface

Science and technology in the coastal zone has a long history. However, the term "Coastal Engineering" did not appear until the First Conference on Coastal Engineering held in 1950 at Long Beach, California, USA. In the preface of its proceedings, Morrough P. O'Brien stated that "Coastal Engineering" is not a new or separate branch of engineering, but the design of coastal works does involve many criteria which are foreign to other phases of civil engineering.

The aim of this book is to provide a comprehensive overview of Coastal Engineering from basic theory to engineering practice. Each chapter deals with an important topic in the field of coastal engineering. The topics are of recent deep concern all over the world motivated by the 2004 Indian Ocean Tsunami, 2005 Hurricane Katrina, 2011 Tohoku Earthquake Tsunami and other natural disasters.

For proper coastal zone management, a broad range of knowledge is necessary. This book provides a basic understanding of the theories behind the diverse natural phenomena within the coastal areas, such as waves, tsunamis and sediment transport. The book also introduces various coastal conservation technologies such as costal structures and beach nourishment. Finally, coastal zone management practices in the USA, Europe, and Japan are introduced. Each chapter is self-standing and readers can begin from any topic depending on their interest

In organizing the authors of this book, the contributions of Dr. Kiyoshi Horikawa, a member of The Japan Academy, have been indispensable. He is the founder of Coastal Engineering in Japan, and has been involved in world-wide research and development within this field for more than half a century. It is because of Dr. Horikawa's established reputation in the costal engineering field, that the editors were able to secure leading authors for this book. We are very grateful for Dr. Horikawa's contributions and would like to dedicate this book to him. The authors of this book are thus world-wide authorities in the field of coastal engineering. Therefore, we believe this book is especially beneficial to coastal engineering professionals, researchers and students.

Shinji Sato
Masahiko Isobe

Contents

Chapter 1

Introduction

Robert L. Wiegel
Professor Emeritus,
University of California, USA

1. Introduction

In its broadest sense, coastal engineering started in antiquity. The name "coastal engineering", however, was probably not used (at least in print) until 1950 — at the Institute on Coastal Engineering held in Long Beach, California, planned by Morrough P. O'Brien and J. W. Johnson, organized by the University of California Extension. The purpose of the program was to help engineers by summarizing the state-of-the-art and science, related to planning and designing coastal works; each of the 35 lectures was by invitation. The written versions were published as the *Proceedings of First Conference on Coastal Engineering*, edited by J. W. Johnson (1951). In the proceeding's preface, O'Brien wrote:

> "A word about the term "Coastal Engineering" is perhaps in order here. It is not a new or separate branch of engineering and there is no implication intended that a new breed of engineer, and a new society is in the making. Coastal Engineering is primarily a branch of Civil Engineering which leans heavily on the sciences of oceanography, meteorology, fluid mechanics, electronics, structural mechanics, and others (such as geology — RLW). However, it is also true that the design of coastal works does involve many criteria which are foreign to other phases of civil engineering and novices in this field should proceed with caution."

Prior to, then there were coastal activities, works, studies and publications on aspects of coastal engineering: some in ancient times such as ports/harbors, lighthouses, fishing, salt water fishponds (piscina), evaporative salt ponds; and some mostly within the past century and a half, such as recreation. People have always lived along the coast owing in part to convenience in transportation (ocean and river) and relatively mild climate, its use as a source of food, and the ease in disposing of wastes (even though sometimes a health hazard). Its use has increased dramatically in the past century owing to population growth, the development of highway and rail systems, low-cost airlines, and great changes in the economic/social environment. Much had been learned in the past about operating in the coastal physical environment; about some characteristics of waves, winds, currents and tides; about the shoaling of harbors; about subsidence/uplift/earthquakes. But, it was not until World War II that major advances were made in analysis techniques and development and use of hydraulic laboratories — the combination of science and engineering — together with advances made in instruments and the increased measurement, analysis and data presentation of components of environmental processes (e.g. Wiegel and Saville, Jr., 1996). Most recent has been the use of digital computers in analysis, data handling and evaluation, and design.

The present book has two purposes: a celebration of coastal engineering for Professor Kiyoshi Horikawa, and looking towards the future beyond its golden anniversary. Professor Horikawa's professional career has spanned nearly the entire half century of modern coastal engineering. He has been an important leader, teacher and researcher in the discipline; his large number of papers on diverse subjects, and his book (1978) are evidences of this. It is appropriate to have this celebration. I first worked with Professor Horikawa when he was a visiting assistant professor at the University of California at Berkeley, November 1957–April 1959 (Horikawa and Wiegel, 1959) and have met with him many times since then, often at the ICCEs. We have worked together, first at Berkeley, then on two US Japan Cooperative Science Program seminars: coastal engineering in 1964, and tsunamis in 1965. We wrote a paper about the first one for *Civil Engineering* (December 1964): "Coastal Engineering in Japan". I value my 40 yrs of association with Professor Horikawa, and I am pleased to participate in honoring him.

The second purpose — the future of coastal engineering beyond its golden anniversary — is difficult. We have never been good in foreseeing the future, and when we have done so correctly, it has often been by luck. I will make a few predictions, but mostly will write what I think should be done in the future, as well as now. Certainly, the future will include more cooperative efforts to live better with the environment, to minimize harmful effects, and to plan and implement more regional rather than local projects. Multiple use of the coast is increasing, so more time and effort must be spent in finding solutions that will satisfy (perhaps only to some extent) a diversity of opinions by affected persons on what can (or should) be done, economically and socially. I think we will continue to recognize that although we have generalized theories, the combination ("mix") of processes, forcings and boundary conditions makes nearly all projects different, each requiring a unique solution, and that the project performance should be monitored quantitatively. This has been recognized by others (e.g. O'Brien, 1972), but should be emphasized again. The more experience I gain, the more I believe it to be true. Also, the recent trend seems to be, at least in the USA, of making little use of hydraulic laboratory studies, and to rely mostly on numerical models and field investigations. I think this is a mistake which hopefully will be recognized and reversed, as all are needed. If a numerical model cannot predict correctly what is observed in a hydraulic model — and the "scale effects" — we should have no confidence in its use for prototype application. Both laboratory and field are part of the "real world", and we learn from each.

Substantial advances have been made in our ability to analyze, design and construct coastal works, as is documented in the present book, and in *History and Heritage of Coastal Engineering* (Kraus, 1996). However, much remains to be done. Sometimes, we have to rectify or mitigate problems caused by the actions of others, such as chronic erosion or narrowing of beaches owing to a decrease in river transport of sand to the coast after the construction of dams, coastal subsidence resulting from withdrawal of oil/gas/water, construction of navigation works, or encroachment on beaches with buildings and infrastructure. Sometimes, engineers are even called upon to rectify a problem caused by a project which was built against their advice. In general, engineers are optimistic, learning from past works, and believing they can do a better job next time; perhaps this is one of the traits that leads us to become engineers — and when we find a societal need, we want to help society by developing a solution for it.

2. Origin and Needs of Coastal Engineering

A few anecdotes from the past, some ancient, some several centuries ago, and some recent, are both interesting and useful in illustrating the origin and needs of coastal engineering. We will recognize types of problems that are still with us, and why there is a continued demand for the resourcefulness of the human mind in coastal engineering — both research and practice. Also, they illustrate how much change can occur, physically and sociologically, and that we must expect and accommodate for these changes.

In antiquity, water-borne transport was often the most convenient manner of conveying cargo and people, and for heavy cargo, probably the only means. For some coastal works, there is only archaeological evidence, but for others, historical information is also available; as an example, ports/harbors have been used for more than 4,000 yrs in the Mediterranean Sea and elsewhere, and have been the subject of much archaeological and historical research. Many Roman ports/harbor sites have been investigated, but even earlier there were Egyptian. Phoenician and Greek coastal works.

Half-way through the time frame mentioned above, about 2,000 yrs ago, an artificial harbor, Caeseria Maritima (Sebastos Harbor), was built from shore into the open Mediterranean Sea by King Herod (Holum *et al.*, 1988). Some archaeologists believe it to be at the site of Strato's Tower, a Phoenician harbor used some two–four centuries prior to the construction of the new harbor. I mention this example, which has been studied extensively, because Roman engineering and construction techniques of great historical interest were used. This included "hydraulic concrete" (rubble stone and "cement" of lime and pozzuolana, a volcanic ash, which sets underwater, discovered in the 3rd Century BC) that was placed underwater in large size wooden form-works that had been weighted and set on the prepared bottom. Quarried limestone blocks were then laid above the large concrete units to form the breakwater. Use was likely made of wave overtopping/water storage/sluicing to dredge the harbor. The breakwaters deteriorated eventually, but their remnant can be seen through the water. They extend further seaward than the present harbor structures, which are modifications of the smaller Byzantine, Arabic, Crusader, and Turkish facilities which followed the Roman period. It appears that the breakwaters caused substantial changes in the transport and deposition of sand in the region, by waves moving it alongshore and by wind blowing great quantities of trapped sand inland. Some of the sand here, travelled from the Nile River, along the littoral of the southeastern Mediterranean Sea (Rohrlich and Goldsmith, 1984).

The ancient port of Ostia, which served Rome, was constructed in 43 AD just north of the mouth of the Tiber River. It was built against the advice of the engineers (Savile, 1940). It filled with sediment as the river delta built seaward, causing difficulties in the use of the harbor by the end of the 1st Century AD. It is no longer a harbor, and is about 2 miles inland (Bradford, 1957 as reported in Flemming, 1969). It is interesting that in the present century, particularly in the recent several decades, the beach at Lido di Ostia has *eroded* owing to the decrease in sand supplied to it by the Tiber River because of dams that had been built. This loss of sand has been mitigated by a beach nourishment and submerged shore-parallel breakwater perched beach project (Ferrante *et al.*, 1992).

Earthquakes have always been with us. Almost the symbol of coastal engineering, is the famous Japanese painting of a tsunami generated by an undersea earthquake. Japan has had a long history of tsunamis, with disastrous ones in the past few years. The most recent, a terrible disaster that occurred on 11 March 2011, was the 2011 Tohoku earthquake (9.0 Mw) and tsunami in Sendai/Onagawa, Japan (e.g. Wikipedia, 2011). In 1996, in Hilo, Hawaii, a symposium was held to mark the 50th anniversary of the 1 April 1946 tsunami which was probably generated by a massive underwater slide triggered by the Aleutian Trench Earthquake. A few years after the tsunami, in 1948, the Seismic Sea Wave Warning System was established. It expanded, with a number of countries on the Pacific Rim participating, and the cooperative effort is now named the International Tsunami Warning System. Much scientific and coastal engineering research and development followed the event (e.g. Kaijura, 1963; Keller, 1963), and is still ongoing. We will learn more from Professor Shuto's chapter in this book.

In another part of the world, the extensive archaeological/sea level changes study by Flemming (1969) of 179 coastal cities in the Western Mediterranean (with definite conclusions on 54 and relevant but inconclusive data on another 67), led him to state: "It is concluded that there has been no net eustatic change in the last 2000 years; that all submerged sites are due to earth movement; and that tectonic movements in the basin are predominantly downward." In other regions of the earth, tectonic movements are upward, or sometimes upward in one part and downward in another part of the area during a single event, such as the 1964 Alaska Earthquake. For ancient events, we have more information on tectonic movement than for subsidence caused by liquefaction — an important problem today, especially where fill is used to create land for the large areas needed for cargo containers. We are all aware of the catastrophe at Kobe, Japan, caused by the 1995 Hyogo-ken Nanbu Earthquake.

There is little, if any, information on the effects of storm waves on ancient coastal structures. However, nearly 500 yrs ago (in the same decade that Columbus made his historic voyage across the Atlantic Ocean), waves of a great storm (in February, 1498) demolished the superstructure of the wall of the "Old Mole" in Genoa, Italy. The inspection and engineering report of the event was made by the artist/scientist/engineer of genius Leonardo da Vinci (D'Arrigo, 1955). Owing to the port's location, the several breakwaters that have replaced the old one have been subject to many severe storms and damages (D'Arrigo, 1955). As a part of engineering studies, one of the first wave height gages made was in Genoa and installed on a section of the vertical concrete wall in water 15 m deep, together with two vertical arrays of pressure cells (7 cells in each array) mounted in the wall to measure wave-induced forces.

Many natural, and some artificial, harbor entrances are difficult, or even dangerous to use. Just south of Biarritz, France is one described in the 16th Century "Sailing Directions of Pierre Garcie" (as translated by Waters, 1967): "Know that when the sea breaks more than two rollers on the Plateau de St. Jean de Luz (Bay of Biscay) you must not attempt to enter Le Boucaut; take heed indeed, because it is not worth it. But if the seas are not breaking you can go in safely." Four centuries later, at the unimproved tidal entrance of Tomales Bay, CA (40 miles north of my University), about one small boat a year capsizes in large waves that break over the entrance bar — often when a strong ebb tidal current is flowing, and northwest winds blowing (Doyle, 1996). Improvement of entrances for safe navigation has been, and is, a major activity of coastal engineers. Sometimes, however, improvements cannot be made economically, and reliance must be placed on public education to improve safety.

The formation of a tidal estuary/lagoon inlet and its changes with time has been documented for Aveiro, Portugal since the 10th Century, with detailed information, including surveys, since the end of the 13th Century (Abecasis, 1955). It was formed by the evolution of a sand spit, and the harbor was used for shipping until the middle of the 18th Century. The entrance nearly closed in 1575, but it opened again. By the end of that century, and during the following century it was unstable, and navigation was difficult. It eventually closed by natural processes, becoming a barrier beach and lagoon, which occasionally opened. The lagoon became a swamp. The population of the city decreased to a small number. Many attempts were made to open and maintain an entrance. During the early part of the 19th Century, an artificial entrance was opened. Entrance works were built and sluicing used during the following years. Other substantial works were constructed in the 20th Century, with a major project in the 1930's, and another in the 1950's. There is probably more detailed long term information on this tidal estuary than any other.

Tidal entrances/estuaries/land reclamation/coastal wetlands are complex coupled nonlinear systems. In problems associated with these, as with many other problems, engineers are asked to serve as expert witnesses in courts of law. John Smeaton was the first to be an expert witness in a court of law in England (other than medical doctors). He was also the first person to call himself a civil engineer (at least in the English language). His testimony based on an extensive study of a tidal estuary in a sediment-rich region was given on 4 May 1782 in regard to the deterioration of a small navigation channel between the North Sea and Wells Harbour (where Admiral Horatio Nelson watched ships when he was a boy), Norfolk, England. One of the questions that had to be resolved by the court was: what were the effects on the navigation of the embankments constructed to enclose some of the salt marshes, together with the condition and operation of a sluice which had been built and rebuilt to help maintain the small channel? Furthermore, what would be the effects, the benefits (if any), and the rate of change with time, of removing the most recently built embankment (Roscoe, 1831; Wiegel, 1993; Golan, 2004). This is the type of problem where a combined numerical and hydraulic model would be useful, with many measured data as input.

Harmful effects of waste disposal (factory and household) on sea water quality, thence oyster and mussel aquaculture and fishing, and the endangerment to public health are not new. As an example, they occurred in the vicinity of Marseille, France, in the early 1700s with refuse being dumped into the harbor and on the beaches (Corbin, 1994). This has been a common problem. One result of the great population increase along the coast in recent decades has been a large increase in use of estuaries and the ocean for waste disposal, including waste heat from thermal-power plants (and the absorption of aircraft noise from airports). Considerable advances have been made in the treatment of these wastes in many countries to reduce pollution to an acceptable level. Pollution control, and remedial work on degraded areas are growing aspects of coastal engineering.

Severe storms have caused major loss of life, and destruction and damages to buildings and infrastructure along the coast. In some areas, it was the combined effects of astronomical tide, storm surge, waves and wind. The hurricane of 1900 that struck Galveston, Texas, killed about 6,000 people, and destroyed nearly 3,600 buildings (Walden, 1990). The great North Sea storm of February 1953 caused extensive loss of life and damages. In The Netherlands 1,800 were killed, 47,300 houses destroyed or damaged and 800,000 acres of land flooded with salt water; in England nearly 300 people were killed (Wemelsfelder, 1954). The Ise-Wan Typhoon of September 1959 was devastating in Japan. There were about 4,600 deaths, and extensive damage to boats, buildings and infrastructure (Otao, 1960). These, and similar natural events, which are unpredictable except on a probability basis, have led to extensive coastal works, building codes, zoning restrictions, and disaster preparedness systems, all requiring coastal science inputs. Owing to improved monitoring of hurricanes, reliability of prediction of paths, and evacuation procedures, loss of life in the USA, Japan, and other countries have been mitigated. Studies of events are available for use in future mitigation planning (e.g. a special issue of Shore and Beach (2012) about Hurricane Irene in 2011).

Recreational use of the seaside is a cultural change that has occurred since about 1750 in Europe (Corbin, 1994), and about the mid-1800's in the USA (Wiegel and Saville, Jr., 1996). (Historical writings tell us that people in ancient times also enjoyed viewing seascapes, and other coastal amenities.) It includes enjoyment of seascapes, sun-bathing, building sand-castles on the beach, and water-sports: sailing, wind-surfing, fishing, surfing, swimming, diving. There is a great demand for beaches. Tourism is a major industry worldwide, and many of the tourists like to visit beaches (Houston, 1995). Coastal regions that were remote, only few decades ago, are being developed, as people like variety and few places are now more than about one day's travel by aeroplane. Also, large numbers of people live along the coast, owing in part to milder climates than inland. Often they like to be right on the shore, and are willing to pay big prices for condominiums, homes and hotels overlooking the sea. There is something therapeutic and almost mystical about looking at the surf. In some regions, such as on the barrier islands of the US. Atlantic and Gulf coasts, this desire has led to a conflict between those wanting to live on the shore and those who believe they should not be permitted to do so owing to beach and building vulnerability to storms and/or environmental reasons.

The state of the art just prior to the modern era was described by Sharp (1927):

> "Conditions vary so widely from place to place that rule-of-thumb methods are sure to give a large percentage of failures, and a structure successful at one place may be a dismal failure at another. On the other hand, the engineer who wishes to attack his problem scientifically finds that science has done very little to help him. He is almost entirely without trustworthy facts, and must work up his data from hasty studies of his own."

3. A Bit of History of Modern Coastal Engineering: Yesterday and Today

The history of modern coastal engineering is basically that of the development of the combined engineering/science approach wanted by Sharp (1927); including data collection/presentation, analysis, hydraulic and numerical model studies. There have been few new project concepts. Modern coastal engineering began to develop shortly after Sharp's comment, with major advances made during and after WWII. I will not include nearly as much history as I originally intended, owing to the new book *History and Heritage of Coastal Engineering* (Kraus, *ed.*, 1996) of the 15 countries that have hosted an ICCE, each chapter written by engineers in the country. Also, I will not write about a major current activity of coastal engineering, beach rehabilitation and nourishment, as a 3-year study has recently been made by the US. National Research Council's Committee on Beach Nourishment and Protection, and a 336 page book (1995) published on its findings, which includes history and many references. This section is "yesterday and today", and much of "today" is included in other chapters of this book.

What is included in modern coastal engineering? The following subjects which were in the *Proceedings of the First Conference on Coastal Engineering* (Johnson, editor, 1951) are reasonably inclusive: beach fills; breakwater/ seawall design and construction; case histories of coastal projects; coastal geology/geomorphology/ petrology; dredging; groins; harbor location/ layout; harbor surging; hydraulic models; hydrographic and topographic surveys, charts and maps; jetties; littoral barriers; littoral processes on sandy coasts; material corrosion and erosion; nearshore currents; project planning; river/storm drain outlets; sand sources; sea and swell data, wave climate; sand bypassing; sediment transport to coast by rivers; sewer outfalls; tidal inlets; tide measurements and prediction; wave forces on piles; wave generation, including forecasting, hindcasting; wave recorders; wave theory, including shoaling, refraction and diffraction; weather maps, winds and temperature data.

Important in the development of coastal engineering has been the formation of relevant organizations (for example, in the USA, the Beach Erosion Board (replaced by the Coastal Engineering Research Board) of the US. Army Corps of Engineers, and the Coastal Engineering Research Council of the ASCE), academic programs in universities, conferences/symposia/workshops, and the preparation of case histories of projects. Bathymetric and topographic surveys, measurements of tidal elevations and currents, and meteorological measurements (temperatures, visibility, rainfall) made for navigation needs have been of great value for the study of changes. The value of long-term accurate measurements is immense (Bowman, 1934; Seymour *et al.*, 1993). The development of setback lines and building codes requires technical input from coastal engineers and scientists, and is an integral part of professional practice. However, these subjects will not be included herein, nor will many others owing to space limitation; only a few will be. I apologize for using more US references than others, but this is because I am more familiar with them.

Waves are important in most coastal processes and works. They fascinate me, and I still watch them on beaches by the hour, when I have the opportunity — transforming, breaking and moving up and down the porous beach face. What I see reminds me of how little I know about their physics after all the years of study. An understanding of their physics, generation by wind (and other mechanisms), characteristics (size and direction), and occurrence (wave climate) is necessary for most coastal engineering activities. They have been studied for more than a century, theory and observations (e.g. Stokes, 1880; Cornish, 1904). It is rather surprising to most coastal engineers today how little was known even in the 1930s of the origin of damaging waves in some regions. As an example, on occasions heavy-surf pounded beaches and buildings in southern California during summer when there were no local storms (Ewing, 1935; Wiegel, 2010). In the 1940s it was determined that some were generated by tropical cyclones off the west coast of Mexico, more than a thousand miles away. Another class, "southern swell"

(Wiegel and Kimberly, 1950), came from thousands of miles away, generated in the 40°–50° degrees south latitude (a classical study has been made of these (Snodgrass *et al.*, 1966)). Major quantitative advances have been made since the early 1940s, in forecasting/hindcasting/nowcasting waves, in measuring and analyzing them, and in developing theory.

The principles of forecasting wind waves and swell, and breakers and surf developed during WW II were given in USN Hydrographic Office publications in 1943, 1944, and 1951. The basic physical relationships and techniques (building on the work of Jeffreys (1925; 1926)) were published by Sverdrup and Munk (1946), and updated by Bretschneider (1951), with additional data. These are the SMB (Sverdrup, Munk, Bretschneider) wave forecasting curves, based on the similarity principle O'Brien (1943) wrote to Sverdrup about. The concept of spectra was known to them, and the term "significant wave" was used, based on an empirical distribution function. A statistical specification of a random sea was developed later by Pierson and by Neumann, with practical methodology given by Pierson, Neumann and James (1955). Two important theories of wave generation by wind were developed by Miles (1957; 1959) and Phillips (1957). Later another similarity relationship, with energy transfer among frequencies by wave–wave interactions, was developed by Hasselman *et al.* (e.g. 1973). There have been many advances since then in theory and in the practice of wave forecasting, with the present techniques being the "3rd generation" WAM (e.g. Komen *et al.*, 1995), requiring the use of a supercomputer. Wave forecasting is done routinely by various government agencies and private companies in many countries, with the US. Navy making forecasts four times a day for most of the world's oceans using the WAM model (e.g. Wittmann and Clancy, 1993) — available on World Wide Web (WWW)/Internet.

Measurement of waves (1D), data analysis and presentation developed in the 1940's and 1950's (e.g. Wiegel, 1955), at first slowly owing to the costs of installation and operation of instruments, and the lengthy process of analyzing the measurements (mostly by hand). Wave height (and wave period) distribution functions were obtained which were used as guides and to verify theoretical models of Longuet–Higgins (1952) and Putz (1954). Putz, using S.O. Rice's theory of random noise, showed the Rayleigh Distribution to be one of a family of functions; the relevant parameter to identify the specific distribution as a best fit being the ratio of the number of maxima and minima to the number of zero-crossings of the mean value. With the advent of digital computers, analysis became fast and inexpensive, and great amounts of data are now available, in some cases almost in "real time". Nearly all analyses are spectral, based on the assumptions of linear superposition, linear wave theory, and random phase. (No more will be written about random waves, as they are treated in this book by Goda.) An important milestone in wave spectral analysis was the Ocean Wave Spectra Conference, held in Easton, MD (1961).

For most coastal processes, wave direction data are important, but difficult to measure. Observations of mean directions were made from ships, reporting separately sea and swell. Observations of breaker height and direction have been made on beaches. Also the combination of 1D wave measurements and hindcasts (with directional information) are made. These are all useful. But, routine measurement and analysis of 2D (directional spectra) is needed, and this is complicated and expensive. The early work on theory of arrays and spectral analysis for obtaining directional spectra was by N. F. Barber in the U.K. The first use of digital recording (punched tape) and analysis of ocean waves was to obtain directional data from an array (Munk *et al.*, 1963). There are now quite a few installations of several types of devices, including arrays, but angular resolution obtained from them is inadequate for many purposes. There are conceptual models of 2D spectra (e.g. Mitsuyasu, 1981). Some background papers on concepts, measurements, and uses were presented at the Conference on Directional Spectra Applications '81 (Wiegel, 1981).

For "today" on ocean wave measurement and analysis, see for example the conference proceedings (New Orleans, LA) on this subject (Edge and Magoon, editors, 1974; Magoon and Hemsley, editors, 1993). See also Wiegel, 1974.

Waves generated at sea are modified in shoaling water (shoaling and refraction) and by islands (diffraction and refraction). Possibly the earliest quantitative use of refraction theory was in regard to sand transport at Santa Barbara, CA, in 1936 by Morrough P. O'Brien (see Wiegel and Saville, Jr., 1996). The technique was extended during the 1940's (Johnson *et al.*, 1948). Theory and techniques have been developed extensively since then (e.g. O'Reilly and Guza, 1993). A state-of–the-art example is: directional spectra from analysis of measurements with a seven-gage array on the Harvest Platform off southern California are transformed by means of a back-refraction (inverse) model using a supercomputer to locations along the coast in almost real time (O'Reilly, 1993; personal communication, 1996). To help in visualizing the wave variability along the coast, the ratios of shore to deepwater heights (linear theory is used) are displayed in color graphics.

Characteristics of breaking waves on a beach (spilling, plunging, surging) were studied in the laboratory and in the ocean, and an approximate relationship among beach slope, wave steepness and breaker type were established (Patrick and Wiegel, 1955). Subsequently, a relationship between these variables and wave runup on breakwaters and beaches was investigated, and a rough relationship found (Hunt, 1959), using the ratio of steepness and slope, what is now referred to as the surf similarity parameter or the Iribarren Number. However, the scatter of data is large (Singamsetti and Wind, 1980). Currently, the terms "dissipative" and "reflective systems" are used (Guza and Inman, 1975). All investigators, starting with J. J. Stoker, knew that crucial to any theory of wave breaking and runup are estimates of energy dissipation (e.g. Sobey, 1993).

The relatively long wave period oscillations (about 2 to 5 min) that occur in the surf zone were named "surf beat" by Munk (1949) in his presentation of theory and measurements. Independently, and nearly simultaneously, Tucker (1950) investigated surf beat. Walter Munk and John Isaacs also mentioned what became known as wave setup (first studied, and named by Saville (1961)), with theory developed by Longuet–Higgins and Stewart (1963; 1964). Generation and characteristics of what are now called infragravity-frequency waves in the surf (we thought of them as subharmonics) are currently being studied theoretically and with large arrays of instruments in the ocean (making use of bi-harmonic spectral analysis) (e.g. Herbers *et al.*, 1995). (Can the surf beat of Munk and of Tucker be distinguished from variations in setup caused by irregular waves breaking?) (These long period waves, combined with trapping by refraction, may well be an important source of the forcing of harbor oscillations.)

Breaker-induced littoral currents are three-dimensional, including seaward flowing rip currents and "undertow", and alongshore currents (e.g. Shepard and Inman, 1951; Horikawa, 1978). An early theory of alongshore currents (it is time we stopped using the word "longshore") was developed by Putnam *et al.* (1949), with numerous variations since. Ocean measurements of setdown, setup, and rip currents were made by Sonu (1972), and theory advanced by Noda (1974), and Horikawa (1978). Rip currents are a major cause of drownings (Brewster, ed., 1995), but little research is done on this phenomenon.

That waves breaking at an angle to a beach, and on the beach face (swash), transport sand along the shore was known to D.D. Gaillard, D.W. Johnson, Morrough P. O'Brien and others during the first half of the previous century. The term "river of sand" appears in one of the first BEB reports (1933). Few quantitative data were available. Semi-empirical formulas were developed of the relationship between waves and sand transport. What is referred to as the SPM formula appears to be derived from the "littoral drift factor" formula (Caldwell, 1956). This was probably based on the section on "average work factor" in the SIO (1947) report to the USACE Los Angeles District. They state that in: "... sand transport by wave action it appears that wave work rather than wave height is the significant parameter." In the SIO report, an equation for the time rate of work (wave power) is developed in terms of "significant wave weight, associated wave period, and the frequency distribution of these in terms of the summation of the product of the square of the significant wave height and the significant wave period, per unit length of beach". Bailard (1984) modified the equation to include the ratio of wave orbital velocity to sand settling velocity (for the importance of settling velocity, see e.g. the article by O'Brien, 1986), and investigated the effect of bottom slope. There is substantial variation in estimates of the empirical coefficient in the formulas (Schooness and Theron, 1995; 1997).

Numerical models have been developed for sand transport and beach deformation (see Watanabe, and others in this book, and also a short review by Komar (1998)). Numerical models make use of Pelnard–Considere's concept (1956), and wave climate data as the forcing (e.g. Hanson and Kraus, 1989). In order to get probabilistic estimates (and sensitivity analysis), Monte Carlo simulation can be used (LeMehaute *et al.*, 1981).

The last subject I will mention in this section is wave-induced forces on structures: breakwaters and piers/offshore platforms. Vertical wall breakwaters were used by the Romans, and even earlier, but no information is available about the forces. Thomas Stevenson, a British engineer measured forces on vertical walls with a dynamometer starting in 1842 (as reported in Gaillard (1904)); Gaillard (1904) reported his own measurements made in the 1890's and early 1900's; D'Arrigo (1955) reported on his and earlier measurements at Genoa, Italy. These were "yesterday"; for "today", I rely on a recent (1995) monograph containing 14 papers from 9 countries: *Wave Forces on Inclined and Vertical Wall Structures*, a product of the ASCE Task Committee on the subject. Iribarren observed rubblemound breakwaters in the Atlantic Ocean at the northwest coast of Spain, and presented a formula for use in their design, relating wave height, breakwater slope and size stone. This was modified by Hudson (1953), making it dimensionally homogeneous. The relationship Hudson developed, with input from hydraulic tests and some prototype observations, has been used for many years. As breakwaters were built in deeper water, and subject to higher waves, concrete armor units were developed (e.g. Danel and Greslou, 1962), with many types invented since. Some engineers became aware that there were scale effects, laboratory to ocean, and the breakage of nearly every cast concrete unit in the breakwater at Sines, Portugal led to a "watershed" change in the way such breakwaters were analyzed (Port Sines Investigating Panel, 1982). Technical workshops have been held on aspects of the subject (e.g. Davidson and Magoon, 1990), and many papers published.

The equation used to calculate wave-induced forces on a pile, or structural component of an offshore platform, is known as the Morison Equation, but should be called the Morison, O'Brien, Johnson, Schaaf (MOJS) Equation, after the four co-authors (1950). Water particle motions in waves are unsteady, and the equation includes both drag and inertia terms. Early estimates of the coefficients of drag and mass (inertia) from ocean tests are given in Wiegel *et al.* (1957). The studies of Keulegan and Carpenter (1959) and McNown and Keulegan (1959) identified a parameter which is as important in unsteady flow as Reynolds Number is in steady flow. It is known as the Keulegan–Carpenter Number. There is an extensive technical literature on the subject, owing to its use in the offshore oil industry (e.g. American Petroleum Institute, 1993). Piers that extend from shore through the surf are subject to breaking waves. Early experiments in the ocean (Snodgrass, *et al.*, 1951) and laboratory tests identified two force components, the slowly varying "normal" forces of the type described by the MOJS Equation, and a quasi-impact load. Few studies have been made of this phenomena (Watanabe and Horikawa, 1974; Wiegel, 1982); many piers have been damaged severely or destroyed.

4. Present and Future Requirements

Substantial advances have been made in our understanding of the physical, chemical and biological coastal environment, and in data collection and analysis. In spite of the great amount of work done during the past half century, there is still need for better understanding of the processes and their mix at a particular site, together with long-term reliable measurements, analyses and data presentation. We must continue to work on this. More planning must be done of coastal systems rather than just components. The value of regional planning and coastal programs has long been recognized, but has not often been implemented for a variety of reasons, including economical and political. Risk analyses are needed of projects and their interactions. (For example, we need to learn more about the probability of occurence and effects of clusters of storms such as those that impacted the USA Pacific Coast during the El Nino winters of 1982/1983 and 1997/1998.) Conditions and demands change, and we must accommodate for these changes, with retrofitting and rehabilitation if possible, but by replacement projects (or perhaps even abandonment in a few instances) if necessary. More information (monitoring) is needed on post-construction (or in some cases non-construction) performance, benefits and costs, and analyses made of indirect or secondary effects/interactions, possibly unintentional, of the projects with other interests of coastal communities.

In Morrough P. O'Brien's (1972) comments at the 13th ICCE in Vancouver, Canada, he quoted Professor William A. Nierenberg, then Director of the Scripps Institution of Oceanography: "Want to make big money on oceanography? Then forget about mining nickel from the ocean floor or refining gold from seawater. Instead, build a harbor that will not silt up or beaches that will not wash away. Learn to clean up pollution." These are still valuable goals. To help reach these goals, and for other future requirements, the following assessments and recommendations are given, with no order of priority implied. (Some of these are treated in chapters of the present book.)

1. Use of the coast will continue to grow, with increasing numbers of people and investments, and more conflict. Multiple demands (beach use, surfing, marinas, fishing, wetlands, accommodations, commercial ports, infrastructure, waste disposal/pollution control) will require better coastal zone management, with substantial technical input from coastal engineers/scientists.
2. Cooperative efforts are required to assure that we live better with the environment, to minimize harmful effects, and to rehabilitate degraded areas.
3. There is need for a more complete understanding of the processes, forcings, and boundary conditions at a particular site, and their combination ("mix"), together with long-term reliable measurements, analyses and data presentation. Quantitative data and models of variability and trends are needed. No two coastal regions are exactly alike, many differences being subtle with observations required over a long period of time to detect and understand them. It is necessary to evaluate the natural processes and the actions by people that led to a condition that requires an engineered solution. The more we understand about what is happening at a site, especially relationships among the components, the more likely planning will be good and projects will perform as desired.
4. Often it is written and said that beaches are eroding, with inference that this is the normal state of nature owing to increasing mean sea level and/or some

other natural cause(s). Many beaches are not eroding, but are stable or even accreting. Many beaches that are eroding are doing so as a result of anthropogenic activities such as dam building, tidal entrance dredging, and/or jetties, which can mitigated. Both what is happening at a site and why it is occurring must be determined if a successful project (or no project) is to be planned, with any harmful effects minimized or mitigated (most activities are in response to economic and social-political-demands, sometimes conflicting).

5. As a corollary to the above statements, do not force a common solution, and do not promulgate regulations that do not recognize diversity.
6. I have never understood why, but have been grateful that, Nature permits the successful use of the combination of linear superposition of waves and linear wave theory in deep water. However, coastal engineers must deal mostly with works in shallow water (relative to wave length), where these approximations are often not valid. Advances must continue to be made in nonlinear analysis of measured data in shallow water, and in the study in general of nonlinear wave mechanics, especially in the surf zone. As an example, how are subharmonic (infragravity) waves generated in the surf? Development is needed of theory of water particle motions in the surf zone, which includes porosity of the sand bottom.
7. Our ability to predict beach erosion/accretion cycles (mostly cross-shore transport) as a function of waves, tide and beach material is poor. It must be improved. The scatter of estimates of the coefficient(s) in alongshore sand transport equations is large, which tells us there is something we do not understand. Furthermore, we have little understanding of how cross-shore and alongshore transports interact, and they often occur simultaneously.
8. A better quantitative understanding of the nonlinear coupled system of tidal entrances/estuaries is required in order to make reliable estimates for project design. A complete two-dimensional model of an estuary and entrance is needed (van de Kreeke, 1992). (Many coastal wetlands are in estuaries, and their value is being increasingly recognized.)
9. Numerical models of coastal (and other earth science) processes have coefficients which are obtained by comparison of predictions with observations. The words calibration, verification, and validation are sometimes used incorrectly (Oreskes *et al.*, 1994). Quantitative studies are needed to determine their reliability, adequacy and probabilistic accuracy.
10. Methods and background studies are needed for use in quantitative risk analysis (e.g. Borgman, 1963), probabilistic design, and estimates of interactions of project components.
11. Use of beach nourishment will continue to expand as the preferred method of rehabilitating beaches, and used to construct artificial beaches where none exist (e.g., Houston and Dean, 2013). Owing to difficulty and/or expense of obtaining natural sand of adequate quality in some locations, sand made of crushed rock (Wiegel, 1993) will be used in more projects. More hybrid projects should be used, combining beach fill with sand-tight terminal groins (or jetties) to form

cells (i.e. "pocket beaches"), and/or shore-parallel submerged breakwaters to form a perched beach.

12. Plan and implement regional systems rather than local projects.
13. Seawalls and revetments are useful for some sites, and should be evaluated on a case by case basis ("time, place, manner"). They are usually built after erosion has occurred, and then are blamed by some people for the erosion. Perhaps they are responsible for some erosion in some circumstances, but there are other conditions where they don't add to erosion (e.g. Kraus, 1988; Griggs, *et al.*, 1996, Wiegel, 2002). Their effects on a beach depends on where they are placed relative to the mean sea level, etc. In some cases they may be used as a backup for an extreme event, and normally covered by sand and vegetation for aesthetics.
14. Quantitative monitoring is needed on post-construction (or in some cases non-construction) performance, benefits and costs, and analyses made of indirect or secondary effects/interactions, possibly unintentional, of the projects with other interests of coastal communities. Has it performed as expected from a technical standpoint? Have the costs and benefits been as predicted? Has it brought about the social benefits claimed for in its presentation to the public and to the decision makers? We must evaluate our ability to plan and design, to learn from both successes and failures, and to continue to develop better procedures.
15. Global warming and mean sea level rise predictions are in flux, with estimates made of the amount and effects changing with each new study (Titus and Narayanan, 1995). Much can be learned of the likely effects of relative mean sea level rise by studies of regions where coastal subsidence has occurred because of withdrawal of oil/gas/water, such at Long Beach, CA and vicinity (Habel, 1978; USACE-SPL, 1995; Wiegel, 2010), and Niigata, Japan (Hirono, 1970).
16. Many people want simple explanations of complex processes, and simple solutions. This is both natural and unfortunate as someone (perhaps many) once said: "For every complex problem, there is a simple solution", and it is wrong. Not only must we improve our understanding of coastal processes and the design of coastal works, but we must improve our methods of educating the affected community.

References

Abecasis, C. K. (1955): The history of a tidal lagoon inlet and its improvement: The case of Aveiro, Portugal, *Proc. 5 th Int. Conf. on Coastal Engrg.* 329–363.

American Petroleum Institute (1993): Recommended Practice for Planning, Designing and Constructing Fixed Offshore Platforms — Working Stress Design, API Recommended Practice 2A-WSD (RP 2A-WSD), 20th Ed.

Bailard, J. A. (1985): A simplified model for longshore sediment transport, *Proc. 19th Int. Conf. on Coastal Engrg.* 1454–1470.

Beach Erosion Board, USACE (1933): *Interim Report of the Beach Erosion Board*, Office of the Chief of Engineers, Washington, D.C.

Borgman, L. E. (1963): Risk criteria, *J. Waterways and Harbors Div., Proc. ASCE* **81**(WW3), 1–36.

Bowman, I. (1934): Address by the chairman of the national research council, *Shore and Beach* **2**(1), 3–4.

Bretschneider, C. L. (1952): Revised wave forecasting relationships, *Proc. 2nd Int. Conf. on Coastal Engrg.* 1–5.

Brewster, B. C. (ed.) (1955): *The United States Lifesaving Association Manual of Open Water Lifesaving* (BRADY/Prentice Hall, Englewood Cliffs, New Jersey), 316 pp.

Caldwell, J. M. (1956): Wave Action and Sand Movement Near Anaheim Bay, California, USACE BEB Technical Report No. 68, 21 pp.

Corbin, A. (translation by J. Phelps) (1994): *The Lure of the Sea. The Discovery of the Seaside in the Western World* 1750–1840: University of California Press, 380 pp.

Cornish, V. (1904): On the dimensions of deep sea waves, and their relations to meteorological and geographical conditions, *Geographical J.* XIII.

Danel, P. and L. Greslou (1962): The 'tetrapod', *Proc. 8th Int. Conf. on Coastal Engrg.* 469–481.

D'Arrigo, A. (1955): The recent damage to the Genoa breakwater, *The Dock and Harbour Auth.* **36**(416), 54–56.

Davidson, D. D. and O. T. Magoon (eds.) (1990): *Stresses in Concrete Armor Units*, ASCE, 421 pp.

Doyle, J. (1996): Motorboat capsizes in Tomales Bay — 4 Killed, *San Francisco Chronicle*, pp. A13 and A17.

Edge, Billy L. and Orville T. Magoon, (editors), 1974, *Proc. of the International Symposium on Ocean Wave Measurement and Analysis*, Sept. 9–11, 1974, New Orleans, Louisiana, USA, ASCE.

Ewing, G. F. (1935): Destructive high waves along the southern California coast in 1934, *Shore and Beach* **3**(2), 61–64.

Ferrante, A., F. Leopoldo and B. Sander (1992): Modelling and monitoring of a perched beach at Lido di Ostia (Rome), *Proc. 23rd Int. Conf. on Coastal Engrg.* 3305–3318.

Flemming, N. C. (1969): Archaeological Evidence for Eustatic Change of Sea Level and Earth Movements in the Western Mediterranean During the Last 2000 Years, Geological Society of America, Special Paper 109, 125 pp.

Gaillard, D. D. (1904): *Wave Action in Relation to Engineering Structures*: The Engineering School, USACE, reprinted in 1935 and 1945, 218 pp.

Golan, T. (2004): *Laws of Men and Laws of Nature*: Harvard University Press, 325 pp.

Griggs, G. B., L. J. Moore, J. F. Tait, K. Scott and D. Pembrook (1996): The effects of the storm waves of 1995 on beaches adjacent to a long-term seawall monitoring site in northern Monterey Bay, California, *Shore & Beach* **64**(1), 34–39.

Guza, R. T. and D. L. Inman (1975): Edge waves and beach cusps, *J. Geophys. Res.* **86**(21), 2997–3012.

Habel, J. S. (1978): *Shoreline Subsidence and Sand Loss*, Internal Report, CA State DNOD, 5 pp. and attach.

Hanson, H. and N. C. Kraus (1989): *GENESIS: Generalized Model for Simulating Shoreline Change, Report I: Reference Manual and Users Guide*, Technical Report CERC-89-19, USACE, WES/CERC, Vicksburg, MS.

Hasselmann, K., T. P. Barnett, E. Bouws, H. Carlson, D. E. Cartwright, K. Enke, J. A. Ewing, H. Gienapp, D. E. Hasselmann, P. Kruseman, A. Meerburg, P. Muller, D. J. Olbers, K. Richter, W. Sell and H. Walden (1973): *Measurements of Wind-Wave Growth and Swell Decay during the Joint North Sea Wave Project (JONSWAP)*, Deutsches Hydrographisches Institute, Hamburg, Germany, 95 pp.

Herbers, T. H. C., S. Elgar, R. T. Guza and W. C. O'Reilly (1995): Infragravity-frequency (0.005-0.05 Hz) motions on the shelf. Part II: Free waves, " *J. Phys. Oceanogr.* **25**(6), 1063–1079.

Hirono, T. (1970): Niigata ground subsidence and ground water change, *Land Subsidence. Proc. of the Tokyo Symposium*, Inter. Assoc. Scientific Hydrology — UNESCO, 144–161.

Holum, K. G., R. L. Hohlfelder, R. J. Bull and A. Raban (1988): *King Herod's Dream: Caesarea on the Sea*: W.W. Norton, New York, NY, 244 pp. and appendices.

Horikawa, K. and R. L. Wiegel (1959): *Secondary Wave Crest Formation*, UCB, College Engrg., Technical Report Series **89**(4), 1959, 62 pp.

Horikawa, K. (1978): *Coastal Engineering. An Introduction to Ocean Engineering*: University of Tokyo Press, 402 pp.

Houston, J. R. (1953): Beach nourishment, *Shore & Beach* **63**(1), 21–24.

Houston, James R., and Robert G. Dean, "Beach Nourishment Provides a Legacy for Future Generations," *Shore & Beach*, **81**(3), 2013. 3–30.

Hudson, R. Y. (1953): Wave forces on breakwaters, *Trans. ASCE* **118**, 653–674.

Hunt, I. A., Jr. (1959): Design of seawalls and breakwaters, *J. Waterways and Harbors Div., Proc. ASCE* **85**(WW3), 123–151.

Jeffreys, H. (1925): On the formation of water waves by wind, *Proc. Royal Soc. Ser. A* **197**, 189–206.

Jeffreys, H. (1926): On the formation of water waves by wind (second paper), *Proc. Royal Soc. Ser. A* **110**, 1926, 241–247.

Johnson, J. W., M. P. O'Brien and J. D. Isaacs (1948): *Graphical Construction of Wave Refraction Diagrams*, USN Hydro. Office Pub. No. 605, 45 pp.

Johnson, J. W. (ed.) (1951): *Proc. 1st Int. Conf. on Coastal Engrg.* 334 pp.

Kaijura, K. (1963): The leading wave of tsunami, *Bull. Earth. Res. Inst.*, Tokyo University, **41**(3), 535–571.

Keller, J. B. (1961): Tsunamis — water waves produced by earthquakes, *Proc. of the Tsunami Meetings Associated with the Tenth Pacific Congress*, IUGG Monograph No. 24, 154–166.

Keulegan, G. H. and L. H. Carpenter (1958): Forces on cylinders and plates in an oscillating fluid, *J. Res., NBS* **60**(5), 423–440.

Komen, G. J., L. Cavaleri, M. Donelan, K. Hasselmann, S. Hasselmann and P. A. E. M. Janssen (1994): *Dynamics and Modelling of Ocean Waves*: Cambridge University Press, NY, xxii, 532 pp.

Komar, P. D. (1998): The modeling of processes and morphology in the coastal zone — reflections on the maturity of our science", *Shore & Beach* **66**(1), 10–22.

Kraus, N. C. (1988): The effects of seawalls on the beach: An extended literature review, *J. Coastal Research* **SI4**, 1–28.

Kraus, N. C., ed. (1996): *History and Heritage of Coastal Engineering*, ASCE, 601 pp.

LeMehaute, B., J. D. Wang and C.-C. Lu (1982): Monte Carlo simulation for shoreline processes, *Proc. Conf. on Directional Wave Spectra Applications*, University of California, Berkeley, CA, ASCE, 269–280.

Longuet-Higgins, M. S. (1952): On the statistical distribution of the heights of sea waves, *J. Marine Res.* **11**(13), 245–266.

Longuet-Higgins, M. S. and R. W. Stewart (1964): A note on wave set-up, *J. Marine Res.* **21**, 4–10.

Longuet-Higgins, M. S. and R. W. Stewart (1964): Radiation stresses in water waves: A physical discussion, with applications, *Deep-Sea Research* **11**, 529–562.

Magoon, O. T. and J. M. Hemsley, eds. (1994): Ocean Wave Measurement and Analysis: Proc. of the Second International Symposium Honoring Professor Robert L. Wiegel, New Orleans, Louisiana, 25–28 July 1993, ASCE.

McNown, J. S. and G. H. Keulegan (1959): Vortex formation and resistance in periodic motion, *J. Engrg. Mech. Div., Proc. ASCE* **85**(EM1), 1–6.

Miles, J. W. (1957): On the generation of surface waves by shear flows, *J. Fluid Mech.* 3(3), 185–204.

Miles, J. W. (1959): On the generation of surface waves by shear flows, part 2, *J. Fluid Mech.* **6**(4), 568–582.

Mitsuyasu, H. (1982): Directional spectra of ocean waves in generating area, *Proc. Conf. on Directional Wave Spectra Applications*, Berkeley, CA, ASCE, 87–102.

Morison, J. R., M. P. O'Brien, J. W. Johnson and S. A. Schaaf (1959): The forces exerted by surface waves on piles, *J. Petroleum Tech., Pet. Branch, Amer. Inst. of Mining and Metal., Engineers* **2**(5), 149–154.

Munk, W. H. (1949): Surf beats, *Trans. Amer. Geophys. Union* **30**(6), 849–854.

Munk, W. H., G. R. Miller, F. E. Snodgrass and N. F. Barber (1963): Directional recording of swell from distant storms, *Philos. Trans. Royal Soc. (London), Ser. A* **255**(1062), 505–584.

National Research Council, Committee on Beach Nourishment and Protection (1995): *Beach Nourishment and Protection*, National Academy Press, Washington, D.C., 1995, 336 pp.

Noda, E. K. (1974): Wave-induced nearshore circulation, *J. Geophys. Res.* **79**(27), 4097–4106.

O'Brien, M. P. (1943): *Letter to H.U. Sverdrup*, 31 December 1943; also an appendix in *Memorandum Concerning Wind, Waves, and Swells. A Basic Method of Forecasting — by H. U. Sverdrup and W. H. Munk*, College on Engrg., UCB, Memo. No. HE 116-7, 5 September 1944, 7 pp.

O'Brien, M. P. (1951): Preface, *Proc. 1st Int. Conf. on Coastal Engrg.* v.

O'Brien, M. P. (1972): Opening Address. Some Comments on Coastal Engineering, *Proc. 13th Int. Conf. on Coastal Engrg.* 3–8.

O'Brien, M. P. (1986): The Voice of Experience I, *Shore and Beach* **54**(2), 23–24.

Ocean Wave Spectra: Proceedings of a Conference, Easton, Maryland, May 1–4, 1961 (1963): *US Navy Oceanographic Office and National Academy of Sciences-National Research Council*: Prentice-Hall, Englewood Cliffs, NJ, 357 pp.

O'Reilly, W. C. and R. T. Guza (1993): A comparison of two spectral wave models in the Southern California Bight, *Coastal Engrg.* **19**, 263–282.

O'Reilly, W. C. (1993): The southern California wave climate: Effects of islands and bathymetry, *Shore & Beach* **61**(3), 14–19.

Oreskes, N., K. Shrader-Frechette and K. Belitz (1994): Verification, validation, and confirmation of numerical models in the earth sciences, *Science* **263**, 641–646.

Otao, H. (1961): A brief outline of the Ise-Wan Typhoon", *Proc. 7th Conf. of Coastal Engrg.*, The Hague, Netherlands, Council on Wave Res., *The Engrg. Found.* 931–941.

Patrick, D. A. and R. L. Wiegel (1955): Amphibian tractors in the surf, *Proc. 1st Conf. on Ships and Waves*, Council on Wave Res., *The Engrg. Found.*, and Amer. Soc. Naval Architects and Marine Engineers, 397–422.

Pelnard-Considere, R. (1956): Essai de theorie d l'evolution des formes de rivate en plage de sable et de galets, *4me Journees de l'Hydraulique, Le Energies de la Mer*, Question III, Report No. 1.

Phillips, O. M. (1957): On the generation of waves by turbulent winds, *J. Fluid Mech.* **2**(5), 417–445.

Pierson, W. J., Jr., G. Neumann and R. W. James (1955): Practical Methods for Observing and Forecasting Ocean Waves by Means of Spectra and Statistics, *USN Hydro. Office*, Pub. No. 603, 284 pp.

Port Sines Investigating Panel (Edge — Chair, Baird, Caldwell, Fairweather, Magoon, Treawell) (1982): *Failure of the Breakwater at Port Sines*, Portugal, ASCE, 278 pp.

Putnam, J. A., W. H. Munk and M. A. Traylor (1949): The prediction of longshore currents, *Trans. Amer. Geophys. Union* **30**(34), 349–357.

Putz, R. R. (1954): Statistical analysis of wave records, *Proc. 4th Conf. on Coastal Engrg.*, Chicago, Illinois., Council on Wave Res., *The Engrg. Found.* 13–24.

Rohrlich, V. and V. Goldsmith (1984): Sediment transport along the southeast Mediterranean: A geological perspective, *Geo-Marine Letters* **4**, 99–103.

Roscoe, H. (1831): *Reports of Cases Argued and Determined in the Court of King's Bench*, In *the 22nd–25th Years of the Reign of George III*, from the manuscripts of The Right Hon. Sylvester Douglas, Baron Glenbervie, and also from the manuscripts of Mr. Justice Lawrence, Mr. Justice Le Blanc, Mr. George Wilson, Vol. III, S. Sweet, and Stevens and Sons, Dublin, 157–161 and 340–343.

Savile, Sir L. H. (1940): Presidential address, *J. Inst. Civil Engineers (London)* (1), 1–26.

Saville, T., Jr. (1962): Experimental determination of wave set-up, *Proc. 2nd Conf. on Hurricanes, Miami Beach, Fla.*, U.S. Dept. Commerce, Weather Bureau, National Hurricane Res. Project, Report No. 50, I, 242–252.

Schooness, J. S. and A. K. Theron, Accuracy and applicability of the spm longshore transport formula, *Proc. 24th Int. Conf. on Coastal Engrg.* 2595–2609.

Schooness, J. S. and A. K. Theron (1997): Improvement of the most accurate longshore transport formula, *Proc. 25th Int. Conf. on Coastal Engrg.* 3652–3665.

Scripps Institution of Oceanography (SIO), Univ. Calif. (1947): *A Statistical Study of Wave Conditions at Five Open Sea Localities Along the California Coast*, Prepared for USACE, Los Angeles Dist., SIO Wave Report No. 68, 34 pp. and appendices.

Seymour, R. J., D. Castel, D. McGehee, J. Thomas and W. O'Reilly (1993): New technology in coastal wave monitoring, in Second International Symposium on Ocean Wave Measurement and Analysis, Waves 1993, ASCE, 105–123.

Sharp, Henry S. (1927): Artificial beach construction in the vicinity of New York, *The Scientific Monthly* XXV, 34–39.

Shepard, F. P. and D. L. Inman (1951): Nearshore circulation, *Proc. 1st Int. Conf. on Coastal Engrg.* 50–59.

Shore & Beach (2012): An issue dedicated to Hurricane Irene, **80**(2), 80 pp.

Singamsetti, S. R. and E. G. Wind (1980): Characteristics of Shoaling and Breaking Periodic Waves Normally Incident to Plane Beaches of Constant Slope, Delft Hyd. Lab. (DHL), Report No. M1371, 67 pp. and appendices.

Snodgrass, F. E., E. K. Rice and M. Hall (1951): Wave Forces on Piling (Monterey Field Test), Technical Report No. 35–4, *Inst. Engrg. Res.*, University of California, Berkeley, CA, 9 pp. and 18 figs.

Snodgrass, F. E., G. W. Groves, K. F. Hasselmann, G. R. Miller, W. H Munk and W. H. Powers (1966): Propagation of ocean swell across the Pacific Ocean, *Philos. Trans. Royal Soc. (London)*, Ser. A, **259**, 431–497.

Sobey, R. J. (1993): Quantifying coastal and ocean processes, *Proc. 11th Aust. Conf. of Coastal and Ocean Engrg.*, Brisbane, Australia, Inst. of Engineers Australia, 1–10.

Sonu, C. J. (1972): Field observation of nearshore circulation and meandering currents, *J. Geophys. Res.* **77**(18), 3232–3247.

Stokes, G. G. (1880): On the theory of oscillatory waves, *Mathematical and Physical Papers*, I: Cambridge University Press, UK.

Sverdrup, H. U. and W. H. Munk (1948): *Wind, Sea, and Swell: Theory of Relations for Forecasting*, USN Hyd. Office, H.O. Pub. No. 601, 44 pp.

Titus, J. G. and V. K. Narayanan (1995): The Probability of Sea Level Rise, US Environmental Protection Agency, 186 pp.

Tucker, M. J. (1950): Surf beat: sea waves of 1 to 5 min. period, *Proc. Roy. Soc. (London)*, Ser. A **202**(1071), 565–573.

USACE, SPL (1995): *Huntington Cliffs Reconnaissance Report, Huntington Beach, Orange County*: California, Technical Appendices.

van de Kreeke, J. (1992): Stability of tidal inlets; Escoffier's analysis, *Shore & Beach* **60**(1), 9–12.

Walden, Don (1990): Raising Galveston, *American Heritage of Invention and Technology* **5**(2), 8–18.

Watanabe, A. and K. Horikawa (1974): Breaking wave forces on a large diameter cell, *Proc. 14th Int. Conf. on Coastal Engrg.* 1741–1760.

Waters, D. W. (1967): *The Rutters of the Sea: The Sailing Directions of Pierre Garcie*: Yale University Press, 478 pp.

Wave Forces on Inclined and Vertical Wall Structures (1995): Task Committee on —, ASCE, 402 pp.

Wemelsfelder, P. J. (1953): The disaster in the netherlands caused by the storm flood of February 1, 1953, *Proc. 4th Int. Conf. on Coastal Engrg.* 258–271.

Wiegel, R. L., ed. (1956): *Proc. 1st Conf. on Coastal Engineering Instruments,* Berkeley, CA, Council on Wave Res., *The Engrg. Found.*, 302 pp.

Wiegel, R. L. (1974): Engineers' concern with waves, and their measurement, *Proc. of the International Symposium on Ocean Wave Measurement and Analysis*, Sept. 9–11, 1974, New Orleans, Louisiana, USA (eds. Billy L. Edge and Orville T. Magoon, ASCE, Vol. II, pp. 1–22).

Wiegel, R. L., ed. (1982): *Proc. Conf. on Directional Wave Spectra Applications,* University of California, Berkeley, CA, ASCE, 495 pp.

Wiegel, R. L. (1983): Forces induced by breakers on piles, *Proc. 18th Int. Conf. on Coastal Engrg.* 1699–1715.

Wiegel, R. L. (1993): Opening remarks, *Ocean Wave Measurement and Analysis, Proc. 2nd Inter. Symposium*, Honoring Professor R. L. Wiegel, New Orleans, LA, ASCE, 194–197.

Wiegel, R. L. (1993): Artificial beach construction with sand/gravel made by crushing rock, *Shore & Beach* **61**(4), 28–29 and cover.

Wiegel, R. L., (2002): Seawalls, seacliffs, beachrock: what beach effects? Part 1, *Shore & Beach* **70**(1), 17–27.

Wiegel, R. L., (2002): Seawalls, seacliffs, bachrock: what beach effects? Part 2, *Shore & Beach* **70**(2), 13–22.

Wiegel, R. L., (2002): Seawalls, seacliffs, bachrock: what beach effects? Part 3, *Shore & Beach* **70**(3), 2–14.

Wiegel, R. L. (2010): San Pedro Bay Delta in Southern California: Shore and shore use changes during the past 1½ centuries from a coastal engineering perspective, *Shore & Beach* **78**(3), 33–63.

Wiegel, R. L., K.E. Beebe and J. Moon (1957): Ocean wave forces on circular cylindrical piles, *J. Hyd. Div., Proc.* ASCE, 83(HY2), Paper 1199, 36 pp.

Wiegel, R.L. and H.L. Kimberly (1950): Southern swell observed at Oceanside, California, *Trans. Amer. Geophysical Union* **31**(5), 717–722.

Wiegel, R. L. and Kiyoshi Horikawa (1964): Coastal Engineering in Japan, *Civil Engineering* **34**(12), 76–77.

Wiegel, R. L. and T. Saville, Jr. (1996): History of Coastal Engineering in the USA, In: *History and Heritage of Coastal Engineering* (Nicholas C. Kraus, ed.), ASCE.

Wikipedia (2011), 2011 Tohoku earthquake and tsunami, Wikipedia, the Free Encyclopedia.

Wittmann, P. A. and R. M. Clancy (1994): Implementation and validation of a global third-generation wave model at Fleet Numerical Oceanography Center, In *Ocean Wave Measurement and Analysis. Proc. 2nd Inter. Symposium Honoring Professor R. L. Wiegel* (O. T. Magoon and J. M. Hemsley, eds.), New Orleans, LA, ASCE, by 406–419.

Chapter 2

Random Waves in the Sea

Yoshimi Goda
Professor Emeritus
Yokohama National University, Japan

1. Introduction

Waves in the sea are characterized by their randomness. Figure 2.1 depicts a crest pattern of wind waves, which is created by a numerical simulation based on the directional wave spectrum described in Sec. 3.2. Because of short extension of individual wave crests, wind waves are often called short-crested waves. Swell from a remote storm area, on the other hand, exhibits much longer extension of wave crests, and thus it is called long-crested waves. Even so, a profile of swell, or the time history of water surface elevation under the swell motion, is not a sinusoidal form of constant height and period but an irregular succession of varying heights and periods. Monochromatic waves, *i.e.* waves made of single component, can never be observed in the sea, though they help us to understand the basic nature of water waves.

Since the introduction of the significant wave concept by Sverdrup and Munk (1946), a number of engineering analyses of wave transformations and wave loading on structures have been executed by means of monochromatic waves, which correspond to significant waves, maximum wave, or some other representative waves. In light of our present knowledge, however, wave analyses with monochromatic approach must be inaccurate. The degree of inaccuracy depends on the category of wave phenomena. For evaluation of wave diffraction by breakwaters, monochromatic wave analysis yields gross error. Wave refraction in the coast of simple geometry analyzed with the monochromatic wave approach may produce an error of minimal percent only, but the same approach to wave refraction in the coast of complicated geometry would yield a large error in the estimate of wave heights. The wave force on a vertical wall exerted by the maximum wave is one of a few cases where the monochromatic wave analysis presents a reliable engineering solution.

In Sec. 2, a brief description is given on how to represent random sea waves mathematically and statistically. There are two basic representations by means of wave spectrum and wave-by-wave analysis. Spectral representation is discussed in

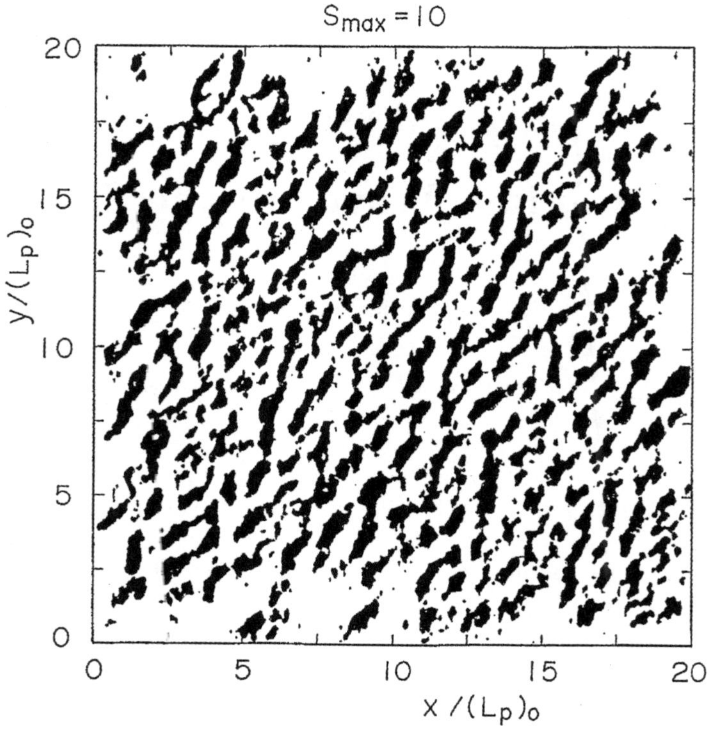

Fig. 2.1 A crest pattern of simulated wind waves with the directional spreading parameter of $s_{\max} = 10$ (Goda, 1994): the reference length $(L_p)_0$ denotes the deepwater wavelength corresponding to the spectral peak frequency.

Sec. 3, while wave statistics are described in Sec. 4 together with the spectral effects on statistics. These representations are based on the assumption of linear behavior of sea waves. In reality, sea waves exhibit a certain degree of nonlinearity. Some features of nonlinear random waves are introduced in Sec. 5. The methodology of incorporating wave randomness into engineering practice is presented in Sec. 6, which includes wave transformations of refraction, diffraction and breaking, wave-induced currents, and wave interactions with structures such as run-up, overtopping and forces.

2. Description of Random Sea Waves

2.1 *Linear model of random sea surface*

Random sea waves are postulated as composed of an infinite number of infinitesimal free waves.

Pierson, Neumann, and James (1955) advocated that postulate, when they proposed the wave forecasting method by means of wave spectra. There are several mathematical expressions to describe the random sea surface, and the following is most frequently referred to:

$$\eta(x, y, t) = \sum_{i=1}^{\infty} a_i \cos(k_i x \cos\theta_i + k_i y \sin\theta_i - 2\pi f_i t + \varepsilon_i), \tag{1}$$

where η denotes the elevation of water surface above the mean water level, x and y are the Cartesian coordinates, and t is the time. The component wave has the amplitude a_i, wave number k_i, direction of propagation, θ_i, frequency f_i, and the random phase angle ε_i that is uniformly distributed in the range $[0, 2\pi]$.

Each component-wave propagates independently as free waves, satisfying the following dispersion relationship between the wave number k and the frequency, f.

$$4\pi^2 f^2 = gk \tanh kh, \tag{2}$$

where g is the acceleration of gravity and h denotes the water depth. Wave motions such as orbital velocities and underwater pressure fluctuations are derived by the small amplitude wave theory, and the motions of component waves are linearly superposed to yield the resultant random wave motions.

The amplitude a_i and the energy $a_i^2/2$ of each component wave is infinitesimally small, but when the energy is summed up in a certain range of frequency and direction it yields a finite value, which defines the wave spectrum discussed in the next subsection.

The surface elevation η defined by Eq. (1) follows the Gaussian distribution by virtue of the central limit theorem. In fact, the quasi-Gaussian distribution has been observed in many field measurements, and that fact supports the validity of Eq. (1). However, there exists a small deviation from the normality of surface elevation. The deviation is attributed to the presence of nonlinear wave components, which are generated by higher order interactions among free wave components. The topic of wave nonlinearity effects is dealt with in Sec. 5.

2.2 *Spectral representation of sea waves*

The energy of component waves when summed up in a small frequency range between $(f, f+df)$ and a small directional range $(\theta, \theta+d\theta)$ can be uniquely expressed with a continuous function $S(f, \theta)$ as in the following:

$$\sum_{f}^{f+df} \sum_{\theta}^{\theta+d\theta} \frac{1}{2} a_i^2 = S(f, \theta) df\, d\theta. \tag{3}$$

Under the condition that the function $S(f, \theta)$ exists, it is called the directional spectral density function or simply the directional spectrum.

The directional spectrum is usually expressed as the product of the frequency spectrum $S(f)$ and the directional spreading function $G(f|\theta)$. That is,

$$S(f,\theta) = S(f)G(f|\theta). \tag{4}$$

The directional spectrum $S(f,\theta)$ has the dimension of the energy density, and this dimension is carried by the frequency spectrum $S(f)$. Thus, the directional spreading function is dimensionless and normalized by the following condition:

$$\int_0^{2\pi} G(\theta|f)d\theta = 1. \tag{5}$$

2.3 *Statistical representation by wave-by-wave analysis*

The spectral representation of random waves can tell us how the wave energy is distributed in the frequency and direction ranges. Nevertheless, it does not explicitly describe the heights and periods of visually observed waves. For this purpose, individual waves are defined from wave records, their heights and periods are measured, and certain statistical parameters are calculated.

For defining individual waves, the current standard is the use of the zero-upcrossing or zero-downcrossing methods (PIANC 1986, IAHR 1989). Both the methods are statistically equivalent, though they yield slightly different values of heights and periods for individual waves.

The significant wave proposed by Sverdrup and Munk (1946) is given the height $H_{1/3}$ being equal to the mean height of the highest one-third waves and the period $T_{1/3}$ being equal to the mean period of the same waves. Several representative wave heights and periods are also defined statistically, among which the mean period $\overline{T}$ of the whole waves is often utilized.

The statistical wave heights and periods can be related with the frequency spectrum. The significant wave height is estimated by the following formula:

$$H_{1/3} \cong H_{m0} = 4.004\sqrt{m_0}, \tag{6}$$

where m_0 is the zeroth moment of frequency spectrum, and the terminology of H_{m0} is employed here so as to clarify its definition based on the spectral moment. The wave height $H_{1/3}$ is the height defined by the wave-by-wave analysis, and it is about 5% smaller than H_{m0}, on the average, for wind waves. The relationship between $H_{1/3}$ and H_{m0} will be further discussed in Sec. 4.3.

The mean period is also estimated as

$$\overline{T} \cong T_{0,2} = \sqrt{m_0/m_2}, \tag{7}$$

where m_2 denotes the second spectral moment as calculated by the following equation for the case $n = 2$:

$$m_n = \int_0^{\infty} f^n S(f)df \tag{8}$$

3. Standard Functional Forms of Wave Spectrum

3.1 *Single-peaked frequency spectrum*

Fully developed wind waves exhibit a single spectral peak. A number of functional forms have been proposed to represent such single-peaked spectra. For engineering applications, it is preferable to express the spectral density in terms of wave height and period parameters. The Bretschneider spectrum with a modification of coefficients by Mitsuyasu (1970) of the following is widely used in Japan, which is called the Bretschneider–Mitsuyasu spectrum:

$$S(f) = 0.257 H_{1/3}^2 T_{1/3}^{-4} f^{-5} \exp[-1.03(T_{1/3} f)^{-4}]. \tag{9}$$

This spectrum has been derived by setting $H_{1/3} = H_{m0}$ and $T_p = 1.05 T_{1/3}$ based on field measurements by Mitsuyasu (1970), where T_p denotes the period corresponding to spectral peak frequency.

Wind waves generated by strong winds over a relatively short fetch often demonstrate sharply peaked spectra. Hasselmann *et al.* (1973) proposed the JONSWAP spectrum by artificially enhancing its peak. The original parameters of wind speed and fetch length have been replaced with the height and period parameters by Goda (1988) on the basis of a large number of Monte Carlo simulation of wave profiles and their zero-upcrossing analysis. The modified JONSWAP spectrum is expressed as follows:

$$S(f) = \beta_J H_{1/3}^2 T_p^{-4} f^{-5} \exp[-1.25(T_p f)^{-4}] \gamma^{\exp[-(T_p f - 1)^2 / 2\sigma^2]}, \tag{10}$$

where

$$\beta_J \cong \frac{0.06238[1.094 - 0.01915 \ln \gamma]}{0.230 + 0.0336\gamma - 0.185(1.9 + \gamma)^{-1}} \tag{11}$$

$$T_{1/3} \cong [1 - 0.132(\gamma + 0.2)^{-0.559}] T_p \tag{12}$$

$$\overline{T} \cong [1 - 0.532(\gamma + 2.5)^{-0.569}] T_p \tag{13}$$

$$\sigma = \begin{cases} 0.07 : f \le f_p \\ 0.09 : f > f_p. \end{cases} \tag{14}$$

The wave height $H_{1/3}$ differs slightly from the spectrally estimated height H_{m0}, depending on the γ value as presented in Table 2.1.

The key parameter of the JONSWAP spectrum is γ, which is called the peak enhancement factor: its value varied from 1 to 7 during the Joint North Sea Wave Project by Hasselmann *et al.* (1973). When γ is 1, the JONSWAP spectrum takes the same functional form with the Bretschneider–Mitsuyasu spectrum. There is a slight difference in the position of spectral peak between the two spectra, partly because of the different period relationship between T_p and $T_{1/3}$. If the modified

JONSWAP spectrum is employed to approximate the swell spectrum, the value $\gamma = 7$ to 15 suffices to model long-traveled swell (Goda, 1983a).

The frequency spectra of Eqs. (9) and (10) are characterized with the spectral density that decreases with the frequency in proportion to f^{-5} in the high frequency range. Phillips (1958) theorized this frequency dependence as the result of wave breaking in deep water which produces the equilibrium range of spectral density. The swell spectra show much faster decrease with the frequency, while waves in shallow water tend to have the spectra in proportion to f^{-4} to f^{-3}. Such a variety of spectral function can be simulated by the Wallops spectrum by Huang *et al.* (1981), by choosing an appropriate exponent m to the frequency f. The Wallops spectrum has been rewritten by Goda (1988) with the input parameters of m, $H_{1/3}$ and T_p.

Phillips' theory of equilibrium wave spectrum has been extended to waves in water of finite depth by Kitaigorodoskii *et al.* (1975), who showed that the equilibrium spectrum there should have a slower rate of decrease of spectral density with the frequency than that in deep water. Bouws *et al.* (1985) applied this equilibrium form to the field data and proposed the following spectral function with the name of TMA spectrum:

$$S(f) = S_J(f)\phi(kh), \tag{15}$$

where $S_J(f)$ is the JONSWAP spectrum and $\phi\ (kh)$ is a function, setting the upper bound of the equilibrium range of spectrum. The function $\phi\ (kh)$ is expressed as below (see Tucker, 1994):

$$\phi(kh) = \frac{\tanh^2 kh}{1 + 2kh/\sinh 2kh}. \tag{16}$$

The significant wave height H_{m0} evaluated for the TMA spectrum gradually decreases as the water depth decreases. As implied by the approach by Kitaigorodoskii *et al.* (1975), the decrease is due to wave breaking in finite water depth. Goda (1998) showed that the decrease in the significant wave height with the TMA spectrum is approximately in agreement with the result of calculation by the random wave breaking model by Goda (1975) for the wave steepness $H_0/L_0 = 0.04$ to 0.08 on the uniform sea bottom slope of 1/100.

3.2 *Directional spreading function*

In comparison with a large amount of frequency spectrum data accumulated since around the 1950s, the field measurements of directional wave spectrum are not so abundant to form the reliable database. Several functional forms of the directional spreading functions $G(\theta|f)$ have been proposed and utilized, but most of them have been derived for convenience of mathematical derivation: an exception is the Mitsuyasu-type spreading function based on field measurement as will be discussed below.

(1) Cosine Power Type Spreading Function

A well-known directional spreading function is of the cosine power type of the following form:

$$G_1 = \begin{cases} \dfrac{2l!!}{\pi(2l-1)!!}\cos^{2l}(\theta-\bar{\theta}) & : |\theta-\bar{\theta}| \le \dfrac{\pi}{2} \\ 0 & : |\theta-\bar{\theta}| > \dfrac{\pi}{2}, \end{cases} \tag{17}$$

in which $\bar{\theta}$ denotes the mean wave direction and $2l!! = 2l(2l-2)(2l-4)\cdots 2$.

The cosine-squared function with $l = 1$ is the oldest formula and has been used widely because of its functional simplicity. The cosine power type spreading function is usually applied for the whole frequency range with the same spreading characteristics.

(2) Half-Cosine Power type and Mitsuyasu-type Spreading Functions

In the field measurement of directional waves, the wave direction cannot be confined in a half plane of $|\theta-\bar{\theta}| \le \pi/2$, but can be extended to the full plane of $|\theta-\bar{\theta}| \le \pi$. Thus, the following half-cosine power type function has been applied for the field data since Longuet-Higgins *et al.* (1963):

$$G_2(\theta|f) = G_0\cos^{2s}\left(\frac{\theta-\bar{\theta}}{2}\right), \tag{18}$$

where G_0 is the constant to satisfy the normalization condition of Eq. (5) as below.

$$G_0 = \left[\int_{\theta_{\min}}^{\theta_{\max}} \cos^{2s}\left(\frac{\theta-\bar{\theta}}{2}\right) d\theta\right]^{-1}, \tag{19}$$

where the range of integration is normally set as $\theta_{\min} = \bar{\theta} - \pi$ and $\theta_{\max} = \bar{\theta} + \pi$, but might be set otherwise when the topographic situation limits the range of wave incidence.

Actual ocean waves usually exhibit narrow directional spreading around a spectral peak frequency and broad spreading at lower and higher frequency ranges. In other words, the spreading parameter s is not constant but variable depending on the frequency. Mitsuyasu *et al.* (1975) derived the following formula based on their own ocean measurements and other data:

$$s = \begin{cases} (f/f_p)^{2.5} s_{\max} & : f \le f_p \\ (f/f_p)^{-5} s_{\max} & : f > f_p. \end{cases} \tag{20}$$

Though, some oceanographers propose different formulas for the frequency dependency of the spreading parameter s, Eq. (20) has been accepted as representative of wind-generated waves.

Mitsuyasu *et al.* (1975) reported the peak value of spreading parameter $s_{\max}$ being in the range of 5 to 30 with the mean of about 10 for wind waves. Then, for the purpose of engineering applications, Goda and Suzuki (1975) tentatively proposed to use the following values inclusive of swell:

$$s_{\max} = \begin{cases} 10 : \text{wind waves} \\ 25 : \text{swell with short decay distance} \\ 75 : \text{swell with long decay distance.} \end{cases} \quad (21)$$

The spreading function $G_2(\theta|f)$ of Eq. (18) together with Eqs. (19) and (20) was named as the Mitsuyasu-type spreading function by Goda and Suzuki (1975), and has been used as the standard directional spectrum coupled with the Bretschneider–Mitsuyasu frequency spectrum in Japan.

(3) Wrapped-Normal Spreading Function

In simulating directional waves in test basins, some laboratory uses the following wrapped-normal function as proposed by Borgman (1984):

$$G_3(\theta) = \frac{1}{2\pi} + \frac{1}{\pi}\sum_{n=1}^{N} \exp\left[-\frac{(n\sigma_\theta)^2}{2}\right] \cos n(\theta - \bar{\theta}), \quad (22)$$

where the parameter σ_θ is the angular standard deviation (in radians) of the spreading function defined as

$$\sigma_\theta^2 = \int_{\bar{\theta}-\pi/2}^{\bar{\theta}+\pi/2} (\theta - \bar{\theta})^2 G_3(\theta) d\theta. \quad (23)$$

The number of terms of the series N can be set as the largest integer not exceeding $10/\sigma_\theta$ or the condition $(\mathrm{N}\sigma_\theta)^2 < 100$ so as to avoid the underflow of exponential function in digital computation.

The wrapped-normal spreading function is uniquely determined with the input parameter σ_θ, which is easily understood by engineers. The angular standard deviation σ_θ can be set as varying with the frequency, but usual practice is to use a single value for the whole frequency range as done by Vincent and Briggs (1989) and Briggs *et al.* (1995).

(4) Intercomparison of Directional Spreading Functions

The pattern of angular spreading of wave energy over the direction depends on the spreading function being employed. It is sometimes required to compare the degree of angular spreading of various functions. For such purposes the frequency-integrated spreading function of the Mitsuyasu-type is defined as

$$\overline{G}_2(\theta) = \frac{G_0}{m_0} \int_0^\infty S(f) \cos^{2s}\left(\frac{\theta - \bar{\theta}}{2}\right) df. \quad (24)$$

Owing to the frequency dependence of the spreading parameter s, the variation of $\overline{G}_2(\theta)$ with the direction θ is slightly affected with the functional form of frequency spectrum employed into Eq. (24).

In order to make inter-comparison of various spreading functions, Goda (1998, 1999) evaluated the angular standard deviations of the cosine power type function of Eq. (17), the half-cosine power type function of Eq. (18), and the frequency-integrated one of Eq. (24). He obtained the following empirical relations from the results of numerical calculation:

$$\sigma_\theta \cong \frac{180^\circ}{\pi}\left(\frac{2}{2.2+4l}\right)^{1/2} \quad : \text{cosine power type,} \tag{25}$$

$$\sigma_\theta \cong \frac{180^\circ}{\pi}\left(\frac{2}{1+s}\right)^{1/2} \quad : \text{half-cosine power type.} \tag{26}$$

The above formulas give the standard deviation of $\sigma_\theta \cong 33^\circ$ for both the $\cos^2\theta$ function ($l = 1$) and the $\cos^{10}(\theta/2)$ function ($s = 5$).

For the Mitsuyasu-type spreading function, the mean spreading parameter $\bar{s}$ has been calculated with Eq. (24) through the evaluation of the angular standard deviation. The value of the mean spreading parameter is affected by the peak enhancement factor γ when the JONSWAP spectrum is coupled with the spreading function as in the following:

$$\bar{s} = \begin{cases} (s_{\max} + 4.0)/2.8 : \gamma = 1.0 \\ (s_{\max} + 1.0)/2.1 : \gamma = 3.3 \\ (s_{\max} + 0.2)/1.7 : \gamma = 10. \end{cases} \tag{27}$$

The directional wave spectrum with the spreading parameter $s_{\max} = 75$ and the peak enhancement factor $\gamma = 10$ is estimated to have the angular spreading equivalent to the half-cosine type function with $\bar{s} = 44.2$. Use of Eq. (26) indicates that the angular standard deviation would be about 12°.

For other types of directional spreading functions such as: the circular-normal and hyperbolic secant-squared types, please refer to the comparative study by Goda (1999).

3.3 *Multi-peaked wave spectrum*

In the severe storm condition, ocean waves tend to show single-peaked spectra around one main direction. In the medium to mild sea states, however, locally-generated wind waves are often superposed on swell, thus producing double-peaked spectra with two distinct directions: sometimes, three spectral peaks can be distinguished.

Because the situation in which multi-peaked wave spectra are generated is so diverse, no effort has yet been made to standardize the multi-peak directional spectra for engineering applications. For the frequency spectra of ocean waves, Ochi and Hubble (1976) have given a six-parameter model.

In numerical and/or laboratory modeling of wave actions and/or transformations, there arises a need to introduce multi-peaked wave spectrum. It can be realized by linearly superposing the spectral density of multiple wave spectra as in the following:

$$S(f,\theta)df\,d\theta = \lfloor S_1(f,\theta|(H_{1/3})_1,(T_p)_1,\theta_1) + S_2(f,\theta|(H_{1/3})_2,(T_p)_2,\theta_2) + \cdots \rfloor df\,d\theta, \tag{28}$$

where S_1, S_2, and so on are the single-peaked spectrum defined by the respective significant wave height $(H_{1/3})_i$, the peak period $(T_p)_i$ (or $T_{1/3}$, $\overline{T}$ etc.), and the mean direction $\bar{\theta}_i$, where $i = 1, 2, \ldots$. Once the input parameters of the height, period, and direction are assigned, appropriate functional forms are chosen to represent individual spectra and then the overall spectral density is easily calculated. There is no guideline for selection of the input parameters and individual spectral functions, but they must be made with due consideration of the field conditions under study.

The overall significant wave height of multi-peaked spectral waves is approximately given as below.

$$(H_{1/3})_{\text{total}} = \sqrt{(H_{1/3})_1^2 + (H_{1/3})_2^2 + \cdots}. \tag{29}$$

The above formula is supported by the fact that the Rayleigh distribution of zero-crossing wave heights is universally applicable regardless of spectral shape. Equation (29) also serves for allocating the component heights of wind waves and swell from a given design wave height.

The situation is different for wave periods, because the period distribution is quite sensitive to the functional shape of frequency spectrum. The only reliable formula for estimating the overall wave period is the following for the mean period:

$$(\overline{T})_{\text{total}} = \sqrt{\frac{(H_{1/3})_1^2 + (H_{1/3})_2^2 + \cdots}{\frac{(H_{1/3})_1^2}{\overline{T}_1^2} + \frac{(H_{1/3})_2^2}{\overline{T}_2^2} + \cdots}}. \tag{30}$$

Tanimoto *et al.* (1986) have proposed an empirical formula to estimate the equivalent significant wave period $T_{1/3}$ for a wave system composed of two wave groups. The formula has been derived for calculation of wave forces on composite breakwaters. Its functional form is similar with Eq. (30) but has an empirical factor, which is a function of the wave height and period ratios of the two wave groups (see OCDI (2002), p. 70).

For some design work, the information of spectral peak period is required. In multi-peaked wave spectra, the definition of spectral peak becomes ambiguous, but some estimate needs to be given to the spectral peak period. With regards to wave run-up and overtopping of seawalls, van der Meer and Janssen (1995) recommend the use of the equivalent peak period of the following by dividing a measured spectrum into two parts of the low and high frequency ranges:

$$(T_p)_{\text{eqv}} = \left[\frac{(m_0)_{\text{low}}}{m_0}(T_p)^4_{\text{low}} + \frac{(m_0)_{\text{high}}}{m_0}(T_p)^4_{\text{high}}\right]^{1/4} : m_0 = (m_0)_{\text{low}} + (m_0)_{\text{high}}, \tag{31}$$

where m_0 denotes the total spectral area (zeroth moment), $(m_0)_{\text{low}}$ and $(T_p)_{\text{low}}$ are the spectral area and peak period of the low frequency part, respectively and $(m_0)_{\text{high}}$ and $(T_p)_{\text{high}}$ are those of the high frequency part, respectively.

4. Statistical Distributions of Wave Heights and Periods

4.1 *Distribution of wave heights*

(1) Rayleigh Distribution

In a number of problems dealing with wave actions and transformations, the wave-by-wave approach rather than the spectral approach provides a better means to solve the problem. Random wave breaking and wave overtopping rate are typical examples.

The distribution of individual wave heights is approximately represented by the Rayleigh distribution of the following form:

$$p(x)dx = 2a^2 x \exp[-a^2x^2]\,dx : x = H/H_*, \tag{32}$$

in which $p(x)$ denotes the probability density function of the dimensionless wave height x normalized with a reference height H_*. The constant a takes the following value depending on the choice of the reference height H_*:

$$a = \begin{cases} 1.416 & : H_* = H_{1/3} \\ 1 & : H_* = H_{\text{rms}} \\ \sqrt{\pi}/2 & : H_* = \overline{H} \\ 1/2\sqrt{2} & : H_* = \sqrt{m_0} = \eta_{\text{rms}}, \end{cases} \tag{33}$$

where the subscript rms stands for the root-mean-square value.

The applicability of the Rayleigh distribution was proven by Longuet-Higgins (1952) for waves with a very narrow spectral bandwidth such that every maximum of surface elevation constitutes a wave crest and every minimum of surface elevation becomes a wave trough. Real ocean waves have a rather broad spectral bandwidth, and wave profiles have a number of maxima and minima between a wave crest and the next crest. Thus, the Rayleigh distribution is not correct theoretically, but its applicability has been confirmed by almost every wave observation so long as individual waves are defined by the zero-upcrossing or zero-downcrossing method, even though the reason for such an applicability has not been clarified yet.

(2) Relationships among Representative Wave Heights

Under the Rayleigh distribution, the following relationships among representative wave heights are easily derived:

$$H_{1/3} = 1.416H_{\rm rms} = 1.597\overline{H} = 4.004\eta_{\rm rms} \tag{34}$$

$$H_{\rm rms} = 1.128\overline{H} = 2.828\eta_{\rm rms} \tag{35}$$

$$\overline{H} = 2.507\eta_{\rm rms}. \tag{36}$$

The relationship in Eq. (6) between the significant wave height and the zeroth spectral moment is based on the assumption of Rayleigh distribution by Eq. (34).

(3) Maximum Wave Height

The maximum wave height among N_0 waves is predicted in an undeterministic but a probabilistic manner. Its probability density function is derived as below (e.g., Goda (2000); Sec. 9.1.3)

$$p^*(x_{\rm max})dx_{\rm max} = 2a^2 x_{\rm max}\xi e^{-\xi}dx_{\rm max} : x_{\rm max} = H_{\rm max}/H_*, \tag{37}$$

where

$$\xi = N_0 \exp[-a^2 x_{\rm max}^2]. \tag{38}$$

The most frequently appearing maximum height is derived from Eq. (37). With the significant height as the reference, it is obtained as

$$(H_{\rm max})_{\rm mod} \cong 0.706H_{1/3}\sqrt{\ln N_0}. \tag{39}$$

The probability μ that $H_{\rm max}$ exceeds a specified value is also calculated as

$$\mu \cong 1 - \exp\left\{-N_0 \exp\left[-2.005\left(\frac{H_{\rm max}}{H_{1/3}}\right)^2\right]\right\}. \tag{40}$$

For example, $H_{\rm max}$ exceeding $2.5H_{1/3}$ among 1,000 waves will appear at the rate of one record per 277 records, while that exceeding $3.0H_{1/3}$ will be one per 69,000 records, if the wave heights strictly follow the Rayleigh distribution.

There exists a slight difference between the Rayleigh and the actual distribution of wave heights. One factor is the effect of spectral shape, which is discussed in Sec. 4.3, and another factor is the effect of wave nonlinearity, which is discussed in Sec. 5.2.

4.2 *Joint distribution of wave heights and periods*

The distribution of wave periods is dependent on the functional shape of frequency spectrum. A swell with a sharp spectral peak has a narrow period distribution, wind waves with a broad spectral spread exhibit a broad range of period variation, and a sea state of combined swell and wind waves may show double modes of the period distribution which correspond to the peak periods of swell and wind waves, respectively. In the following, discussions are limited to the case of waves with single peaked spectra.

(1) Marginal Distribution of Wave Periods

The marginal distribution of wave periods can be approximated by the following function, which was derived by Longuet-Higgins (1975) under the assumption of a very narrow spectral bandwidth:

$$p(\tau) = \frac{\nu^2}{2[\nu^2 + (\tau - 1)^2]^{3/2}} \quad : \quad \tau = T/\overline{T}, \tag{41}$$

where $p(\tau)$ denotes the probability density function of the dimensionless wave period τ, which is normalized with the mean period $\overline{T}$. The symbol ν is a spectral width parameter introduced by Longuet–Higgins and defined as

$$\nu = \left[\frac{m_0 m_2}{m_1^2} - 1\right]^{1/2}, \tag{42}$$

which was assumed very small in theoretical derivation.

On the other hand, actual wind waves yield the ν value ranging from 0.3 to 0.7, which is not appropriate to be substituted into Eq. (41). Instead, an empirical value of ν defined below should be employed to describe the wave period distribution as recommended by Longuet–Higgins (1975).

$$\nu = \frac{\sqrt{3}}{2}(Q_3 - Q_1), \tag{43}$$

where Q_3 and Q_1 denote the upper and lower quartiles of period distribution, respectively.

Equation (41) describes a distribution symmetrical with respect to $\tau = 1$. The distribution of actual waves, however, shows a slight asymmetry with the mode at $\tau = 1.1$ to 1.2 (Goda, 1978).

(2) Joint Distribution of Wave Heights and Periods

The marginal distribution of Eq. (41) is derived from the following joint distribution by Longuet–Higgins (1975):

$$p(x, \tau) = \frac{2a^3x^2}{\sqrt{x}\nu} \exp\left\{-a^2x^2\left[1 + \frac{(\tau - 1)^2}{\nu^2}\right]\right\}, \tag{44}$$

where $p(x, \tau)$ is the joint probability density function of the dimensionless height x and the dimensionless period τ.

The joint distribution by Eq. (44) is symmetrical with respect to $\tau = 1$, so that the correlation between the height and period is zero. However, wave periods in the sea exhibit a positive correlation with wave heights; *e.g.* the correlation coefficient

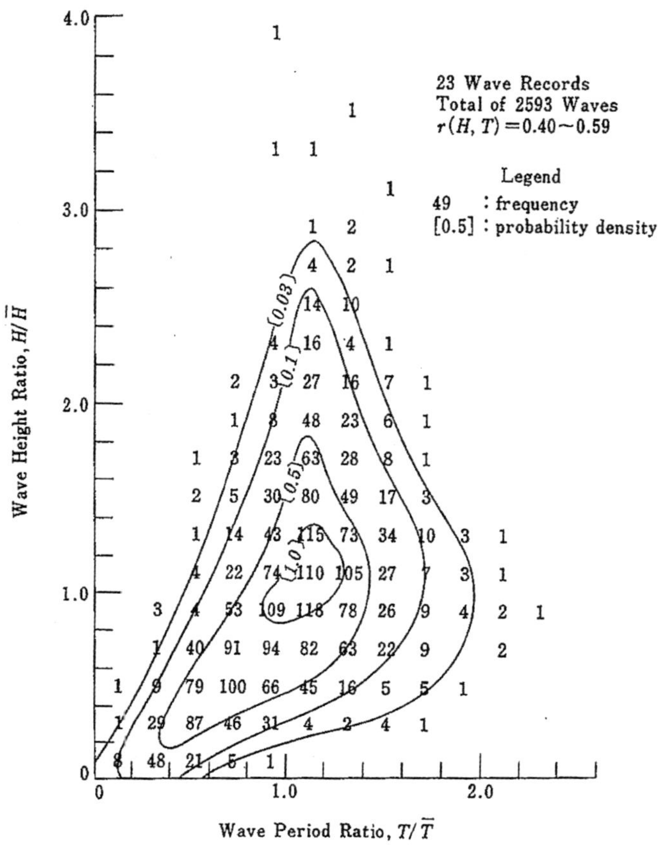

Fig. 2.2 Joint distribution of dimensionless wave heights and periods of 23 wave records having the correlation coefficient between 0.40 and 0.59 (Goda, 1978).

may exceed 0.6 in some case. Figure 2.2 is an example of field data on the joint distribution of wave heights and periods, which represents a compilation of 23 wave records having the correlation coefficient between 0.40 and 0.59: the heights and periods have been normalized with the mean values of respective records. As shown in Fig. 2.2, there is a strong correlation among waves of small heights. On the other hand, waves of large heights, say greater than the mean height, show no correlation between heights and periods and the joint distribution is symmetrical with respect to $\tau \cong 1.2$.

Several theories have been proposed to describe such a tendency of strong correlation among small waves and no correlation among large waves, including a modified version of Eq. (44) by Longuet–Higgins (1983) himself. However, no satisfactory theory has yet been derived. In case that the information of a joint distribution of wave heights and periods is required for waves with a given spectral characteristics, a recommended practice is to make a Monte Carlo simulation of surface elevation, to count the heights and periods of individual waves defined by the zero-crossing method, and to tabulate them in a form of joint distribution.

(3) Relationships between Representative Wave Periods

The wave-by-wave analysis yields several representative periods such as $T_{\max}$ (period of largest wave in a record), $T_{1/10}$ (mean of highest one-tenth waves), $T_{1/3}$ and $\overline{T}$. Relationships between these representative periods vary from a wave record to another, but statistical analysis of many wave data tends to indicate the following relationships:

$$T_{\max} \cong T_{1/10} \cong T_{1/3} \cong 1.2\overline{T}. \tag{45}$$

The approximate equality among $T_{\max}$, $T_{1/10}$ and $T_{1/3}$ reflects the tendency of no correlation between heights and periods among large waves.

The relationships between the spectral peak period T_p and the statistical wave periods of $T_{1/3}$ and $\overline{T}$ are dependent on the functional shape of frequency spectrum. Equations (12) and (13) provide such relationships for the case of JONSWAP spectrum as a function of the peak enhancement factor.

4.3 *Spectral effects on wave heights and periods*

The statistical distributions of wave heights and periods are slightly affected by the functional shape of wave spectrum. The spectral effect has been demonstrated by Goda (1988) through the analysis of numerically simulated wave profiles by means of the Monte Carlo technique. Tables 2.1 and 2.2 list the mean values of the ratios among several heights and periods as a function of spectral shape parameter. The simulation was carried out by producing 2,000 wave profiles for each wave spectrum under the condition that the sampling time interval was set at $\Delta t = T_p/12$ or $f_{\max} = 6.0 f_p$.

The spectrum with the parameter $m = 3$ refers to the Wallops type spectrum of the following function:

$$S(f) \propto H_{1/3}^2 T_p^{1-m} f^{-m} \exp\left[-\frac{m}{4}(fT_p)^{-4}\right]. \tag{46}$$

Table 2.1 Mean values of the ratios of representative wave heights to the rms surface elevation.

Parameter	$m = 3$	$\gamma = 1$	$\gamma = 3.3$	$\gamma = 10$	Rayleigh
$[H_{\max}/\eta_{\rm rms}]$	0.893*	0.925*	0.936*	0.937*	1.000
$H_{1/10}/\eta_{\rm rms}$	4.658	4.781	4.849	4.921	5.090
$H_{1/3}/\eta_{\rm rms}$	3.736	3.836	3.866	3.910	4.004
$H_{\rm rms}/\eta_{\rm rms}$	2.647	2.734	2.757	2.783	2.828
$\overline{H}/\eta_{\rm rms}$	2.359	2.448	2.461	2.482	2.507

Note: The supersript * refers to the ratio of the observed mean value to the theoretical prediction under the Rayleigh distribution.

Table 2.2 Mean values of the ratios among representative wave periods.

Parameter	$m = 3$	$\gamma = 1$	$\gamma = 3.3$	$\gamma = 10$
$T_{\max}/T_p$	0.834	0.872	0.922	0.953
$T_{1/10}/T_p$	0.824	0.889	0.933	0.962
$T_{1/3}/T_p$	0.782	0.883	0.934	0.965
$\overline{T}/T_p$	0.579	0.740	0.779	0.873
$T_{1/3}/\overline{T}$	1.350	1.192	1.162	1.106

Note: The above ratios are based on the simulation with the condition $f_{\max} = 6f_p$.

The parameter γ refers to the JONSWAP spectrum of Eq. (10). The result of $H_{1/3}/\eta_{\rm rms}$ has been utilized in the formulation of Eq. (11), and the ratios $T_{1/3}/T_p$ and $\overline{T}/T_p$ have been formulated as Eqs. (12) and (13).

As observed in Table 2.1, the ratios of various wave heights to the root-mean-square surface elevation $\eta_{\rm rms}$ are smaller than those derived from the Rayleigh distribution. The commonly observed value of $H_{1/3}/\eta_{\rm rms} \cong 3.8$ among the field data on the average corresponds to the Bretschneider–Mitsuyasu spectrum and the JONSWAP spectrum with $\gamma = 1$. Deviations of the wave height ratios from those of Rayleigh distribution indicate the wave height distribution being narrower than the Rayleigh. In fact, Forristall (1978) fitted a Weibull distribution to the wave height data recorded in a storm in the Gulf of Mexico and obtained the following empirical formula:

$$P(\xi) = \exp[-\xi^{2.126}/8.42] : \xi = H/\eta_{\rm rms} \tag{47}$$

where $P(\xi)$ denotes the distribution function of the dimensionless wave height ξ.

There are several more field reports of the wave height distribution being narrower than the Rayleigh. Theoretical approaches have also been made by a number of researchers such as Tayfun (1990b) to derive the wave height distribution for waves of broad spectral bandwidth. However, no theory has yet been successful in fully predicting the wave spectral effects. It should be mentioned here that, as the wave nonlinearity becomes appreciable, the distribution of wave heights becomes broad as will be discussed in Sec. 5.2.

The effect of wave spectrum also appears as the decrease of representative wave periods relative to the peak period with a broadening of frequency spread (decrease in the value of γ or m), as listed in Table 2.2.

4.4 *Wave grouping*

In a train of random waves, high waves are not always isolated but appearing in groups. Such a phenomenon is called the wave grouping. An indicator of wave grouping is the correlation coefficient between successive wave heights: a high correlation is associated with a pronounced wave grouping. The correlation coefficient is about

0.3 among wind waves and goes up to 0.7 in the case of long-traveled swell (Goda, 1983a).

Quantitative analysis of wave grouping is made by several methods. The run length analysis by Goda (1970) is one of the earliest. A succession of waves exceeding a threshold height is called the run of wave height and the number of such successive waves is called the run length, A total run is also defined as a succession of waves after the first wave exceeds the threshold height until the next upward exceedance takes place: the definition is similar with the method of defining zero-upcrossing waves. The probability of the length j of the run of wave height is given by

$$p(j) = p_{22}^{j-1}(1 - p_{22}), \tag{48}$$

where p_{22} denotes the probability that, under the condition of the first wave exceeding the threshold, the second wave also exceeds the threshold. The mean and standard deviation of the run length are derived as follows:

$$\bar{j} = \frac{1}{1 - p_{22}} \tag{49}$$

$$\sigma(j) = \frac{\sqrt{p_{22}}}{1 - p_{22}}. \tag{50}$$

The probability p_{22} can be evaluated by the theory of Kimura (1980) under the assumption of the Rayleigh distribution for wave heights as below:

$$p_{22} = \frac{\int_{H_c}^{\infty}\int_{H_c}^{\infty} p(H_1, H_2)dH_1 dH_2}{\int_{H_c}^{\infty} p(H_1)\, dH_1}, \tag{51}$$

where $p(H_1)$ is the marginal probability density function of wave heights (Rayleigh distribution), and $p(H_1, H_2)$ is the joint probability density function of successive wave heights H_1 and H_2.

The joint probability density function $p(H_1, H_2)$ has been derived by Kimura (1980) as follows:

$$p(H_1, H_2) = \frac{4H_1H_2}{(1-\kappa^2)H_{\rm rms}^4} \exp\left[-\frac{H_1^2 + H_2^2}{(1-\kappa^2)H_{\rm rms}^2}\right] I_0\left[\frac{2_k H_1 H_2}{(1-\kappa^2)H_{\rm rms}^2}\right] \tag{52}$$

where I_0 is the modified Bessel function of the first kind and κ is the parameter defined by

$$\kappa(\tau)^2 = \left|\frac{1}{m_0}\int_0^{\infty} S(f)\cos 2\pi f\tau\, d\tau\right|^2 + \left|\frac{1}{m_0}\int_0^{\infty} S(f)\sin 2\pi f\tau\, d\tau\right|^2. \tag{53}$$

The parameter τ is the time lag and is set equal to the mean wave period $\overline{T}$ in the analysis of run length statistics.

The correlation coefficient between successive wave heights is governed by the above parameter κ, which takes the value between 0 and 1. Kimura (1980) derived the relationship between them and estimated the κ value from the correlation coefficient between successive wave heights. Battjes and van Vledder (1984) as well as Longuet–Higgins (1984) promoted the use of Eq. (53) for evaluation of κ from spectral information.

When a wave spectrum has a sharp peak, the parameter κ takes a value close to 1 and so is the wave height correlation coefficient. The situation yields a large value of the probability p_{22}, which results in a conspicuous wave grouping. When a wave spectrum shows a broad peak, then the κ value is close to 0 and wave grouping is not well established. Thus, the wave grouping as described with the run length is controlled by the frequency spectrum through the parameter κ of Eq. (53).

Another method of describing the degree of wave grouping is by means of the wave groupiness propcsed by Funke and Mansard (1980). The surface elevation η of a wave record is squared and smoothed with a triangular filter to yield a smoothed instantaneous wave energy history (SIWEH). The coefficient of variation of the SIWEH profile is computed and defined as the groupiness factor GF. Funke and Mansard (1980) insisted that GF is a vital parameter to realistically simulate sea waves in the laboratory. However, the GF value exhibits a variation during wave propagation in the sea, because the phase differences between spectral wave components constantly change with wave propagation and so is the time history of surface elevation at various locations. A large GF value indicates that a particular wave record has a large degree of energy variation, which can be interpreted as the presence of large waves much higher than the rest of waves. In other words, the wave height distribution is broad and has a long tail. The groupiness factor is a random variable describing one feature of a sample of wave data, *i.e.*, wave profile record, with regard to the fluctuation of wave heights.

5. Properties of Weakly Nonlinear Waves

5.1 *Parameters of wave nonlinearity*

(1) Parameters Characterizing the Distribution of Surface Elevation

Representation of random wave field with Eq. (1) is based on the assumption of wave linearity or the Gaussian (normal) distribution of surface elevation, as stated in Sec. 2.1. A departure from the Gaussian is measured with the skewness $\sqrt{\beta_1}$ and the kurtosis β_2, defined below.

$$\sqrt{\beta_1} = \frac{1}{\eta_{\rm rms}^3} \cdot \frac{1}{N} \sum_{i=1}^{N} (\eta_i - \bar{\eta})^3 \tag{54}$$

$$\beta_2 = \frac{1}{\eta_{\rm rms}^4} \cdot \frac{1}{N} \sum_{i=1}^{N} (\eta_i - \bar{\eta})^4 \tag{55}$$

where N denotes the number of data points in a wave record and $\bar{\eta}$ is the mean of the surface elevation η.

The Gaussian distribution yields $\sqrt{\beta_1} = 0$ and $\beta_2 = 3.0$. Waves in the sea exhibit a positive skewness up to around 1.0 and the kurtosis greater than 3.0 up to around 4.0. A positive skewness corresponds to the tendency of water waves having sharp and high crests associated with flat and shallow troughs. A kurtosis greater than 3.0 indicates a presence of long tails of the distribution at the extreme sides compensated by a concentration around $\eta = 0$. In other words, a large kurtosis suggests the existence of quite large waves in a wave record. The wave height ratios such as $H_{\max}/H_{1/3}$, $\eta_{\max}/H_{1/3}$ and $H_{1/3}/\overline{H}$ are strongly correlated with the kurtosis β_2 as seen in the data by Goda and Nagai (1974). Mori and Yasuda (1996) have derived a theory of wave height distribution with the non-Gaussian surface elevation to clarify the effect of kurtosis on large waves.

The skewness is a measure of asymmetry with respect to the mean water level. In the analysis of the action of large waves upon a structure, the crest-to-height ratio is sometimes called the horizontal asymmetry (PIANC 1986, IAHR 1989). The ratio varies from about 0.6 in deep water to about 0.8 in very shallow water.

Waves at the stage of breaking, either under strong wind actions or in shoaling water, generally exhibit a forward tilting in their profiles. Such a feature can be described with the following parameter, which is called the atiltness by Goda (1986):

$$\beta_3 = \frac{\frac{1}{N}\sum_{i=1}^{N-1}(\dot{\eta}_i - \bar{\dot{\eta}})^3}{\left[\frac{1}{N}\sum_{i=1}^{N-1}(\dot{\eta}_i - \bar{\dot{\eta}})^2\right]^{3/2}}, \tag{56}$$

where $\dot{\eta}$ denotes the time derivative of surface elevation.

A positive atiltness indicates the forward tilting, while a negative atiltness is for the backward tilting. When the atiltness takes a large positive value as occurring within the surf zone, the representative periods of large waves defined by the zero-upcrossing method tend to be longer than those defined by the zero-downcrossing although no difference is found with regard to the representative heights (Goda, 1986).

(2) Parameters Governing Wave Nonlinearity

The parameter directly related to wave nonlinearity is the wave steepness H/L in deep water and the relative height H/h in shallow water. The steepness of random waves is expressed as $H_{1/3}/(L_{1/3})_0$, $\eta_{\rm rms}/(L_p)_0$, and others, where $(L_{1/3})_0$ and $(L_p)_0$ denote the deepwater wavelengths corresponding to the significant and spectral peak periods, respectively.

In the analysis of regular waves in shallow water, the Ursell parameter $\eta L^2/h^3$ is often employed. In order to fill the gap between the wave steepness in deep water and the Ursell parameter in shallow water, Goda (1983b) has proposed a unified nonlinearity parameter of the following form though not based on theory:

$$\Pi = \frac{H}{L\tanh^3 kh}, \tag{57}$$

in which the wave number k and the wavelength L are to be calculated by Eq. (2) based on the small amplitude wave theory. For random waves, the Π-parameter is evaluated with the significant or maximum wave height and the wavelength corresponding to the significant or spectral peak period.

5.2 *Wave nonlinearity effects on wave heights and periods*

(1) Wave Nonlinearity Effects on Wave Heights

The skewness of surface elevation increases as the wave nonlinearity becomes strong. Figure 2.3 shows the relationship between the skewness $\sqrt{\beta_1}$ and the parameter $\Pi_{1/3}$ based on the significant wave height for the field data in deep to shallow water (Goda, 1983b). The symbols L_A and k_A appearing in the parameter $\Pi_{1/3}$ in the abscissa denote the wavelength and the wave number calculated by the small wave theory by Airy. As stated before, a positive skewness is associated with a raise of crest elevation relative to the wave height. Theoretical prediction of the crest rise

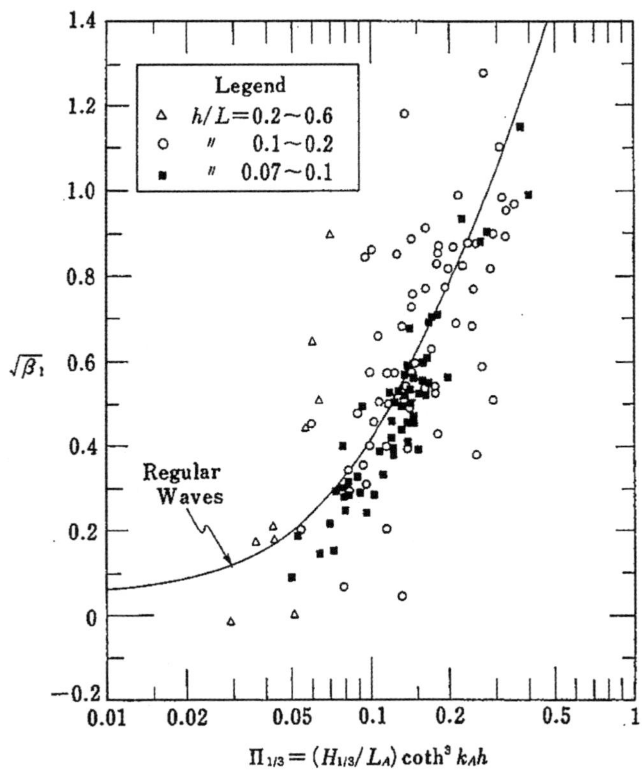

Fig. 2.3 Field data on the relationship between the skewness $\sqrt{\beta_1}$ and the nonlinearity parameter $\Pi_{1/3}$ (Goda, 1983b): L_A and k_A refer to the wavelength and wavenumber of small amplitude waves, respectively.

of random waves by wave nonlinearity has been made by Tayfun (1983) based on the calculation of secondary interaction between spectral components. Vinje (1989) presented the wave data in the North Sea in which the maximum crest elevation exceeds the expected value according to the Rayleigh distribution as the wave height increases. Recently, Tayfun (2004) presents a theory of statistics of largest crests in random waves and shows good agreement with storm wave data in the North Sea.

Wave heights are affected by the nonlinearity at the third order, and thus the nonlinearity effect on wave heights is less pronounced than that on crest elevation as demonstrated by Vinje (1989). Nevertheless, the ratio such as $H_{\rm rms}/\eta_{\rm rms}$ in deep water increases several percent with the increase in wave steepness. Longuet–Higgins (1980) predicted the increase of $H_{\rm rms}$ by calculating the value of $\eta_{\rm rms}$ of nonlinear regular waves and estimating the expected value of $H_{\rm rms}/\eta_{\rm rms}$ under the assumption of Rayleigh distribution of wave heights. Yasuda and Mori (1994) have evaluated the nonlinearity effect on wave heights by using their third-order interaction theory.

In the water of intermediate to shallow depth, nonlinearity effects are more pronounced. Figure 2.4 shows the variation of the ratio $H_{1/3}/\eta_{\rm rms}$ with the nonlinearity parameter $\Pi_{1/3}$ (Goda, 1983b). The line marked as "Theory" was drawn by making

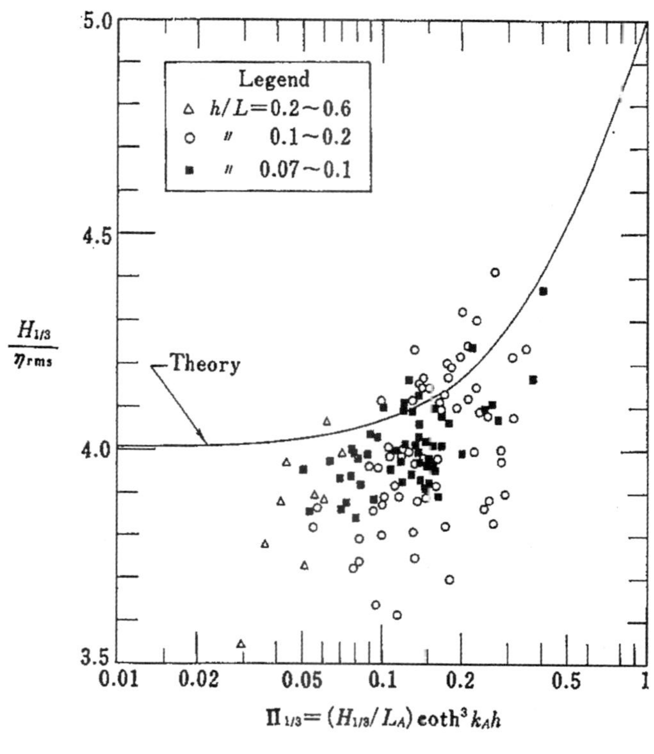

Fig. 2.4 Variation of wave height ratio $H_{1/3}/\eta_{\rm rms}$ with the nonlinearity parameter $\Pi_{1/3}$ (Goda, 1983b).

calculation similar with that by Longuet–Higgins (1980). In the region $\Pi_{1/3} < 0.1$, the ratio $H_{1/3}/\eta_{\rm rms}$ indicates a mean of about 3.8 which agrees with many field data. In the region $\Pi_{1/3} > 0.1$, however, a clear tendency of the increase of the ratio $H_{1/3}/\eta_{\rm rms}$ with the $\Pi_{1/3}$ parameter is observed: some data show the ratio exceeding the value 4.0 by the Rayleigh distribution.

An increase in the occurrence probability of large waves owing to wave nonlinearity has been addressed as the problem of freak waves or unexpectedly large waves in the ocean, which are thought to be a cause of marine accidents. Some researcher defines freak waves as the situation under which the maximum wave height exceeds 10 m and its ratio to $H_{1/3}$ is greater than 2.0. Yasuda and Mori (1997) have examined wave records around Japanese islands, which were measured with ultrasonic-type (inverted echo-sounder type) sensors. Their conclusion was such that the occurrence probability of freak waves is well described with the prediction derived from the Rayleigh distribution as given in Eq. (40). Tayfun (2004) also shows that two exceptionally large waves observed in the data he analyzed are simple outliers within the context of his extreme-value statistics. Because the waves in the low to medium sea state have the height distribution narrower than the Rayleigh, the return to the Rayleigh distribution should be attributed to the nonlinearity effect.

The above discussions of wave nonlinearity effects on the skewness and wave heights are limited to nonbreaking waves. When waves begin to break in finite depth water, the height of maximum wave is limited by the water depth and the range of wave height distribution becomes narrow. The surface elevation approaches to the Gaussian distribution again, and the skewness begins to take a small value.

(2) Wave Nonlinearity Effect on Wave Periods

The wave nonlinearity does not affect the statistics of period derived by the wave-by-wave analysis. When the mean wave period is estimated with the spectral moments as by Eq. (7), however, the spectral mean period $T_{0,2}$ becomes smaller than the zero-crossing mean period $\overline{T}$; in shallow water $T_{0,2}$ may become smaller than 80% of $\overline{T}$. It is due to the contribution of phase-locked nonlinear waves to wave spectra, as will be discussed in the next sub-section. The spectral density at the frequency around twice the peak frequency and beyond is mostly composed of second order waves. Such second order components superficially increase the second spectral moment m_2 and decrease the spectrally-evaluated mean period $T_{0,2} = \sqrt{m_0/m_2}$. Thus, for better application of spectrally-derived theories of wave statistics to field wave data, the portion of frequency spectrum beyond the frequency (1.5–1.8) f_p is often cut out so as to remove the effect of nonlinear wave components.

5.3 *Nonlinear spectral components and infragravity waves*

In a train of regular waves, a single component of sinusoidal profile with the fundamental frequency alone cannot satisfy the nonlinear surface boundary condition, and the higher order nonlinear components with the twice, thrice, and higher harmonic frequencies must accompany the linear component. The result is a train of Stokes waves. When two trains of regular waves with different frequencies are superposed, there appear the second order nonlinear waves with the frequencies equal to the sum and difference of the two linear frequencies. In a random wave system, each pair of spectral components interacts and generates the sum and difference secondary waves. Tick (1963) was the first in presenting the theory of spectral second order interactions, and Hamada (1965) gave a correction to it.

The theory was applied by Goda (1983a) to the frequency spectrum of long-traveled swell in order to resolve it into the first and second order components by iteration. Figure 2.5 is an example of spectral resolutions. The observed spectrum had the main peak at the frequency $f_p \cong 0.055\,\text{Hz}$ and a secondary peak at $f_p \cong 0.11\,\text{Hz}$. The second order spectrum coincides with this secondary peak, indicating that no free wave component exist beyond $f > 0.1\,\text{Hz}$. According to Tayfun (1990a),

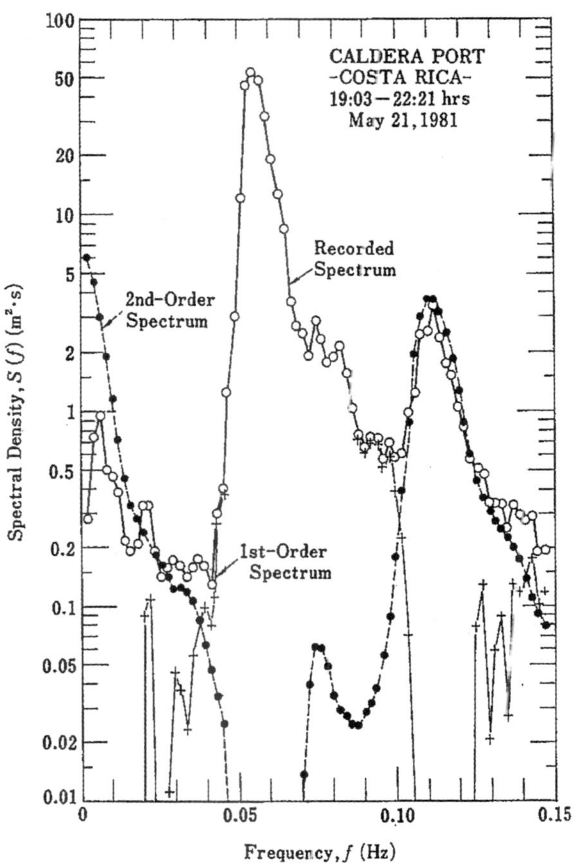

Fig. 2.5 Resolution of linear (first-order) and nonlinear (second-order) spectral components for the spectrum of long-traveled swell (Goda, 1983a): the water depth being about 17 m.

the linear spectral components of wind waves in the high frequency range attenuate in proportion to f^{-8} to f^{-9}, even though the overall shape of wind wave spectra indicates the f^{-5} attenuation as represented by the standard spectral functions of Eqs. (9) and (10) etc.

Another feature of the observed spectrum is a hump at the low frequency range. This hump represents the secondary components of the difference frequency, and corresponds to the infragravity waves bounded by a wave group. The existence of group-bounded long waves was predicted by Longuet–Higgins and Stewart (1964) with the concept of radiation stress in connection with the surf beat phenomenon. An explicit form of group-bounded long waves in the unidirectional wave system was given by Ottesen Hansen (1978).

The predicted spectral density of low frequency waves in Fig. 2.5 is greater than the observation. The difference is partly attributed to the effect of multi-directionality of ocean waves. The theory of multi-directional infra-gravity waves has been presented by Sand (1982), and then Okihiro *et al.* (1992), Sekimoto *et al.* (1992), Herbers *et al.* (1994), and others.

The group-bounded long waves constitute a major part of infra-gravity waves observed in the nearshore. The problems of infra-gravity waves are also discussed in relation with long-period motions of moored ships in harbors.

6. Incorporation of Wave Randomness in Engineering Practice

6.1 *Spectral transformation of refraction, diffraction, and reflection*

(1) General Principles

The phenomena of wave refraction, diffraction, and reflection are essentially linear processes. Each component of directional wave spectrum undergoes transformations without interference by other component waves. Therefore, the wave spectra after these transformations can be computed by applying the transfer function to the incident wave spectrum. The generalized formula is expressed as below.

$$S_{tt}(f,\Theta)df\,d\Theta = K_{tt}^2 S_{\eta\eta}(f,\theta)\,df\,d\theta, \tag{58}$$

where S_{tt} and $S_{\eta\eta}$ refer to the transformed and incident wave spectra, respectively, K_{tt} is the transfer function or the ratio of transformed height to incident height for monochromatic waves, and Θ is the direction of wave propagation after transformation while θ stands for the incident wave direction.

For wave refraction, the transfer function is defined as the product of the shoaling coefficient K_s (kh) and the refraction coefficient K_r (kh,θ). Diffraction by breakwaters is analyzed by using the diffraction coefficient K_d (k_x, k_y, θ) as the transfer function K_{tt}. Refraction–diffraction problems such as wave transformations over a shoal can be solved by employing the ratio of wave heights at respective locations to incident height as the transfer function. For wave reflection problems in a wave flume, the reflection coefficient $C_r(f)$ of a test structure is taken as the transfer function. However, for evaluation of the 2D wave field in front of a reflective structure, the diffraction of reflected waves must be included in the transfer function (see Goda (2000), Sec. 3.6.2).

Once the directional spectrum of transformed waves is evaluated by Eq. (58), its spectral moments are calculated by Eq. (8) and the wave height H_{m0} and the wave period $T_{0,2}$ are evaluated by Eqs. (6) and (7), respectively: the information of T_p can directly be obtained from the shape of transformed spectrum. In the calculation of spectral moments, care should be taken for the change of wave direction and its differentiated interval.

(2) Diffraction of Directional Random Waves

Transformation of directional random waves can be very different from that of monochromatic waves. Figure 2.6 compares the diffraction diagrams by monochromatic and directional spectral waves through a breakwater gap having the opening equivalent to five times the wavelength (Nagai, 1972): the directional spreading function of the cosine squared type by Eq. (17) with $l = 1$ was used in the computation. The left-hand side of Fig. 2.6 shows the diffraction coefficient of monochromatic (sine) waves, while the right diagram shows that of directional spectral waves. The diffraction coefficient of monochromatic waves shows an intensive concentration along the center axis, while that of directional spectral waves exhibits a rapid

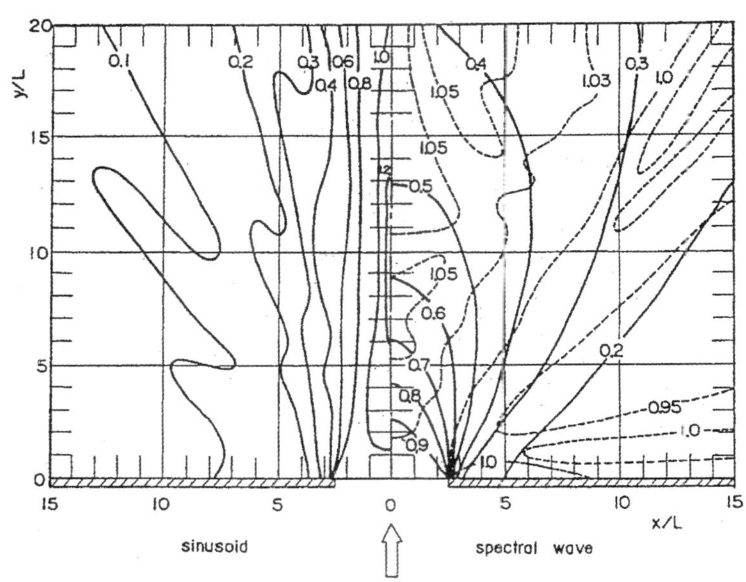

Fig. 2.6 Diffraction diagram of monochromatic waves (left) and that of directional random waves (right) through a breakwater gap of $B/L = 5$ computed by Nagai (1972): solid lines for the ratio of wave height and dashed lines for that of wave period.

attenuation along the center axis but a deep penetration into the sheltered area. Laboratory experiments by Briggs *et al.* (1995) have confirmed such a large difference between monochromatic and directional spectral waves. Because field measurements have demonstrated the superiority of directional diffraction analysis to monochromatic wave analysis as exemplified by Goda (2000; Sec. 3.2), engineering analysis of wave diffraction and reflection must be carried out by introducing directional wave spectra.

Numerical methods for evaluation of transfer functions will be discussed in Chapter 7. An overview of wave transformation analysis with directional spectra is also found in Goda (1997; 2000).

6.2 *Wave breaking and generation of longshore currents*

(1) Features of Random Wave Breaking

Wave breaking is another example in which wave randomness exercises profound influence. In the monochromatic wave system, waves break at a fixed location with the same breaker height and thus we can easily define the wave breaking point. In the random wave system, however, large individual waves break in the water of greater depth, while small individual waves do not break until they reach to the water of smaller depth. If the coast has a pronounced longshore bar, many waves break at the longshore bar which may define the outer edge of surf zone: even so, many small waves will pass over the bar without breaking until they reach to shallow water where they break.

The first theoretical model of random wave breaking process was given by Collins (1970). Then, Battjes (1972), Goda (1975), and others presented improved versions, which were all based on the modification of the wave height distribution by introducing an upper limit due to depth-controlled wave breaking. Various statistical wave heights are calculated from the derived probability density function of wave heights in the surf zone. Another approach was initiated by Battjes and Janssen (1978) who evaluated the rate of energy dissipation by breaking in analogy to a hydraulic bore and took the root-mean-square wave height $H_{\rm rms}$ as the reference wave height. Thornton and Guza (1983) presented a slightly different version of energy dissipation model with $H_{\rm rms}$. There are many other models, which are contesting each other for the accuracy of prediction and the versatility of applications. Goda (2004a) has presented a versatile 2D spectral wave transformation model with the concept of gradational breaker index, which can predict spatial variations of $H_{\max}$, $H_{1/20}$, $H_{1/3}$ and other definitions of wave heights.

Despite differences among random wave breaking models, they are common in negating the concept of a single location of wave breaking and a single value of breaker height. Many coastal sediment problems so far have been investigated with the breaker index curves of monochromatic waves. However, it is necessary to introduce the random wave breaking process in the analysis of coastal sediment problems so as to yield more reliable solutions. Research efforts being made in this direction will produce a new development in coastal morphological studies.

(2) Longshore Currents Induced by Random Waves

The difference between the prediction of longshore currents by regular waves and that by random waves is a typical example that illustrates the validity of random wave approach. As well known, Longuet–Higgins (1970) clarified the generating mechanism of longshore currents as being caused by the spatial gradient of the radiation stress. However, he had to use the breaker index of monochromatic waves at that time and obtained the cross-shore profile of longshore current velocity which jumps to the maximum speed from zero at the wave breaking point and then linearly decreases toward zero at the shoreline. Longuet–Higgins introduced the artifice of the horizontal mixing coefficient so as to yield a smooth variation of longshore current speed that would compare well with laboratory data.

Computation of longshore currents induced by random breaking waves was first undertaken by Battjes (1974). His work remained mostly unnoticed until Thornton and Guza (1986) analyzed the field measurement data and compared it with their model prediction. Goda and Watanabe (1991) formulated empirical expressions for the longshore current profile with the input of deepwater wave height, period, and incident angle, based on the random wave breaking model by Goda (1975). All these models of longshore currents induced by random waves do not require the employment of the horizontal mixing coefficient, because the random breaking process itself produces a smoothly varying profile of the radiation stress. Kuriyama *et al.* (1992) also report a minor role of the horizontal mixing effect by analyzing a large number of longshore current profiles measured along the HORF pier at Hazaki Coast, Japan. In a bar-trough bathymetry, however, the lateral mixing term seems necessary as introduced in the model by Kuriyama (1994).

Another factor to be considered is the effect of surface rollers on longshore and cross-shore currents induced by breaking waves. A recent paper by Tajima and Madsen (2003) demonstrates the surface roller effect. Goda (2004b) also shows the computation of longshore currents on a planar beach with inclusion of the surface roller effect.

Computed results of longshore currents are very sensitive to the random wave breaking model being employed, as demonstrated by Goda (2004b). Examination of various models through calibrations with laboratory and field data of longshore currents is the task to be undertaken at present.

6.3 *Wave actions on structures*

(1) Wave Run-up and Overtopping

Wave run-up on a sloped structure such as a seawall is a highly nonlinear phenomenon, because waves run-up a slope after breaking. It is also a random process with the run-up height varying from wave to wave. If the toe depth of slope is small, the mean water level would fluctuate owing to the presence of infra-gravity waves or surf beats, and the run-up height would be affected by the infra-gravity waves. Numerical simulation of wave run-up on beaches is being tried by several researchers, especially for prediction of infra-gravity waves; *e.g.*, Watson and Peregrine (1992), Roelvink (1993), and others. However, the simulation has not yet been able to produce detailed information of run-up height on seawalls for design purposes.

Presently, the wave run-up is evaluated by laboratory tests using random waves. For representation of random run-up heights, the 2% exceedance run-up height is often employed. That is to say, the cumulative distribution of run-up heights is constructed and the threshold height at the 2% exceedance level is interpolated. Van der Meer and Janssen (1995) discuss the design practice in the Netherlands for estimation of the 2% exceedance run-up height $R_{2\%}$ based on the surf similarity parameter and other factors. The significant wave height is used as the parameter for estimation of $R_{2\%}$. Wang and Grüne (1996) also report results of large-scale tests of wave run-up on revetments with composite slopes. Several empirical formulas for estimation of run-up height have been proposed, which include one by van der Meer (1998).

Wave overtopping of a seawall must also be evaluated by laboratory tests using random waves. In this case, the mean rate of overtopping is a major concern because the drainage system behind the seawall must have a sufficient capacity to deal with the overtopped water to prevent flooding there. If the toe depth of seawall is not so small and the effect of infra-gravity waves is minimum, then a laboratory test with regular waves of various height levels may be carried out so that the expected rate of overtopping is calculated by assuming the Rayleigh distribution of wave heights.

There are several design diagrams for estimating random wave overtopping rate based on laboratory data. For seawalls of bulkhead type with and without sloped mounds of energy dissipating concrete blocks, Goda *et al.* (1975) have prepared a set of 12 diagrams for the seabed gradient of 1 on 10 and 1 on 30 with three values of wave steepness (see Goda, 2000: Sec. 5.1). For gently-sloped seawalls, van der Meer and Janssen (1995) provide two diagrams of dimensionless overtopping rate for different ranges of surf similarity parameter.

One of the topics related with wave overtopping is the fluctuation of overtopping rate; grouping of several high waves may produce the overtopping rate more than 10 times the mean rate. Wave grouping is enhanced when waves have a sharply-peaked spectrum. Along a long stretch of seawall, the overtopping also exhibits significant spatial variations, especially under the attack by directional random waves. Sakakiyama and Kajima (1996) report a model test of 2D wave overtopping.

(2) Wave Forces Exerted on Breakwaters

Vertical breakwaters are monolithic structures such as concrete caissons rested upon rubble mound foundations. If the wave forces exerted upon them are excessive, they might be slid and/or the rubble mound or the bottom ground might fail by having circular slips. What is critical in the stability of vertical breakwaters is the largest wave force among storm waves. Though, the greatest wave loading may not be caused by the largest wave in individual wave trains, it would be natural to assume that the greatest force is brought by the largest wave on the statistical average. Thus, the design formula for vertical breakwaters is mostly prepared with the largest wave in the design storm waves.

A question remains however, as to how to select the height $H_{\max}$ from the design significant wave height $H_{1/3}$, because the largest wave height is a random variable, the value of which cannot be assigned deterministically as described by Eq. (37). However, design practice requires a simple methodology to select $H_{\max}$. Goda (1974) has chosen the average height of highest 1/250 waves or $H_{1/250}$ as $H_{\max}$ for design of caisson breakwaters, based on the analysis of 21 slid and 13 non-slid prototype breakwaters in Japan. The height $H_{1/250}$ is about $1.8H_{1/3}$ according to the Rayleigh distribution outside the surf zone, but its ratio to $H_{1/3}$ may become as small as 1.3 within the surf zone.

In the case of mound-type breakwaters, the main design consideration is the stability of armor units at the top layer. Wave forces upon individual armor units placed on inclined surfaces govern the stability, but there are many other factors such as the inter-locking between armor units, permeability of filter layers, and duration of storm waves. Therefore, the stability of armor units is examined by scale model tests in laboratories using random waves. The test results are usually expressed with the parameters of $H_{1/3}$ and T_p.

Stability of mound-type breakwaters is affected by the degree of randomness of wave heights of wave trains under tests. Because of statistical variability of random waves, each realization of wave trains under the same input condition produces different combinations of wave heights and periods. Some trains contain very large waves, while others exhibit a relatively small $H_{\max}$. Galland and Manoha (1991) demonstrated that the stability of armor stones slightly deteriorates as the groupiness factor GF of a wave train under test increases. However, the groupiness factor is a random variable and its value varies with the propagation of wave trains as discussed in Sec. 4.4. Thus, the dependency of armor unit stability on the wave groupiness should be regarded as a cause of its statistical variability rather than a deterministic factor.

7. Summary

Waves in the sea are essentially random. The randomness has been clarified to a certain extent through various research efforts in the past. The present knowledge on random sea waves can be summarized as below.

(1) The composition of random sea waves is well described with the directional wave spectrum, and several functional expressions for directional spectrum are available.

(2) The wave-by-wave analysis of wave records is necessary to identify individual waves, the information of which becomes important when dealing with wave breaking, wave actions on structures, and highly nonlinear phenomena.
(3) The statistical distribution of individual wave heights is well approximated with the Rayleigh distribution regardless of spectral shapes, while the statistics of wave periods are much affected by wave spectra.
(4) Waves in the sea exhibit nonlinear behaviors. The nonlinearity of random waves has been analyzed to a certain degree but not fully. Infra-gravity waves are one of the important but unsolved phenomena.
(5) Wave diffraction and 2D wave reflection must be analyzed with the input of directional wave spectra, because use of regular wave analysis yields a gross error in the estimate of wave height distribution.
(6) Analysis of wave transformations in the surf zone must be made by incorporating the random breaking process, and the near-shore current system should also be predicted for random waves. Use of regular wave trains for any phenomenon in the surf zone will remain as of academic interest only.
(7) Wave actions on structures are analyzed either with the largest wave height $H_{\max}$ as for vertical breakwaters, or with the significant wave height $H_{1/3}$ as the reference for testing of mound-type breakwaters and estimation of wave run-up and overtopping rate.

References

Battjes, J. A. (1972): Set-up due to irregular waves, *Proceedings of the 13th International Conference on Coastal Engineering*, Vancouver, ASCE, 1993–2004.

Battjes, J. A. (1974): Computation of set-up, longshore currents, Run-up and overtopping due to wind-generated waves, Department of Civil Engineering, Delft University Technical Report 74-2, 1–244.

Battjes, J. A. and J. P. F. M. Janssen (1978): Energy loss and set-up due to breaking of random waves, *Proceedings of the 16th International Conference on Coastal Engineering*, Hamburg, ASCE, 569–587.

Battjes, J. A. and G. Ph. van Vledder (1984): Verification of Kimura's theory for wave group statistics, *Proceedings of the 19th International Conference on Coastal Engineering*, Houston, ASCE, 642–648.

Bouws, E., H. Gunther, W. Rosenthal and C. L. Vincent (1985): Similarity for the wind wave spectrum in finite depth water, Part I: Spectral form, *J. Geophys. Res.* **90**(C1), 975–986.

Borgman, L. E. (1984): Directional spectrum estimation for the S_{xy} gauges, Technical Report, Coastal Engineering Research Center, USAE Waterways Experimental Station, Vicksburg, 1–104.

Briggs, M. J., E. T. Thompson and C. L. Vincent (1995): Wave diffraction around breakwater, *J. Wtrwy. Port, Coast., and Oc. Engrg.*, ASCE, **121**(1), 23–35.

Collins, J. I. (1970): Probabilities of breaking wave characteristics, *Proceedings of the 12th International Conference on Coastal Engineering*, Washington, D.C., ASCE, 399–414.

Forristall, G. Z. (1978): On the statistical distribution of wave heights in a storm, *J. Geophys. Res.* **83**(C5), 2353–2358.

Funke, E. R. and E. P. D. Mansard (1980): On the synthesis of realistic sea state, *Proceedings of the 17th International Conference on Coastal Engineering*, Sydney, ASCE, 2974–2991.

Galland, J. C. and B. Manoha (1991): Influence of wave grouping on the stability of rubble mound breakwaters, *Proceedings of the XXIVth Congress International Association Hydraulic Research*, Madrid, B41–B49.

Goda, Y. (1970): Numerical experiments on wave statistics with spectral simulation, *Rept. Port and Harbour Res. Inst.* **9**(3), 3–57.

Goda, Y. (1974): New wave pressure formulae for composite breakwater, *Proceedings of the International Conference on Coastal Engineering, Copenhagen*, ASCE, 1702–1720.

Goda, Y. (1975): Irregular wave deformation in the surf zone, *Coastal Engrg. in Japan*, JSCE, **18**, 13–26.

Goda, Y. (1978): The observed joint distribution of periods and heights of sea waves, *Proceedings of the 16th International Conference on Coastal Engineering*, Hamburg, ASCE, 227–246.

Goda, Y. (1983a): Analysis of wave grouping and spectra of long-travelled swell, *Rept. Port and Harbour Res. Inst.* **22**(1), 3–41.

Goda, Y. (1983b): A unified nonlinearity parameter of water waves, *Rept. Port and Harbour Res. Inst.* **22**(3), 3–30.

Goda, Y. (1986): Effect of wave tilting on zero-crossing wave heights and periods, *Coastal Engrg. in Japan*, JSCE, **29**, 79–90.

Goda, Y. (1988): Statistical variability of sea state parameter as a function of wave spectrum, *Coastal Engrg. in Japan*, JSCE, **31**(2), 39–52.

Goda, Y. (1994): Statistics of wave crest lengths based on directional wave simulations, *J. Offshore Mech. and Arctic Engrg.*, ASME **9**(116), 239–245.

Goda, Y. (1997): Directional wave spectrum and its engineering applications, *Advances in Coastal and Ocean Engrg.* (P. L.-F. Liu, ed.): World Scientific, Singapore, **3**, 67–102.

Goda, Y. (1998): A comparison of several functional forms of directional wave spectrum, *Proceedings of the International Workshop on Modeling of Ocean Environments in Waves & Current Basin, Korean Research Institute Ships and Ocean Engineering*, Taejon, Korea, 1–21.

Goda, Y. (1999): A comparative review on the functional forms of directional wave spectrum, *Coastal Engineering J.* **40**(4), 1–20.

Goda, Y. (2000): *Random Seas and Design of Maritime Structures*, 2nd edn.: World Scientific, Singapore, 1–443.

Goda, Y (2004a): A 2-D random wave transformation model with gradational breaker index, *Coastal Engrg. J.* **46**(1), 1–38.

Goda, Y. (2004b): Examination of various factors affecting longshore currents induced by random waves, Draft Contribution to Professor Battjes' Symposium at Delft University of Technology, 13–14 May, 1–22.

Goda, Y., K. Kishira and Y. Kamiyama (1975): Laboratory investigation on the overtopping rate of seawalls by irregular waves, *Rept. Port and Harbour Res. Inst.* **14**(4), 3–44 (in Japanese).

Goda, Y. and K. Nagai (1974): Investigation of the statistical properties of sea waves with field and simulation data, *Rept. Port and Harbour Res. Inst.* **13**(1), 3–37 (in Japanese).

Goda, Y. and Y. Suzuki (1975): Computation of refraction and diffraction of sea waves with Mitsuyasu's directional spectrum, Technical Note, *Port and Harbour Res. Inst.* **230**, 1–45 (in Japanese).

Goda, Y. and N. Watanabe (1991): A longshore current formula for random breaking waves, *Coastal Engrg. in Japan*, JSCE, **34**(2), 159–175.

Hamada, T. (1965): The secondary interactions of surface waves, *Rept. Port and Harbour Res. Inst.* **10**, 1–28.

Hasselmann, K. *et al.* (1973): Measurements of wind-wave growth and swell decay during the Joint North Sea Wave Project (JONSWAP), *Deutsche Hydr. Zeit*, **A8**(2), 1–95.

Herbers, T. H. C., S. Elgar and R. T. Guza (1994): Infragravity-frequency (0.005–0.05 Hz) motions on the shelf, Part I: Forced waves, *J. Phys. Oceanogr.* **25**(6), 917–927.

Huang, N. E., S. Long, C. C. Tung, Y. Yuen and L. F. Bilven (1981): A unified two-parameter wave spectral model for a general sea state, *J. Fluid Mech.* **112**, 203–224.

IAHR Working Group on Wave Generation and Analysis (1989): List of sea-state parameters, *J. Wtwry., Port, Coast., and Oc. Engrg.*, ASCE, **115**(6), 793–808.

Kimura, A. (1980): Statistical properties of random wave group, *Proceedings of the 17th International Conference on Coastal Engineering*, Sydney, ASCE, 2955–2973.

Kitaigorodoskii, S. A.,V. P. Krasitskii and M. M. Zaslavskii (1975): On Phillips' theory of equilibrium range in the spectra of wind-generated gravity waves, *J. Phys. Oceanogr.* **5**, 410–420.

Kuriyama, Y., K. Katoh and Y. Ozaki (1992): Longshore current distributions, bottom profiles and waves, *Proceedings of the 39th Japanese Conference on Coastal Engineering*, JSCE, 196–200 (in Japanese).

Kuriyama, Y (1994): Numerical model for longshore current distribution on a bar-trough beach, *Proceedings of the 24th International Conference on Coastal Engineering*, Kobe, ASCE, 2237–2251.

Longuet-Higgins, M. S. (1952): On the statistical distributions of the heights of sea waves, *J. Marine Res., IX* (3), 245–266.

Longuet-Higgins, M. S. (1970): Longshore currents generated by obliquely incident sea waves, 1 & 2, *J. Geophys. Res. IS* (33), 6778–6789 & 6790–6801.

Longuet-Higgins, M. S. (1975): On the joint distribution of the periods and amplitudes of sea waves, *J. Geophys. Res.* **80**(15), 2688–2693.

Longuet-Higgins, M. S. (1980): On the distribution of the heights of sea waves: Some effects of nonlinearity and finite band width, *J. Geophys. Res.* **85**(C3), 1519–1523.

Longuet-Higgins, M. S. (1983): On the joint distribution of wave periods and amplitudes in a random wave field, *Proc. R. Soc. London, Ser. A* **339**, 241–258.

Longuet-Higgins, M. S. (1984): Statistical properties of wave groups in a random sea state, *Philos. Trans. Royal Soc. London, Ser. A* **312**, 219–250.

Longuet-Higgins, M. S., D. E. Cartwright and N. D. Smith (1963): Observations of the directional spectrum of sea waves using the motions of a floating buoy, *Ocean Wave Spectra*: Prentice-Hall, Englewood Cliffs, N.J., 111–132.

Longuet-Higgins, M. S. and R. W. Stewart (1964): Radiation stresses in water waves; a physical discussion, with applications, *Beep-Sea Res.* **11**, 529–562.

Mitsuyasu, H. (1970): On the growth of spectrum of wind-generated waves (2) — spectral shape of wind waves at finite fetch–, *Proceedings of the 17th Japanese Conference on Coastal Engineering*, JSCE, 1–7 (in Japanese).

Mitsuyasu, H. *et al.*, (1975): Observation of the directional spectrum of ocean waves using a clover leaf buoy, *J. Phys. Oceanogr.* **5**(4), 750–760.

Mori, N. and T. Yasuda (1996): Weakly non-Gaussian model of wave height distribution for nonlinear random waves, *Proceedings of the 25th International Conference on Coastal Engineering*, Orlando, Florida, ASCE, 850–863.

OCDI (2002): Technical Standards and Commentaries for Port and Harbour Facilities in Japan, The Overseas Coastal Area Development Institute of Japan, Tokyo, 1–599 with index.

Ochi, M. K. and E. N. Hubble (1976): On six-parameter wave spectra, *Proceedings of the 15th International Conference on Coastal Engineering*, Hawaii, ASCE, 301–328.

Nagai, K. (1972): Diffraction of the irregular sea due to breakwaters, *Coastal Engrg. in Japan*, JSCE, **15**, 59–67.

Okihiro, M., R. T. Guza and R. J. Seymore (1992): Bound infragravity waves, *J. Geophys. Res.* **97**(C7), 11453–11469.

Ottesen Hansen, N. E. (1978): Long period waves in natural wave trains, Inst. Hydrodyn. Engrg., Tech. Univ. of Denmark, Progress Report 46, 13–24.

Phillips, O. M. (1958): The equilibrium range in the spectrum of wind-generated waves, *J. Fluid Mech.* **4**, 426–434.

PIANC (1986): List of Sea State Parameter, Supplement to Bulletin No. 52, Permanent Int. Assoc. Navigation Congresses, Brussels.

Pierson, W. J., Jr., G. Neumann and R. W. James (1955): Practical Methods for Observing and Forecasting Ocean Waves by Means of Wave Spectra and Statistics, US Navy Hydrog. Office Pub. No. 603.

Reoelvink, J. A. (1993): Surf beat and its effect on cross-shore profiles, Ph. D. thesis, Delft University of Technology.

Sakakiyama, T. and R. Kajima (1966): Wave overtopping of armor units under multidirectional waves, *Proceedings of the 25th International Conference on Coastal Engineering*, Orlando, Honda, ASCE, 1862–1873.

Sand, S. E. (1982): Long waves in directional seas, *Coastal Engrg.* **6**(3), 195–208.

Sekimoto, T., T. Shimizu, K. Kondo and Y. Kubo (1992): Field observation of surf beats outside the surf zone, *Proceedings of the 23rd International Conference on Coastal Engineering*, Venice, ASCE, 804–817.

Sverdrup, H. U. and W. H. Munk (1946): Empirical and theoretical relations between wind, sea, and swell, *Trans. Amer. Geophys. Union*, **27**(6), 823–827.

Tajima, Y. and O. S. Madsen (2003): Modeling near-shore waves and surface roller, *Proceedings of the 2nd International Conference Asian and Pacific Coasts (ABAC 2003)*, Makuhari, Chiba, Japan, Paper No. 28 in CD-ROM, 12 p.

Tanimoto, K., K. Kimura and A. P. dos Santos Pinto (1986): Random wave forces and design wave periods of composite breakwaters under the action of double peaked spectral waves, *Rept. Port and Harbour Res. Inst.* **25**(2), 3–25 (in Japanese).

Tayfun, M. Aziz (1983): Nonlinear effects on the distribution of crest-to-trough wave heights, *Ocean Engrg.*, **10**(2), 97–106.

Tayfun, M. Aziz (1990a): High wave-number/frequency attenuation of wind-wave spectra, *J. Wtrwy., Port, Coast., and Oc. Engrg.*, ASCE, **116**(3), 381–398.

Tayfun, M. Aziz (1990b): Distribution of large wave heights, *J. Wtrwy., Port, Coast, and Oc. Engrg.* ASCE, **116**(6), 686–707.

Tayfun, M. Aziz (2004): Statisitcs of wave crests in storms, *J. Wtrwy., Port, Coast, and Oc. Engrg.* ASCE, **130**(4), 155–161.

Thornton, E. B. and R. T. Guza (1983): Transformation of wave height distribution, *J. Geophys. Res.* **88**(C10), 5925–5938.

Thornton, E. B. and R. T. Guza (1986): Surf zone longshore currents and random waves: Field data and models, *J. Phys. Oceanogr.* **16**, 1165–1178.

Tick, L. J. (1963): Nonlinear probability models of ocean waves, *Ocean Wave Spectra*: Prentice-Hall, Englewood Cliffs, NJ, 163–169.

Tucker, M. J. (1994): Nearshore wave height during storms, *Coastal Engrg.* **24**, 111–136.

Van der Meer, J. W. and J. P. F. M. Janssen (1995): Wave run-up and wave overtopping at dikes, Wave Forces on Inclined and Vertical Wall Structures, by Task Committee of Waterway, Port, Coastal and Ocean Division, ASCE, 1–27.

Van der Meer, J. W. (1998): *Wave Run-up and Overtopping*, Dikes and Revetments, eds. K. W. Pilarczky, A. A. Balkema, Rotterdam, 145–160.

Vincent, C. L. and M. J. Briggs (1989): Refraction-diffraction of irregular waves over a mound, *J. Wtrwy., Port, Coast., and Oc. Engrg.*, ASCE, **115**(2), 269–284.

Vinje, T. (1989): The statistical distribution of wave heights in a random seaway, *Applied Ocean Res.* **11**(3), 143–152.

Wang, Z. and J. Grüne (1996): Wave run-up on revetments with composite slopes, *Proceedings of the 25th International Conference on Coastal Engineering*, Orlando, Florida, ASCE, 1008–1021.

Watson, G and D. H. Peregrine (1992): Low frequency waves in the surf zone, *Proceedings of the 23rd International Conference on Coastal Engineering*, Venice, ASCE, 818–831.

Yasuda, T. and N. Mori (1994): High order nonlinear effects on deep-water random wave trains, *Proceedings of the International Symposium: Waves-Physical and Numerical Modelling*, IAHR, University British Columbia, Vancouver, 823–832.

Yasuda, T. and N. Mori (1997): Occurrence properties of giant freak waves in sea area around Japan, *J. Wtrwy., Port, Coast., and Oc. Engrg.* ASCE, **123**(4), 209–213.

Chapter 3

Tsunamis — Their Coastal Effects and Defense Works

Nobuo Shuto
Professor Emeritus
Tohoku University, Japan

1. Introduction

1.1 *Causes of tsunamis*

The Sanriku District, Japan, was shaken by a weak earthquake on 15 June 1896, around 7 pm. Its largest intensity was 3 in the JMA intensity scale. Coastal residents were celebrating the Boys' Festival. They paid no attention to the earthquake and did not attempt to evacuate. Thirty minutes later, a giant tsunami suddenly attacked. More than 22,000 lives were lost in an instant. The highest run-up was measured as high as 40 m. This earthquake was a typical "tsunami earthquake".

The most general cause of tsunami is a submarine earthquake. In case of an ordinary tsunamigenic earthquake, a large earthquake generates a large tsunami. The magnitude of a tsunami, however, has no direct relation to the magnitude of an earthquake but to the magnitude of the vertical displacement of sea bottom caused by the earthquake. In case of a "tsunami earthquake", the vertical displacement of sea bottom is exceptionally large by some reasons which are not yet fully understood.

On 27 August 1883, a giant tsunami was generated by a volcanic activity of Krakatau Island in the Sunda Strait, Indonesia. Its run-ups were reported higher than 30 m. More than 36,000 lives were lost. Because of thick falling ashes, no one could witness how the tsunami was generated. A comparison of the topography before and after the eruption indicates that two-third of the original island was lost, leaving an area 200 m deep. The horizontal length of this area is of the order of 10 km (Simkin and Fiske, 1983). Three mechanisms of tsunami generation are assumed; the impact of falling rocks and/or island itself after ejected into the air, the underwater vapor explosion and the caldera formation (Kawamata *et al.*, 1993).

On 10 July 1958, an earthquake of Ms = 7.9 generated at least six separate tsunamis by landslides, submarine slides and icefall in southern Alaska. The most catastrophic events occurred in Lituya Bay, 11 km long, 3 km wide and 200 m deep. About 300 million cubic meters of rock collapsed and fell into the bay. The water was surged up to a height of 525 m. The tsunami, however, rapidly decreased its height, after emitted from the bay entrance. (Lander and Lockridge, 1989).

1.2 *Far- and near-field tsunamis*

On 22 May 1960, an earthquake of Ms = 8.6, the largest in the history of earthquake in the Pacific, occurred off the coast of Chile. In terms of moment magnitude Mw, which has no saturated upper value different from Ms ordinarily used, its magnitude is 9.5. The estimated fault length, width and dislocation are 800 km, 200 km and 24 m, respectively (Kanamori and Cipar, 1974). As a "local" or "near-field" tsunami, its run-ups measured as high as 25 m and claimed 1,000 lives in Chile.

It traveled over the Pacific, passed Hawaii on causing 61 deaths, concentrated toward Japan, as a "distant" or "far-field" tsunami, and claimed 142 lives. The affected area in Japan was longer than 1,800 km from Hokkaido to Okinawa. Tsunami heights were 5 m to 6 m on the Sanriku coast, and 3 m to 4 m on the other coasts.

The initial profile of a tsunami consists of many components of different frequency. If the tsunami hits the nearby shore as a near-field tsunami, all the components arrive without attenuation and with negligible phase lag. The dominant tsunami period of a near-field tsunami is 5 to 10 min.

During the propagation over the ocean, the phase-dispersion effect makes the longer wave components propagate faster while the entrapment and reflection effects of islands decrease the uniformity and energy of the short wave components. A far-field tsunami tends to have a long wave period of 40 to 100 min. Whether or not the dispersion effect should be taken into consideration is judged by the Kajiura p_a value (1970)

$$p_a \cong (6h/R)^{1/3}(a/h),$$

where h is the water depth, a the horizontal dimension of the tsunami source and R the distance from the source. For $p_a < 4$, the assumption of the non-dispersive long waves is not applicable.

1.3 *Major amplification mechanisms*

Tsunami is a long wave. The phase velocity C is the same as the energy propagation velocity. They are given by the square root of the product of water depth, h, and gravitational acceleration, g. Approaching the land, a tsunami increases its height, H, owing to the shoaling effect, because the energy transmission rate proportional to $H^2C_G = H^2(gh)^{1/2}$ is kept constant.

After a tsunami enters a bay with a wide entrance, the tsunami increases its height on proceeding to the narrow bay bottom. This is the concentration effect.

Shoaling and concentration effects are expressed in the Green formula for the linear long waves, where b is the width of the bay.

$$H \propto b^{-\frac{1}{2}} h^{-\frac{1}{4}}.$$

A container of water has its natural oscillation. A bay is not an exception. The oscillation which occurs most easily has its node at the entrance of the bay and its loop at the bottom of the bay. The period of the first-mode natural oscillation, T, is given as follows, where l is the length of the bay,

$$T = 4l/\sqrt{gh}.$$

When a tsunami of similar period to this T enters the bay, its height is amplified by resonance. In case of a long wave, resonance is completed if three waves consecutively come into the bay. (Kajiura, 1963).

The dispersion effect is another mechanism to increase the tsunami height, often observed in rivers and shallow seas. A series of solitons develop at the front of a tsunami (for example, Shuto, 1985). This is also called as cnoidal wave train or undular bore. The period of solitons, about 10 s, is similar to that of wind waves. The highest crest height reaches twice the height of the original tsunami.

2. Tsunamis, Observed and Experienced

2.1 *The 1983 Nihonkai-Chubu earthquake Tsunami*

At noon on 26 May 1983, an earthquake of Ms = 7.7 occurred in the Japan Sea, 100 km off Akita and Aomori Prefectures, Japan: A tsunami was generated. It was a perfectly fine day with no wind and no waves. The tidal range in the Japan Sea is small, less than 50 cm at spring tide. If there were abnormal water waves under these conditions, all of them were surely correlated with the earthquake.

The highest run-up of 15 m was measured on a sandy dune at Minehama, located at the middle of the north Akita coast, which has a smooth shoreline 55 km long, bounded by rocky cliffs projecting perpendicularly to the shoreline at the north and south ends. Different from a Ria coast, there is no topography which suggests the tsunami amplification mentioned in Sec. 4.1.3).

Collecting and analyzing witnesses, photos and videos, the tsunami on the north Akita coast was reconstructed (Shuto, 1985). Figure 3.1 shows the area of concern and wave rays drawn from the tsunami source (Shuto *et al.*, 1986).

The southern end of the coast is bounded by the Oga Peninsula. Along the northern shore of this peninsula, evolution of solitons was recorded by photos and a video. Near the tip of the peninsula, beginning of the soliton fission was photographed in six pictures in sequence, on the last picture of which two short-period waves were recognized. At a point 3 km further east, a video was taken to show the later stage of the evolution. There were five short-period waves. Four of them were white-capped. The solitons had wave length of 150 m, wave height of 3 m to 4 m, and wave period of 10 s.

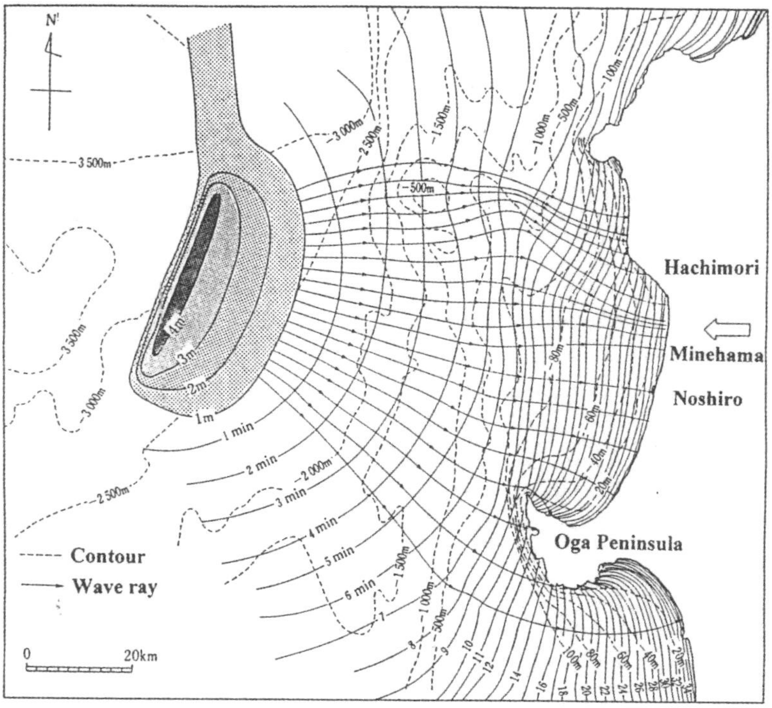

Fig. 3.1 The Nihonkai–Chubu earthquake tsunami computed with the Aida Model-10. The tsunami source is shown by the dotted area with contours of vertical displacement. Solid lines with arrows are wave rays, solid lines with numeral attached by min are the position of the wave front at one minute interval and dotted lines are bottom contours.

At the northern end of the coast, a fisherman was surprised by a series of solitons from the land. The first wave from the sea was so small and gentle that he did note recognize it on the deep sea. The tsunami had enough time to evolve solitons after they were reflected from the shore.

On this north Akita coast, many riddles were left after the tsunami. The first riddle was that the initial tsunami profile determined from seismic data could not explain the total energy of the tsunami. Among several models, the Aida Model-10 (Aida, 1984) is now accepted as the best. It assumes a fault gently dipping eastward. The initial profile has a crest 4 m high, about 2.5 times that determined from seismic data, near the western end, and a shallow trough eastern side.

The second riddle is the reason why the tsunami ran up to a height of 15 m at Minehama. Many researchers tried to simulate with the shallow-water theory, and obtained the maximum run-up of about 6 m. Considering that the reason might be due to the evolution of undular bores at the front, Aida (1984) used the Boussinesq equation and obtained an estimate of the maximum run-up height of 9 m, still leaving 6 m difference. Shuto *et al.* (1986) solved this problem, starting with the Aida Model-10. They used much finer spatial grids than those usually adopted.

As a result, the refraction due to bottom topography was well simulated and the maximum run-up height was, too. Figure 3.1 is the result. Due to refraction, the tsunami concentrated on to Minehama. In their simulation, a high tsunami corresponding the initial crest, 4 m high, invaded the land, preceded by a small ebb.

The solution above is not perfect. According to witnesses, the first wave which was not so high came around 12:15. About 10 min later, there was the huge second wave higher than 20 m. The second wave, however, ran up lower than the first wave, because it met the receding currents induced by the first wave. This observation contradicts the simulation above. In addition, the computed arrival time shown in Fig. 3.1 is the same as the arrival time of the observed second wave. A possible generation mechanism of the small first wave is discussed by Shuto *et al.* (1995), leaving, however, its magnitude undetermined.

The tsunami was trapped along the coast. Near the tips of the southern and northern boundaries, one or two incident tsunamis were recognized. In contrast, seven huge waves were experienced off Noshiro at the middle of the coast. On the shore north of Noshiro, the tsunami repeatedly came as bores along the shore. The tsunami could not easily return to the sea but was refracted again and again toward the land. Near the shoreline, the front of the tsunami propagating alongshore showed a violent breaking. This edge bore propagates sometimes following the ordinary wave refraction law and sometimes neglecting the topography. In hydraulic experiments by Uda *et al.* (1988), a small difference in the side boundary condition introduced a big difference in wave profile, thus suggesting also a big difference in wave force. There is no theory and no simulation method applicable to edge bores.

2.2 *The 1933 showa great Sanriku tsunami*

In early morning on 3 March 1933, an earthquake of Ms = 8.1 occurred in the neighborhood of the Japan Trench off the Sanriku District. This was due to an innerplate normal fault. A giant tsunami hit the Sanriku coast, a typical Ria coast rich in the indented coastline. Different from the former Meiji tsunami in 1896 mentioned in Sec. 4.1.1, this fault motion generated violent shaking of ground. More than 3,000 lives were lost by the tsunami. The highest run-up measured as high as 30 m. The dominant wave period was between 5 and 10 minutes. This was a typical, giant local tsunami on the Sanriku coast. Imamura (1942) defined a tsunami magnitude m, letting $m = 3$ for this Showa tsunami and $m = 4$ for the Meiji tsunami.

By using 160 examples reported by witnesses (ERI, 1934), profiles of this tsunami near the shoreline are estimated (Shuto, 1993) in terms of the "tsunami height". Although, the definition and the method of measurement are not stated in the ERI report, the "tsunami height" might be one of the followings: the tsunami run-up height, (most probably) the height of tsunami trace near the shoreline, or the estimated tsunami crown height in the offing. A general idea of profiles of a giant local tsunami might be formed by this analysis.

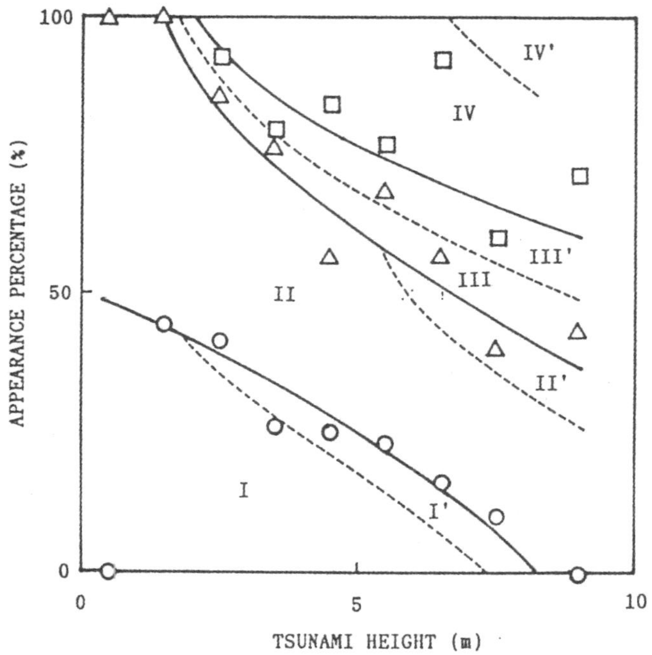

Fig. 3.2 Appearance percentage of different wave profiles of the 1933 Showa Sanriku tsunami. (With kind permission from Kluwer Academic Publishers.)

Figure 3.2 shows the appearance percentage of four types: Type I is somewhat like a tide. It is described as, "the water level made a slow rise like a tide", "the tsunami was like a rapid tide" or "the tsunami quietly advanced shoreward and rose suddenly at the breakwaters". Some of them had a wavy front like "ridges in the field". These short waves sometimes break (Subtype I'). Type I is for a tsunami of gentle wave steepness on a steep bottom slope.

Type II is characterized by "a rapid growth near the shoreline, although the tsunami is not recognized in the offing". This expression means that the bottom slope is relatively steep. It is described as "the water level rapidly swells near the shore", "the water swells from the bottom" or "the water level is raised by a train of short waves, the succeeding waves overtaking and lying upon the preceding ones". Subtype II' shows spilling breaking.

Type III is characterized by "a rapid growth, near the shoreline, of the tsunami that is well recognized in the offing". This is a tsunami on a relatively gentle bottom slope. It is described as "a tsunami like a bank", "a tsunami like a stretched curtain" or "a wall with splash at its crest in the offing". Even if they are not high, half of them shows spilling breaking (Subtype III').

Type IV is for the tsunami of plunging breaking. The second, third or later waves of relatively small height (the smallest is about 2 m) may show plunging breaking when they meet the receding current of the preceding waves. The first wave higher than 7 m shows the plunging breaker (Subtype IV').

3. Initial Profile, An Enigma At The Start Line

3.1 *Method of determination*

Characteristics of a fault motion are described by fault parameters. From analysis of seismic wave records, seismic moment, strike direction, dip angle, slip direction and depth of fault are estimated. The seismic moment is the product of the dislocation, the area of fault plane and the rigidity constant which is $(3–6) \times 1011\,\mathrm{dyne/cm^2}$. If the fault plane area is estimated by empirical formulas or by the spatial extension of aftershocks, the dislocation is determined.

With these fault parameters, the vertical displacement of sea bottom is computed by the Mansinha–Smylie method (1971), on assuming a rectangle fault plane in an isotropic homogeneous semi-infinite medium. For a thorough discussion and misprint correction, refer Okada (1985).

A method introduced by Fukuyama and Irikura (1986) can be used, to estimate heterogeneity of a fault motion, if seismograms of smaller events after a main shock are available. The fault plane of the main shock is divided into segments, the size of which is determined from the scaling relations between the main shock and the aftershocks. Unknown displacements are given on each segment. The seismic waves due to these displacements are evaluated by using the seismograms of smaller events as empirical Green's functions and are superposed to yield synthesized seismograms of the main shock. Unknowns are solved to minimize the cross-correlation error functions between the observed and synthesized seismograms.

Satake (1989) introduced a similar technique of wave-form inversion, using tsunami data instead of seismograms. The fault plane is arbitrarily divided into small segments. Unit dislocation is given on each segment and the corresponding vertical displacement of sea bottom (= initial rise of water surface) is computed by the Mansinha–Smylie method. Starting with this initial condition, tsunami is computed. The time series of the computed tsunami at tide gauge stations are output and are used as Green's functions.

3.2 *The 1993 Hokkaido Nansei-Oki earthquake tsunami*

An earthquake of Ms = 7.8 occurred off the southwest coast of Hokkaido, Japan at 22:17 on 12 July 1993 (local time). A giant tsunami was generated. Its highest run-up measured as high as 31.7 m. Figure 3.3 shows initial profiles determined with the Mansinha–Smylie method for different faults models (Takahashi *et al.*, 1995). Differences in fault models are caused by differences in the seismic data used and in the analysis procedure. Figures 3.3(a) to 3.3(e) assume one fault plane. Figure 3.3(f) proposed by Kikuchi assumes three fault planes. Figure 3.3(g) shows the initial profile determined from two tide records at Iwanai and Esashi by use of Satake's inversion method mentioned above. A trial to satisfy the seismic data and measured ground subsidence in Okushiri Island leads to a model of three fault

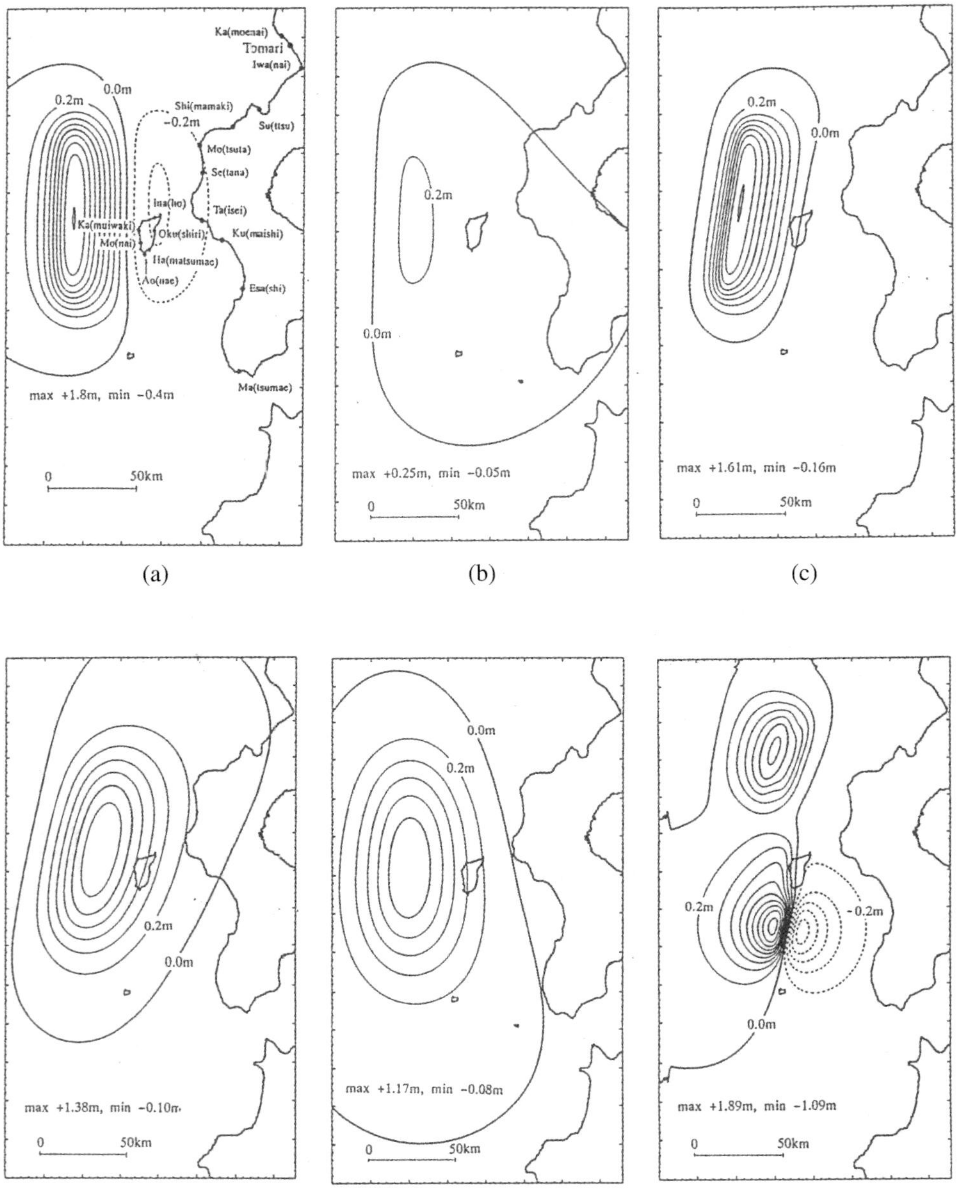

Fig. 3.3 Initial profile of the 1993 Hokkaido–Nansei–Oki earthquake tsunami. (a) Harvard, (b) USGS, (c) ERI, the University of Tokyo, (d) Hokkaido University, (e) JMA, (f) Kikuchi, (g) Takahashi *et al.* by tsunami inversion, (h) Hashimoto *et al.*, and (i) Latest model of Takahashi *et al.* (With kind permission from Birkhauser, Verlag, AG.).

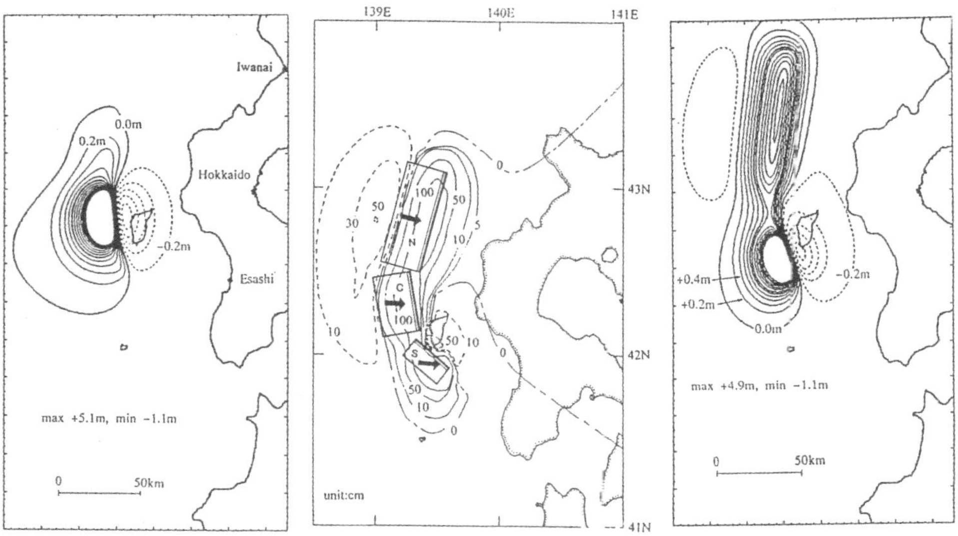

(g) (h) (i)

Fig. 3.3 (*Continued*)

planes shown in Fig. 3.3(h) (Hashimoto *et al.*, 1994). All of these initial profiles do not satisfy the tsunami trace heights in the neighborhood of the tsunami source such as Okushiri Island and Hokkaido.

Another effort to satisfy the seismic data, ground subsidence and tsunami run-up heights in the near-field gives the initial profile shown in Fig. 3.3(i) on assuming three fault planes (Takahashi *et al.*, 1995).

As shown in this example, the initial profile of a tsunami is not uniquely determined even for an ordinary tsunamigenic earthquake. No method is developed to estimate the initial profile of a "tsunami earthquake" from seismic information. Use of accurate tsunami data is becoming more and more important in estimation of the initial profile and fault motion. For this topic, refer Secs. 4.5.2 and 3).

4. Numerical Simulation, A Powerful Means

4.1 *Equations*

When born in the sea several km deep, a tsunami is several tens to 200 km long and several meters high. At its infancy and during the near-field propagation in deep sea, the linear long wave theory in the Cartesian coordinates is applicable. For the transoceanic propagation, the dispersion effect and the Coriolis force should not be neglected (refer Sec. 4.1.2). The linear Boussinesq equation including the Coriolis force expressed in the longitude–latitude coordinates is applied.

Approaching the shore where the bottom friction is an important factor, the tsunami height increases and becomes comparable with or larger than the water depth. The shallow-water theory with the bottom friction included should be used.

In a river or in a shallow sea on a very gentle slope, a tsunami often shows evolution of undular bores at its front. The Boussinesq equation is needed for the analysis. If one wishes to analyze the run-up of undular bores, he needs such equations of higher approximation as the Goto equation (1983), until bores become breaking.

Under some conditions, the shallow-water theory should be replaced by other theories. As mentioned in Sec. 4.2.1), edge bores cannot be described by any of existing theories. In this case, a new theory should be developed and established. In case of the 1993 Hokkaido Nansei-Oki earthquake tsunami, the highest run-up of 31.7 m was measured at a small V-shaped bay, 50 m wide at the entrance. The tsunami trace height on the cliff at the valley entrance was 23 m to 25 m. The horizontal distance from the entrance to the location of 31.7 m high run-up was only 50 m, quite short compared with the original tsunami tide. The bottom slope is estimated to be 3/5. Long wave assumption is no longer applicable to this situation.

The MAC (mark-and-cell) method may be used for the breaking and after-breaking-regrowth of undular bores, with verification by hydraulic experiments to formulate the energy loss due to breaking.

4.2 *Numerical error control*

(1) Origin of numerical errors

Simulation by a difference equation introduces numerical errors which are closely related with the original differential equation, difference scheme adopted, grid sizes, size of tsunamis, and characteristic size of the sea bottom. In order to ensure stable computation and reliable results, these errors should be controlled and evaluated.

The long wave theory belongs to the wave equation. For a stable computation, the CFL condition should be satisfied first. The propagation velocity in the numerical computation, i.e. spatial grid length divided by time step interval, should be taken faster than the physical propagation velocity.

In the followings, other conditions to be considered are briefed, on assuming the leap-frog scheme and the up-wind difference for convection terms, because these schemes are widely used and their mechanism of error generation is examined best.

(2) Wave length and grid size

The more spatial grids are in one wave length of a tsunami, the more smoothly the wave profile is expressed. The first term of the truncation error has the effect of dispersion. A coarse grid size has a large dispersion effect and the computed wave reduces its height as it travels. An example is shown in Fig. 3.4. The uppermost figure is the initial profile. Although the wave should propagate without any change because it is computed with the linear long wave theory, the wave profile changes

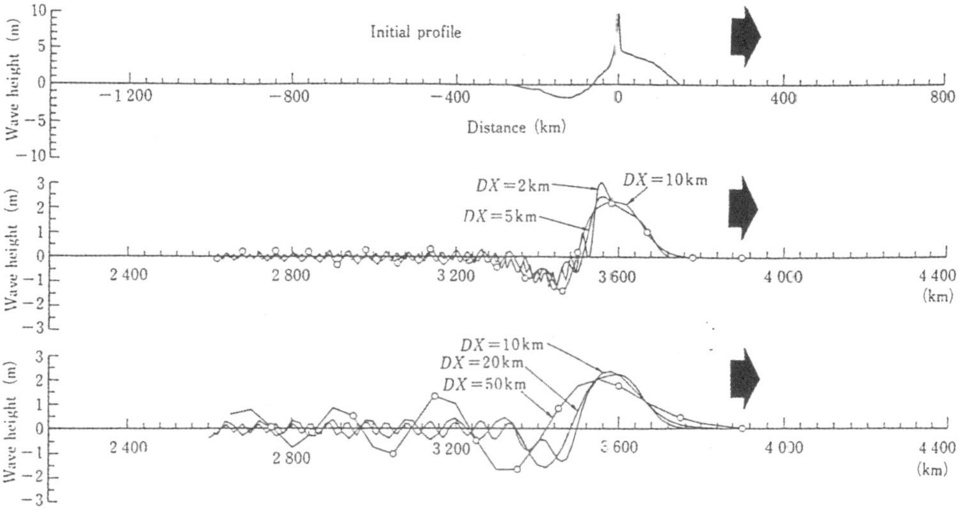

Fig. 3.4 Numerical results with the linear long wave theory for different grid length. (With kind permission from Kluwer Academic Publishers.)

with the size of spatial grids and the travel distance. A criterion is proposed by Shuto *et al.* (1986) that 20 grid points, preferably 30 grid points, are necessary within one wave length in order to ensure null reduction in wave height during when a tsunami propagates over a distance equivalent to four wave lengths.

(3) Run-up front and grid size

With equations in the Eulerian description, it is not easy to handle a run-up front in simulation, because the moving boundary changes its location with time. There are, of course, exact methods to include this boundary condition by use of a variable transform at the front (Takeda, 1984) or by use of equations in the Lagrangian description (Shuto and Goto, 1978). They are, however, inconvenient for practical application where the topography is complicated.

Instead, approximate moving boundary conditions are proposed as shown in Fig. 3.5. In the leap-frog scheme, grid points are alternatively located for velocity and water level. Assume that the water level is already computed at a computation cell. Then, compare this water level with the bottom height of the next landward cell. If this water level is higher than the bottom height, the water may flow into the landward cell.

Iwasaki and Mano (1979) assume that a line connecting the water level at a cell and the bottom height of the landward cell gives the surface slope of the first-order approximation and the inflow is determined according to this water surface slope.

Hibberd and Peregrine (1979) give a provisional water level in the dry cell on a linearly extrapolated water surface. Then, the discharge calculated with this provisional water level gives the total amount of water into the landward dry cell

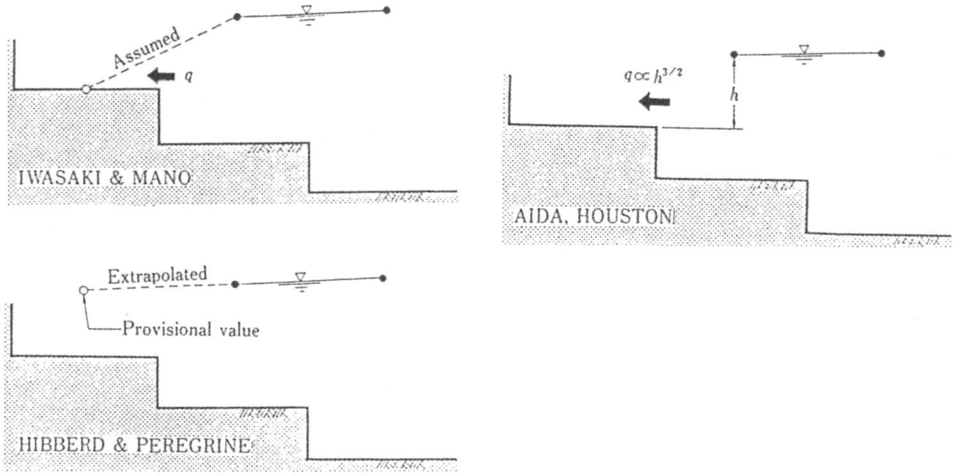

Fig. 3.5 Approximate moving boundary conditions in numerical simulation. (With kind permission from Kluwer Academic Publishers.)

and, therefore, the water level in the cell too. If necessary, the computation will be repeated with the water level thus modified until the expected accuracy is attained.

Aida (1977) and Houston and Butler (1979) evaluate the discharge into the landward dry cell with broad-crested weir formulas in which the water depth above the bottom of the dry cell is substituted.

Use of an approximate moving boundary condition introduces numerical errors which deteriorate the solution. In order to limit the error within 5% and ensure stability of computation for the Iwasaki and Mano condition, the grid size at the front should satisfy the following relationship, where α is the angle of slope, g the gravitational acceleration, and T the wave period (Goto and Shuto, 1983). This means that the grid size is smaller than 1/50 of one local wave length at the front.

$$\Delta x/\alpha g T^2 \leq 4 \times 10^{-4}.$$

(4) The smallest water depth

Near the run-up and run-down fronts, the water depth is very small and easily leads to instability through the friction and convection terms. Both of them have the water depth, D, in the denominators. The friction term expressed with the Manning roughness, n, is given by $gn^2M|M|/D^{7/3}$, where M is the discharge per unit width. The convection term is expressed as $\partial(M^2/D)/\partial x$. In order to suppress the instability in a practical application, the smallest value of D is set, for example, to be 5 cm.

(5) Grid size and topography

The sea bottom configuration is digitized and reproduced in a simulation model. This discretization procedure deteriorates refraction, because the direction of tsunami propagation does not change in a computation cell.

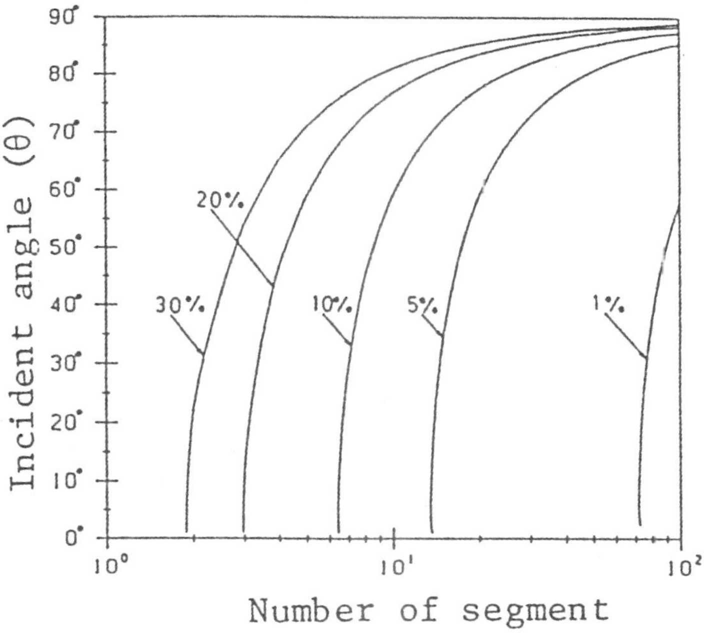

Fig. 3.6 Error in the location of wave rays at the shore as a function of the number of cells on a uniform slope.

In case of a uniform slope, a criterion is obtained by Sayama *et al.* (1988). A wave ray on the digitized slope is estimated as a result of subsequent refraction at the boundaries between the computation cells, and is compared with the analytical solution. The error is defined by the difference between locations of wave rays on the original and digitized slopes. Figures 3.6 give the error percentage as a function of the incident angle and the number of cells on the slope.

In order to determine the spatial grid size for an actual topography like a seamount or an island, Lee (1998) introduced three different kinds of FRACTAL relating sea-bottom topography, tsunami travel time and refraction effect.

(6) Use of numerical dispersion

When the linear long wave theory is differenced, it is reduced to the first equation in the followings, leaving the first term of truncation error. Compare it with the linear Boussinesq equation, the second equation shown below, which has the physical dispersion term. The truncation error term has a similar form to the physical dispersion term.

$$\frac{\partial^2 \eta}{\partial t^2} - C_0^2 \frac{\partial^2 \eta}{\partial x^2} - \frac{C_0^2 \Delta x^2}{12}\left[1 - \left(C_0 \frac{\Delta t}{\Delta x}\right)^2\right] \frac{\partial^4 \eta}{\partial x^4} = 0,$$

$$\frac{\partial^2 \eta}{\partial t^2} - C_0^2 \frac{\partial^2 \eta}{\partial x^2} - \frac{C_0^2 h^2}{3} \frac{\partial^4 \eta}{\partial x^4} = 0.$$

Imamura used the linear long wave equation with the grid size determined by equating the coefficients of the third terms of the two equations and obtained the same results as those by the linear Boussinesq equation, thereby saving the CPU time (Refer Appendix of Nagano *et al.*, 1991).

4.3 *Credibility of bathymetry data*

Hydrographic charts often used as the source of bathymetry data for numerical simulation are originally made for the navigation purpose. If the water is sufficiently deep for navigation, no details are usually surveyed. An example is found in case of the 1983 Nihonkai–Chubu earthquake tsunami mentioned in Sec. 4.2.1. The highest run-up of 15 m was concluded as a result of refraction due to sea topography which was only simulated with fine spatial grids. This conclusion was obtained in 1986, by using the bathymetry shown in the left-hand side of Fig. 3.7. After this tsunami, the Japanese Maritime Safety Agency measured the area in detail and published the result in 1995, as shown in the right-hand side of Fig. 3.7. To our embarrassment, the water depth which was believed to give a big influence to the refraction was reduced to 42 m from 91 m. This change may affect much the refraction of the tsunami.

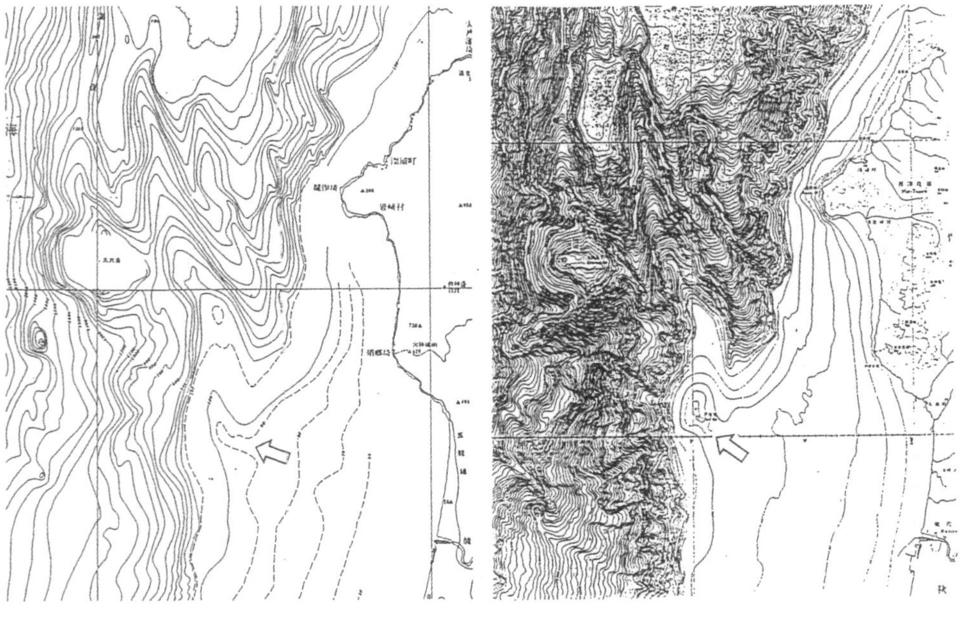

Fig. 3.7 Comparison of an old chart (left, Chart No.1195, JMSA) and the recent result (right, Chart No.6659, JMSA). The water depth of a shoal (indicated by an arrow) which was considered to be important in refraction of the 1983 Nihonkai–Chubu earthquake tsunami is reduced from 91 m to 42 m. (Permission No. 100035, Japan Maritime Safety Agency.)

4.4 *Verification of computed results*

A tsunami simulation starts with an initial profile which is only ambiguously known. Data of sea bottom topography is sometimes not accurate enough for computation of tsunami propagation and refraction. Numerical errors will be accumulated during the propagation and run-up computation. It is always a problem whether or not the computed results reproduce an actual tsunami and are reliable enough for practical design. A comparison with the tide records has a difficulty discussed in Sec. 4.5.1. The most popular way to verify the computed results is a comparison with the tsunami run-up heights. Aida (1978) introduced the K value, a geometric mean of the ratio of the recorded to computed run-up heights. If the K value is close to unity, the total tsunami energy is well estimated. His κ value, the corresponding variance, judges whether the spatial distribution of the initial profile is reasonable or not.

For a practical use, the initial tsunami profile under consideration is adjusted until K and κ values take a given accuracy. After this adjustment, run-up heights of a historical tsunami is reproduced within an error of 15% as far as the maximum run-up height is concerned.

5. Measurements

5.1 *Tide gauge*

A tide gauge is installed in a stilling well connected to the outer sea by an orifice or by an intake pipe, both of which are designed to reduce the influence of wind waves and swells of period of several to several ten seconds. A long and narrow intake pipe also affects the recording of tsunamis of period of several minutes to several ten minutes.

When the 1983 Nihonkai–Chubu earthquake tsunamis hit the Japan Sea, Japan Meteorological Agency (JMA) immediately collected the tsunami heights measured by tide gauges through its microwave network and reported them to public. Most of the reported tsunami heights were less than 1 m. It was the first impression that this tsunami might give almost no damage to the coast. Then, local autonomies began to report damages of unexpectedly large magnitude. There are two reasons of this gap: the first is that tide gauge stations did not cover the area of the most affected area; the second is that the tsunami heights recorded by the tide gauges were strongly reduced by hydraulic filtering.

Satake *et al.* (1988) measured the response characteristics of 40 tide gauges, and found that recorded tsunamis should be doubled in the worst case. The response characteristics vary with respect to time, because the in-take pipe will be narrowed by sedimentation as well as by bio-effects such as algae breeding.

Other defects of tide gauges are; (i) scaling-out when a giant tsunami overflows the tide well or when the in-take pipe opens to air at a big ebb, and (ii) the poor time resolution due to slow recording paper speed, such as one centimeter per hour.

For comparison with the computed tsunamis, tide records are usually corrected for hydraulic filtering. This is not recommended because records of short-period components which are once lost in the tide records can never be reproduced. Inversely, the computed tsunami time histories should be filtered by use of the response function and should be compared with the original tide records.

5.2 *Ultrasonic wave gauge*

Among several types of wave gauge, an ultrasonic wave gauge installed on or near the sea bottom supplies good data. It measures the vertical distance to the water surface with no attenuation. The Bureau of Harbor, Ministry of Transport, Japan has a wave measurement network, NOWPHAS, in which ultrasonic-type wave gauges installed at the water depth of 50 m are used. No scaling-out is expected even in case of a giant tsunami. Its time-resolution is excellent, because the main purpose is the recording of wind waves and swells. The sampling time interval is 0.5 s.

Only one, but vital, defect is the way of operation. It works for 10 to 20 min every two hours. After the measurement and data-processing to yield wave characteristics, the gauge returns to another silence for two hours. If a tsunami appears during this silence, the gauge will miss it. The Bureau of Harbor is changing the operation method from the present intermittent operation to the 24 h continuous operation (Nagai *et al.*, 1994).

5.3 *Post-tsunami field survey aided by numerical simulation*

Tsunami data rich in number is the tsunami trace heights measured after tsunamis. The difficulty in the reproduction of a tsunami from the trace heights lies in the fact that there are many ambiguous factors in the life of a tsunami.

It is not possible to exactly determine the initial profile of the tsunami from seismic data. In an extreme case, a tsunami hit the shore at Hamada, Shimane Prefecture, Japan on the Japan Sea coast, on 6 February 1872, before an earthquake occurred (Imamura, 1977). In case of the 1983 Nihonkai–Chubu earthquake tsunami, an early arrival of the first wave is left unsolved as described in Sec. 4.2.1. In order to improve the initial tsunami profile estimated by seismic data, geological data and tsunami trace heights are inevitable.

A tsunami is translational. It can be progressive, partially or perfectly reflected, breaking or non-breaking, dispersive or non-dispersive. Some of traces might be marked by splashes but not by waves. At the time of a post-tsunami field survey, these characteristics are not known. In addition, it is rare to be informed of how many waves hit the site and which wave left the trace. Local topography assists to make this situation more complicated. A several-meter difference in height is often measured at a several-ten-meter horizontal distance.

At present, the first approximation of a fault can be given within a few hours after an earthquake. Then, the corresponding tsunami is calculated within three

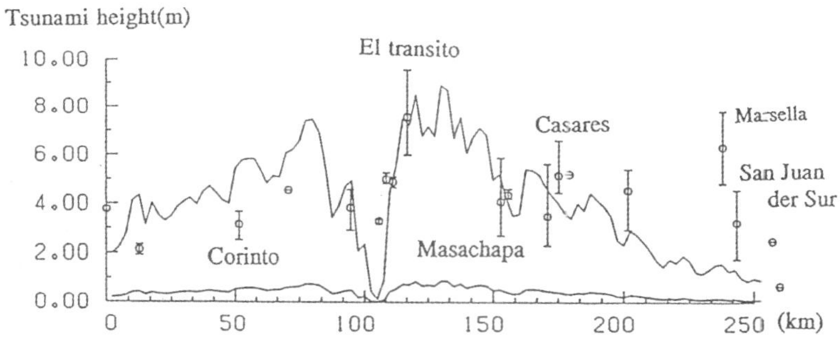

Fig. 3.8 Comparison of the computed and measured tsunami height distributions of the 1992 Nicaraguan tsunami. The lower line assumes fault parameters estimated from seismic data. The upper line is modified by making the fault dislocation 10 times larger.

days. The most time-consuming work is the preparation from finding good charts to digitizing the topography data. The computation itself will be finished within an hour.

It was in the 1992 Nicaraguan tsunami when the international cooperation of field survey and also the tsunami-simulation-aided field survey began. A numerical simulation was finished before the survey team entered the damaged area. Its results, Fig. 3.8 (Imamura *et al.*, 1993), showed that the computed tsunami was unexpectedly small compared with the tsunami data reported by local journalists. On expecting a possibility of a "tsunami earthquake", the survey team prepared well inquires in Spanish on the earthquake intensity felt by residents. They found that the maximum intensity was 2 in the modified Mercalli Scale and concluded that the earthquake was a tsunami-earthquake.

Another outcome of the pre-simulation was the maximum run-up at El Transito, a town at the middle of the damaged area. Although, nothing was reported by newspapers and TVs, the simulation suggested the occurrence of the highest run-up at this town. The survey team forced themselves to reach the town through very bad road condition, and found the damaged town and the maximum run-up height in the affected area.

After this experience, the present author's group carried out tsunami simulations as soon as possible and distributed the results to any field survey teams of any nationality at their requests. When there is a difference between the computed results and that measured in the field, it is important to confirm this difference by increasing accuracy and density of measurements. These data are inevitable to improve the initial tsunami profile.

5.4 *Tsunami measurement in deep sea*

Efforts have been made to measure tsunamis in deep sea for early warning. JMA installed seismographs and tsunami gauges at the water depth of 2,200 m, 110 km south–southwest of Omaezaki, Central Honshu, in August 1978 (MRI, 1984). Signals are transmitted through a submarine cable 160 km long on real time. Another set was installed at the water depth of 4,000 m off Boso Peninsula near Tokyo.

NOAA, USA operates the Pacific Tsunami Observation Program (PacTop) since 1986. They employ buoy stations, from which signals of pressure gauges are telemetered via satellites (Gonzalez *et al.*, 1987).

The deep-sea tsunami gauges are useful not only for warning but also for the detailed study of the initial profiles because the recorded tsunamis are expected to be less affected by topography. The number of such deep-sea gauges are, however, very limited at present.

6. Disasters and Countermeasures

6.1 *Disasters*

(1) Kinds of disasters

In addition to the well-known damages to human lives, houses and boats, there are many kinds of tsunami disasters. Table 3.1 summarizes them collected from documents. Figure 3.9 shows a rough estimate of damages in terms of a tsunami intensity i defined by

$$i = \log_2 H,$$

where H (m) is the tsunami height (Shuto, 1993).

A natural disaster occurs as an interaction of a natural force and the human society. Even if the natural force is the same, the magnitude and type of the disaster is quite variable, because of the change and difference of the coastal society.

Under a special condition, a tsunami gives a formidable damage which is never mended. For example, the island of Miyako, Okinawa, Japan, was hit by a giant tsunami in 1771. Its highest run-up was 85 m. When the tsunami receded, it eroded and washed away the fertile soil from agricultural fields. The barren areas were left and are not yet recovered (Makino, 1968).

(2) Loss of lives

As described in Sec. 4.1.1, the Meiji Great Sanriku tsunami hit after a weak earthquake. None of coastal residents tried to evacuate. As a consequence, many lives were lost. Figure 3.10 shows the relationship between the number of houses washed away and/or destroyed and the number of loss of lives (Shuto, 1991). About seven lives were lost per house.

Miyano and Ro (1992) obtained the following relations for the number of the dead, P, and the number of the injured, Q, in terms of the number of houses washed away and/or destroyed, W, in case of the Tonankai earthquake tsunami of 1944.

$$P = 0.072\, W^{1.018}, \quad Q = 5.584 \times 10^{-4}\, W^{1.961}.$$

This gives 0.07 dead per house.

Table 3.1 Kinds, types and causes of tsunami disaster.

Human Lives

Drowned. Injured hit by debris etc. Disease caused by swallowing alien substances during drifting.

Houses

Washed away. Destroyed. Flooded.

Coastal Structures

Toe erosion, displacement and overturning of sea walls, sea dikes, breakwaters and quay walls.
Scattering and subsidence of concrete blocks.

Traffic

- Railway
  - Erosion of embankments.
  - Displacement of rails and bridges. Rails buried by sands.
- Highway
  - Displacement and falling down of bridges.
  - Overturning of bridge abutment by erosion.
  - Erosion of embankment.
  - Closure of traffic by debris on roads.
- Harbor
  - Change in water depth (erosion and accumulation).
  - Closure of port area due to transported debris and cars.
  - Collision of ships in harbors.

Lifelines

- Water supply
  - Destruction of hydrants by collision of debris.
- Electricity
  - Overturning and washed-away of electric poles.
- Telephone
  - Damage to telephone lines and poles.
  - Cut-off of underground telephone line at the junction to the aerial lines.
  - Submergence of telephone receivers.

Fishery

Damage to fishing boats.
Destruction and loss of rafts, fishes and shells in aquaculture.
Loss of fishing nets and other fishing gears.
Closure of port entrance by fishing gears washed-away.

Commerce and industry

Depreciation of goods by submergence.

Agriculture

Physiological damage to crops due to submergence.
Damage to farms buried by sands.
Closure of irrigation channels filled by sands and debris.

Forest

Physical damage (breaking and overturning of trees.) Soil erosion.
Physiological damage by sea water and sands.

Oil spill

Environmental pollution.
Spread of fires.

Fire (causes)

Kitchen fire. Heating. Engine room of fishing boats. Submerged batteries of fishing boats.
Collision to gasoline tanks. Electricity leakage.

TSUNAMI INTENSITY	0–1	1–2	2–3	3–4	4–5
TSUNAMI HEIGHT (m)	1–2	2–4	4–8	8–16	16–32
WAVE PROFILE On a mild slope	Sudden rise near the shore.	Wall of water in the offing. Second wave breaks.	Wave front often breaks.	First wave may show plunging breakers.	
On a steep slope	Rapid tide.				
SOUND		Continuous sound caused by spilling breaking at wave front. (Like a sea roar, a storm, a locomotive or big trucks.)			
			Sudden big sound caused by plunging breaking on the beach. (Like a thunderbolt. Only hearable at the beaach of breaking.)		
			Big sound caused by impact of tssunamis on cliffs. (Like a distant thunder or an explosion. Hearable at distant places.)		
HOUSE Wooden house	Partial damage	Demolished			
Stone house	Withstand		(No data)	Demolished	
R. C. building	Withstand			(No data)	Demolished
FISHING BOAT		Damage begins.	Damage > 50%	Damage = 100%	
TSUNAMI CONTROL FOREST	Negligible damage. Drifts stopped. Reduction of tsunamis.		Partial damage. Drifts stopped.	Damaged. No effect.	
AQUACULTURE RAFT	Damage begins.				

Fig. 3.9 Tsunami intensity, tsunami phenomena and degree of damage.

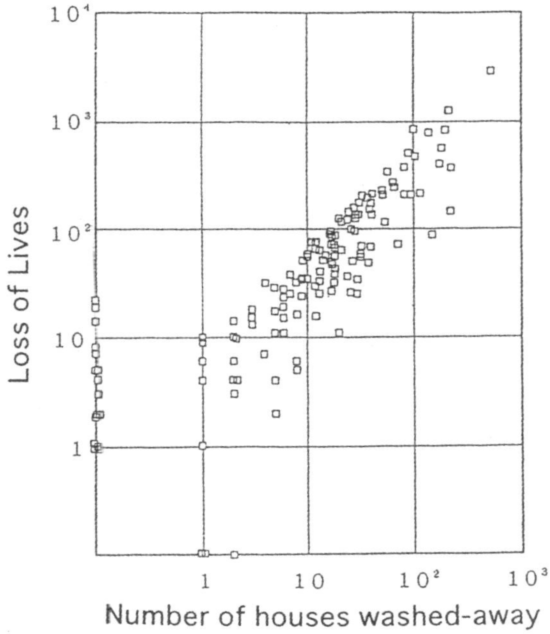

Fig. 3.10 Loss of lives as a function of destroyed and washed-away houses in case of the 1896 Meiji Sanriku Great tsunami.

This comparison suggests that evacuation is the most crucial action to save human lives.

(3) Damage to houses

In addition to hydrodynamic forces such as buoyancy, drag force and impact of water, collision of floating materials is the most dangerous force to damage houses. A rough estimate of damage to individual houses is given in terms of tsunami height and type of house, in Fig. 3.9.

Damage percentage of houses in a flooded area, R_{HD}, is defined by Hatori (1984) as follows.

$$R_{HD} = (a + 0.5b)/(a + b + c),$$

where a, b and c are numbers of houses washed away & completely destroyed, partially damaged, and only flooded, respectively. This R_{HD} is expressed as a function of the tsunami height (for example, see Shuto, 1993). Aida (1977) introduced a tsunami hydraulic force defined as the product of water depth and square of current velocity, both estimated from numerical simulation, and correlated it with R_{HD}. This method does not give much difference from the relation in terms of tsunami height (Shuto, 1993). Matsutomi and Shuto (1994) tried to evaluate the drag force from field data, but the data are scarce yet.

Matsutomi *et al.* (1995) carried out hydraulic experiments of impact forces caused by driftwood. The results are not yet matured for practical design.

(4) Oil-related fires

A fire starts from such a source as kitchen fire when houses are destroyed by earthquakes and tsunamis. If the fire spreads by oils transported by tsunamis, the result is disastrous.

In March 1964, three towns in Alaska suffered this kind of damage. Whittier was the developed community closest to the epicenter of the 1964 Great Alaska earthquake. Locally generated sea waves occurred during and after the quake. Three major sea waves hit the community. The second and third waves crested at 13 m and 10 m. The waves toppled Union Oil and U.S. Army storage tanks, causing an oil spill that caught fire and burned over 3 square miles. The total volume of the oil spill was estimated to be 2.8 × 104 kl consisting of jet fuel and diesel oil, among others. In June of the same year, the city of Niigata, Japan, was hit by an earthquake and its low land was flooded by the tsunami. Oil leaked from a tank damaged by the earthquake and spread over the sea water. Five hours after the earthquake, the oil caught fire, other tanks ignited, the fire continued for 15 days, and 300 residences were burnt (see Shuto, 1987).

Goto (1985) developed a method to simulate the spread of oil transported by tsunamis. His result approaches the Fay formula (1969) with the lapse of time.

The burnt area might be estimated from the boundary between the gravity-viscous and surface tension-viscous regimes of the oil spreading law. Taking typical values of the density of oil, capillary force at the air-oil-water interface and empirical coefficients given by Fay, the burnt area A_B (m^2) is related to the volume of oil V (kl) by $A_B = 324V$ (Shuto, 1991). This relationship clearly explains the Whittier case.

6.2 *Countermeasures*

(1) Design tsunami and countermeasures

Tsunami is a rare natural phenomenon. Enough data are not available for a sound statistical analysis. The basis of the recurrence year of a design tsunami is always ambiguous. If a tsunami that occurs once per 100 yrs is selected as the design tsunami, a seawall may become very tall. On the Ria coast where the topography amplifies tsunamis, there is no wide area between the sea and hills behind. It is hard for coastal residents to accept tall structures which deteriorate the living condition. They cannot bear inconvenience for 100 yrs in order to secure their safety against a giant tsunami which may continue only half a day at the longest. In addition, these tall structures are powerless for an excess tsunami of larger recurrence year. To the farther past we go, the details on largest tsunamis are found. At present, there is no sure means to determine geophysically the upper bound of tsunami.

A good way of countermeasures is not to rely upon structures only but to combine structures, regional planning and soft-wares in a harmony with daily activity. Structures provide a direct means to protect human lives and properties against tsunamis lower than their crown height. Regional planning strengthens coastal villages although perfect protection is not expected. Soft-wares should be introduced in order to save human lives in any case.

(2) Structures

A typical structure is a seawall made of concrete. A sea dike made of soil should be covered by concrete on fore-, top- and rear-faces, otherwise it will be easily eroded if a tsunami overflows. Their crown height is determined by taking into account the tsunami heights estimated from historical data and/or computed with numerical simulation. No method is yet established to design their sections, because of the lack of knowledge of the tsunami forces. The sections are, therefore, designed by referring to examples which were safe against tsunamis in the past. Once these structures are overflowed by a tsunami, the back toe might be scoured. The mechanism of this scouring is not yet understood.

A tsunami gate is constructed at the mouth of a river to prevent the intrusion of tsunamis into the river. Hydraulic experiments are the only means to evaluate tsunami-induced forces on tsunami gates. A tsunami breakwater is constructed at the entrance of a long bay. It does not completely prevent tsunami invasion. It decreases tsunami height in the bay, by reducing the sea water discharge into the bay and by changing the resonance characteristics of the bay. This effect can be estimated by numerical simulation. A tsunami breakwater is an expensive structure, because its site is usually very deep compared to ordinary breakwaters against wind waves and swells.

It is important but difficult to maintain the function of structures against deterioration such as uneven subsidence over a long interval of the appearance of giant tsunamis.

(3) Regional planning

A coastal village changes before the next huge tsunami hits. The tsunami resistance ability of the village should not be deteriorated but much more improved by regulation of land-use.

Movement of residences to the tsunami-free high ground is often used in the past. According to Yamaguchi (1972), in case of the Sanriku coast, Japan, where the major industry is fishing, the distance and height between the new residence and the beach should not exceed 400 m and 15 m, respectively, for a successful relocation of residence in the days when no automobile was used.

Important facilities such as town office and fire station as well as facilities for the disaster weak such as hospital and nursery school should not be located in the tsunami-prone low ground. The tsunami resistant building zone, i.e. alignment of substantial buildings along the shoreline, is a hopeful countermeasure in the future. This idea has no conflict to daily activity. From the experiences in the past, all the reinforced concrete buildings could withstand tsunamis and protect wooden houses behind, if the inundation height was not larger than 6 m (refer Fig. 3.9). The corresponding tsunami height might be about 8 m, because buildings are usually located on the ground 1–2 m higher than sea water level. Only one example of destruction is found in case of the 1946 Aleutian tsunami. A light house 18 m high on the ground, 10 m above sea water level was destroyed by the tsunami 30 m high (inundation height is 22 m) (Lander and Lockridge, 1989). For the inundation height between 6 m and 22 m, no data is found in documents.

A tsunami control forest has a similar effect, but with limit due to strength and thickness of forest trees and undergrowth (Shuto, 1987).

(4) Soft-wares

Evacuation according to forecasting and warning is the most effective means to save human lives. Section 4.7 explains this problem. Disaster culture concerning tsunamis and tsunami disasters should be learnt by and continued to the future generation. Difficulty in education is indifference to other person's experience and a wrong optimistic belief that "I am exceptionally safe". It is important to prepare rescue operations beforehand. If a rescue operation is put in action within 24 h (Golden 24) after a disastrous event, many persons can be saved alive. On the second day, the number of the sufferers found dead exceeds the number of the sufferers found alive. After 72 h, there is almost no possibility to relieve any survivors (Kawata, 1995). This rule is believed universal in any natural disasters. In case of a giant tsunami, the first difficulty in rescue operations is to open the coastal roads covered by debris, houses and boats.

When a violent shaking occurs, some drivers on coastal roads may not be aware of the shaking and unprepared for the approaching tsunami. How to control the coastal traffic is another problem. Readers may find an example of road preparation before, traffic management at, and road opening after a tsunami in Okazaki *et al.* (1995).

7. Forecasting and Evacuation

7.1 *Tsunami forecasting and warning*

Tsunami forecasting is developed and used in many countries as well as in an international cooperation in the Pacific Ocean (see e.g. IOC UNESCO, 1989; Bernard, 1997). Most of them is based upon a general rule that the stronger earthquake in the nearer location generates the larger tsunami which results in the larger coastal effect. The Japan Meteorological Agency determines the magnitude of earthquake from the maximum amplitude of seismic waves recorded on the standard seismograph and the distance to the epicenter from the arrival time difference between the P and S waves (see e.g. Ichikawa and Watanabe, 1983; Uchiike and Hosono, 1995). This method was established in 1957 and replaced, later in 1999, by a new method based upon numerical simulation (Tatehata, 1997). The new method which will give very detailed information is still based upon the principle mentioned above.

This principle is applicable to ordinary tsunamis but not to "tsunami earthquake", a typical example of which is the 1896 Sanriku tsunami mentioned in Sec. 4.1.1. In the sea around Japan, 10% of submarine earthquakes is "tsunami earthquake" (Abe, 1988). Among nine tsunamis in the Pacific Ocean which damaged human lives and properties, from 1992–1996, three were generated by "tsunami earthquake". They are; Nicaraguan tsunami of 1992; East Java tsunami of 1994 and; Peruvian tsunami of 1996. The Nicaraguan earthquake was the first example that the broadband body waves (seismic waves) were useful to clarify characteristics of "tsunami earthquake", slow rupture propagation and slow dislocation motion (Kikuchi and Kanamori, 1995).

A solution to cover all the type of tsunami in forecasting may be the use of broadband seismographs which do not miss the energy of long-period components responsible in generating "tsunami earthquake", as being operated in French Polynesia (Reymond *et al.*, 1993).

When a tsunami is forecasted, a warning is issued and is disseminated to local authorities and coastal residents by every communication means such as telephone, facsimile, radio, TV, information transmission systems specially arranged by civil defense or by similar authorities. The best will be the direct transmission via satellites, excluding any possible errors and delays on the line of information flow.

7.2 *Evacuation*

Evacuation is the most important and effective method to save human lives. Tourists in Hawaii should have a look at the special pages in telephone books placed in every rooms in Hawaiian hotels at their check-in. The pages show the possible tsunami inundation maps and tell, "When the ATTENTION/ALERT SIGNAL is sounded, the EMERGENCY BROADCAST SYSTEM will direct evacuation of threatened areas", adding another important warning in case of a local tsunami, "Any violent earthquake — one that causes you to fall or hold onto something to keep from falling — is a natural tsunami warning. Immediately evacuate beaches and low-lying coastal areas as soon as the shaking stops. Go to an area that is safe from flooding." This rule is widely adopted around the world (e.g. Farreras and Sanchez, 1991). Recent tragic examples as a result of the violation to this rule are found in Shuto (1995).

Another empirical rule that a loud booming noise means an approaching tsunami is analyzed and classified into three major types (Shuto, 1997). Drills make coastal residents aware of evacuation routes and shelters, and keep them informed about how to act in an emergency.

8. Research Needs In The Near Future

8.1 *Reduction of false alarm*

In case of the Aleutian tsunami of May 1986, a false alarm cost a loss of 30 million dollars and resulted in massive traffic gridlock in Honolulu (Adams, 1988). According to Yanagi (1996), among 20 tsunami warnings issued for distant tsunamis since 1946, 15 were considered false alarms (Bernard, 1997).

Gusiakov and Blackford (1998) showed the warning statistics in the Pacific for the years 1991–1997. From all the 30 warnings issued by all tsunami warning centers including regional, national and the Pacific Tsunami Warning Center (PTWC) , at least 20 should be considered as false from the practical point of view.

In case of a far-field tsunami, numerical simulations may be a good means to reduce false judgement, because there is enough time between the generation and the arrival of the tsunami, if the computation is well designed and if the results are well supported and/or well corrected by tsunami measurements on the way of propagation of the tsunami.

8.2 *Offshore tsunami measurement*

From two points of view, increase of offshore tsunami measurement systems in number is in an urgent need. The first need is to make forecasting more reliable as mentioned above. If a tsunami is measured just after its birth and sufficiently before its arrival at the coast, the warning will become more accurate. This improves much, the present incompetent forecasting for a tsunami earthquake.

The second need of offshore measurement is to understand the mechanism of tsunami generation, which is not yet clearly explained as described in Sec. 4.3. The only way to solve this problem is by measuring tsunami's which were less affected by topographical effects.

8.3 *New theories*

As stated in Sec. 4.2.1, edge bores were observed in case of the 1983 tsunami. A theory should be developed to describe this phenomenon. Another necessary theory is for the growth, breaking and regrowth of cnoidal bores at a tsunami front. Energy transfer from behind to the frontal wave can be evaluated with either of the Boussinesq, Peregrine or Goto equations. The most essential importance of this phenomenon is how the energy loss due to breaking is evaluated and formulated.

8.4 *CG animation for a better understanding of tsunami movement*

A tsunami shows a complicated movement under the effect of topography. Refraction and diffraction determine the major course of the tsunami. The tsunami height in the early stage is a result of converging and shoaling. Reflection and entrapment change the type of the tsunami from progressive to standing, then the tsunami in the later stage is determined.

In order to understand which mechanism leads to the final run-up distribution, a CG animation by use of the whole data created in a numerical simulation is a good means. This method is a powerful replacement of hydraulic experiments which require a wide basin and are not adequate for a good understanding because of the scale-effect.

8.5 *Public education aided by "virtual-reality" graphics*

The greatest difficulty in tsunami defence is caused from the infrequent attack of major tsunamis. Coastal residents easily forget severe experiences of their grand-parents. Even if told again and again, other person's experiences never become their owns. Young generation cannot realize the actual faces of major tsunamis.

A method of improvement may be realized by making people have a quasi-experience of a giant tsunami. In the Tsunami Experience Hall, Karakuwa Town, Miyagi Prefecture, Japan, provides a special program by which visitors can experience an earthquake by vibration of chairs and then can see a tsunami on screen. If the "virtual-reality" graphics are introduced, visitors can select several types of reactions and can learn the results: drowned; injured or; safely evacuated, then this expression will help them to recollect the necessary reaction when they encounter a giant tsunami.

References

Abe, K. (1988): Tsunami magnitude and the quantification of earthquake tsunamis around Japan, *Bull. Earthq. Res. Inst.* **63**, 4289–303 (in Japanese).

Adams, W. M. (1988): Basic improvement of tsunami warning systems on previous experiences, Science of Tsunami Hazards, **6**, No. 1.

Aida, I. (1977): Numerical experiments for inundation of tsunami -Susaki and Usa, in Kochi Prefecture, *Bull. Earthq. Res. Inst.* **56**, 441–460 (in Japanese).

Aida, I. (1978): Reliability of a tsunami source model derived from fault parameters, *J. Phys. Earth* **26**, 57–73.

Bernard, E. N. (1997): Reducing tsunami hazards along US Coastlines, *Advances in Natural and Technological Hazards Res.* **9**, 189–203.

Farreras, S. F. and A. J. Sanchez (1991): The tsunami threat on the Mexican West Coast: A historical analysis and recommendations for hazard mitigation, In: Tsunami Hazard (E. N. Bernard, ed.), 301–316.

Fay, J. M. (1969): The spread of oil slicks on a calm sea, In *Oil on the Sea*: Plenum Press, 53–63.

Fukuyama, E. and K. Irikura (1986): Rupture process of the 1983 Japan Sea (Akita-oki) earthquake using a waveform inversion method, *Bull. Ser. Soc. America*, **76**, 1623–1640.

Gonzalez, F. I., E. N. Bernard, H. B. Miburn, D. Castel, J. Thomas and J. M. Hemley (1987): The Pacific Tsunami Observation Program (PacTop), Proceedings of the International Tsunami Symposium, 3–19.

Goto, C. (1983): Numerical simulation of spread of floating timbers, *Proc. Coastal Engrg., JSCE,* **30**, 594–597 (in Japanese).

Goto, C. (1984): Equations of nonlinear dispersive long waves for a large Ursell number, *Proc. JSCE,* 351/II-2, 193–201 (in Japanese).

Goto, C. (1985): A simulation model of oil spread due to tsunamis, *Proc. JSCE,* No. 357/II-3, 21–223 (in Japanese).

Goto, C. and N. Shuto (1983): Numerical simulation of tsunami propagations and run-up, *Tsunamis-Their Science and Engrg.*, Terra Science Pub. Co., Tokyo/Reidel, Dordrecht, 551–525.

Gusiakov, V. (1998): Report on V. Gusiakov's trip to Chile, Tsunami Bulletin Board <TSUNAMI@itic.noaa.gov> on internet, 18 May 1998.

Hashimoto, M. *et al.* (1994): Crustal movement and fault model of the Hokkaido Nansei-Oki earthquake, Kaiyo Monthly, Extra No. 7, 55–61 (in Japanese).

Hatori, T. (1984): On the damage to houses due to tsunamis, *Bull. Earthq. Res. Inst.* **59**, 433–439 (in Japanese).

Ichikawa, M. and H. Watanabe (1983): A new system for tsunami warning in the Japan Meteorological Agency, In *Tsunami-Their Science and Engrg.* (K. Iida and T. Iwasaki, eds.): Terra Science Publication, Tokyo, pp. 51–60.

Imamura, A. (1942): History of Japanese tsunamis, Kaiyo-no Kagaku (Oceanography), **2**, 74–80 (in Japanese).

Imamura, A. (1977): Data of precursory phenomena of big earthquake, Manuscripts left by the late Dr. Aritsune Imamura, Association for Prevention of Earthquake Disasters, 170 p., (in Japanese).

Imamura, F., N. Shuto, S. Ide, Y. Yoshida and Ka. Abe (1993): Estimate of the tsunami source of the 1992 Nicaragua earthquake from tsunami data, *Geophys. Res. Lett.* **20**, 1515–1518.

Intergovernmental Oceanographic Commission (1989): Second International Tsunami Workshop on the Technical Aspects of Tsunami Warning Systems, Tsunami Analysis, Preparedness, Observation and Instrumentation, UNESCO, IOC Workshop Report No. 58, 286 p.

Kajiura, K. (1963): Effects of a breakwater on the oscillations of bay water, *Bull. Earthq. Res. Inst.* **41**, 403–418 (in Japanese).

Kajiura, K. (1970): Tsunami source, energy and the directivity of wave radiation, *Bull. Earthq. Res. Inst.* **48**, 835–869.

Kanamori, H. and J. J. Cipar (1974): Focal process of the great Chilean earthquake May 22, 1960, *Phys. Earth Planet Interiors* **9**, 128–136.

Kawamata, S., F. Imamura and N. Shuto (1993): Numerical simulation of the 1883 Krakatau tsunami, Proceedings of the XXV Congress IAHR, Vol. IV, 24–31.

Kawata, Y. (1995): Delay of response just after earthquake and emergency management, *J. Japan Society for Natural Disaster Science*, Special Issue, 7–17 (in Japanese).

Kikuchi, M. and H. Kanamori (1995): Source characteristics of the 1992 Nicaragua tsunami earthquake inferred from teleseismic body waves, PAGEOPH, 144, Nos. 3/4, 441–453.

Lander, J. F. and P. Lockridge (1989): United States Tsunamis 1690–1988, *National Geophysical Data Center*, 265 p.

Lander, F. J. and P. A. Lockridge (1989): United States Tsunamis, 1690–1988, *National Geophysical Data Center*, 94–95.

Lee, Ho Jun (1998): Study on Characteristics of Tsunami Propagation and Hazards in the Japan Sea, Doctoral Thesis submitted to Tohoku University, 150 p. (In Japanese).

Makino, K. (1968): The Great Meiwa Tsunami at Yaeyama, published by the author, 462 p. (In Japanese).

Mansinha, L. and D. E. Smylie (1971): The displacement fields of inclined faults, *Bull. Seis. Soc. America* **61**, 1433–1440.

Matsutomi, H. and N. Shuto (1994): Tsunami inundation depth, current velocity and damage to houses, Proceedings of the Coastal Engineering, JSCE, **41**, 246–250 (in Japanese).

Matsutomi, H. and K. Ikeda (1995): Impulsive force of drift-woods with consideration of its plasticity, Proceedings of the Coastal Engineering, JSCE, **41**, 806–810 (in Japanese).

Matsutomi, H. and M. Shiratorri (1995): Laboratory study on maximum impulsive force of timbers drifted by surge front of tsunamis, Proceedings of the Coastal Engineering, JSCE, **41**, 811–815 (in Japanese).

Miyano, M. and K. Ro (1992): A study of damages to human lives caused by earthquakes, based upon data in the past, Papers Annual Conference, *Inst. of Social Safety Science*, No. 2, 129–135 (in Japanese).

MRI (Meteorol. Res. Inst.) (1984): Bottom Pressure Observation South Off Omaezaki, Central Honshu, Technical Report MRI, No. 9, 90 p. (in Japanese with an English abstract).

Nagai, T. *et al.* (1994): Introduction of Japanese NOWPHAS System and its recent topics (Nationwide Ocean Wave information network for Ports and Harbours), HYDRO-PORT'94, Proceedings of the International Conference on Hydro-Technical Engineering for Ports and Harbor Construction, 67–82.

Nagano, O., F. Imamura and N. Shuto (1991): A numerical model for far-field tsunamis and its application to predict damage done to aquaculture, Tsunami Hazard: Kluwer Academic Publishers, 235–255.

Okada, Y. (1985): Surface deformation due to shear and tensile faults in a half-space, *Bull. Seis. Soc. America*, **75**, 1135–1154.

Okazaki, S., K. Shibata and N. Shuto (1995): A road management approach for tsunami disaster planning, Tsunami Progress in Prediction, Disaster Prevention and Warning, Advances in Natural and Technological Hazards Research: Kluwer Academic Press, Vol. 4, 223–234.

Reymond, D., O. Hyvernaud and J. Talandier (1993): An integrated system for real time estimation of seismic source parameters and its application to tsunami warning, Advances in Natural and Technological Hazards Res., (Tinti, S., ed.), **1**, 177–196.

Satake, K. (1989): Inversion of tsunami waveforms for the estimation of heterogeneous fault motion of large submarine earthquakes: the 1968 Tokachi-oki and 1983 Japan Sea Earthquakes, *J. Geophysical Res.* **94**, 5627–5636.

Satake, K., M. Okada and Ku. Abe (1988): Tide gauge response to tsunamis: Measurements at 40 tide gauge stations in Japan, *J. Marine Res.* **46**, 557–571.

Sayama, J., N. Shuto and C. Goto (1988): Errors induced by refraction in tsunami numerical simulation, *Proceedings of the 6th Congress, Asian and Pacific Regional Division*, IAHR, Vol. IV, 257–264.

Shuto, N. (1985): The Nihonkai-Chubu earthquake tsunami on the North Akita coast, *Coastal Engrg. in Japan* **28**, 255–164.

Shuto, N. (1987): Effectiveness and limit of tsunami control forests, *Coastal Engrg. in Japan*, **30**(1), 143–153.

Shuto, N. (1991): Historical changes in characteristics of tsunami disasters, *Proceedings of the International Symposium on Natural Disaster Reduction and Civil Engineering*, JSCE, 77–86.

Shuto, N. (1993): Tsunami intensity and disasters, Tsunamis in the World: Kluwer Academic Publishers, 197–216.

Shuto, N. (1995): Tsunami, disasters and defense works in case of the 1993 Hokkaido-oki earthquake tsunami, *Advances in Natural and Technological Hazard Res.* **4**, 263–276.

Shuto, N. (1997): A natural warning of tsunami arrival, *Advances in Natural and Technological Hazards Res.* **9**, 157–173.

Shuto, N. and C. Goto (1978): Numerical simulation of tsunami run-up, *Coastal Engrg. in Japan* **21**, 13–20.

Shuto, N. and H. Matsutomi (1995): Field Survey of the 1993 Hokkaido Nansei–Oki Earthquake Tsunami, PAGEOPH, **144**(3/4), pp. 649–663.

Shuto, N., T. Suzuki, K. Hasegawa and K. Inagaki (1986): A study of numerical technique on the tsunami propagation and run-up, *Science of Tsunami Hazards* **4**(2), 111–124.

Shuto, N., C. Goto and F. Imamura (1990): Numerical simulation as a means of warning for near-field tsunami, *Coastal Engrg. in Japan* **33**(2), pp. 173–193.

Shuto, N., K. Chida and F. Imamura (1995): Generation mechanism of the first wave of the 1983 Nihonkai Chubu earthquake tsunami, Tsunami: Progress in Prediction, Disaster Prevention and Warning: Kluwer Academic Publishers, 37–53.

Simkin, T. and R. S. Fiske (1983): Krakatau 1883, Eruption and its Effects: Smithonian Institution Press, 464 p.

Takahashi, To., Ta. Takahashi, N. Shuto, F. Imamura and M. Ortiz (1995): Source models for the 1993 Hokkaido Nansei–Oki earthquake tsunami, *Pure and Appl. Geophys.* **144**, 747–767.

Takeda, H. (1984): Numerical simulation of run-up by variable transformation, *J. Oceanogr. Soc. Japan* **40**, 271–278.

Tatehata, H. (1997): The new tsunami warning system of the Japan Meteorological Agency, *Advances in Natural and Technological Hazards Res.*, **9**, 175–188.

Uda, T. *et al.* (1988): Two-dimensional deformation of nonlinear long waves on a beach, Report No. 2627, *Public Works Res. Inst.*, Ministry of Construction, Japan (in Japanese).

Uchiike, H. and K. Hosono (1995): Japan tsunami warning system; Present status and future plan, *Advances in Natural and Technological Hazards Res.* **4**, 305–322.

Yamaguchi, Y. (1972): Movement of villages in the tsunami-prone, Sanriku Coast, in The Selected Papers of Yamaguchi Yaichiro, Sekai Bunko, **6**, 427 (in Japanese).

For tsunami data, the following publications written in English and in Japanese are available.

Soloviev, S. L. and Ch. N. Go (1984): Catalogue of Tsunamis on the Eastern Shore of the Pacific Ocean, *Canada Inst. for Scientific and Technical Information, National Res. Council*, 285 pp.

Soloviev, S. L. and Ch. N. Go (1984): Catalogue of Tsunamis on the Western Shore of the Pacific Ocean, *Canada Inst. for Scientific and Technical Information, National Res. Council*, 439 pp.

Lockridge, P. A. (1985): Tsunamis in Peru-Chile, Report SE-39, National Geophysical Data Center, U.S. Department of Commerce.

Lander, J. F. and P. A. Lockridge (1989): United States Tsunamis 1690–1988, Publication 41-2, *National Geophysical Data Center, U.S. Department of Commerce*, 265 pp.

Lander, J. F., P. A. Lockridge and M. J. Kozuch (1993): Tsunamis Affecting the West Coast of the United States 1806–1992, NGDC Key to Geophysical Records Documentation No. 29, *National Geophysical Data Center, U.S. Department of Commerce*, 242 pp.

Sanchez, A. J. and S. F. Farreras (1993): Catalog of Tsunamis on the Western Coast of Mexico, Report SE-50, *National Geophysical Data Center, U.S. Department of Commerce*, 79 pp.

Lander, J. F. (1996): Tsunamis Affecting Alaska 1737–1996, NGDC Key to Geophysical Records Documentation No. 29, 242 pp., National Geophysical Data Center, U.S. Department of Commerce, 195 pp.

Watanabe, H. (1998): *Comprehensive List of Tsunamis to Hit the Japanese Islands*, 2nd edn.: University of Tokyo Press, 238 pp. (In Japanese).

To know development in tsunami research and tsunami prevention practice, the following books are recommended.

Bernard, E. N., ed. (1991): *Tsuanami Hazard, A Practical Guide for Tsunami Hazard Reduction*: Kluwer Academic Press, 336 p.

Tinti, S., ed. (1993): *Tsunamis in the World, Fifteenth International Tsunami Symposium, 1991, Advances in Natural and Technological Hazards Research,* Vol. 1: Kluwer Academic Press, 228 pp.

Satake, K. and F. Imamura, eds. (1995): Tsunamis: 1992–1994, Their Generation, Dynamics, and Hazard, *Pure and Appl. Geophys.* **144**(3/4), 890 pp.

Tsuchiya, Y. and N. Shuto, eds. (1995): Tsunami: Progress in Prediction, Disaster Prevention and Warning, *Advances in Natural and Technological Hazards Res.* Vol. 4: Kluwer Academic Press, 336 p.

Hebenstreit, G., ed. (1997): Perspectives on Tsunami Hazard Reduction, Observation, Theory and Planning, *Advances in Natural and Technological Hazards Res.* **9**: Kluwer Academic Press, 218 p.

Chapter 4

Basic Coastal Sediment Transport Mechanisms

Peter Nielsen
Associate Professor
University of Queensland, Australia

4.1. Introduction

For an introductory overview of the sediment transport modes on a beach consider Fig. 4.1.

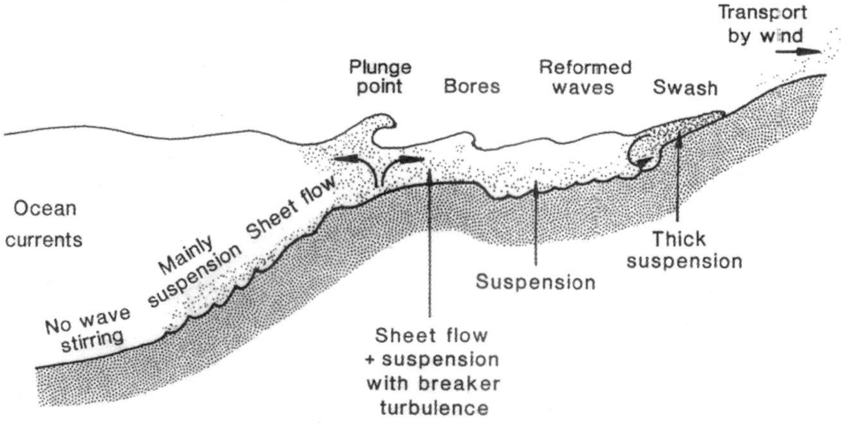

Fig. 4.1 Sediment transport modes on a beach, a simplified 2D view. 3D features like rip currents and longshore currents will increase the vertical scale of entrainment and transport sand. On fine.g.rained beaches, the step at the bottom of the swash zone may be absent. From Nielsen, 2009.

In deep water, the bed is unaffected by waves and consequently any sediment transport will be due to tidal or other ocean currents.

In somewhat shallower depths ($h < 25\,\mathrm{m}$), the wave motion will often dominate flow near the bed and the sediment transport. As an indication of this, the bed will often be covered by sharp crested, shore-parallel ripples (vortex ripples), over which most of the sediment is transported in suspension.

Close to the breakpoint, the wave induced water motion at the bed is often so vigorous that the vortex ripples disappear and the bed will have megaripples or may almost be perfectly flat. In this area, the transport occurs essentially within a few millimetres of the bed level. The transport mode which is called "sheet flow" involves a mixture of bedload and suspended load.

Around the breakpoint, the sediment transport occurs throughout the depth because the breaking waves can generate strong vertical flows, bringing large amounts of sediment toward the surface, see Fig. 4.37. This sediment, being suspended for several wave periods, is then transported by the currents including wave generated circulation. This area still holds many challenges including description of the water motion, which is not at all like that of progressive waves, and of the breaking induced, largely convective sediment entrainment mechanisms.

In the surf zone the bed may be rippled or flat, but because of the breaker generated turbulence a considerable amount of sand will be suspended throughout the water column. In this area, the wave generated undertow, rip currents or longshore currents play a major role in transporting the sediment.

On many beaches a step exists at the base of the swash zone on which vigorous wave breaking occurs, e.g. Pearl Beach, Nielsen (2009). Surprisingly, despite the strong turbulence in this area, very sharp horizontal gradients in sediment size are often found at the step. The flow and sediment transport processes that maintain this step are yet to be fully understood and modeled.

Landward of the step, in the swash zone, the water motion is quite different from ordinary wave motion (Figs. 4.2 and 4.47) and very little is known about the boundary layer structure and bed shear stresses. Infiltration/exfiltration effects may be significant parts of the swash zone sediment transport mechanisms. While the sand surface in the swash zone is usually flat, sandwaves similar to antidunes in supercritical steady flow sometimes develop during long lasting backwash events, see Fig. 4.2.

The sediment transport in the swash zone is very significant in the shore parallel as well as the shore normal direction.

From the top of the swash zone the sand may be moved on by onshore winds.

4.2. Boundary Layers and Bed Shear Stresses

4.2.1 *Coastal bottom boundary layers*

Sediment particles are entrained by the flow via the bed shear stresses $\tau_b(t)$. In general, both waves and currents contribute to the bed shear stress. However, since the wave boundary layers are thinner, the waves tend to generate greater bed shear stresses for a given velocity magnitude, see Fig. 4.3.

The vertical scale of laminar oscillatory boundary layers is the Stokes length $\sqrt{2\upsilon/\omega}$. For turbulent oscillatory boundary layers it is $0.09\sqrt{Ar}$, where r is the hydraulic roughness of the bed, cf. Nielsen and Guard (2011).

Fig. 4.2 Backrush of strong surf beat. South Beach Brunswick Heads, 17 March 1992. Such events contribute a very large part of the beach erosion during storms.

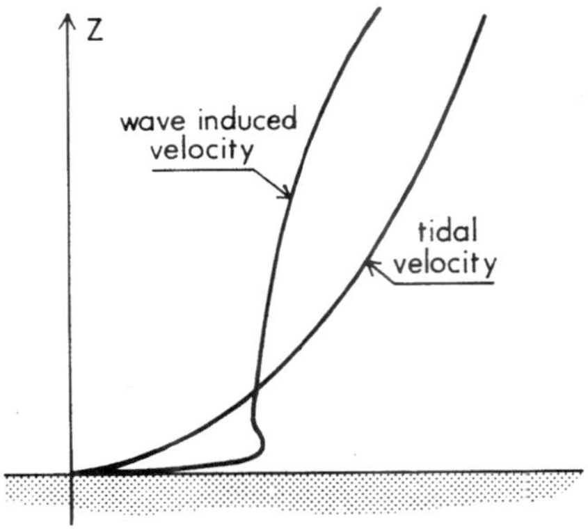

Fig. 4.3 Rapidly accelerated flows induced by waves have thinner boundary layers than longer period flows like tides.

The boundary layer thickness δ is proportional to the square root of the wave period,

$$\delta \propto (v_t T)^{1/2}, \quad (4.2.1)$$

where v_t is the eddy viscosity. Consequently, since

$$\tau_b \propto v_t u_\infty / \delta, \quad (4.2.2)$$

wave motions generate greater bed shear stresses for a given free stream velocity (u_∞) magnitude than say tidal flows.

Experimental evidence indicates that the wave boundary layer structure and the wave shear stresses are practically unaffected by the presence of currents (cf.

Nielsen, 1992, Fig. 1.5.8 and Simons *et al.*, 2000). Hence, for practical purposes, the sediment entrainment can usually be contributed to the wave-generated stresses and these can be calculated as if the current was not there.

The laminar solution for simple harmonic oscillatory boundary layer flow, (cf. Nielsen, 1992, pp. 19–23) has been known for a long time, and it is a good guide to some of the important features of natural wave boundary layers, for example the velocity overshoot indicated in Fig. 4.3. However, some features, like the phase lead of the bed shear stress relative to the free stream velocity, are different in turbulent boundary layers, typically 15°–30° compared with the laminar 45°.

For more complicated, unsteady (but still x-uniform) boundary layer flows the laminar bed shear stress can be expressed in terms of a convolution of previous free stream accelerations

$$\tau_b(t) = \rho\sqrt{v}\int_{t'=-\infty}^{t} \frac{\frac{du}{dt'}}{\sqrt{t-t'}}dt'. \tag{4.2.3}$$

This expression has been used successfully to explain why the bed shear stress changes to negative after the passage of the crest of a solitary wave although the velocity remains positive, Liu *et al.* (2007). Liu (2006) also pointed out that similar convolutions apply if the time-invariant eddy viscosity varies as a power of z:

$$v_t(z) = v_t(z_0)\left(\frac{z}{z_0}\right)^p, \quad 0 \le p < 1. \tag{4.2.4}$$

However, it must be stressed that this superposition law (the convolution) is only valid when v_t is time-invariant, and that is usually not the case in transient and turbulent boundary layer flows.

To sort out how the eddy viscosity scales on the bed roughness r and fluid viscosity in an arbitrary unsteady flow remains an urgent and important challenge.

The qualitative concepts of velocity and shear stress distributions are illustrated in Fig. 4.4, over beds of increasing complexity. The roles of "THE bed shear stress" — perhaps the stress that would be measured by a shear-plate at $z = 0$ (?) — and the stream-wise pressure gradients, with respect to moving sediment throughout a moving layer, need serious considerations.

4.2.2 *Shear stresses on rippled beds*

Unsteady bed shear stresses are very difficult to measure over movable sand beds. Consequently, knowledge about them was for a long time limited to two experimental studies: Carstens *et al.* (1969) and Löfquist (1986). Carstens *et al.* only measured the time-averaged energy dissipation and hence they provide no information about the shape of $\tau_b(t)$. Löfquist did measure time-series of $\tau_b(t)$, but only on rippled beds where the behavior of $\tau_b(t)$ is quite complicated because of the periodic vortex formation, see Fig. 4.5.

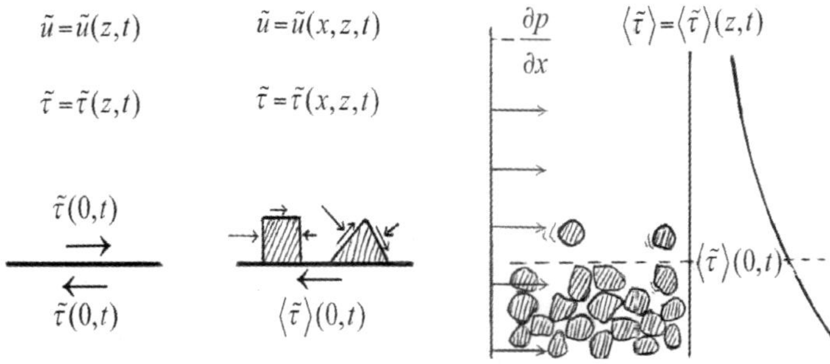

Fig. 4.4 Shear stress distributions $\tilde{\tau}(x, z, t)$ and "bed shear stresses" on three different types of beds: left, a flat impermeable bed; middle, a rough impermeable bed; and right a movable sand bed. For the left-hand scenario, smooth bed, the definition of THE bed shear stress is obvious. For the right-hand scenario of a movable bed it is not.

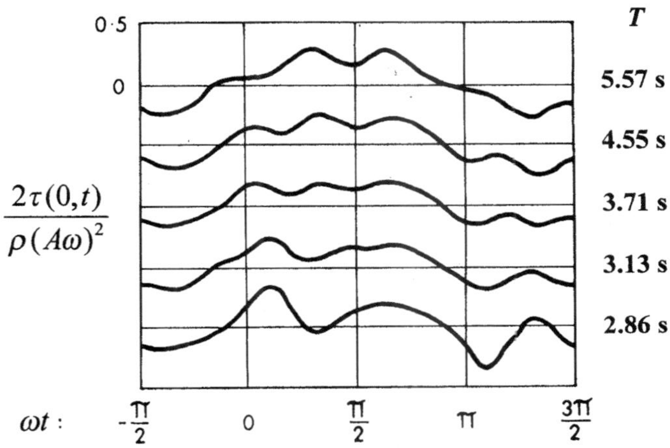

Fig. 4.5 Bed shear stress measured over natural, rippled sand beds by Löfquist (1986). The free stream velocity varied as sin ωt. The multi-peaked behavior is due to the rhythmic formation of vortices at the ripple crests.

For simple harmonic wave motion over flat beds, $\tau_b(t)$ displays a well defined phase-lead compared with the free stream velocity. For laminar flow, this phase-lead is 45°, cf. Nielsen (1992), p. 21. For turbulent flows it is less, typically 15°–30°. Recent detailed measurements under sheet-flow conditions have opened the window on the boundary layer structure over flat movable beds.

For sediment transport purposes, the total bed shear stress on a bed with bed-forms, e.g. ripples, is often split into the form drag $\tau''(t)$ and the skin friction $\tau'(t)$.

$$\tau(t) = \tau'(t) + \tau''(t). \tag{4.2.5}$$

This is because the form drag, which is related to the front-and-back pressure differences on the bed forms, is considered irrelevant to the movement of sediment particles. The sediment is moved by the skin friction.

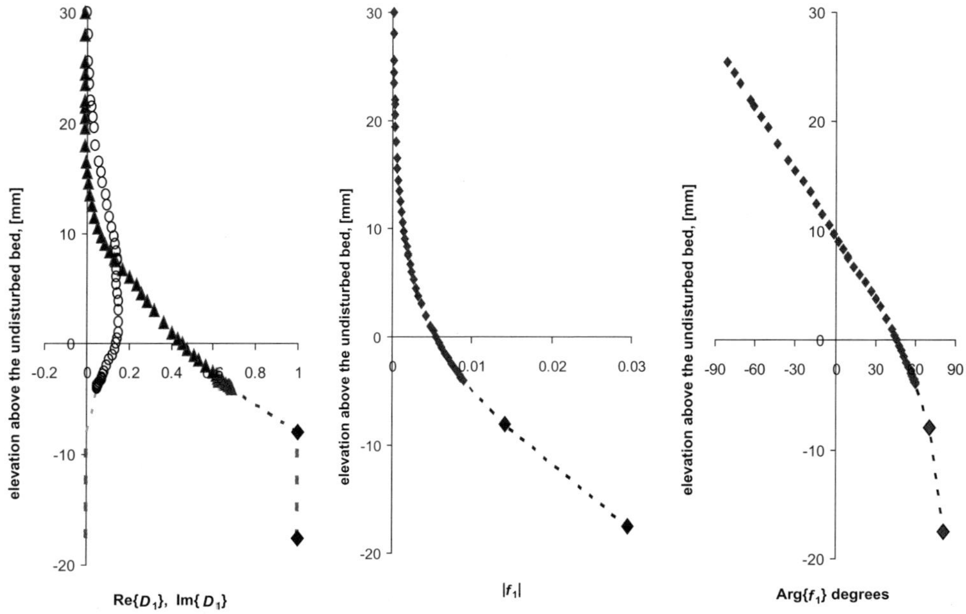

Fig. 4.6 Left: Velocity defect function $D_1(z)$ (in terms of its real and imaginary parts). Center: magnitude of the friction factor $f_1(z)$ corresponding to the local shear stress magnitude and right: argument of $f_1(z)$ corresponding to the phase-lead of the local shear stress relative to the free stream velocity. Data from O'Donoghue *et al.* (2004), $T = 6\,\text{s}$, $A = 1.2\,\text{m}$, $d_{50} = 0.15\,\text{mm}$. $Z = 0$ is the undisturbed bed level. From Nielsen and Guard, 2011.

4.2.3 *Shear stresses over flat movable beds*

Sleath (1984) pointed out that, because of the horizontal pressure gradients, which drive unsteady flows, the shear stress will generally increase in magnitude downwards, and the variation is often considerable through the moving sand layer, as shown by the data in Fig. 4.6. Along with the shear stress increase through the moving sediment layer goes an increase of the phase-lead from around 45° at the undisturbed bed level toward an asymptotic value of 90° deep in the bed, Fig. 4.6 (right).

There is thus no obvious definition of "THE bed shear stress" on a movable sand bed.

4.2.4 *The log law for steady flow*

In a steady uniform flow with depth h and surface slope S, the bed shear stress must balance the downhill pull by gravity. This gives

$$\overline{\tau_b} = \rho g h S \tag{4.2.6}$$

and correspondingly, the steady friction velocity u_* is given by

$$\overline{u_*} = \sqrt{\overline{\tau_b}/\rho} = \sqrt{ghS}. \tag{4.2.7}$$

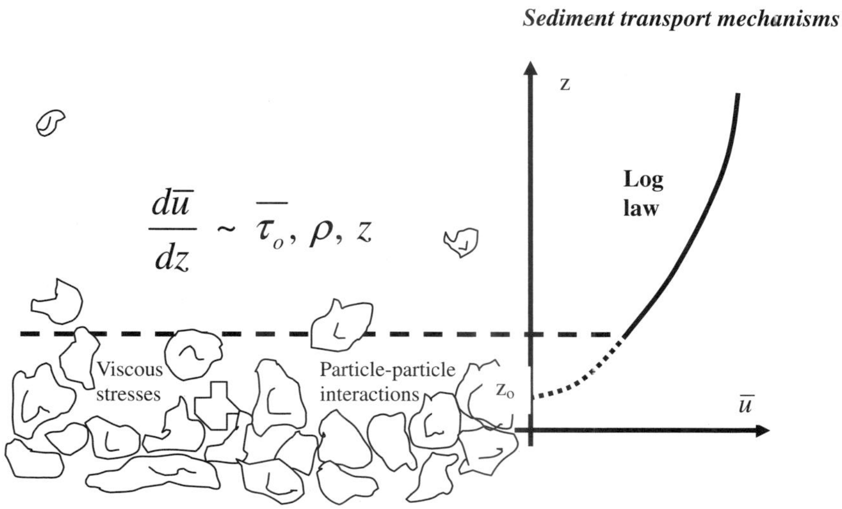

Fig. 4.7 In a steady, uniform flow, an important layer exists where the shape of the velocity profile is governed by the bed shear stress and the fluid density. In this layer the velocity distribution must be logarithmic. In the layer below, viscous stresses and/or intergranular forces from moving sediment will also influence $\bar{u}(z)$. From Nielsen, 2009.

A reasonable fit to steady flow velocity profiles and a simple, yet useful theoretical model is provided by the logarithmic velocity profile also known as "the law of the wall",

$$\bar{u}(z) = \frac{\overline{u_*}}{\kappa} \ln \frac{z}{z_o} \approx \frac{\overline{u_*}}{\kappa} \ln \frac{z}{r/30}, \tag{4.2.8}$$

where κ is von Karman's constant and r is the hydraulic roughness (Nikuradse roughness) of the bed.

The *law of the wall* was developed by Theodore von Karman in 1920s but its validity or otherwise is perhaps most succinctly summarized by the derivation given by Landau and Lifshitz (1987). That is: assume as in Fig. 4.7, that a layer exists where the velocity gradient is determined completely by the bed shear stress, the fluid density and the elevation.

If indeed $\frac{d\bar{u}}{dz}, \overline{\tau_o}, \rho, z$ are the only relevant physical variables, it follows from dimensional analysis that only one dimensionless combination is possible and that it must be a constant. One way of expressing this is

$$\frac{z \frac{d\bar{u}}{dz}}{\sqrt{\overline{\tau_o}/\rho}} = \text{constant} \tag{4.2.9}$$

where the constant is written as $1/\kappa$, and $\kappa \approx 0.4$ is called von Karman's constant.

Integrating and introducing the friction velocity $\bar{u}_* = \sqrt{\overline{\tau_o}/\rho}$, this gives (4.2.4) where the constant of integration z_o depends on the conditions in the *inner layer* below the log layer. The value $z_o = r/30$ implied by (4.2.8) applies to fully rough

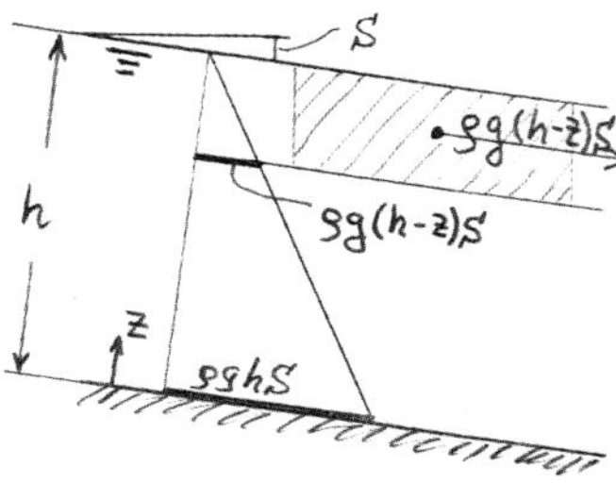

Fig. 4.8 Steady equilibrium with the downhill pull of gravity requires $\overline{\tau}(z)$ to vary linearly with the distance from the surface.

turbulent conditions which are usually assumed in vigorous sediment transport scenarios, see Schlichting and Gersten (2000) for details.

Note that z_o is also the level at which the downward extrapolation of the log velocity profile hits zero velocity.

Steady equilibrium with the downhill pull of gravity requires the shear stress distribution to be linear: $\overline{\tau}(z) = \tau_b(1 - z/h) = \rho u_*^2(1 - z/h)$, see Fig. 4.8. Hence, the logarithmic velocity profile requires a parabolic eddy viscosity distribution

$$\overline{v_t}(z) = \frac{\overline{\tau}/\rho}{\frac{\partial \bar{u}}{\partial z}} = \kappa \overline{u_*} z(1 - z/h) \tag{4.2.10}$$

4.2.5 *The wave friction factor*

Since the time dependence of $\tau_b(t)$ is excessively complicated for rippled beds, Fig. 4.5, $\tau_b(t)$ is often only modeled as far as its peak value $\hat{\tau}_b$ is concerned, through the definition,

$$\hat{\tau}_b = \frac{1}{2}\rho(A\omega)^2 f_w, \tag{4.2.11}$$

this then involves prediction of the wave friction factor f_w, which is a function of the relative bed roughness r/A and of the boundary layer Reynolds number $A^2\omega/v$. A is the water particle semi excursions just above the boundary layer.

The behavior of $f_w(r/A, A^2\omega/v)$ was investigated comprehensively by Riedel (1972) for fixed beds with variable sand grain roughness and his results are shown in Fig. 4.8, which is analogous to the Moody diagram for pipe flow friction factors.

Note that $\tau_{b,\max}$ always occurs ahead of $u_{\infty,max}$ even if the bed is not rippled as for the data in Fig. 4.5. For laminar flow the phase-lead is 45°. For turbulent flows it is less, decreasing with increasing Reynolds number.

For most field applications, the Reynolds number dependence of f_w can be ignored, i.e. the flow is considered "fully rough turbulent", and the friction factor is considered to be a function of the relative roughness only, e.g.,

$$f_w = \exp\left[5.5\left(\frac{r}{A}\right)^{0.2} - 6.3\right] \tag{4.2.12}$$

which is within a factor 1.5 of the available fixed bed measurements for $0.0002 < r/A < 1$, cf. Nielsen (1992), p. 26. Thus, calculation of the friction factor, for total bed friction, is not a problem, provided the hydraulic roughness r of the bed is known. Knowing what r is for movable sediment beds is, however, a very complex and largely unresolved issue. Further uncertainty arises with respect to the sediment transport because it is a function of the skin friction τ' only, rather than of the measurable total.

The f_w-formalism reported above, is based on the notion of "THE bed shear stress" which is shown by Fig. 4.5 to be somewhat naïve. Nevertheless, it is included here for the possible use in simplistic operational sediment transport models.

At the current state-of-the-art, these complications are usually bypassed in sediment transport estimations, and the wave shear stresses and resulting sediment transport are dealt with in terms of the nominal *grain roughness Shields Parameter* $\theta_{2.5}$ and the corresponding friction factor $f_{2.5}$, which is determined from Eq. (4.2.12) with $r = 2.5\, d_{50}$:

$$f_{2.5} = \exp\left[5.5\left(\frac{2.5 d_{50}}{A}\right)^{0.2} - 6.3\right]. \tag{4.2.13}$$

4.2.6 *Time dependent bed shear stresses*

A perfectly sinusoidal $u_\infty(t)$ will, due to its symmetry, not generate any net sediment transport. However, other wave motions with zero net flow are known to generate considerable net sediment transport. For typical finite-amplitude wave shapes (see e.g. Nielsen, 2009, p. 46), this can be understood by steady flow reasoning: Since the bed shear stress magnitude varies more or less as u^2, the larger velocities associated with the wave crest will win, despite their shorter duration. For such wave shapes, the net sediment transport rates, which are of not too fine sand, follow the simple rule $\overline{Q_s} \propto \overline{u_\infty^3}$. However, most surf zone waves have a degree of saw-tooth asymmetry like the wave in Fig. 4.9 and for these waves the simple $\overline{Q_s} \propto \overline{u_\infty^3}$ rule does not work. Waves with saw-tooth asymmetry often generate significant shoreward sediment transport even though $\overline{u_\infty^3} = 0$ as well as $\bar{u}_\infty = 0$.

The reason for this landward transport is the strong asymmetry in the bed shear stress, which is indicated in Fig. 4.10 and can be evaluated by the convolution (4.2.3) for laminar flow. That is, when the flow is accelerated very abruptly, the boundary layer has had little time to grow, and the shear stresses become very large for a given value of the free stream velocity. In addition, the stream-wise pressure gradients, which drive the abrupt acceleration act directly on the sediment particles.

While, as mentioned above, the time dependent shear stresses in a laminar flow over a flat impermeable bed can be evaluated by the convolution (4.2.3), this method has not been generalized to scenarios with time-varying eddy viscosity.

Instead, more ad hoc methods, based on the friction factor concept with influence from fluid accelerations have been suggested. For example, several kinds of indirect

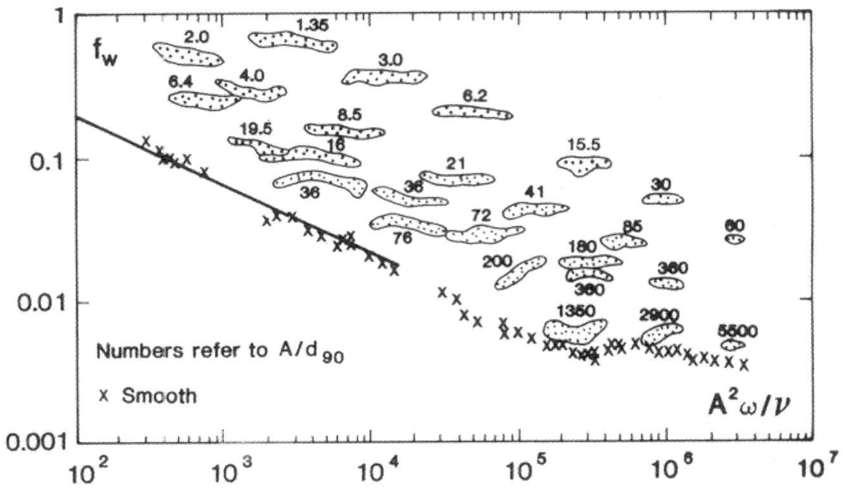

Fig. 4.9 Wave friction factors for fixed flat sand beds, i.e. a single layer of sand grains glued to the top of a shear plate, measured by Riedel (1972). The numbers refer to A/d_{90}. The solid line corresponds to the theoretical, smooth laminar flow result, $f_w = \frac{2}{\sqrt{RE}} = \frac{2}{\sqrt{A^2\omega/v}} =$, cf. Nielsen (1992), p. 21.

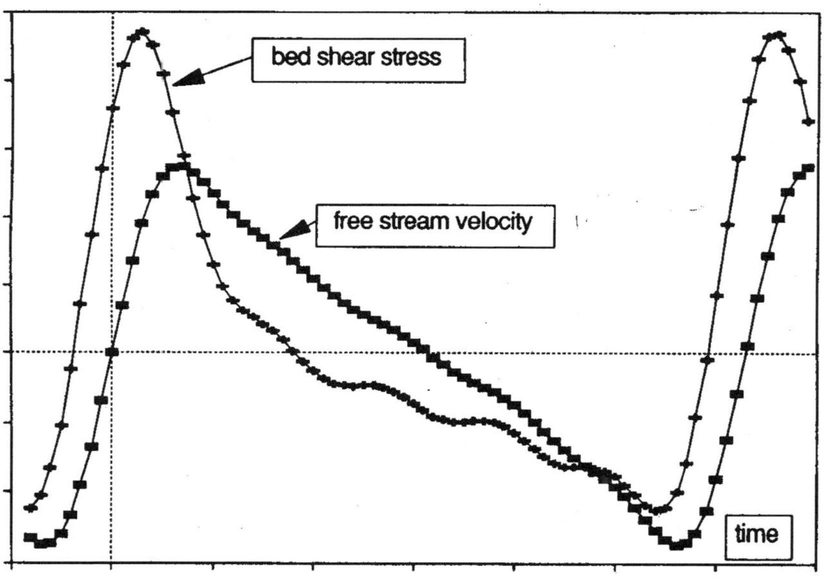

Fig. 4.10 The thinner boundary layer during abrupt acceleration, leads to greater bed shear stresses for a given value of the free stream velocity.

evidence (Nielsen, 2006) support the formula

$$\tau_{2.5}(t) = \frac{1}{2}\rho f_{2.5}\left[\cos\varphi_\tau u_\infty(t) + \sin\varphi_\tau \frac{u_\infty(t+\delta_t) - u_\infty(t-\delta_t)}{2\omega_p\delta_t}\right]^2 \operatorname{sign} u_*(t)], \tag{4.2.14}$$

where δ_t is the time step of the free stream velocity $u_\infty(t)$ time series and φ_τ is the phase-lead of the bed shear stress at the peak angular frequency ω_p. Its best-fit value for sheet flow, including swash zone, applications seems to be $\varphi_\tau \approx 45°$. The corresponding instantaneous friction velocity is

$$u_*(t) = \sqrt{\frac{1}{2}f_{2.5}}\left[\cos\varphi_\tau u_\infty(t) + \sin\varphi_\tau \frac{u_\infty(t+\delta_t) - u_\infty(t-\delta_t)}{2\omega_p\delta_t}\right]. \tag{4.2.15}$$

For the purpose of calculating the grain roughness wave friction factor, $f_{2.5}$, via Eq. (4.2.8), for an arbitrary $u_\infty(t)$, the equivalent water particle semi excursion is taken to be

$$A = \frac{\sqrt{2\mathrm{Var}\{u_\infty(t)\}}}{\omega_p}, \tag{4.2.16}$$

where ω_p is the peak frequency of the u_∞ spectrum. The subscript "2.5" refers to the bed roughness being taken as 2.5 d_{50}.

4.3. Sediment Parameters and Sediment Mobility

4.3.1 *Introduction*

Natural flows and natural sediments are complex. That is, the flow is usually unsteady and non-uniform and, if it is separated into time average, periodic and random components, these may have different directions. Indeed, oscillatory components with different periods may have quite different directions. At the same time, the sediment may contain a mixture of densities, sizes and shapes, which respond differently to the flow. For any given grain, its response to the flow depends on its position on an uneven bed. A grain at the crest of a ripple is much more exposed than one in the trough. For fine-grained (cohesive) sediments, the mobility may even depend on the history of consolidation and biological activity. All of these possible complications are to be kept in mind while developing formulae for simplified scenarios.

4.3.2 *Sediment characteristics*

The mode and rate of sediment transport are determined by the parameters of the flow and of the sediment. Among the parameters of the sediment one thinks usually first of the size, which is an explicit part of most sediment transport formulae. The

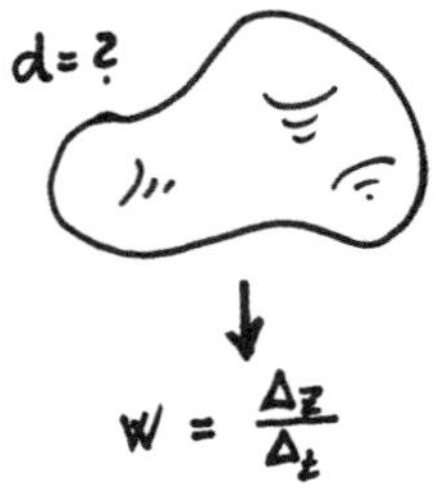

Fig. 4.11 The settling velocity is easier to define experimentally than the diameter for a natural sand grain.

definition of "the size" is however not obvious for a mixture of grains with different shapes, Fig. 4.11.

To this end, odd-shaped particles are usually assigned a diameter d, which is either the sieve diameter d_s, the volume diameter d_v or the fall diameter d_f, the latter two being the diameters of spheres with respectively the same volume and the same settling velocity. For relations between these different diameters, see e.g., Fredsøe and Deigaard (1992), p. 197. The *dimensionless grain size* can be defined as $d\sqrt{(s-1)gd}/v$ which is a particle Reynolds number based on the nominal settling velocity $\sqrt{(s-1)gd}$. Mixtures of different diameters are primarily described in terms of statistics like the mean diameter $\overline{d}$ or the median diameter d_{50}.

The gradation, i.e. the size variation is most simply quantified as the ratio between two size fractiles, e.g., d_{90}/d_{10}. This ratio is large for well-graded sediments and close to unity for well-sorted sediments.

The sediment behavior will also depend on its density ρ_s or the specific weight, $s = \rho_s/\rho$. Typical beach sands consisting of quartz or carbonate have $\rho_s \approx 2650\,\text{kg/m}^3$ corresponding to $s \approx 2.65$.

Some sediment transport formulae include the angle of repose φ, which is also a friction coefficient between layers of sediment particles. It depends on the grain shape, gradation and packing of the sediment.

The solids concentration [solids volume/total volume] in the non-moving bed $c_{\max}$ depends on grain shape, gradation and packing. For spheres of uniform size, the loosest packing is the cubic configuration for which the solid concentration is $c \approx 0.52$, and the densest is the tetrahedron configuration for which $c \approx 0.74$. For natural sands the range is $c_{\max} \in [0.6; 0.9]$.

The settling velocity w_o of a single particle in still water is easier defined than "the size". It does, however, depend on water temperature (via the fluid viscosity v) as well as on particle size and shape. For a spherical particle, the settling velocity is related to the diameter through

$$w_o = \sqrt{\frac{4(s-1)gd}{3C_D}}, \tag{4.3.1}$$

where the drag coefficient C_{D} for spheres may be estimated from Fig. 4.12.

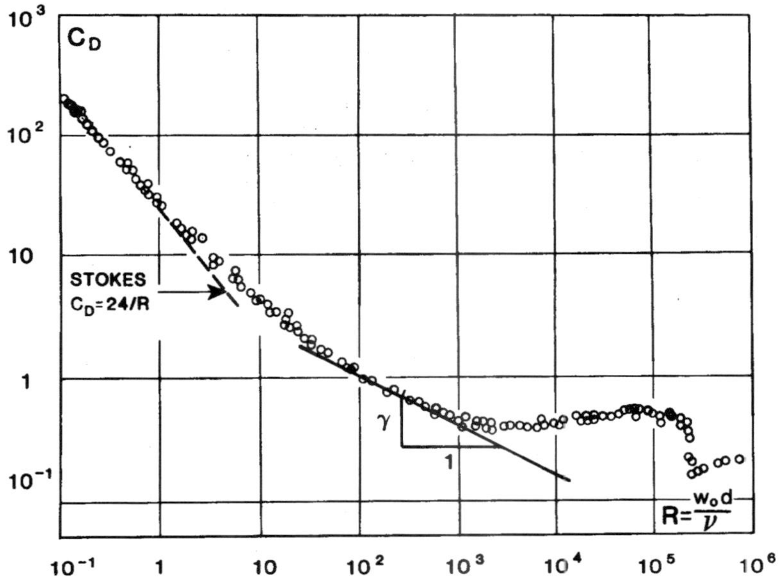

Fig. 4.12 Drag coefficients for spheres as functions of Reynolds number. The dashed line corresponds to Stokes Law, which gives $w_o \sim d^2$ for laminar conditions.

An example of the settling velocity-diameter relationship for a natural beach sand sample, including odd-shaped particles, is shown in Fig. 4.13.

She *et al.* (2005) showed that C_{D} for natural sand is typically 2 to 3 times larger than for spheres and suggested the formula

$$C_D = 2.1 + 40/R, \tag{4.3.2}$$

based on the sieve diameter.

Settling velocity depends on sediment concentration, the clear water value w_{o} being the limit for $c \to 0$. The other limit is the fluidization velocity w_f (see below), which can be seen as the settling velocity corresponding to the concentration $c_{\max}$ in the non-moving bed.

In the absence of fluid velocity shear it may be assumed that $w = w(c)$ varies as a simple power function between these two limits, i.e.,

$$\frac{w}{w_o} = (1-c)^n \quad \text{for } 0 < c < c_{\max}. \tag{4.3.3}$$

This form has no theoretical basis but it is convenient and not seriously compromised by the available data from settling columns. Such experiments yield the typical exponent range $n \in [3;4]$ for beach sand. Equation (4.3.3) does, however, not apply in boundary layer flow scenarios where the effects of velocity shear are dominant, as pointed out by Nielsen *et al.* (2002).

The porosity of a given sediment sample is the ratio

$$\varepsilon = \frac{\text{void volume}}{\text{solids volume}}. \tag{4.3.4}$$

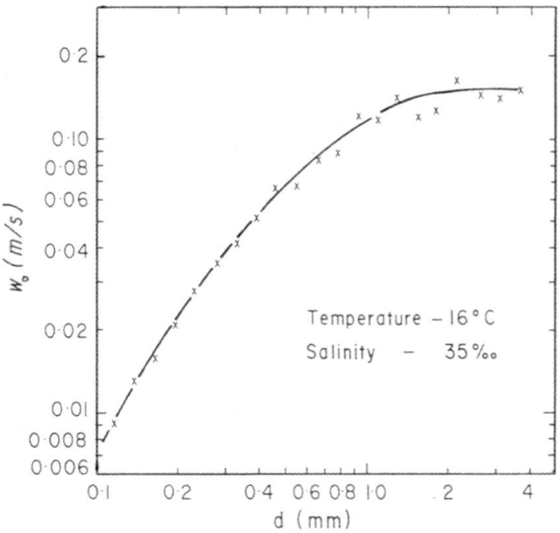

Fig. 4.13 Settling velocity versus sieve diameter for the various sieve fractions of one natural surf zone sand sample. The flattening of the curve for $d > 1\,\text{mm}$ is due to the larger particles being disk shaped shell fragments. The behavior of the smaller particles (<0.5 mm) is fairly typical for quasi-spherical quartz sand. From Nielsen (1983).

The *permeability* P of a granular medium is usually an increasing function of the porosity but also a function of particle shapes and grading. It is usually determined experimentally via the *hydraulic conductivity* K which, for the x-direction is defined by

$$u = \frac{dq_x}{dA} = -K_x \frac{\partial}{\partial x}\left(\frac{p}{\rho g} + z\right) \tag{4.3.5}$$

and similarly for the y and z directions. K_x, K_y, and K_z will be mutually different unless the porous medium is isotropic, in which case the simple vector relation,

$$\mathbf{u} = \frac{d\mathbf{q}}{dA} = -K\overrightarrow{\text{grad}\left(\frac{p}{\rho g} + z\right)} \tag{4.3.6}$$

called Darcy's Law is valid. The hydraulic conductivity is not a property of the porous medium alone since it depends on the density ρ and the kinematic viscosity v of the fluid. Based on a measured hydraulic conductivity, the permeability can be determined from

$$P_x = K_x \frac{\mu}{\rho g} \tag{4.3.7}$$

and similarly for the other directions, i.e. the permeability has dimension of area while hydraulic conductivity has dimension of velocity.

As a rule of thumb, for uniform almost spherical particles, the hydraulic conductivity can be estimated roughly as

$$K \approx 10^{-3}\frac{gd^2}{v} \tag{4.3.8}$$

but note that K varies strongly with particle shape, gradation and packing.

The fluidization velocity w_f is the upward seepage velocity at which quicksand is formed, that is for which the upward pressure force equals the weight. Considering a unit volume on which the vertical pressure force is $-\frac{dp}{dz}$, and noting that according to Darcy's law for the vertical direction $\frac{w}{K_z} = -\frac{\partial}{\partial z}(\frac{p}{\rho g} + z)$, where K_z is the "vertical" hydraulic conductivity, this means,

$$-\frac{dp}{dz} = \rho g \left(\frac{w_f}{K_z} + 1\right) = \rho g (s c_{\max} + [1 - c_{\max}]) \tag{4.3.9}$$

and hence

$$w_{\rm f} = (s-1) c_{\max} K_z. \tag{4.3.10}$$

The erodability and settling behavior of very fine sediments ($d < 50\,\mu$m) is also dependent on *cohesion*, which originates from electrical attraction between the particles and depends on the chemical composition of the water as well as on the mineralogy and consolidation history of the sediment, see e.g., Winterwerp and van Kesteren (2004). A more detailed discussion of sediment parameters can be found in Allen (1984).

4.3.3 *Forces on sediment particles*

The forces which act to move the sediment are mainly drag forces $\overrightarrow{F_{\rm D}} \sim d^2|\mathbf{u}|\mathbf{u}$, lift forces directed perpendicular to $\mathbf{u}$ with magnitude $F_L \sim u^2 d^2$, and pressure forces $\overrightarrow{F_{\rm p}} \sim -\overrightarrow{\text{grad } p} \times d^3$.

The definition of "The velocity" is not a trivial matter in the presence of the boundary layer and sheltering from other particles. We shall be using the free steam velocity $u_\infty(t)$ as a nominal drag velocity. Similar issues arise with respect to the horizontal pressure gradient. Here the practical, and quite reasonable, approach is to assume hydrostatic conditions throughout the layer of moving sediment. The horizontal pressure gradient is then proportional to the free stream fluid acceleration, $\frac{dp}{dx} = -\rho \frac{du_\infty}{dt}$ and hence

$$F_{p,x} \propto -\rho \frac{du_\infty}{dt} d^3. \tag{4.3.11}$$

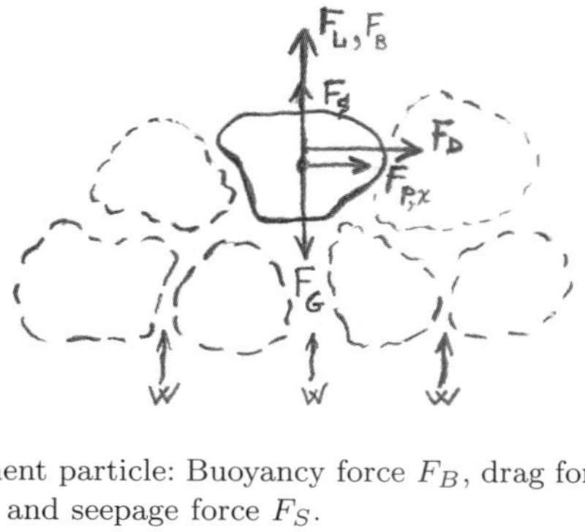

Fig. 4.14 Forces on a sediment particle: Buoyancy force F_B, drag force F_D, gravity force F_G, lift force F_L, pressure force F_P, and seepage force F_S.

Vertical pressure gradients give rise to the familiar buoyancy force $F_B = F_{p,z} \propto \rho g d^3$.

A vertical seepage velocity w corresponds to a vertical force, which may generate quicksand or alternatively stabilize the bed. Through Darcy's law ($\mathbf{u} = -K\, \overrightarrow{\text{grad}}\, p/\rho g$), the vertical seepage force per unit volume is $F_s = \rho g w/K$, where K is the (vertical) hydraulic conductivity of the bed material.

Stabilizing forces are usually the gravity force $F_G \sim \rho s g d^3$ and the cohesive forces.

For a sloping sand surface, slope $\tan\beta$, gravity delivers a downhill tangential force of magnitude $F_G \sin\beta$ and the normal force due to gravity is reduced to $F_G \cos\beta$.

4.3.4 *The mobility number*

The mobility number

$$\psi = \frac{|u|^2}{(s-1)gd}, \tag{4.3.12}$$

is the simplest measure of the ratio between moving forces and gravity on a sediment particle. The moving forces are, implicitly, represented by the canonical form $F \propto \rho d^2 |u|^2$ of the drag and lift forces, while the magnitude of the gravity force, corrected for buoyancy, is $\rho(s-1)gd^3$. In a pure wave motion, the mobility number is expressed in terms of the velocity amplitude just above the boundary layer

$$\psi = \frac{(A\omega)^2}{(s-1)gd}. \tag{4.3.13}$$

Simplicity makes the mobility number very useful. However, the neglect of the different natures of the drag versus pressure forces is sometimes significant. For oscillatory flows, e.g., $u(t) = A\omega\cos\omega t$, the relative importance of the pressure force, i.e., F_p/F_D is measured by d/A, the inverse of the Keulegan–Carpenter number.

4.3.5 *The Shields parameter*

The Shields parameter

$$\theta(t) = \frac{\tau_b(t)d^2}{\rho(s-1)gd^3} = \frac{\tau_b(t)}{\rho(s-1)gd} = \frac{u_*^2(t)}{(s-1)gd}, \tag{4.3.14}$$

is a more sophisticated measure of sediment mobility which, in principle, accounts for both pressure gradients and drag, in as much as the acceleration effects, are incorporated in $\tau_b(t)$.

Due to a shortage of information about the details of $\theta(t)$, other related quantities like the peak value

$$\hat{\theta} = \frac{\hat{\tau}}{\rho(s-1)gd} = \frac{\frac{1}{2}f_w(A\omega)^2}{(s-1)gd} = \frac{u_{*\max}^2}{(s-1)gd} \tag{4.3.15}$$

are often used for simplicity.

Like the shear stress in Eq. (4.2.5), the Shields parameter is often split into a form drag component and a skin friction component, $\theta = \theta'' + \theta'$ where only θ' is assumed to generate sediment transport. The questions about defining θ', and the hydraulic roughness for calculating it, are usually bypassed by using a nominal grain roughness Shields parameter, based on $r = 2.5d_{50}$ instead. That is, instead of θ', the grain roughness Shields parameter $\theta_{2.5}$ with the peak (maximum) value,

$$\hat{\theta}_{2.5} = \frac{\frac{1}{2}f_{2.5}(A\omega)^2}{(s-1)gd}, \tag{4.3.16}$$

is often used. The grain roughness friction factor $f_{2.5}$ is calculated from Eq. (4.2.13).

The complex issue of hydraulic roughness of natural seabeds is comprehensively discussed in Nielsen (1992), Sec. 3.6.

4.3.6 *Sleath's acceleration parameter*

For sediment transport scenarios where the fluid accelerations ($\sim$ the sreamwise pressure gradients) are more important than drag, like coarse sediment under bore fronts, as observed by Khezri and Chanson (2012), Sleath (1999) suggested the acceleration parameter

$$S = \frac{A\omega^2}{(s-1)g}, \tag{4.3.17}$$

for simple harmonic flow or more generally $S = \frac{a_{\max}}{(s-1)g}$.

4.3.7 *Infiltration effects*

In the swash zone, in particular, but possibly in the nearshore area in general (Conley and Inman, 1992), the seepage forces associated with in- or out-flow of water from the bed may influence the sediment mobility.

First, to thicken the boundary layer and hence to reduce the bed shear stress.

Second, to destabilize the particles through the upward seepage force F_s.

For infiltration ($w < 0$), both effects are reversed.

Flow which is perpendicular to the bed has two opposing effects on sediment mobility, see Fig. 4.15. That is, inflow will make the boundary layer thinner and thereby increase the shear stress. At the same time however, the inflow exerts a downward drag, which helps to stabilize the sediment.

Martin (1970) found experimentally that either of the two effects may be dominant, depending on the sediment size and density. For large or very dense particles, the mobilizing effect of the increased shear stress dominates. For finer and/or lighter sediments the stabilizing effect of the downward drag dominates.

Nielsen *et al.* (2001) suggested incorporating the influence of a vertical seepage velocity w (positive upwards) on sediment stability through a modified Shields

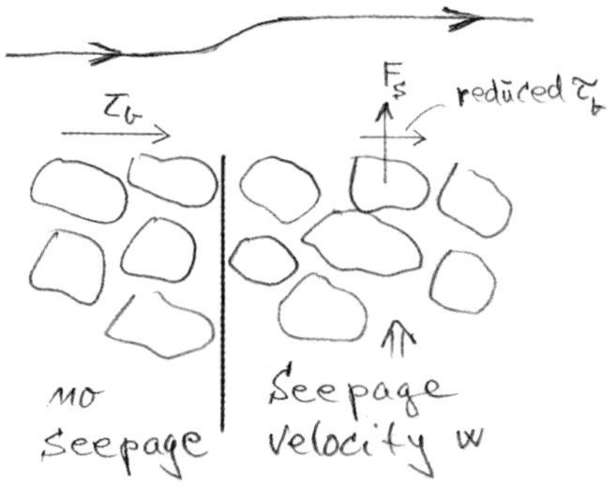

Fig. 4.15 The effects of a positive seepage velocity are to decrease shear stress and to exert an upward drag.

parameter

$$\theta = \frac{u_{*o}^2\left(1 - \alpha\frac{w}{u_{*o}}\right)}{\left[s - 1 - \beta\frac{w}{K}\right]gd}, \tag{4.3.18}$$

where u_{*o} is the shear velocity with no seepage, α and β are dimensionless coefficients and K is the hydraulic conductivity of the sand. Based on experiments with non-breaking waves over a horizontal bed, Nielsen *et al.* (2001) recommended $(\alpha, \beta) = (16, 0.4)$.

The infiltration effects in the experiments of Nielsen *et al.* (2001) $[(h, H) = (0.45\,\text{m}, 0.15\,\text{m})]$ were relatively weak. However, Baldock and Holmes (1998) performed smaller scale $[(h, H) = (0.1\,\text{m}, 0.038\,\text{m})]$ experiments with similar (0.2 mm quartz) sand and found that the stabilizing infiltration effect could dominate over these waves. They found that vertical head gradients greater than 0.1 completely prevented ripple formation. They also found that exfiltration had a strong enhancing effect on the net sediment transport rate by waves, e.g. a factor 4 for head gradients of −0.2. No such increase was found with a steady current. Baldock and Holmes, therefore, concluded that exfiltration enhances the bulk sediment motion driven by the strong horizontal pressure gradients in the wave motion but has no net effect on the transport by currents, which is driven by steady surface shear.

4.3.8 *Initiation of sediment motion*

The Shields parameter has been used to formulate the most commonly used criterion for the initiation of sediment motion.

Shields (1936) postulated that the critical Shields parameter θ_c at which the sediment starts moving, should be a function of the grain Reynolds number $u_* d/v$. This is reasonably well supported by data as indicated by Fig. 4.16, as long as the specific gravity s does not vary too much.

However, if vastly different s-values are considered, e.g., 2.65 for quartz in water and 2200 for quartz in air, the dependence of θ_c upon s is quite strong. For coastal

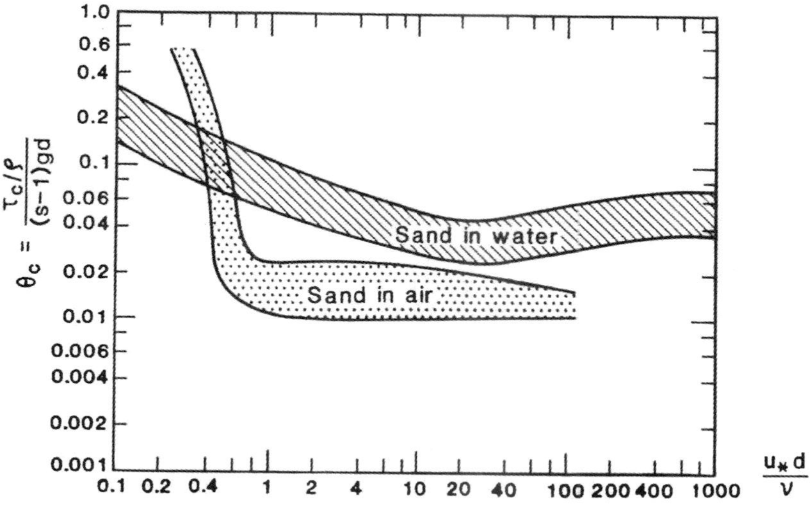

Fig. 4.16 Within a given, reasonably narrow, range of specific gravity, the critical Shields parameter θ_c is a function of the grain Reynolds number only. For sand in water the 'rule of thumb' value is 0.05. From Nielsen, 2009.

applications, recent work by Terrile *et al.* (2006) indicates that wave shape has strong and as yet not fully understood influence on the initiation of sediment motion.

Cohesive sediments are much more resilient to erosion than indicated by the Shields diagram. However, the presence of sand may enhance the erosion process very significantly as described by Mitchener and Torfs (1996).

4.4. Bedload and Sheet Flow

4.4.1 *Introduction*

Sediment transport over flat beds under non-breaking waves is concentrated in a very thin layer (of the order 100 grain diameters) around the level of the undisturbed bed, see Fig. 4.17. The process is called sheet flow and may consist partly of bedload and partly of suspended load. Measurements from within this layer were first provided by the pioneering work of Horikawa *et al.* (1982) but more details have been added by the work of Dick and Sleath (1991a, 1991b); Zala-Flores and Sleath (1997); Sleath (1999) and Ribberink *et al.* (1994, 2000); O'Donoghue and Wright (2004); and Watanabe and Sato (2004).

Particularly interesting with respect to future model developments are the time series of concentrations at and below $z = 0$, see Fig. 4.18. These ought to help bring the state-of-the-art beyond the current use of separate bedload models and suspension models with nominal reference concentrations and into the era of continuous descriptions covering all moving sediment from the lowest level of sediment motion to the water surface. A number of detailed models of the granular flow near $z = 0$

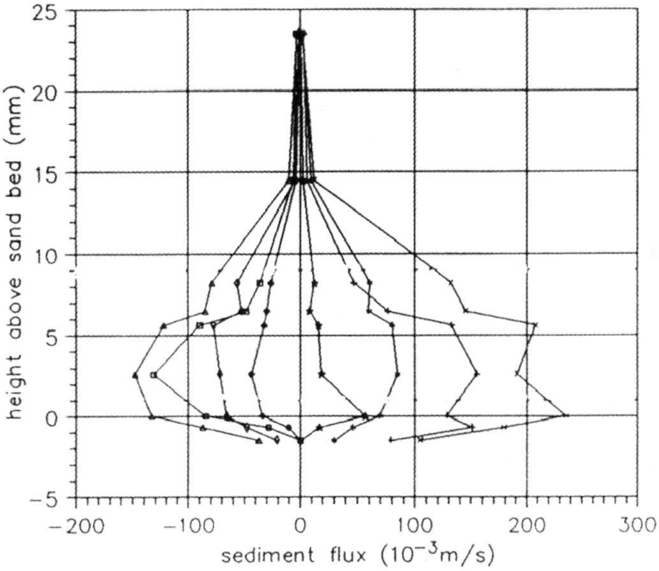

Fig. 4.17 Instantaneous sediment flux densities measured in a water tunnel by Ribberink *et al.* (1994). The grain diameter is 0.21 mm, the wave period is 7.2 s and the peak velocities are 1.8 m/s and 1.4 m/s in the landward and seaward directions, respectively.

have been developed with this aim, e.g., Hanes and Bowen (1985) and; Kaczmarek and Ostrowski (2002). However, a great many simplifying and sometimes unrealistic assumptions are still required in order to make them workable.

4.4.2 *What is bedload?*

In the following, the term bedload refers to that part of the sediment load, which is carried by inter-granular forces, as opposed to suspended load and wash load, which are carried by vertical fluid drag. This definition which is due to Bagnold (1956) is not convenient from a measuring point of view since a given sand particle may be supported by a combination of the two kinds of forces and hence be partly bedload, partly suspended load. It is however the most convenient definition with respect to conceptual model building.

4.4.3 *Dispersive stress and the amount of bedload*

Consider a sand bed, which is exposed to a bed shear stress τ_b capable of mobilizing the top layer of sand. Given that the shear stress on the next layer is then at least as great as τ_b, one might expect that it will also be mobilized, and so will the next, and so on. This is, however, in conflict with observations: Usually only a finite number of layers will be mobilized.

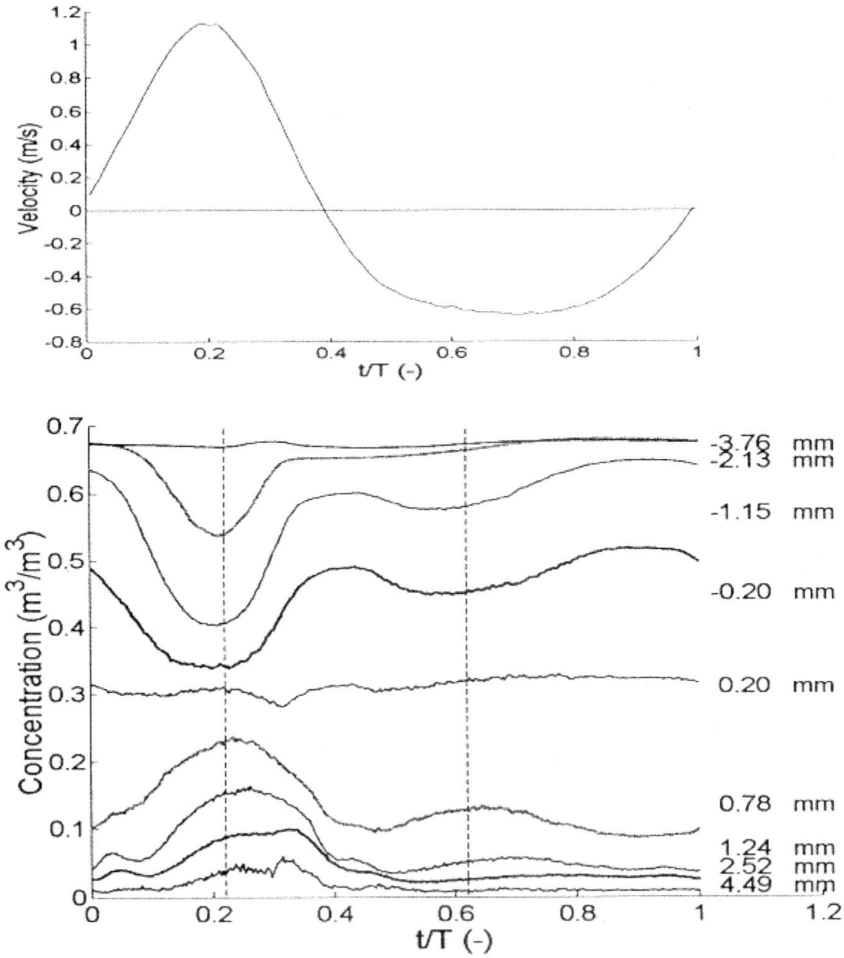

Fig. 4.18 Free stream velocity and sediment concentrations in sheet flow under waves in a flume, measured by Ribberink *et al.* (2000). The wave period is 6.5 s and $d_{50} = 0.245$ mm. At levels below the undisturbed bed, the peak velocities create a dip in concentrations. Above the undisturbed bed level they create peaks. No sediment movement occurs at $z = -3.76$ mm and below.

This represented a paradox until Bagnold (1956) introduced the concept of dispersive stress. That is, the weight of the bedload is transferred as an effective normal stress σ_e to the non-moving bed, and this extra effective stress, enables the top-most non-moving layer to withstand a shear stress greater than τ_c.

Consider the situation in Fig. 4.19 and assume that the maximum shear stress, which can be withstood by the top-most non-moving layer of particles is given by the Coulomb criterion:

$$\tau_{\max} = \tau_c + \sigma_e \tan\phi = \tau_c + \tan\phi \int_{z_b}^{\infty} \rho g(s-1)c_b dz \tag{4.4.1}$$

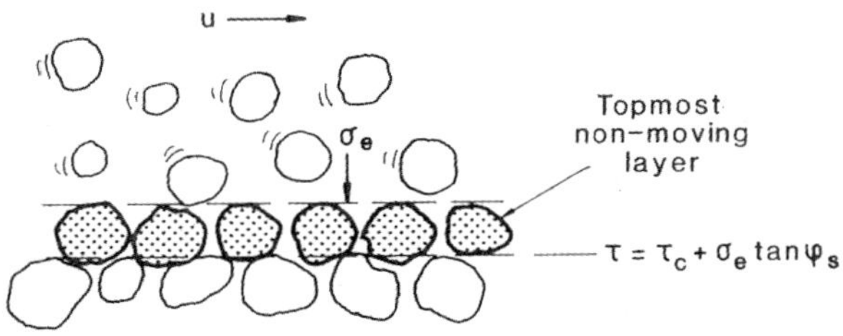

Fig. 4.19 Due to the weight of the bedload above, which is transferred as the dispersive stress σ_e (Bagnold 1956), the topmost non-moving layer can withstand a shear stress that is greater than τ_c.

Then, quantifying the total amount of bedload (m^3/m^2) by its equivalent resting thickness at concentration, $c_{\max}$,

$$L_b = \frac{1}{c_{\max}} \int_{z_b}^{\infty} c_b dz, \tag{4.4.2}$$

we find

$$L_b = \frac{\tau - \tau_c}{\rho g(s-1)c_{\max} \tan\phi} \tag{4.4.3}$$

or introducing the Shields parameter and using Bagnold's (1956) general ballpark value:

$$c_{\max} \tan\varphi \approx 0.4 \tag{4.4.4}$$

we get,

$$L_b \approx 2.5(\theta' - \theta_c)d. \tag{4.4.5}$$

4.4.4 *Meyer–Peter and Müller type bedload formulae*

Meyer–Peter and Müller (1948) found that steady bedload was well described by,

$$\Phi_b = \frac{q_b}{d\sqrt{(s-1)gd}} = 8(\theta' - \theta_c)^{1.5} \tag{4.4.6}$$

and a number of similar formulae have since been derived from physical considerations, cf Madsen (1991) and Fredsøe (1993). Correspondingly, the net transport from half a sine wave $\Phi_{T/2}$ was found, by Nielsen (1992), to behave similarly, see Fig. 4.20.

4.4.5 *The average speed of the bedload*

The semi-empirical transport rate q_b corresponding to (4.4.6) can be used together with the bedload amount given by (4.4.5) to find the typical velocity $\langle u_b \rangle$ for bedload particles. One finds

$$\langle u_b \rangle = \frac{q_b}{c_{\max} L_b} \approx 7u_* \quad \text{for } \theta' \gg \theta_c. \tag{4.4.7}$$

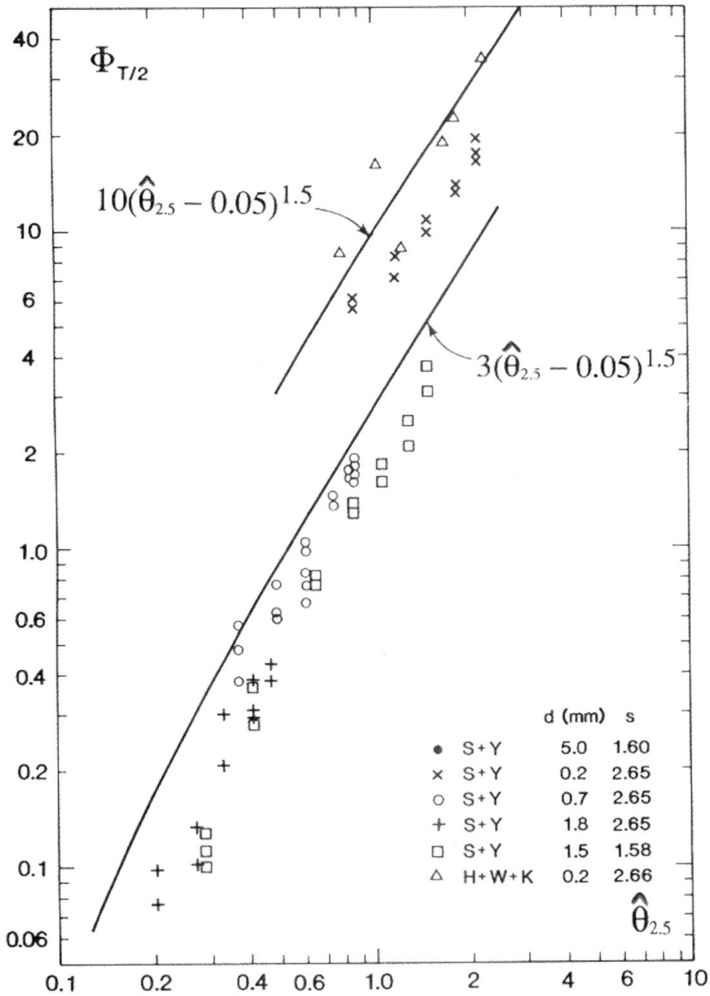

Fig. 4.20 The total net transport from half a sine wave can be described in terms of $\hat{\theta}_{2.5}$ and a Meyer–Peter and Müller type of transport formulae. The coefficient will however depend on the amount of suspended load, which is greater for fine sediments at a given $\hat{\theta}_{2.5}$. The value 3 corresponds directly to Meyer–Peter and Müller's value (8) for steady bedload via the fact that $\overline{|\sin|^3} = 3/8$.

4.4.6 *Energy dissipation and bedload*

In order to bypass many of the difficult details of the sediment transport modeling process, Bagnold (1963) suggested the shortcut of relating the sediment transport rate q_s [m^2/s] directly to the energy dissipation rate D_E [W/m^2]. The basic idea was that the power, P_b per unit bed area expended on dragging the bedload along, should be a certain fixed fraction ε of D_E. The required drag power P_b is estimated as:

$$P_b \approx \rho(s-1)gc_{\max}L_b \tan\varphi\langle u_b\rangle = \rho(s-1)g\tan\varphi q_b, \tag{4.4.8}$$

where the first part is recognized as the frictional force due to the submerged weight of bedload per unit area while $\langle u_b \rangle$ is, somewhat arbitrarily but conveniently, taken as the relevant dragging velocity. Inserting the Meyer–Peter expression for q_b then gives,

$$P_b \approx \rho(s-1)g\tan\varphi\sqrt{(s-1)gd}\,d8(\theta'-\theta_c)^{1.5}. \tag{4.4.9}$$

The boundary layer energy dissipation rate, on the other hand, is $D_E = \overline{\tau_b u_\infty}$. We can now find the value of the sediment transport coefficient $\varepsilon = P_b/D_E$, which corresponds to those experiments that support the Meyer–Peter formula, by combining the latter two expressions. We find that

$$\varepsilon \approx 8\tan\phi\sqrt{\frac{f}{2}} \quad \text{for } \theta' \gg \theta_c, \tag{4.4.10}$$

where f is the friction factor. That is, the Meyer–Peter formula (and other formulae of the form $\Phi \propto \theta^{1.5}$) are consistent with the "energetics hypothesis" $P_b = \varepsilon D_E$. Taking $\langle u_b \rangle$ as the typical dragging velocity, $\tan\varphi \approx 0.6$ and $f \approx 0.02$ we are lead to

$$\varepsilon \approx 8 \times 0.6 \times \sqrt{0.02/2} \approx 1/2. \tag{4.4.11}$$

That is, the drag work required for moving the sediment along is usually of the same order of magnitude as the energy dissipation in the bottom boundary layer. However, the energetics approach has no way of determining the direction of $\overline{q_s}$ if u changes direction as in most coastal flows.

4.4.7 *Velocity moment formulae for sediment flux*

Adapting a formula like the Meyer–Peter and Müller formula to an arbitrary unsteady flow is difficult because it requires calculation of an instantaneous skin friction Shields parameter $\theta'(t)$. Several shortcuts have therefore been tried in order to avoid this hurdle. One class of models in this vein are the velocity moment formulae where an empirical relation of the form $\overline{q_s} = K\overline{|u_\infty|^{p-1}u_\infty)}$ is sought. K is a constant whose dimension depends on p, u_∞ is the velocity just above the bottom boundary layer, and the overbar denotes time averaging. Ribberink and Al-Salem (1994) showed that for 0.21 mm sand and typical prototype wave-current (collinear) flows, such a model with $p = 3$ and $K = 0.00018\,\text{s}^2/\text{m}$ gave good predictions. However, similar experiments by Ribberink and Chen (1993) using 0.13 mm sand could not be modeled in this way. For the finer sand, the assumption of quasi steadiness $q_s(t) \propto |u_\infty(t)|^{p-1}u_\infty(t)$, which seems valid for 0.21 mm (and probably coarser) sand, breaks down as it does for most sand sizes over rippled beds. The reason for the breakdown of the quasi-steady approach is the presence of a considerable amount of suspended sand and that suspended sand does not move in step with $u_\infty(t)$. The Ribberink and Al-Salem sheet flow formula:

$$\overline{q_s} = 0.00018\,\overline{u_\infty^3} \quad \text{(SI units)}, \tag{4.4.12}$$

can be tied in with the Meyer–Peter and Müller type formulae by considering a particular velocity variation, e.g., half a sine wave, and sand of quartz density: s =2.65. In this case it corresponds to

$$\Phi = 0.003\,\overline{\psi^{3/2}} \tag{4.4.13}$$

and, if a constant wave friction factor of $f_w \equiv 0.008$ is assumed, it is equivalent to the Meyer–Peter and Müller like formula $\Phi = 12\overline{\theta^{3/2}}$, which was shown to fit the 0.2 mm sand data in Fig. 4.20.

4.4.8 *Bedload and sheetflow transport in arbitrary flows*

For quartz sand which is not too fine ($d_{50} \geq 0.2$ mm), Eq. (4.4.12) has been shown to apply for water tunnel experiments with combinations of currents and sine waves and for currents plus waves with Stokes wave asymmetry. However, as pointed out by Ribberink *et al.* (2000), it under-predicts the sediment transport rates measured under real waves by about a factor 2, see Fig. 4.21.

We see that the flume waves generate about twice the sediment transport rate for a given value of $\overline{u_\infty^3}$ compared with the "sine waves plus current" or the second-order Stokes waves in the tunnel. There are two possible reasons for this difference:

First, the flume waves generate a steady drift at the bed and a corresponding mean bed shear stress in the shoreward direction. This streaming is driven by wave Reynolds stress terms of the form, $-\rho\overline{\tilde{u}\tilde{w}}$, which are non-zero in real waves with a

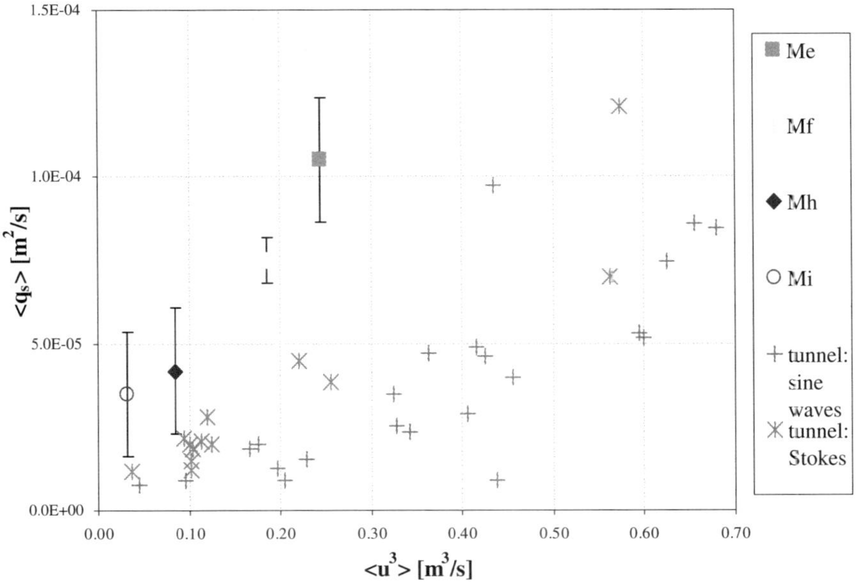

Fig. 4.21 Measured net sediment transport rates versus third moment of the free stream velocity for real waves in a flume (symbols with error bars) and from water tunnel experiments with "sine waves plus currents", +, and second-order Stokes waves, $*$. From Nielsen (2009).

bottom boundary layer as explained by Longuet-Higgins (1956). These stresses do not exist in the water tunnels where $\tilde{w} \equiv 0$.

For the flume situation, the streaming related extra bed shear stress can be estimated from:

$$-\rho(\overline{\tilde{u}\tilde{w}})_\infty = \rho\frac{1}{4\sqrt{2}}kA^3\omega^2 f_w \tag{4.4.14}$$

with f_w based on the hydraulic roughness expression (4.5.17), and the corresponding contribution to the Shields parameter is

$$\Delta_\theta = \frac{-(\overline{\tilde{u}\tilde{w}})_\infty}{(s-1)gd}. \tag{4.4.15}$$

Second, the flume waves have a bit of saw-tooth asymmetry with more abrupt acceleration before the wave crest. The effect of this saw-tooth asymmetry on the bed shear stress was discussed in connection with Fig. 4.8 and accounted for by Eq. (4.2.10). The corresponding instantaneous grain roughness Shields parameter is

$$\theta_{2.5}(t) = \frac{\frac{1}{2}f_{2.5}}{(s-1)gd_{50}}\left[\cos\phi_\tau u_\infty(t) + \sin\phi_\tau\frac{u_\infty(t+\delta_t) - u_\infty(t-\delta_t)}{2\omega_p\delta_t}\right]^2 \operatorname{sign}(u_*(t)), \tag{4.4.16}$$

where the instantaneous friction velocity is taken as

$$u_*(t) = \sqrt{\frac{1}{2}f_{2.5}}\left[\cos\phi_\tau u_\infty(t) + \sin\phi_\tau\frac{u_\infty(t+\delta_t) - u_\infty(t-\delta_t)}{2\omega_p\delta_t}\right]. \tag{4.4.17}$$

When the total grain roughness Shields parameter $\theta_{2.5}(t)$, obtained as the sum of (4.4.15) and (4.4.16), is inserted into the Meyer–Peter and Muller style transport formula we get:

$$\begin{aligned}\Phi(t) &= \frac{q_s(t)}{\sqrt{(s-1)gd_{50}^3}} \\ &= \begin{cases} 0 & \text{for } \theta_{2.5} < 0.05 \\ 12[\theta_{2.5}(t) - 0.05]\sqrt{\theta_{2.5}(t)}\operatorname{sign}(u_*(t)) & \text{for } \theta_{2.5} > 0.05 \end{cases}\end{aligned} \tag{4.4.18}$$

The above equation is in good agreement with the flume measurements of Ribberink *et al.* (2000), see Fig. 4.22. For further details and comparisons with data see Nielsen (2006).

The method outlined above also gives reasonable agreement with measured swash zone net sediment transport rates, cf. Nielsen (2002).

4.5. Bedforms and Hydraulic Roughness

4.5.1 *Introduction*

Bedforms have a strong effect of sediment transport through modifications to the flow and to the mechanisms of sediment entrainment. The nature of the influence is

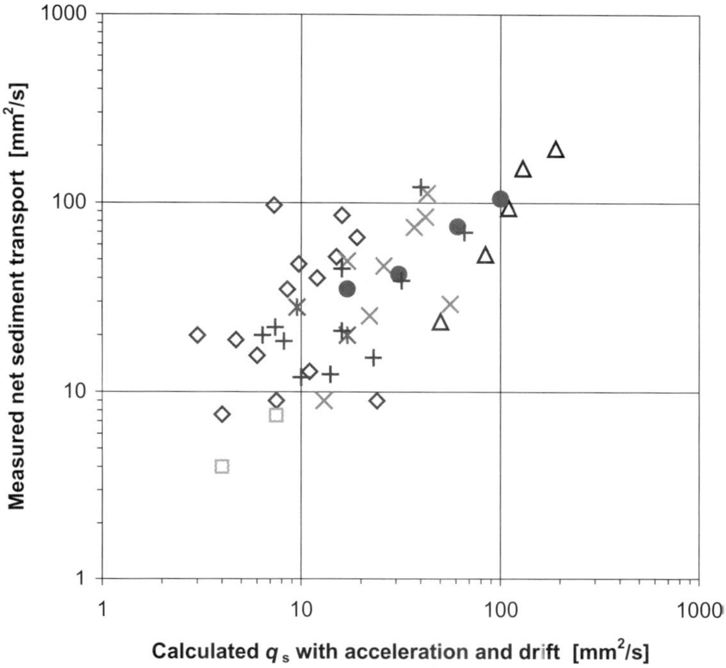

Fig. 4.22 Measured versus calculated net sediment transport rates for the same data as in Fig. 4.21.

different for different types of bedforms. Thus, bars will determine the position of the breakpoint and the longshore location of rip currents while vortex ripples influence the vertical structure of the flow and of sediment concentrations. A schematic overview of typical coastal bedforms on one type of beach morphology is provided by Fig. 4.23. For an overall description of coastal bedforms see also Clifton (1976).

4.5.2 *Bedform growth and migration*

In order to model coastal erosion and beach changes in general, we need to establish the relation between local sediment transport rates and bedlevel changes. This is done in the following section. It turns out that the obvious requirement of conservation of sediment volume leads to a simple relationship between erosion rates and the spatial gradients of sediment transport. The same simple continuity equation is also used to gain qualitative insights into the conditions for bedform growth and migration.

4.5.2.1 *Migration of natural bedforms*

From analysis of beach profiles from Duck, Larson and Kraus (1992) found offshore bar migration speeds up to 18 m/day and onshore bar speeds up to 2.9 m/day. They noted that the offshore speeds, in particular, were low estimates because surveys

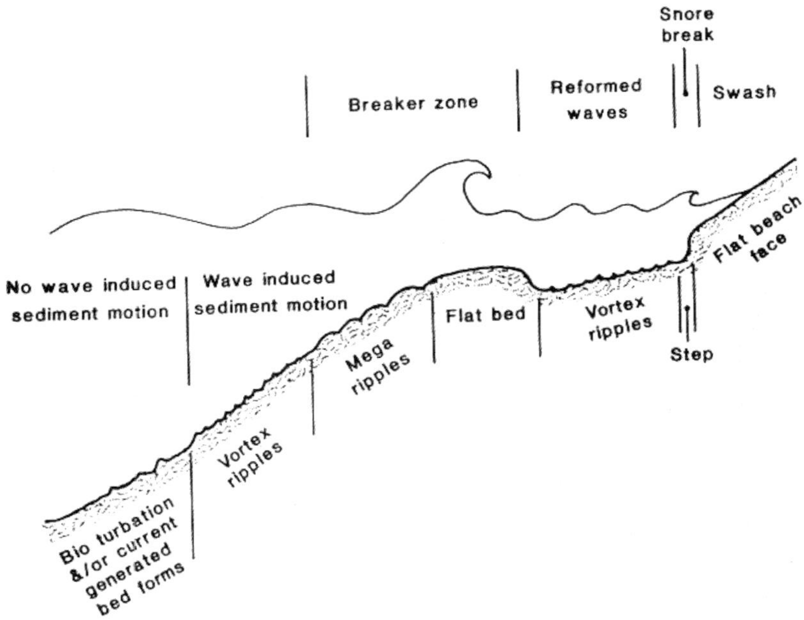

Fig. 4.23 Typical shore normal bedform distribution for an accreting beach. During storms the situation may be quite different and special bedforms are present in rips and strong feeder channels. From Nielsen (1992).

were usually not done immediately after high energy events. In other words, their speeds were averages between surveys rather than peak values during storms. While Larson and Kraus reported no long-term trend, Wijnberg (1996) described how, on the multi-barred Dutch coast, all bars are moving offshore in the long term with speeds of the order 50 m/year. Elgar *et al.* (2001) demonstrated that onshore bar migration during moderate wave conditions can be understood in terms of wave acceleration skewness. When storm waves arrive however, the breakpoint moves seaward and strong undertow makes the bar move seaward toward the new breakpoint. Megaripples have been observed to move landward with typical speeds of 0.5 cm/min ranging up to about 2.5 cm/min by Gallager *et al.* (1998) and Gallagher (2003). Inman and Bowen (1963) observed both seaward and landward ripple migration at up to 2 mm/min in the laboratory, and similar speeds, i.e. typically 0.5–1 mm/min, were observed by Farachi and Foti (2002) and Davis *et al.* (2004). In the field, Dingler (1974) observed ripples migrating landwards at rates up to a few cm/min, while Traikovsky *et al.* (1999) found a peak value of 80 cm/day with an average of 24 cm/day over a 25 day period. This ripple migration is due to a combination of wave asymmetry, $\overline{u^3} \neq 0$, acceleration asymmetry, $\overline{(\frac{du}{dt})^3} \neq 0$, and boundary layer streaming. All of these are greater in shallower depths for given (H, T) which explains the much greater migration rates observed by Dingler at $h \approx 2$ m compared with Traikovsky *et al.* at $h \approx 11$ m.

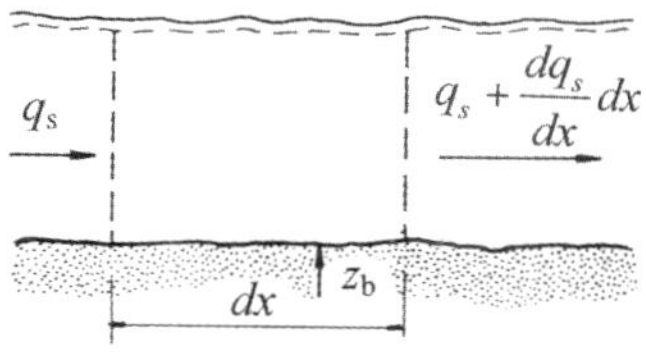

Fig. 4.24 2D control volume. The solids fraction of the porous bed is n. From Nielsen (1992).

4.5.2.2 *The continuity equation applied to bedforms*

The relation between changes in the sediment transport rate and changes in bed elevation is derived by expressing the conservation of sediment volume for a control volume like the one in Fig. 4.24.

If there is more sediment leaving than entering, the bed level of the control volume must go down and if the solids fraction of the bed is n, the simple 2D continuity equation is:

$$n\frac{dz_b}{dt} = -\frac{dq_s}{dx}, \tag{4.5.1}$$

while the more general 3D version, where the sediment transport rate is a vector in the xy-plane, $\mathbf{q}_s = (q_{s,x}, q_{s,y})$ reads:

$$n\frac{dz_b}{dt} = -\text{div}\,\mathbf{q}_s = -\frac{dq_{s,x}}{dx} - \frac{dq_{s,y}}{dy}, \tag{4.5.2}$$

i.e. the local erosion rate is $\frac{1}{n}\text{div}\,\mathbf{q}_s$. Both of these equations are approximate in that they ignore changes to the amount of sediment suspended above z_b inside the control volume.

4.5.2.3 *Bedforms migrating with constant form*

The continuity principle can be used very easily to derive sediment transport rates from the shape and speed of (2D) bedforms if these are assumed to migrate with constant form. The argument is as follows. Consider two-dimensional bedforms of arbitrary shape f, which propagate with speed c so that the sand level can be described by:

$$z_b(x,t) = f(x - ct). \tag{4.5.3}$$

Inserting this into the left-hand side of (4.5.1) we get:

$$n(-c)\mathrm{f}'(x - ct) = -\frac{dq_s}{dx} \tag{4.5.4}$$

and integration with respect to x gives:

$$q_s = q_o + nc\mathrm{f}(x - ct) = q_o + ncz_b(x,t), \tag{4.5.5}$$

where the constant of integration q_o is the sediment transport rate at the point(s) where the bed level is zero.

Equation (4.5.5) shows that if the bedforms move downstream ($c > 0$) the sediment transport rate varies in step with z_b, i.e. the sediment transport rate is maximum over the bedform crest and minimum over the trough. Conversely, for bedforms which move upstream like anti-dunes, and thus have $c < 0$, the maximum sediment transport rate occurs over the trough and the minimum over the bedform crest.

Experiments show that both ripples and megaripples migrate with typical speeds around $c \approx 1\,\text{cm/min}$ in surf zones with skewed velocities. This fairly constant value of the migration speed c indicates that there is a fairly constant ratio between bedform height and bedform related sediment transport for these types of bedforms. That is, Eq. (4.5.5) gives

$$q_{s,\text{crest}} - q_{s,\text{trough}} = nc\eta \approx \frac{2}{3} \times (1\,\text{cm/min}) \times \eta \tag{4.5.6}$$

for ripples and megaripples under skewed nearshore waves.

4.5.2.4 *Migration and growth of sinusoidal bedforms*

Making use of the simple continuity principle, we saw above that if bedforms migrate with constant form, sediment transport rates and bed shape are exactly in step for $c > 0$ and exactly opposite if $c < 0$. The same simple continuity principle can be used to show that if q_s is out of step with z_b as in Fig. 4.25, the bedforms must be either growing or eroding. We shall see that, if q_s peaks before the crest, i.e. $\delta_x < 0$, the bedform will steepen. Conversely, if $\delta_x > 0$ the crest will be eroding and the bedform becomes flatter.

The qualitative principle is quite general but the mathematics are only simple for a simple harmonic bedform, so consider,

$$z_b(x,t) = A\cos\left[\frac{2\pi}{\lambda}(x - ct)\right] = A\cos k(x - ct) \tag{4.5.7}$$

or by introducing the complex exponential $e^{ix} = \cos x + i\sin x$ and attaching physical meaning to the real part only

$$z_b(x,t) = Ae^{ik(x-ct)}. \tag{4.5.8}$$

Assuming then that the sediment transport varies in a similar way to the bed elevation, as indicated in Fig. 4.25:

$$q_s(x,t) = q_o + q_1 e^{ik(x-ct-\delta_x)}, \tag{4.5.9}$$

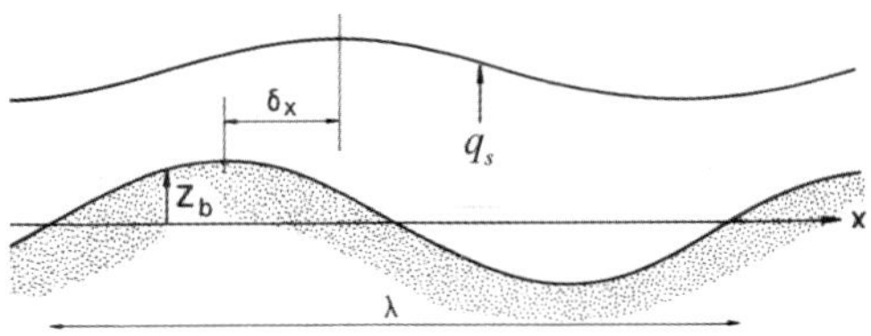

Fig. 4.25 If the peak in q_s occurs after the crest ($\delta_x > 0$), the bedform will become flatter. Conversely if $\delta_x < 0$.

the steepening or flattening rate of the bedform is determined by the continuity equation (4.5.1). Inserting the expressions for z_b and q_s we get:

$$nAik(-c)e^{ik(x-ct)} = -q_1 ike^{ik(x-ct-\delta_x)}, \tag{4.5.10}$$

which we can solve for the speed of propagation and get

$$c = \frac{q_1}{nA}e^{-ik\delta_x} = \frac{q_1}{nA}[\cos k\delta_x - i\sin k\delta_x]. \tag{4.5.11}$$

This result is most easily interpreted when c is real-valued, i.e. if $\delta_x = 0$, leading to $c = \frac{q_1}{nA}$ and if $\delta_x = \lambda/2$ leading to $c = -\frac{q_1}{nA}$. That is, the bedforms move downstream with speed $c = \frac{q_1}{nA}$ for $\delta_x = 0$ and upstream (against the sediment transport direction) with speed $c = -\frac{q_1}{nA}$ for $\delta_x = \lambda/2$.

We can however also interpret the complex c-values, which result for $\delta_x \neq 0, \lambda/2$. Then, the imaginary part of c expresses the growth rate. This is seen by inserting the solution $c = \frac{q_1}{nA}[\cos k\delta_x - i\sin k\delta_x]$ into $z_b(x,t) = Ae^{ik(x-ct)}$ which leads to

$$z_b(x,t) = Ae^{-k\frac{q_1 \sin k\delta_x}{nA}t}e^{ik\left(x-\frac{q_1\cos k\delta_x}{nA}t\right)} \tag{4.5.12}$$

or

$$z_b(x,t) = Ae^{-k\text{Im}\{c\}t}e^{ik(x-\text{Re}\{c\}t)} = Ae^{-k\text{Im}\{c\}t}\cos k(x - \text{Re}\{c\}t), \tag{4.5.13}$$

where Re{} and Im{} denote real and imaginary parts of a complex number, respectively.

The analysis above is of a unidirectional flow scenario. However, the derived relative growth rate for the amplitude A

$$\frac{1}{A}\frac{dA}{dt} = -k\frac{q_1 \sin k\delta_x}{nA} = -\frac{2\pi}{\lambda}\frac{q_1 \sin\frac{2\pi\delta_x}{\lambda}}{nA} \tag{4.5.14}$$

can be used to see what it takes for symmetrical bedforms in a symmetrical oscillatory flow to grow. The general indication is that the bedforms will grow if $\sin\frac{2\pi\delta_x}{\lambda} < 0$ corresponding to $-\lambda/2 < \delta_\mathrm{x} < 0$. That is, if the sediment transport rate peaks along the uphill slope, i.e. where the flow is contracting. This may be the case for both half-cycles of a symmetrical wave motion and the growth rate will then be proportional to the gross sediment transport rate $q_{\mathrm{s,gross}} = |q_{\mathrm{s,landward}}| + |q_{\mathrm{s,seaward}}|$, not to the net transport rate. A corollary to this is that the direction of bedforms tends to be such that the gross transport rate across the crest is maximized. This is an experimentally verified fact from Aeolian (Rubin and Hunter, 1987) as well as from coastal flows, (Gallagher *et al.*, 1998).

4.5.2.5 *Numerical issues with bed updating*

The nature of the continuity Equation (4.5.1) is the source of much trouble for numerical modelers. This is related to the fact that, it becomes a so-called hyperbolic equation, if the sediment transport rate is proportional to the bed elevation, $q_s \propto z_b$ which, as we saw above, corresponds to bedforms which propagate with constant form.

Hyperbolic equations are often solvable analytically, but they are notoriously difficult to solve numerically. Simple numerical schemes give rise to spurious oscillations and smoothing with the aim of getting rid of these will often smooth out most of the real detail as well.

The "centered in space and forward in time" numerical scheme, which is often applied in coastal morphodynamics modeling, is a dispersive scheme and when applied to non-linear systems it becomes unconditionally unstable. To overcome this, higher order Lax–Wendroff schemes and/or smoothing have been applied with limited success. For a recent review see, e.g. Callaghan *et al.* (2006).

4.5.3 *Vortex ripples*

Vortex ripples are symmetrical bedforms which are present in wave dominated flows. Under low energy conditions, $\theta_{2.5} < 0.5$, the ripples have sharp crests which cause regular vortex shedding, Fig. 4.28. Their length λ is then of the order $1.3\,A$ and their height η about $0.2\,A$. As the flow gets more vigorous the ripples get relatively shorter and flatter, they vanish at around $\theta_{2.5} = 1$.

The vortex ripples are larger (in terms of the relative measures λ/A and η/A) and more regular under regular waves than under random waves. Asymmetrical wave motion generates asymmetrical ripples with a milder slope facing the largest peak velocity. Quantitatively, the behavior for field conditions (irregular waves) is

Fig. 4.26 Small vortex ripples in fine sand exposed at low tide. The profile of a vortex ripple in somewhat coarser sand in a wave flume is shown in Fig. 4.28.

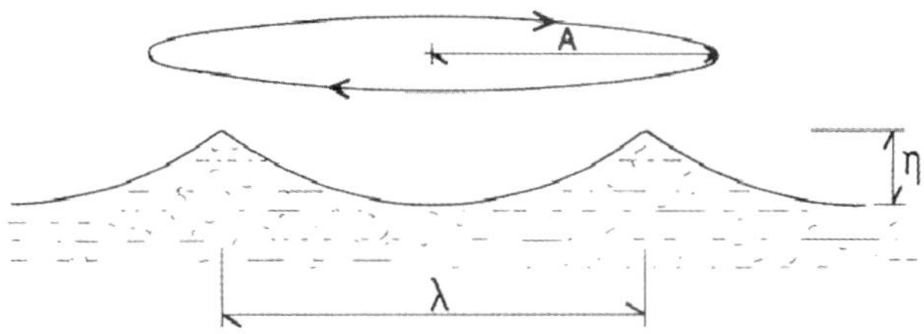

Fig. 4.27 The length λ of wave ripples is related to the amplitude A of the water motion. Such a flow scale is not available for current ripples.

Fig. 4.28 Sharp crested vortex ripples over which sand is entrained into the lee vortices, which form periodically at the crests.

described by Eqs. (4.5.15) and (4.5.16):

$$\frac{\lambda}{A} = \exp\left(\frac{693 - 0.37(\ln\psi)^8}{1000 + 0.75(\ln\psi)^7}\right), \tag{4.5.15}$$

$$\frac{\eta}{A} = 21\psi^{-1.85} \quad (\psi > 10). \tag{4.5.16}$$

Vortex ripples are sometimes found on top of larger bedforms as in Fig. 4.29.

The modeling of vortex ripple formation and their adjustment to varying flow conditions, as measured by Löfquist (1978) is still a challenge. For a recent modeling attempt, see Marieu *et al.* (2008).

4.5.4 *Megaripples*

Megaripples are larger than vortex ripples, typically $\lambda > 2A$, and have rounded crests which do not usually cause flow separation and vortex shedding. The flow boundary layer over megaripples is therefore, for the most part, similar to that

Fig. 4.29 Bedforms, very similar to river dunes in a rip feeder channel. At the top of the feeder, where the current is weak, the bedforms are shore-parallel vortex ripples. Further down where the current becomes dominant, one finds these dune-like bedforms, often with ripples on top.

over a flat bed. However, large (up to 1 m high) clouds of suspended sand are often observed over particularly active spots where the three-dimensional flow structure generates strong vortices.

4.5.5 *Bedforms in combined wave current flows*

Coastal flows are usually combinations of waves and currents with a variety of relative orientations. These different flow combinations result in a great variety of bedforms on which there is a substantial literature.

Within certain limits, the bedforms can be so dominated as to be wholly determined by either the waves or the current, see e.g. Sleath (1984), p. 169 or van Rijn (1993), p. 5.45. Details for steady, intermediate conditions and for transitional states can be found in several experimental studies, e.g. Amos *et al.* (1988), Arnott and Southard (1990) and Hay and Wilson (1994).

4.5.6 *Hydraulic roughness*

Natural sand beds are never perfectly flat and hence the Nikuradse roughness $r \approx d_{50}$ is usually an underestimate. For a nominally flat sand bed with little sediment movement we use $r = 2.5d_{50}$. As the sand starts to move, but while ripples have not yet formed, the momentum transfer and the corresponding bed roughness increase as indicated by the triangles in Fig. 4.30.

For further details about boundary layer structure and roughness in fully developed sheet-flow see Fig. 4.5 above and Nielsen and Guard (2011).

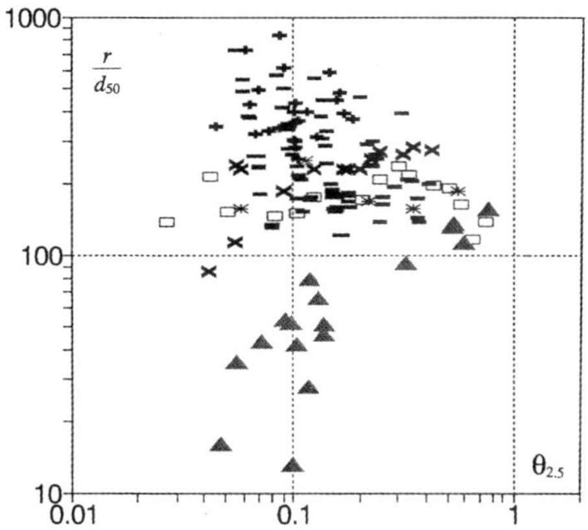

Fig. 4.30 Observed hydraulic roughness for oscillatory flows over flat beds = triangles and rippled beds = all other symbols. Data from energy dissipation measurements by Carstens *et al.* (1968).

When the bed is rippled and/or when a considerable amount of sediment is in motion, and thus contribute to the momentum transfer from the fluid to the bed, the roughness is much greater as indicated by Fig. 4.30. Those "measured" r-values are inferred from measurements of energy dissipation D_E from which a wave friction factor f_w can be calculated, see Nielsen (1992), p. 27. The roughness r is then found by using Eq. (4.2.8) in reverse.

We note that under flow conditions, which correspond to important sediment transport scenarios, the hydraulic roughness is usually of the order 100 d_{50}. Based on the energy dissipation data set in Fig. 4.30 and the corresponding bedform data, Nielsen (1992) suggested the formula

$$r = 8\frac{\eta^2}{\lambda} + 170\sqrt{\hat{\theta}_{2.5} - 0.05}\, d_{50}. \tag{4.5.17}$$

4.6. The Motion of Suspended Particles

4.6.1 *Sediment velocities and sediment orbits*

In order to model suspended sediment transport which occurs under breaking waves, over steep bedforms and in the swash, it is necessary to understand the behavior of suspended particles in various types of flows. In particular, we need to know

1. how fast do the particles settle and
2. how closely do they follow the horizontal velocity of the water.

It turns out that the former question is very complex and has attracted several erroneous theories along with a few useful ones. We shall see how the dominant effects have little to do with the micro-dynamics of the flow (non-linear drag, Magnus effect etc) but all to do with the coherent structures of the flow. The latter question

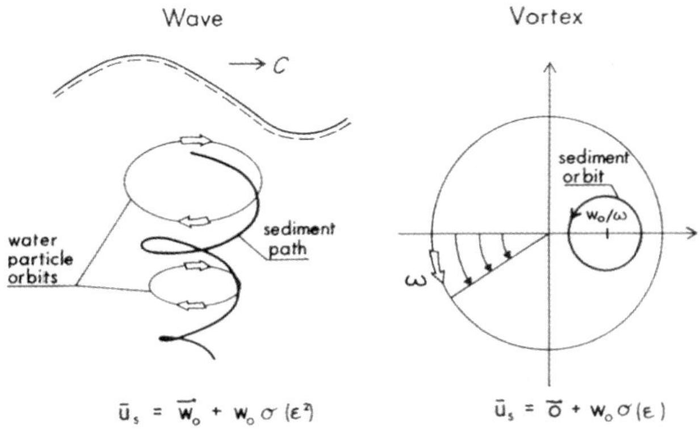

Fig. 4.31 Wave motions and vortex motions are similar in the sense that both have elliptical water particle orbits. Yet, sediment particles behave very differently in the two. The vortex will trap particles (and bubbles) on closed orbits, while the wave motion has no such trapping ability. Settling is unhindered by waves (without turbulence). After Nielsen 2009.

has a simple answer: For all practical purposes, sediment particles have the same horizontal velocity as the surrounding fluid. Most of the results in the following hold analogously for buoyant particles and bubbles and for sinking sediments.

The overwhelming importance of the overall flow structure and the, sometimes surprising, nature of its influence is illustrated in Fig. 4.31.

4.6.2 *Equation of motion for a suspended particle or bubble*

The following discussion will neglect effects of particle–particle interactions, and of particle-boundary interactions but these simplifications are deemed to be of little consequence for sediment moving more than about 1 cm above the bed. We shall also neglect the viscous Basset history term and effects of particle spin. Experience shows that these effects contribute little to the most important phenomena. With these simplifications, the velocity vector $\mathbf{u}_s$ of a spherical sediment particle with diameter d and density ρ_s in a flow with velocity $\mathbf{u}$ is given by

$$\rho s\frac{\pi}{6}d^3\frac{d\mathbf{u}_s}{dt} = \rho s\frac{\pi}{6}d^3\mathbf{g} - \frac{\pi}{6}d^3\overrightarrow{\text{grad }p} + \frac{1}{2}\rho\frac{\pi}{4}d^2C_D|\mathbf{u}-\mathbf{u}_s|(\mathbf{u}-\mathbf{u}_s)$$
$$+\rho\frac{\pi}{6}d^3C_M\frac{d}{dt}(\mathbf{u}-\mathbf{u}_s) \tag{4.6.1}$$

where the driving terms on the right-hand side are respectively due to gravity; pressure gradients; drag, and added hydrodynamic mass.

This equation has (apart from very short-lived transients) solutions of the form

$$\mathbf{u}_s = \mathbf{u} + \mathbf{w}_o + w_o\mathbf{o}(\varepsilon) \tag{4.6.2}$$

where

$$\varepsilon = \frac{1}{g}\left|\frac{d\mathbf{u}}{dt}\right| \tag{4.6.3}$$

meaning that, as long as fluid accelerations are small compared with g, the sediment (or bubble) will quickly obtain a velocity which is well described by the simple superposition law:

$$\mathbf{u}_s = \mathbf{u} + \mathbf{w}_o. \tag{4.6.4}$$

This simple approximation, combined with the kinematics of the two flows, is sufficient for understanding the differences displayed in Fig. 4.31.

4.6.3 *The time scale of particle acceleration*

The time scale δ_t for the decay of the transients, i.e. the time until (4.6.4) becomes an accurate representation, can be found by considering the case of a particle accelerating from rest towards its terminal settling (rise) velocity w_o in a still fluid. The result, corresponding to Eq. (4.6.1) is

$$\delta_t = \frac{s + C_M}{s-1}\frac{w_o}{g}, \tag{4.6.5}$$

which for typical beach sand with $w_\mathrm{o} \approx 2.5\,\mathrm{cm/s}$ is as little as 5 ms (taking $C_M = 0.5$ as for a sphere, Liggett (1994), p. 114), and hence without practical consequences.

An estimate of the velocity variance for a particle or bubble can be derived using Nielsen (1992), Eq. (4.5.13) which, in the simple case of a simple harmonic, horizontal fluid velocity $u = U\mathrm{e}^{i\omega t}$ gives the first-order approximation for the particle velocity:

$$u_p(t) = U_p e^{i\omega t} = \left[1 - \frac{\alpha}{1 - i\frac{\alpha g}{w_o \omega}}\right] U e^{i\omega t} \tag{4.6.6}$$

where $\alpha = \frac{s-1}{s+C_\mathrm{M}}$. For heavy particles ($s > 1 \sim \alpha > 0$) this indicates $\mathrm{Var}\{u_p\} < \mathrm{Var}\{u\}$. However, for light particles and bubbles we can have $\mathrm{Var}\{u_p\} > \mathrm{Var}\{u\}$ with the limit of $\mathrm{Var}\{u_p\} = 3\mathrm{Var}\{u\}$ for a mass-less sphere ($\alpha \to -2$) at high frequencies where $g/w_\mathrm{o}\omega \ll 1$.

4.6.4 *Suspended particles in wave flows*

For a particle submerged in a uniform oscillatory velocity field like (approximately) pure wave motion, e.g.,

$$\mathbf{u}(x, z, t) = \omega \begin{pmatrix} A\cos\ \omega t \\ B\sin\ \omega t \end{pmatrix}, \tag{4.6.7}$$

it can similarly be shown that

$$\mathbf{u}_s = \mathbf{u}(t - \delta_t) + \mathbf{w}_o + w_o\mathbf{o}(\varepsilon^2). \tag{4.6.8}$$

The delay represented by δ_t has no net effect over a wave period and the order ε^2 terms include the settling velocity reduction which occurs if the drag force is non-linear. For a purely vertical oscillatory flow [$A = 0$ in (4.6.7)], and quadratic drag, this settling velocity reduction is given by

$$\frac{\overline{w}}{w_o} = 1 - \frac{1}{16}\left(\frac{\omega^2 B}{g}\right)^2, \tag{4.6.9}$$

cf Nielsen (1992), p. 177. Since the acceleration amplitude $\omega^2 B \ll g$ in most natural flows, this non-linear drag correction is not important in praxis.

4.6.5 *Particles or bubbles suspended in vortices*

Vortices play an important role for the transport of suspended sediment because they are able to trap sediment particles or bubbles, which have settling (rise) velocity smaller than the typical vortex velocities. This was, most strikingly, pointed out by Tooby *et al.* (1977), see their photograph in Fig. 4.32.

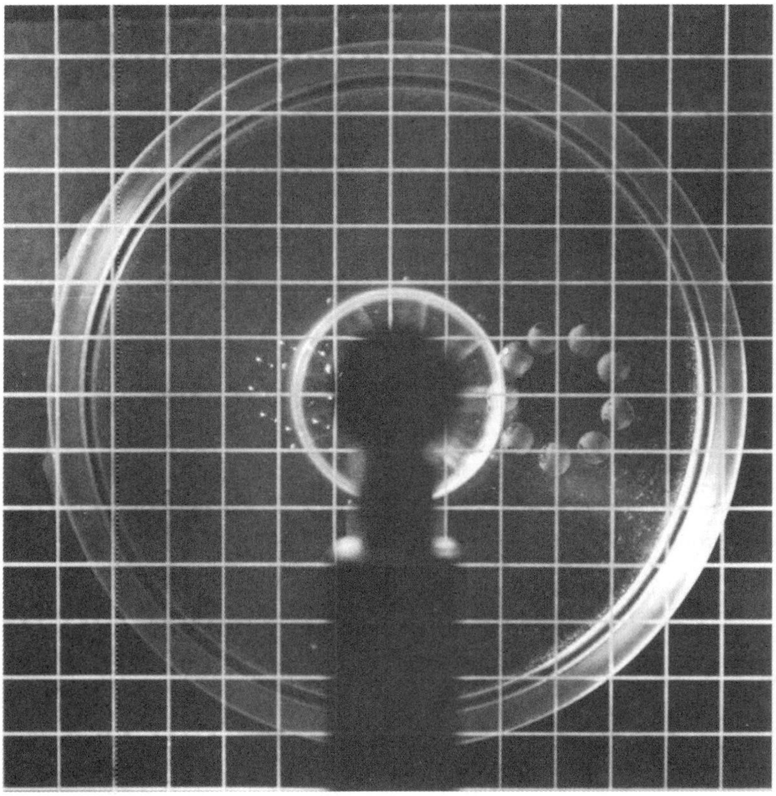

Fig. 4.32 Stroboscopic photograph, by Tooby *et al.* (1977), of a heavy particle and some bubbles in a water filled drum. The drum rotates anti-clockwise so that water velocities are upwards on the right and downwards on the left.

The heavy particle on the right and the bubbles on the left all move on (nearly) closed orbits and hence have zero average velocity,

$$\overline{\mathbf{u}_s} \equiv \mathbf{o} \tag{4.6.10}$$

just like the water particles.

In order to end up with zero net velocity, any bubble or particles must sample water velocities with a particular bias. That is, the identity (4.6.9) with (4.6.4) inserted is

$$\overline{\mathbf{u} + \mathbf{W}_o|_{\text{following a particle}}} \equiv \mathbf{0} \tag{4.6.11}$$

and hence

$$\overline{\mathbf{u}_{\text{following a particle}}} \equiv -\mathbf{W}_o. \tag{4.6.12}$$

In other words: In a steady vortex flow, both heavy particles and bubbles will move along orbits where the average fluid velocity eliminates their settling (rise) velocity. This is true at the same time as every water particle has $\overline{\mathbf{u}} \equiv \mathbf{0}$ and the volume averaged water velocity at any time is also zero: $\langle \mathbf{u}(t) \rangle \equiv \mathbf{0}$.

This truly remarkable result is useful for its direct applications, and for understanding why the inference of particle velocity statistics from the statistics of a surrounding turbulent flow field remains a barrier for researchers. The relevant Lagrangian particle velocity statistics can in praxis only be found by particle tracking.

The particle orbits in Fig. 4.32 can easily be derived using the simple superposition law (4.6.4): Let the drum be centered at $(x, z) = (0, 0)$ and rotate with angular velocity ω. The fluid velocity field is then given by:

$$\mathbf{u}(x, z) \equiv \omega \begin{pmatrix} -z \\ x \end{pmatrix} \tag{4.6.13}$$

and the particle velocity field, obtained by superposition of the settling velocity is

$$\mathbf{u}_s(x, z) = \omega \begin{pmatrix} -z \\ x \end{pmatrix} + \begin{pmatrix} 0 \\ -w_o \end{pmatrix} = \omega \begin{pmatrix} -z \\ x - w_o/\omega \end{pmatrix}. \tag{4.6.14}$$

The analogy between the two expressions is obvious: when $\mathbf{u}(x, z)$ corresponds to the circular fluid orbits around the origin, $\mathbf{u}_s(x, z)$ corresponds to circular orbits around $\binom{0}{w_o/\omega}$. The angular velocity of the orbital motion is in both cases, ω.

4.6.6 *Turbulence effects on settling of particles or the rise of bubbles*

An intriguing and, until recently unanswered, question is whether turbulence, with zero mean velocity has an effect on the mean settling velocity of particles and/or the rise velocity of bubbles? And, is the velocity variance the same for fluid and particles?

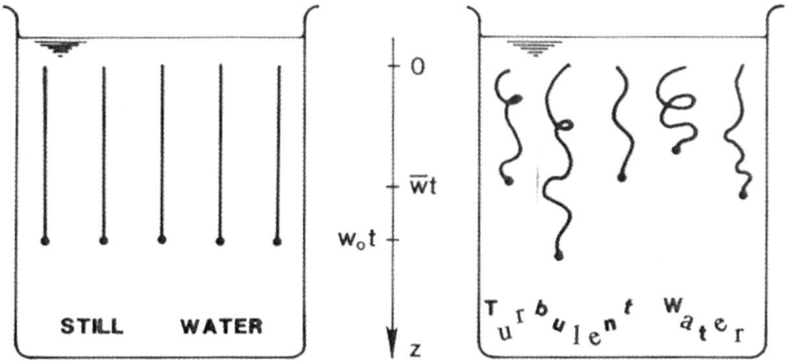

Fig. 4.33 Although the turbulence has zero mean velocities, it changes the average settling velocity of particle and the average rise velocity of bubbles.

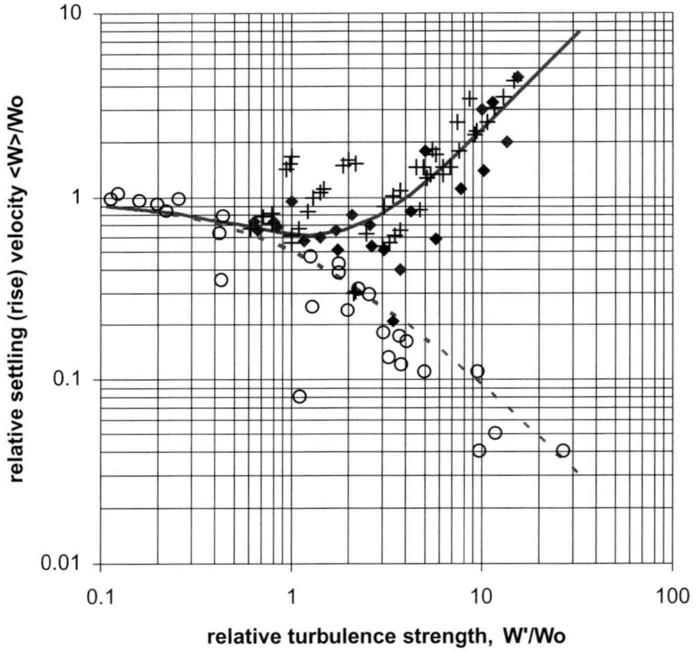

Fig. 4.34 Measured settling velocities of dense particles (♦), rise velocities of light particles and bubbles (○), and rise velocities of diesel droplets (+) in turbulence. In these experiments, the turbulence intensity is taken as the rms vertical particle velocity, w_s'.

There is now comprehensive experimental data, see Fig. 4.34, which shows that in fairly weak turbulence, $\sqrt{\text{Var}\{w_s\}} = w_s' < 5w_o$, both heavy (solid symbols) and buoyant (open symbols) particles are delayed. This can be understood in terms of the vortex trapping mechanism discussed in connection with Fig. 4.32.

For very buoyant particles and bubbles, the delay keeps getting stronger with increasing turbulence strength. Surprisingly however, the trend reverses for heavy

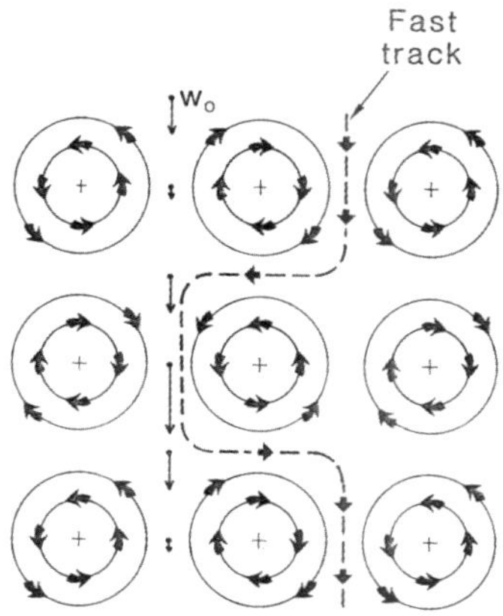

Fig. 4.35 In a field of vortices, heavy particles will, by centrifugal effects become concentrated along the vortex boundaries and get swept downwards along the "fast tracks" by the vortex flow. Bubbles and very light particles are however expected to spiral inwards, and get trapped like the bubbles in Fig. 4.32. A particle following the fast track will experience accelerated settling. It will also experience a reduced velocity variance compared with the flow field as a whole, indeed, the particle sees only downward fluid velocities.

particles (♦) around $w_s'/w_o = 3$. In stronger turbulence, the heavy particles speed up again and end up settling at several times their still water rate for $w'/w_o > 10$.

The increased settling rate for heavy particles in strong turbulence can be qualitatively understood in terms of the fast tracking concept of Nielsen (1992, 1993), see Fig. 4.35.

Nielsen (1992, 1993) also pointed out that this scenario predicts unbounded increase of the settling rate for very strong turbulence:

$$\overline{w_s} \propto w' \quad \text{for } w'/w_o \to \infty \tag{4.6.15}$$

The solid curve, which mimics the trend for heavy particles and diesel droplets in Fig. 4.34 is given by:

$$\frac{\overline{w_s}}{w_o} = \frac{1 + \frac{1}{4}\left(\frac{w'}{w_o}\right)^2}{1 + \frac{w'}{w_o}}. \tag{4.6.16}$$

The dotted curve, which mimics the trend for light particles and bubbles, is given by

$$\frac{\overline{w_s}}{w_o} = \frac{1}{1 + \frac{w'}{w_o}}. \tag{4.6.17}$$

The measurements of Friedman and Katz (2002) on diesel droplets in water showed some dependence of $|\overline{w_s}|/w_o$ upon the Stokes number, *St*, which is the ratio between the response time scale of the particle and the time scale of the turbulence, with the strongest tendency for accelerated rise for particles with $St \approx 1$. Expressing

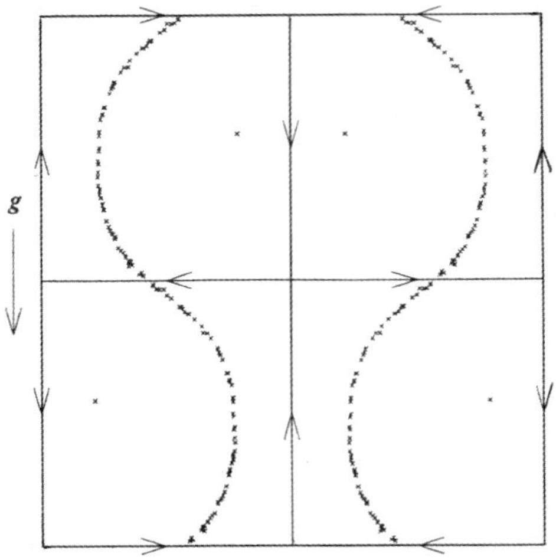

Fig. 4.36 Pattern of concentrated bubbles in a cellular flow field calculated by Maxey (1990). The bubbles were initially uniformly scattered. The isolated "bubble" in each cell is at the stable neutral point, where $\mathbf{u}_f = -\mathbf{w}_o$, into which a great number of particles have actually converged. The curves are rising fast tracks which are pieced together from arcs, which within each cell are inward spirals toward the neutral point. From Nielsen (2009).

the Stokes number as w_o/gT_L, where T_L is the Lagrangian integral time scale of the turbulence we may thus expect that the influence of turbulence on the settling/rise velocity is essentially given by:

$$\frac{\overline{w_s}}{w_o} = \Phi\left(\frac{w'}{w_o}, \frac{w_o g}{T_L}\right), \tag{4.6.18}$$

where both acceleration ($\Phi > 1$) and delay ($\Phi < 1$) are possible for both buoyant and sinking particles. The enhanced rising of some of the diesel droplets of Friedman and Katz around $St \approx 1$ can be understood in terms of the light particle fast track in Fig. 4.36.

4.6.7 *Reduced particle velocity variance due to selective sampling*

The concentration of sediment particles and bubbles along certain preferred trajectories, e.g. the "fast tracks" in Figs. 4.35 and 4.36, will also affect the particle velocity variance. That is, since velocity variance taken along the "fast track" is less than the velocity variance for the velocity field as a whole, one often finds the Lagrangian velocity variance from particle tracking to be smaller than the Eulerian fluid velocity variance measured by a fixed probe. A classical example is the dataset of Snyder and Lumley (1971). They found a significant reduction: $\text{Var}\{w_s\} \sim 0.6\text{Var}\{w_{\text{fluid}}\}$ by tracking particles, which were so small ($w_o/gT_L \approx 0.03$), that Eq. (4.6.6) predicts only insignificant variance reduction.

4.7. Sediment Suspensions

4.7.1 *What is suspended load?*

In Sec. 4.4.2, bedload was defined as that part of the sediment load, which is supported by inter-granular forces. The rest, the suspended load, is then that which is supported by upward fluid drag.

The amount of suspended load is determined either by supply or by a balance between settling and entrainment at the bed, if supply is plentiful. Consider gradually adding fine sediment to a recirculating flow over a concrete bed. Initially, all the sediment will be suspended, but after a while, the flows carrying capacity is reached and any excess sediment settles out.

Sandy sediments are usually carried at capacity, while silt and clay is often carried at loads below capacity. Sediment carried below capacity is sometimes called *wash load*. By its nature wash load is "migratory" rather than "resident" and is therefore only found in rivers and estuaries.

A large part of the sediment transport in surf zones occurs as suspended transport near the breakpoint and in the swash zone. Figure 4.37 shows large plunging waves stirring up sand as the plunging jets impinge on the bar. The sand is subsequently brought to the surface by the rising plumes of entrained bubbles, which drive the "fountains" of Fig. 4.39. After the plumes reach the surface and the air escapes, the sand may travel some distance along the surface before it gradually settles out.

Clouds of suspended sand on the smaller scale of the vortices over vortex ripples were shown in Fig. 4.28.

The nature of sediment suspensions depends strongly upon the bed geometry. The simplest situation is of course that of a flat bed, where the sediment concentrations can be modeled as functions of elevation and time only: $c = c(z, t)$. Examples of $c(z, t)$-time-series from sheet flow were shown in Fig. 4.18.

The extensive laboratory experiments by Ribberink *et al.* (1994, 2000) with $d_{50} = 0.21\,\text{mm}$, $0.32\,\text{mm}$, and $0.13\,\text{mm}$ shows that upwards from about 1 cm from the undisturbed bed, the corresponding time averaged sediment concentrations vary as $\overline{c}(z) \propto (z/L)^{-2}$, where the vertical scale L is of the order 5 mm.

If the bed consists of vortex ripples the vertical scale of the concentration profiles is of the order of the ripple height.

While sediment concentration time series from different levels above a flat bed (e.g. Fig. 4.18) show a time variation which is entirely due to the variation of $u_\infty(t)$, the concentration time series from different levels above rippled beds show two or four concentration peaks per wave period, which are due to the passage of sand clouds. These clouds are the remainders of the sediment laden vortices which are released from the lee of the ripples at each flow reversal, cf. Fig. 4.28.

Also over flat or almost flat beds will concentration time series usually show peaks due to the passage of sediment clouds. These clouds will be generated by

Fig. 4.37 Large breakers bringing suspended sand to the surface. Whale Beach, Sydney, Australia. Just seaward of the break point, concentrations are bigger at the surface than at mid depth in this scenario.

occasional, more or less randomly spaced vortex shedding and are therefore not periodic. They do not show up in Fig. 4.18 because they have been averaged out by ensemble averaging over many (regular) wave periods. In field conditions however, where the waves are not periodic and ensemble averaging not possible, these random clouds (the spotted carpet effect) make it very difficult to relate $c(z,t)$ to $u_\infty(t)$.

For both flat and rippled beds the profile shape, $\overline{c}(z)$ may vary considerably depending on the size of sediment considered. An example of this is shown by the data in Fig. 4.38, which was measured by suction sampling over vortex ripples under waves. Each curve represents one sieve fraction.

The settling velocity varies by a factor 9 from the finest fraction to the coarsest. The fact that the difference in slope is nowhere near this magnitude indicates strong differential diffusion (Sec. 6.7.5 below). The change in shape from upward convex for the fine sand to upward concave for coarser sand is also an interesting modeling challenge.

For the data in Fig. 4.38 the wave period T is obviously fixed while w_o varies by a factor 9 between the sieve fractions. Examples with fixed w_o and variable T,

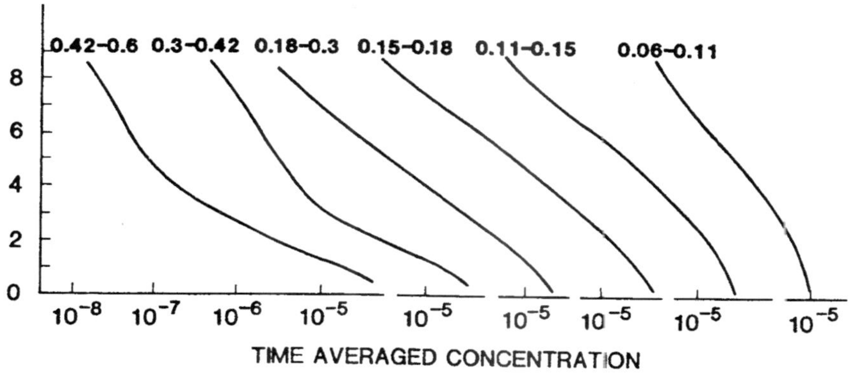

Fig. 4.38 $\bar{c}(z)$-profiles for different sand fractions in the same flow over a rippled sand bed under non-breaking waves. The numbers on the curves are sand size in mm. The $\bar{c}$-axes are logarithmic. The wave and bedform conditions were: $(h, T, H, d_{50}, \lambda, \eta) = (0.3\,\text{m}, 1.51\,\text{s}, 0.13\,\text{m}, 0.19\,\text{mm}, 0.078\,\text{m}, 0.011\,\text{m})$.

i.e. same sand in different waves, are shown by Nielsen (1992), p. 217. In that case, similar $\bar{c}(z)$-shape changes occur, which correlate well with the shape parameter $S = w_o T/\sqrt{Ar}$, where r is the hydraulic roughness estimated by (4.5.17). That is, for small S the profiles are upward convex, while for larger S, they become more or less straight ($c \sim \mathrm{e}^{-z/L}$) and subsequently upward concave with further increase of S.

Wave irregularity leads to smaller bedforms in general and hence to smaller vertical scale of the $\bar{c}(z)$-profiles.

Currents influence the bedforms and add turbulence (compared with non-breaking waves). Hence the $\bar{c}(z)$-profiles are usually stretched vertically due to the presence of a current.

Wave breaking has a strong influence on the $\bar{c}(z)$-profiles because it provides mechanisms for transporting suspended sediment convectively from the bottom to the surface (Figs. 4.37 and 4.39) and because it enhances the turbulent mixing. A set of laboratory data is shown in Fig. 4.40.

The turbulence, which provides the mixing to keep the sediment in suspension, is to a variable extent dampened by stratification. This effect is visible in the data in Fig. 4.41.

The ordinate is essentially the vertical scale $L_c(\varepsilon_m = w_o L_c)$ of the concentration profiles obtained over ripple beds under non-breaking waves.

We see that the concentration scale initially increases with increasing $A\omega/w_o$ but then rather abruptly decreases for $15 < A\omega/w_o < 25$ and then decreases at a slower rate for $A\omega/w_o > 25$. We believe that the reason for the decline is stratification as the concentrations and total sediment loads increase beyond some critical values occurring at about $A\omega/w_o = 15$. The same threshold has recently been obtained from a stratification model for sheet flow by Conley *et al.* (2008).

Fig. 4.39 Fountains of water rising 4.5 m in the air after the plunging of swell waves. Figure 4.37 offers a view from the back of some of these fountains. In such situations, the plunger jet may impact on the bed and entrain large amounts of sand which is easily carried straight to the surface (and into the air) by the vertical flow velocities which generate the fountain. The vertical velocities w_{fountain} in the "roots" of these fountains can be inferred from the height $z_{\max}$ to which the water rises: $w_{\text{fountain}} = \sqrt{2gz_{\max}}$, in this case around $\sqrt{2 \times 9.8 \times 4} \approx 9\,\text{m/s}$.

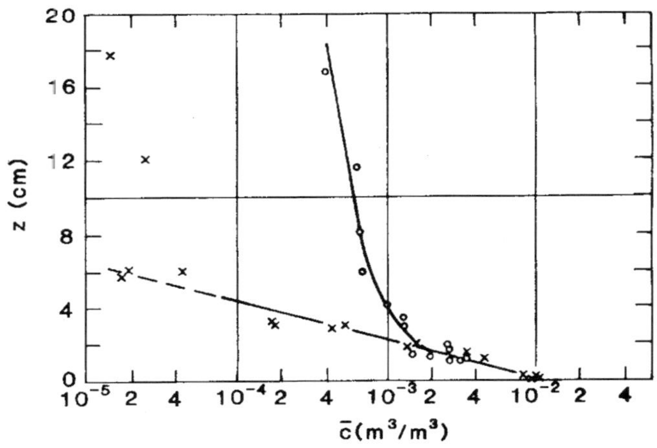

Fig. 4.40 Suspended sediment concentrations under non-breaking waves (×) and spilling breakers (o) of the same height and period over the same sand bed. $T = 1.3\,\text{s}$, $h = 0.38\,\text{m}$, $H = 0.19\,\text{m}$, $d_{50} = 0.08\,\text{mm}$.

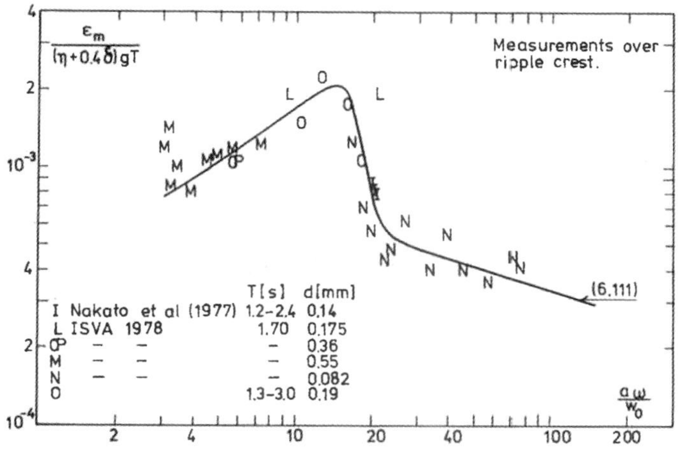

Fig. 4.41 Relative vertical scale of suspended sediment concentration profiles versus the relative wave velocity $A\omega/w_{\rm o}$. Rippled beds under non-breaking waves. From Nielsen (1979).

4.7.2 *The modeling framework for sediment suspensions*

The modeling framework for the suspended sediment concentrations consist of a field equation and boundary conditions at the model boundaries.

The field equation expresses the conservation of sediment:

$$\frac{\partial c}{\partial t} = -\nabla \cdot \mathbf{q}_s = -\nabla \cdot \{\mathbf{u}_p(x,y,z,t)c(x,y,z,t)\}, \tag{4.7.1}$$

i.e. the local rate of change of sediment concentration equals minus the divergence of the sediment flux field. (The operator "$\nabla\cdot$" is the divergence operator on a vector field $\mathbf{R} = (R_x, R_y, R_z)$: $\nabla \cdot \mathbf{R} = \frac{\partial R_x}{\partial x} + \frac{\partial R_y}{\partial y} + \frac{\partial R_z}{\partial z}$).

In the applications it is only the bottom boundary condition that requires detailed attention. It has traditionally been handled in terms of a reference concentration, but this approach is increasingly abandoned in favor of pickup functions, i.e. specification of the upward sediment flux rather than the concentration at the base of the modeling domain.

4.7.3 *Near-bed reference concentrations*

Many previous suspension models have relied on the concept of a reference concentration by using a boundary condition of the form $c(z_b, t) = C_b(t)$ at the bottom of the model domain, $z = z_b$.

This approach is however at the most applicable for steady and uniform conditions (even there it may lead to unrealistic results, cf. Parker 1978). Data sets like the one in Fig. 4.18, which show the concentration at the undisturbed bed level to vary less than those 5 or 10 grain diameters above, also show the inappropriateness of $c(0,t)$ as a driver of the sediment concentrations at higher levels. See the discussion of Nielsen *et al.* (2002) for further details.

4.7.4 *Sediment pickup functions*

The alternative to a prescribed concentration at the model base is a prescribed upward flux or pickup function $p(t)$, which may plausibly be related to the instantaneous skin friction Shields parameter or for practical purposes to the grain roughness Shields parameter $p(t) = p(\theta_{2.5}[t])$ with $\theta_{2.5}(t)$ given by (4.4.16). For an equilibrium situation the pickup flux must equal the settling flux: $\overline{p} = w_s\overline{c_b}$ but for an unsteady or non-uniform situation a difference may exist, resulting in a sand level change:

$$c_{\max}\frac{dz_s}{dt} = w_s c(0,t) - p(t).$$

Van Rijn (1984) reviewed the pickup function concept for steady flow. He provided new data and a steady flow formula which was adapted by Nielsen (1992), Nielsen *et al.* (2002) to oscillatory sheet-flow in the form

$$p(t) = \begin{cases} 0.017w_o|\theta_{2.5}(t) - 0.05| & \text{for } \theta_{2.5}(t) > 0.05 \\ 0 & \text{for } \theta_{2.5}(t) < 0.05 \end{cases}. \tag{4.7.2}$$

Nielsen (1979) introduced pickup functions for waves over vortex ripples. For these rippled beds the pickup function has two peaks per wave period, one at each flow reversal where the sand laden lee vortices (Fig. 4.24) are released into the main flow.

4.7.5 *Sediment distribution modeling*

The concentration distribution in the interior of the modeling domain is determined by the conservation of sediment:

$$\frac{\partial c}{\partial t} = -\nabla \cdot \mathbf{q}_s = -\nabla \cdot \{\mathbf{u}_p(x,y,z,t)c(x,y,z,t)\}, \tag{4.7.1}$$

where $\mathbf{u}_p$ is the sediment particle velocity.

This equation may of course be attacked directly with a large enough computer but usually simplification through various kinds of averaging is preferred.

Horizontal averaging is commonly applied over an area which is small compared to the scales of wave deformation and beach morphology but greater than the scale of ripples or of turbulent bursts over a flat bed.

Time averaging of c and u separately is usually not recommendable because the transport contribution from $\overline{\tilde{u}_p\tilde{c}}$ is quite often greater than that from $\overline{u_p}\,\overline{c}$. Thus, the periodic component $\tilde{c}$ must be calculated, as well as $\overline{c}$.

In order to get started on modeling, a horizontally uniform situation is often assumed. This leads to the following, simplified version of (4.7.1):

$$\frac{\partial c}{\partial t} = -\frac{\partial q_{sz}}{\partial z} = -\frac{\partial}{\partial z}\{w_p(z,t)c(z,t)\} \approx -\frac{\partial}{\partial z}\{w(z,t)c(z,t) - w_s c(z,t)\}, \tag{4.7.3}$$

where the issues of time dependence may be illustrated by considering a periodic flow where $w = 0 + \tilde{w} + w'$ and $c = \overline{c} + \tilde{c} + c'$ ($\overline{w} \equiv 0$ for a horizontally uniform flow). From phase averaging we then get,

$$\frac{\partial \tilde{c}}{\partial t} = -\frac{\partial}{\partial z}\{\tilde{w}\overline{c} + \tilde{w}\tilde{c} + w'c' - w_s\tilde{c}\} \tag{4.7.4}$$

and, by further averaging over a few wave periods

$$0 = \frac{\partial}{\partial z}\{\overline{\tilde{w}\tilde{c}} + \overline{w'c'} - w_s\overline{c}\}. \tag{4.7.5}$$

These equations have traditionally been dealt with in terms of gradient diffusion, e.g. $\overline{\tilde{w}\tilde{c}} + \overline{w'c'} = -K_{\text{Fick}}\frac{\partial \overline{c}}{\partial z}$ for (4.7.5). That is however an oversimplification, which misses the differential diffusion displayed in Fig. 4.38 and by several other classical data sets. A more satisfactory representation is obtained with a finite mixing length model as shown below.

4.7.5.1 *The mixing length approach*

Rather than trying to account for all the details of $w = \overline{w} + \tilde{w} + w'$ and $c = \overline{c} + \tilde{c} + c'$ it is usually more appropriate to consider the sediment flux in terms of the mixing flux

$$q_m = w_m\left[c\left(z - \frac{l_m}{2}\right) - c\left(z + \frac{l_m}{2}\right)\right], \tag{4.7.6}$$

which is illustrated in Fig. 4.42, and the settling flux $w_s c$: $q_{sz} = q_m - w_s c$.

The conservation equation (4.7.3) then becomes,

$$\frac{\partial c}{\partial t} = -\frac{\partial}{\partial z}(q_m - w_s c) = -\frac{\partial}{\partial z}\left[w_m\left[c\left(z - \frac{l_m}{2}\right) - c\left(z + \frac{l_m}{2}\right)\right] - w_s c(z)\right], \tag{4.7.7}$$

which by time averaging (permitted when the wave period is much longer than the turbulent time scale) becomes,

$$0 = \frac{\partial}{\partial z}\left[-w_m\left[\overline{c}\left(z - \frac{l_m}{2}\right) - \overline{c}\left(z + \frac{l_m}{2}\right)\right] - w_s\overline{c}(z)\right]. \tag{4.7.8}$$

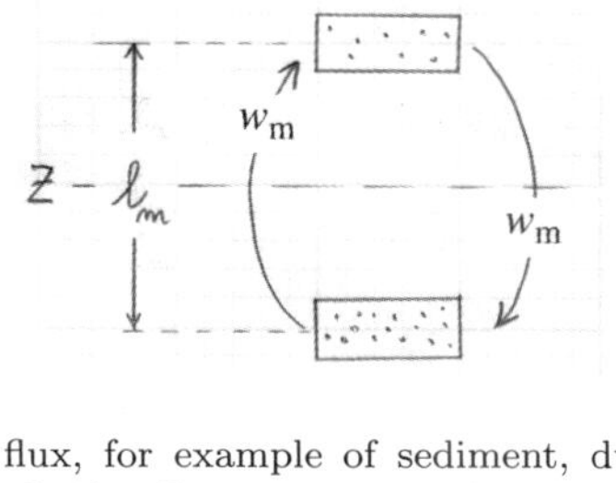

Fig. 4.42 The down-gradient flux, for example of sediment, due to turbulent mixing can be understood in terms of swapping fluid-sediment-mix parcels over a vertical distance l_m, the mixing length.

Further simplification is achieved by assuming that the [ ], which represents the total vertical flux of sediment, is zero at the free surface and hence is zero everywhere as $\partial/\partial z = 0$:

$$w_m \left[\overline{c}\left(z - \frac{l_m}{2}\right) - \overline{c}\left(z + \frac{l_m}{2}\right) \right] - w_s \overline{c}(z) = 0. \tag{4.7.9}$$

In Sec. 5.4.7 of Nielsen (2009) this equation was solved for the case of homogeneous turbulence, i.e. constant w_{m}, l_{m}. It was found that the solution to this complete equation is analogous to that of the Fickian approximation (obtained by assuming $\overline{c}(z - \frac{l_m}{2}) - \overline{c}(z + \frac{l_m}{2}) = -l_m \frac{\partial \overline{c}}{\partial z}$), but that only the complete equation explains differential diffusion. That is, coarser sediment (greater w_{s}) has a greater apparent Fickian diffusivity than fine sediment:

$$K_{\mathrm{Fick}}(w_s) = w_m l_m \left[1 + \frac{1}{24}\left(\frac{w_s}{w_m}\right)^2 + \cdots \right], \tag{4.7.10}$$

which is indeed the trend shown by experimental data where the concentration profiles of different sediment fractions have been measured in the same flow, cf. Fig. 4.38 for oscillatory flows and Coleman (1970) for steady flow. In other words, a simple gradient diffusion approach with a sediment diffusivity, which is independent of w_{s} is inadequate.

Van Rijn (1984b) suggested an ad hoc approach expressed as $K_{\mathrm{Fick}}(w_s) = \beta v_t = v_t[1 + 2(\frac{w_s}{u_*})^2]$ to mimic experimental findings in line with (4.7.10), if $w_m \sim u_*$. This ad hoc fix may be a reasonable engineering approach as long as it is realized that *Reynolds hypothesis*: $K_{\mathrm{Fick}} = v_{\mathrm{t}}$ does not hold for suspended sediment. It does, however, not enable modeling the different profile shapes displayed in Fig. 4.38.

We shall see below that similar results to (4.7.10) are obtained with various different pairs (w_m, l_m), which are relevant to natural flows, and that the details of Fig. 4.38 are, at least qualitatively, captured by the mixing length approach with appropriate choices of $l_m(z)$ and $w_m(z)$.

4.7.5.2 *Mixing length model of $u(z)$ and $c(z)$ for $l_{\mathrm{m}} = \lambda z$ and $w_{\mathrm{m}} = \gamma u_*$*

First consider the flow near the bed in a river or under an ocean current. That is, a flow which has traditionally been modeled as a constant shear stress layer ($\tau \approx \tau_{\mathrm{o}} = \rho u_*^2$) with a linearly increasing eddy viscosity, $v_t = \kappa u_* z$. (This corresponds to the bottom part of the logarithmic velocity profile discussed in Sec. 4.2.2). More precisely, we consider a flow where the mixing velocity is constant $w_m = \gamma u_*$, while the mixing length grows linearly with distance from the bed: $l_m = \lambda z$.

First, we find the parameter space (λ, γ), which corresponds to the experimentally verified logarithmic velocity distribution: Writing the upward flux of x-momentum $(-\tau = -\rho u_*^2)$ as a mixing flux in the form of Eq. (4.7.6) with $w_m = \gamma u_*$, and $l_m = \lambda z$, we get:

$$-\rho u_*^2 = \gamma u_* \left(\rho u \left\{ z - \frac{1}{2}\lambda z \right\} - \rho u \left\{ z + \frac{1}{2}\lambda z \right\} \right), \tag{4.7.10}$$

where insertion of the log-law,

$$u(z) = \frac{u_*}{\kappa} \ln \frac{z}{z_o}, \tag{4.2.4}$$

gives

$$\gamma \ln \frac{1 + \lambda/2}{1 - \lambda/2} = \kappa. \tag{4.7.11}$$

This equation, together with the physical limits $0 < \lambda < 2$ (cf. Nielsen, 2009, Fig. 5.4.9), defines our limits of choice for (λ, γ). Experiments, e.g. Graf and Cellino (2002) indicate $\sqrt{\overline{w'^2}} \approx 0.5u_*$ and adopting this as the mixing velocity: $w_m = 0.5u_*$ then gives $\lambda \approx 0.7$ and $l_m \approx 0.7z$.

Turning now to the sediment, we have to solve (4.7.9) with $w_m = \gamma u_*$, and $l_m = \lambda z$, i.e.,

$$\gamma u_* \left(\overline{c} \left\{ z - \frac{1}{2}\lambda z \right\} - \overline{c} \left\{ z + \frac{1}{2}\lambda z \right\} \right) - w_s \overline{c}(z) = 0. \tag{4.7.12}$$

As for homogeneous turbulence the solution to this complete, finite-mixing-length equation, is similar to that of the Fickian, infinitesimal-mixing-length (iml) approximation. The iml approximation,

$$-w_m l_m \frac{\partial \overline{c}}{\partial z} - w_s \overline{c} = -\gamma u_* \lambda z \frac{\partial \overline{c}}{\partial z} - w_s \overline{c} = 0, \tag{4.7.13}$$

which is obtained from (4.7.12) by applying the linearization $\overline{c}\{z - \frac{1}{2}\lambda z\} - \overline{c}\{z + \frac{1}{2}\lambda z\} = -\lambda z \frac{\partial \overline{c}}{\partial z}$, has the solution:

$$\overline{c}(z) = \overline{c}(z_o) \left(\frac{z}{z_o} \right)^{-\frac{w_s}{\lambda \gamma u_*}}, \tag{4.7.14}$$

which corresponds to a classical model by Rouse (1937) for $\lambda\gamma = \kappa$.

Inserting an analogous power function expression, e.g. $c(z) = c(z_o)(z/z_o)^p$ into (4.7.12), leads to the following equation for the power p

$$\left(1 - \frac{\lambda}{2}\right)^p - \left(1 + \frac{\lambda}{2}\right)^p = \frac{w_s}{\gamma u_*}. \tag{4.7.15}$$

The behavior of p according to the above equation is illustrated by Fig. 4.43.

We see that the finite mixing length effects are negligible for very fine sediment, i.e. $p \to -w_s/\kappa u_*$ and $\beta = K_{\text{Fick}}/v_t \to 1$ for $w_s/u_* \to 0$.

The increasing trend for $\beta = \beta(w_s/u_*)$ is in good qualitative agreement with flume studies, e.g. Coleman (1970) and Graf and Cellino (2002). However, some

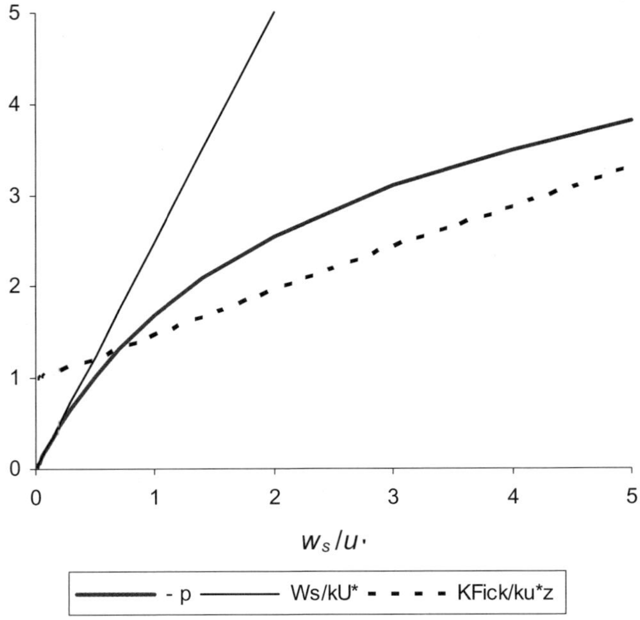

Fig. 4.43 The exponent p compared with the "infinitesimal mixing length" value $-w_s/\kappa u_*$ for $(\lambda, \gamma) = (1, 0.36)$. The finite mixing length behavior quantified by $w_s/ku_* - p$, is stronger for larger λ. Also shown is $\beta = K_{\text{Fick}}/v_t = K_{\text{Fick}}/\kappa u_* z$, where $-K_{\text{Fick}} \frac{\partial c}{\partial z} - w_s c = 0 \Leftrightarrow K_{\text{Fick}} = \frac{-w_s z}{p}$.

details are still not understood, including the occurrence of $\beta < 1$ for some flat bed experiments by Coleman (1970).

4.7.5.3 *Suspension under waves*

Application of Eq. (4.7.9) with $l_m = \lambda z$ and $w_m = \gamma u_* \exp(-z/L_w)$ can also explain the qualitative different $\overline{c}(z)$-profiles for fine and coarse suspended sediment over wave ripples, cf. Fig. 4.38. For full details see Teakle and Nielsen (2004). The traditional Fickian (gradient diffusion) approach could never explain the different shapes of the $c(z)$-profiles displayed by fine and coarse sand in a given flow of this type.

Under breaking waves or in combined wave current flows, where the turbulence is stronger at greater distance from the bed (cf. Fig. 4.39), the model should probably be modified to $l_m = \lambda z$ and $w_m = \gamma u_*/(1 + z/L_w)$.

4.7.6 *Suspensions of cohesive sediments*

Suspensions of cohesive sediments are more complicated because the effective settling velocity varies due to flocculation, which depends on salinity and turbulence intensity as described by Winterwerp (1998).

Salinity enhances flocculation by cushioning electrically charged particles against the electrostatic repulsion, allowing the van der Waal's forces to bind the particles together. Hence, fine sediment which encounters salt water seaward of the null point of the salt wedge, will sink and return upstream with the intruding saltwater. There is hence a convergence of fine sediment near the null point which gives rise to the frequently observed turbidity maxima in this area. Open University (1999), p. 164 shows details of the turbidity maximum in the Seine estuary and Ellis *et al.* (2008) present a recent model of turbidity maxima in the Irish Sea.

4.8. Basic Sediment Transport Model Building

4.8.1 *Introduction*

The aim of the following is, where possible, to deliver the building blocks for morphological change models. The task is easy in the case of sheet flow under a pure wave motion where the Meyer–Peter and Müller type formulae (4.4.17) may be applied. However, when considerable amounts of sediment moves in suspension, the timing of entrainment and settling must be given an detailed consideration.

When currents are present they may have quite different directions and vertical distributions from the wave induced velocities. Many of these issues are still beyond the state-of-the-art, and most of the data, available for model verification is 2DV only.

4.8.2 *Two different families of transport models*

Models of the short term averaged sediment transport rate come in two distinct varieties: The traditional $u \times c$-integral models with the general form $\mathbf{q}_s = \int_{\text{bed}}^{\text{surface}} c\mathbf{u}_s dz$ and the much less common particle trajectory models which are based on the pickup function concept and the average trajectory of moving sediment particles, Fig. 4.44.

The literature contains a vast number of $u \times c$-integral models and a recent guide to the building of such models can be found in van Rijn (2007, I–IV). One important general consideration about these models is that the net transport due to oscillatory components is not to be ignored, i.e. the term $\overline{\tilde{u}_s\tilde{c}}$ is often at least as important as $\overline{u_s}\,\overline{c}$.

A very simple and very successful model of the particle trajectory type for shore normal transport over vortex ripples was developed by Nielsen (1988) and shown

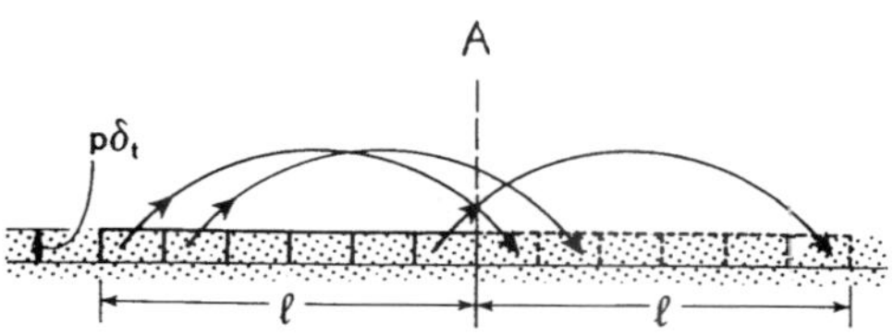

Fig. 4.44 The amount of sand picked up per unit time is the pickup function $p(t)$ and on the average $\overline{p}$. Then, if the average jump length is l, the flux density through a vertical cross-section (A) is seen to be $\overline{q}_s = \overline{p}l$ (where l and then q may be vectors in the horizontal plane).

to be more reliable across different sand sizes than more complicated models of the $u \times c$-integral type.

4.9. Sediment Transport Outside the Surf Zone

4.9.1 *Shorenormal transport*

Outside the surf zone, the experience from detailed large scale laboratory studies with non-breaking waves or oscillating water tunnels applies. Hence, Meyer–Peter and Müller type transport formulae (4.4.15)–(4.4.18) are fairly reliable for sheet flow conditions.

For rippled beds, the accuracy of available models is not as good. The trouble starts with the considerable uncertainty in predicting the ripple geometry. This has impact on the estimation of near-bed concentration magnitude (higher, sharper ripples → greater stresses at the crest → larger concentrations) as well as on the distribution of suspended sediment concentrations and fluid velocities. For a given ripple geometry, the finer details of the suspended sediment concentrations, mentioned in connection with Fig. 4.38 are only modeled with computationally expensive models and the modeling of the stratification effects, clearly visible in Fig. 4.7.5, is currently an area of new research. Simple models, corresponding to conventional diffusion modeling and two alternative approaches, were developed by the writer in the 1980s and tested against small scale laboratory data, cf. Nielsen (1992), p. 266. These models can be calibrated for field use, given suitable large scale measurements.

It is worth a special note here that, the "grab and dump" model, Nielsen (1992), p. 273 which is based on the concept of Fig. 4.44, was the most successful when tested against the small scale experimental data, available at the time, and has not been dethroned by later validations against larger scale experiments.

4.9.2 *Shoreparallel transport outside the surf zone*

The shoreparallel sediment transport outside the surf zone is usually insignificant or dominated by strong currents of tidal or oceanic origin. Cherlet *et al.* (2007) is a recent example of modeling this sediment transport and the resulting bedform dynamics.

4.10. Surf Zone Sediment Transport

4.10.1 *Shorenormal transport*

The modeling of shore normal surf zone sediment transport q_x in a purely 2D situation, such as a wave flume, has progressed to a stage where the resulting beach profile developments have been modeled with reasonable accuracy, at least for irregular waves where the morphological features are not too sharp, see e.g. Roelvink and Broeker (1993). Some of these good end results are however somewhat fortuitous in that none of the models deal with acceleration effects or streaming which have since

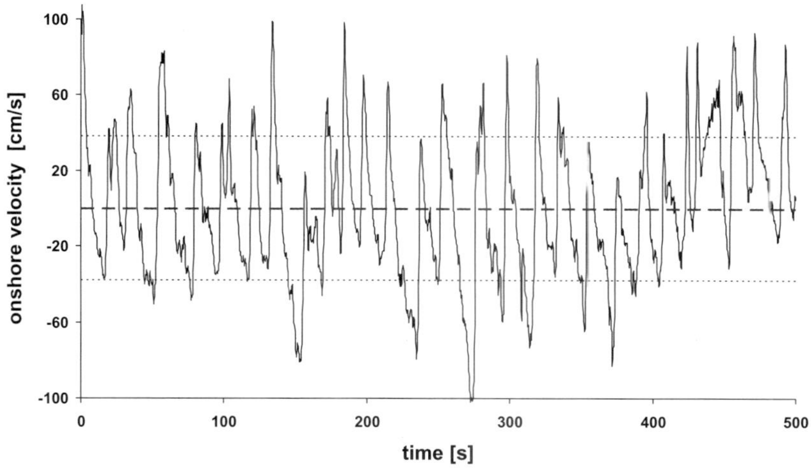

Fig. 4.45 An example of near-bed ($z \sim 20$cm) shorenormal velocity in the inner surf-zone. The mean value $\overline{u}$ has been set to zero in this plot by default. In reality the depth averaged value $\langle\overline{u}\rangle$ is strongly negative in a 2D situation where $\overline{\langle u\rangle(h+\nu)} = 0$. Correspondingly $\overline{u^3}$ is strongly negative and, if there is equilibrium with $q_s = 0$, this must be due to the strong positive acceleration skewness. In this case: $\overline{(\frac{\partial u}{\partial t})^3}/(\overline{(\frac{\partial u}{\partial t})^2})^{3/2} = 1.3$ acting through Eq. (4.4.16).

been shown to be very important, see Sec. 4.4.8. Other challenges are in the details of the flow near the breakpoint and in the distribution of the undertow, which may or may not overpower the bottom boundary layer streaming.

The suspended load is strongly influenced by wave breaking as indicated by Fig. 4.40, and the breaking induced mixing is highly variable between areas of strong organized upward motion, Figs. 4.7.1 and 4.39, and gently spilling breakers.

Neglecting the unsolved problems with the suspended load, i.e. just trying to model the transport closest to the bed with a Meyer–Peter and Müller type transport formula like (4.4.15)–(4.4.18) is not even sure to succeed. The problem is that wave shape is so variable in the surf zone, Fig. 4.45, and that surf beat is usually present.

Application of a single typical wave shape like the one parameterized by Elfrink *et al.* (2006) may then give quite meaningless sediment transport rates due to the highly nonlinear sediment transport process. Comprehensive calibration with detailed large scale data is needed.

Practical beach erosion models still often use the simple approach of Kriebel and Dean (1985), where the offshore sediment transport rate is assumed proportional to the excess energy dissipation rate per unit volume compared with equilibrium conditions.

4.10.2 *Shoreparallel transport*

For the shore parallel sediment transport $q_{sy}(x)$ in the surf zone, there are now also a number of detailed models of the $u \times c$-integral type mentioned above. They are however still not very successful with respect to modeling the distribution of the sediment transport across the surf zone, cf. Bayram *et al.* (2001). It is of course not expected, that the distribution of the longshore sediment transport can be modeled with confidence, when the distribution of the longshore current still cannot, see e.g. Nielsen (2009), Sec. 2.5.4. Even if the longshore currents could be modeled, there are however many unresolved questions with respect to the sediment entrainment and bedform geometry across the surf zone.

A reasonable practical alternative is still the empirical formulae for the overall transport rate Q_{sy}. For example the so-called CERC formula

$$Q_{sy} = \frac{K}{16(s-1)\sqrt{\gamma_b}}\sqrt{g}H_b^{2.5}\sin 2\alpha_b, \tag{4.10.1}$$

where γ_b is the breaker index H_b/h_b. This formula is remarkable in that, it does not contain the grain size. This is counter-intuitive, but reasonably supported by data within the most common range of grain sizes: $0.18\,\text{mm} < d_{50} < 0.6\,\text{mm}$. For coarser sediment, the transport rate tends to be a decreasing function of grain size as indicated by the data for K in Fig. 4.46, but so far the field data does not define a clear trend.

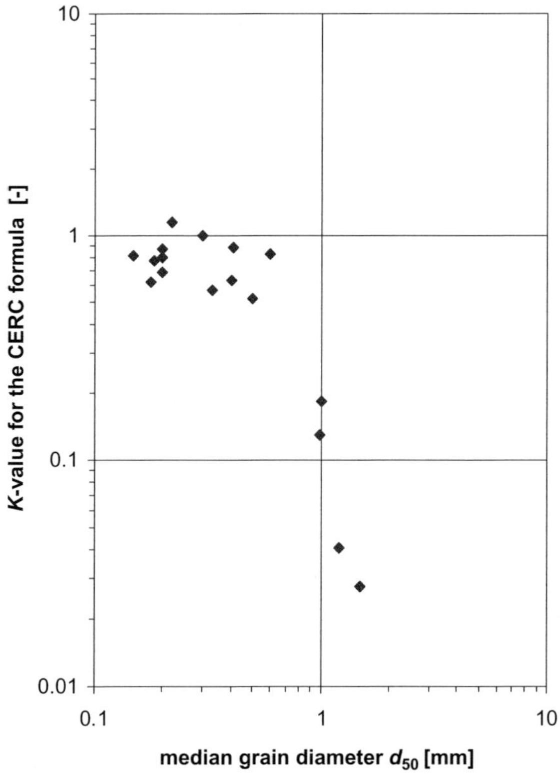

Fig. 4.46 Field experimental values of the K coefficient in the CERC formula. Data from Del Valle *et al.* (1993).

An alternative formula proposed by Kamphuis (1990) on the basis of both laboratory and field data has only very weak grain size dependence:

$$\frac{Q_{\rm sy}}{H^3_{\rm b,rms}/T_{\rm p}} = 2.6 \times 10^{-3} \left(\frac{H_{\rm b,rms}}{L_{\rm o,p}}\right)^{-1.25} m_b^{0.75} \left(\frac{H_{\rm b,rms}}{d_{50}}\right)^{0.25} (\sin 2\alpha_b)^{0.6} \tag{4.10.2}$$

where m_b is the beach slope near the breakpoint. Note that the Kamphuis' formula, which is based on data spanning a larger range of wave heights, has a weaker wave height dependence: $Q_{sy} \sim H^2$ compared with $Q_{sy} \sim H^{2.5}$ for the CERC formula. Given that both formulae have H-exponents near 2, it is acceptable to use $H_{\rm rms}$ as the effective wave height for irregular waves.

4.11. Swash Zone Sediment Transport

Sediment transport in the swash zone is an area of great challenge for present and future research. It is obvious that all the sediments which move onshore and eventually on to the coastal dunes on accreting beaches must be transported through the swash zone, yet we do not understand the details of how it happens. Similarly, we lack understanding of the details of longshore transport in the swash zone although this has long been recognized as one of the major contributing areas together with the area around the breakpoint, see e.g. Bayram *et al.* (2001). The challenges lie in all areas from the overall water motion, Fig. 4.47, through the boundary layer flow structure to the sediment dynamics. A special concern with respect to the swash zone sediment motion is the possible influence of in- or out-flow of water through sand surface, see Sec. 4.3.6. It is also possible, that the horizontal pressure gradients in the bed under a swash bore fronts (cf. Fig. 4.17) are strong enough to cause fluidization, which enhances the sediment transport rate. For a review of recent developments see, e.g. Weir *et al.* (2006) and; Guard and Baldock (2007).

Additional complications are related to surf beat and the associated, occasionally very violent backwash, see Fig. 4.16. These events are undoubtedly very significant contributors to the erosion during storms.

At the very top of the swash zone conditions are qualitatively different depending on whether the slope is continuous or ends at the foot of an erosion scarp, cf. Fig. 7.8 of Nielsen, 2009. The process of scarp retreat on an eroding beach is yet to given a sensible physical description.

4.11.1 *Swash sediment transport and the beach watertable*

The relations between beach watertable heights and beach erosion have been the subject of much research in recent years. It is obvious that if the beach is very "thirsty", some sand will get deposited near the run-up limit because the water drains away around it. It may also be the case that a lower watertable and hence

Fig. 4.47 Swash zone hydrodynamics and the associated sediment transport is an area of vigorous ongoing research. In recent years, advances have been made by Guard and Baldock (2007) on the hydrodynamics by including the incoming momentum from the bore as opposed to the momentum-free start of a dam-break wave. This leads to more realistic depths and flow rates in the swash zone.

relatively more in- than ex-filtration through the beach face makes the beach more resilient to erosion. The forces on sediment particles corresponding to in- and ex-filtration are discussed in Sec. 4.3.6.

The potential for beach stabilization through watertable manipulation has prompted the emergence of several commercial systems with claimed benefit in terms of erosion protection. Turner and Leatherman (1997) reviewed the

Fig. 4.48(a) Beach drains made like 900 mm wide egg cartons wrapped in filter cloth ready for deployment on Dee-Why Beach, Sydney.

performance of pumped dewatering systems deployed in the field. They found scant evidence of positive effects and several pumping systems have been lost during storms on open coasts. The writer's experience with pumped dewatering systems at smaller scales is that spectacular effects can be obtained at small laboratory scales $h \sim 10\,\text{cm}$ with strong pumps, while the effects are difficult to identify experimentally already at $h \sim 50\,\text{cm}$.

Due to the considerable cost of pumping systems and inspired by the fact that the watertable in the beach is higher than the MWS of the inner surf zone, as explained in Sec. 8.2 of Nielsen (2009), a pump free dewatering system was tested at Dee Why Beach near Sydney in 1991, and the results were reported in detail by Davis *et al.* (1992), see Fig. 4.48.

The pump free dewatering system shown above had a clearly measurable effect on the beach watertable, which is visible via the seaward extension of the dry-looking beach face above the drains in Fig. 4.48(d). There were however no significant morphological effects. That is, any morphological effect of the drains was overpowered by the natural signal due to rip system migration etc.

4.12. Concluding Remarks on Sediment Transport Modeling

The state-of-the-art of hydrodynamics and sediment transport modeling is still not progressed to the level where the development of complex topography can be predicted. Such capabilities may be achievable in a few years for simple 2D beach

Fig. 4.48(b) Shorenormal trench for filter at low tide.

Fig. 4.48(c) Outflow of beach groundwater from the bottom end of one of the drains at low tide.

Fig. 4.48(d) The presence of two of the drains is clearly shown by tongues of dryer sand at mid tide.

profiles, but the morphodynamics of 3D sand banks and deltas lies many years ahead.

The most reliable predictive tool is thus still extrapolation of historical data where the main climatic and geological conditions have not changed.

Where these overall defining circumstances are changing: one should use a combination of physics based models and intuition based on the study of morphodynamics in nature and in cyberspace.

References

Bagnold, R. A. (1956): The flow of cohesionless grains in fluids, *Philos. Trans. R. Soc. Lond.* **249**, 235–297.

Bagnold, R. A. (1962): Auto-suspension of transported sediment; turbidity currents, *Proc. R. Soc. Lond. A* **265**, 315–319.

Bagnold, R. A. (1963): Mechanics of marine sedimentation, in: *The Sea*, Vol. 3, ed. N. M. Hill: Interscience, NY, pp. 507–528.

Bailard, J. A. (1981): An energetics total load sediment transport model for a plane sloping beach, *J. Geophys. Res.* **86**(C11), 10938–10954.

Bakker, W. T. and W. G. M. van Kesteren (1986): The dynamics of oscillatory sheet flow, *Proc. 20th Int. Conf. on Coastal Engrg.*, Taipei, ASCE, 940–954.

Baldock, T. E. and P. Holmes (1998): Seepage effects on sediment transport by waves and currents, *Proceedings 26th International Conference on Coastal Engineering*, Copenhagen, ASCE, 3601–3614.

Bascom, W. (1980): *Waves and Beaches*: Anchor Press/Doubleday, New York.

Barry, D. A, S. J. Barry and J-Y Parlange (1996): Capillarity correction to periodic solutions of the shallow flow approximation, in: *Mixing in Estuaries and Coastal Seas, Coastal and Estuarine Studies*, ed. C. B. Pattiaratchi: AGU, Washington DC, pp. 496–510.

Bayram, A, M. Larson, H. C. Miller and N. C. Kraus (2001): Cross shore distribution of longshore sediment transport: Comparison between predictive formulas and field measurements, *Coastal Engrg.* **44**(2), 79–99.

Bijker, E, W. E. van Hijum and P. Vellinga (1976): Sand transport by waves, *Proceedings 15th International Conference on Coastal Engineering*, Honolulu, 1149–1167.

Bosman, J. J. and H. Steetzel (1986): Time and bed averaged sediment concentrations under waves, *Proceedings 20th International Conference on Coastal Engineering*, Taipei, ASCE, 986–1000.

Carstens, M. R, F. M. Neilson and H. D. Altinbilek (1969): *Bedforms generated in the laboratory under an oscillatory flow*, CERC Tech Memo 28.

Chanson, H. (2004): *The Hydraulics of Open Channel Flows: An Introduction*: Butterworth-Heinemann, Oxford, UK, 2nd edn., 630 pp.

Cherlet, J, G. Besio, P. Blondeaux, V. van Lancker, E. Verfaillie and G. Vittori (2007): Modeling sand wave characteristics on the Belgian continental shelf and in the Calais-Dover Strait, *J. Geophys. Res.* **112**(C6), Article Number: C06002.

Clifton, H. E. (1976): Wave formed sedimentary structures — A conceptual model, SEPM Special publication, No 24, 126148.

Coleman, N. (1970): Flume studies of the sediment transfer coefficient, *Water Resour. Res.* **6**(3), 801809.

Conley, D. C. and D. L. Inman (1992): Field observations of the fluid-granular boundary layer flow under near breaking waves, *J. Geophys. Res.* **97**(C6), 9631–9643.

Conley, D. C. and D. L. Inman (1994): Ventilated oscillatory boundary layers, *J. Fluid Mech.* **273**, 261–284.

Conley, D. C, S. Falchetti, I. P. Lohman and M. Brocchini (2008): The effects of flow stratification by non-cohesive sediment on transport in high-energy wave-driven flows, *J. Fluid Mech.* **610**, 43–67.

Dean, R. G. (1991): Equilibrium beach profiles: Principles and applications, *J. Coastal Res.* **7**(1), 53–84.

Del Valle, R, R. Medina and M. A. Losada (1993): Dependence of coefficient on grain size, *J. Waterway Port & Coastal Engrg.* **119**(5), 568–574.

Dibajnia, M. and A. Watanabe (1998): Transport rate under irregular sheet flow conditions, *Coastal Engrg.* **35**, 167–183.

Dick J. E. and J. F. A. Sleath (1991): Velocity and concentrations in oscillatory sheet flow over beds of sediment, *J. Fluid Mech.* **223**, 165–196.

Dick J. E. and J. F. A. Sleath (1991): Sediment transport in oscillatory sheet flow, *J. Geophys. Res.* **97**(C4), 5745–5758.

Dingler, J. R. (1974): *Wave formed ripples in nearshore sands*, PhD thesis, University of California San Diego, 136 pp.

Dixen, M, F. Hatipoglu, B. M. Sumer and J. Fredsøe (2008): Wave boundary layer over a stone covered bed, *Coastal Engrg.* **55**(1), 1–20.

Dohmen-Janssen, C. M. and D. M. Hanes (2002): Sheet flow dynamics under monocromatic non-breaking waves. *J. Geophys. Res.* **107**(C10), 3149, doi:10.1029/2001JC001045.

Dyer, K. R. (1997): *Estuaries: A physical introduction*: Wiley, 195 pp.

Ellis, K. M, C. E. Binding, D. G. Bowers, S. E. Jones and J. H. Simpson (2008): A model of turbidity maximum maintenance in the Irish Sea, *Estuarine, Coastal & Shelf Science* **76**(4), 765–774.

Farachi, C. and E. Foti (2002): Geometry, migration and evolution of small scale bedforms generated by regular and irregular waves, *Coastal Engrg.* **47**, 35–52.

Fredsøe, J. (1984): Turbulent boundary layers in wave current motion, *J. Hydraul. Engrg. ASCE,* **110**(Hy10), 1103–1120.

Fredsøe, J. and R. Deigaard (1992): *Mechanics of Coastal Sediment Transport*: World Scientific, 369 pp.

Fredsøe, J, K. H. Andersen and B. M. Sumer (1999): Wave plus current over a ripple covered bed, *Coastal Engrg.* **38**, 177–221.

Friedman, P. D. and J. Katz (2002): Mean rise rate of droplets in isotropic turbulence, *Phys. Fluids* **14**(9), 3059–3073.

Gallagher, E. L. (2003): A note on megaripples in the surf zone. Evidence for their relation to steady flow dunes. *Marine geology* **193**, 171–176.

Gallagher, E. L, S. Elgar and E. B. Thornton (1998): Megaripple migration in a natural surf zone, *Nature* **394**, 165–168.

Gourlay, M. R. (1968): *Beach and dune erosion tests*, Delft Hydraulics Lab, Report No M935/M936 (unpublished).

Graf, W. H. and M. Cellino (2002): Suspension flows in open channels; experimental study, *J. Hydraul. Res.* **40**(4), 435–447.

Grassmeier, B. T. and L. C. van Rijn (1999): Transport of fine sand by current and waves, Part II: Breaking waves over barred profile with ripples, *J. Waterway, Port, Coastal & Ocean Eng.* **125**(2), 71–79.

Guard, P. A. and T. E. Baldock (2007): The influence of seaward boundary conditions on swash zone hydrodynamics, *Coastal Engrg.* **54**, 321–331.

Hamm, L. and C. Migniot (1994): Elements of cohesive sediment deposition, consolidation and erosion, in: *Coastal Estuarial and Harbour Engineers' Reference Book* (M. B. Abbott and A. Price, eds.), E. Spon and F. N. Spon, pp. 93–106.

Hancock, M. W. and N. Kobayashi (1994): Wave overtopping and sediment transport over dunes, *Proceedings of the 24th International Conference on Coastal Engineering*, Kobe, ASCE, 2028–2042.

Hanes, D. M. (1990): The structure of events of intermittent suspension of sand due to shoaling waves, in: *The Sea*, Vol. 9: John Wiley & Sons, 941–952.

Hanes, D. M. and A. J. Bowen (1985): A granular fluid model for steady, intense bedload transport, *J. Geophys. Res.* **90**(C5), 9149–9158.

Hay, A. E. and D. J. Wilson (1994): Rotary side scan images of nearshore bedform evolution during a storm, *Marine geology* **119**, 57–65.

Ho, H. W. (1964): Fall velocity of a sphere in an oscillating fluid, PhD Thesis, University of Iowa.

Horikawa, K. (1988): *Nearshore dynamics and coastal processes*, University of Tokyo press, 522 pp.

Horikawa, K., A. Watanabe and S. Katori (1982): Sediment transport under sheet flow condition, *Proceedings of the 18th International Conference on Coastal Engineering*, Capetown, ASCE, 1335–1352.

Huppert, H. E, J. S. Turner and M. A. Hallworth (1995): Sedimentation and entrainment in dense layers of suspended particles stirred by an oscillating grid, *J. Fluid Mech.* **289**, 263–293.

Inman, D. L. and R. A. Bagnold (1963): Littoral processes, in: *The Sea*, Vol. 3 (N. M. Hill, ed.), Interscience, NY, pp. 529–553.

Inman, D. L. and A. J. Bowen (1963): Flume experiments on sand transport by waves and currents, *Proceedings of the 8th International Conference on Coastal Engineering*, ASCE, 137–150.

Jensen, B. L., B. M. Sumer and J. Fredsoe (1989): Turbulent oscillatory boundary layers at high Reynolds numbers, *J. Fluid Mech.* **206**, 265–297.

Jobson, H. E. and W. W. Sayre (1970): Vertical transfer in open channel flow, *J. Hydraul. Div.* ASCE, **96**(Hy3), 703–724.

Jonsson, I. G. (1980): A new approach to rough turbulent oscillatory boundary layers, *Ocean Engrg.* **7**, 109–152.

Jonsson, I. G. (1990): Wave-current interactions. *The Sea*, Vol. 9 (eds. B. Le Mehaute and D. M. Hanes): Wiley, New York.

Justesen, P. (1988): *Turbulent wave boundary layers*, Series Paper 43, ISVA, TU Denmark, 226 pp.

Kajiura, K. (1968): A model of the bottom boundary layer in water waves, *Bull. Earthq. Res. Inst.* **46**, 75–123.

Kamphuis, J. W. (1990): Littoral sediment transport rate, *Proceedings of the 22nd International Conference on Coastal Engineering*, Delft, ASCE, 2402–2415.

Kamphuis, J. W. (1991): Alongshore sediment transport rate distribution, *Proceedings Coastal Sediments* '91, ASCE, 170–183.

Kaczmarek, L. M. and R. Ostrowski (2002): Modeling intensive near bed sand transport under wave-current flow versus laboratory and field data, *Coastal Engrg.* **45**(1), 1–18.

Khezri, N. and H. Chanson (2012): Inception of bed load motion beneath a bore, *Geomorphology* **153–154**, 39–47.

King, D. B. (1991): *Studies in oscillatory flow bedload sediment transport*, PhD Thesis, University of California, San Diego, 184 pp.

Klopman, G. (1994): Vertical structure of the flow due to waves and currents, Laser Doppler flow measurements for waves following or opposing a current. Tech Rep H840.32 Part 2, Delft Hydraulics.

Komar, P. D. (1998): *Beach Processes and Sedimentation*: Prentice Hall.

Kraus, N. C, M. Larson and D. L. Kriebel (1991): Evaluation of beach erosion and accretion predictors, *Proc. Coastal Sediments* '91, ASCE, 572–587.

Kriebel, D. L. and R. G. Dean (1985): Numerical simulation of time-dependent beach profile response, *Coastal Engrg.* **9**(N 3), 221–245.

Landau, L. D. and E. M. Lifschitz (1987): *Course of Theoretical Physics*, Vol. 6: Fluid Mechanics, 2nd edn., Butterworth-Heineman, 539 pp.

Liu, P. F-L. (2006): Turbulent boundary layer effects on transient wave propagation in shallow water, *Proc. R Soc. Lond. A* **462**, 3481–3491.

Liu, P. F-L, Y. S. Park and E. A. Cowen (2007): Boundary layer flow and bed shear stress under a solitary wave, *J. Fluid Mech.* **574**, 449–463.

Löfquist, K. E. B. (1978): Sand ripple growth in an oscillatory-flow tunnel, Tech Paper No 78-5, US Army Corps of Engineers Coastal Engrg. research center, 101 pp.

Löfquist, K. E. B. (1986): Drag on naturally rippled beds under oscillatory flows, Misc Paper CERC-86-13, US Army Corps of Engineers, 121 pp.

Longuet-Higgins, M. S. (1953): Mass transport in water waves, *Philos. Trans. R Soc. Lond. A* **245**, 535–581.

Longuet-Higgins, M. S. (1956): The mechanics of the boundary layer near the bottom in a progressive wave, Proceedings of the 6th International Conference on Coastal Engineering, Miami, ASCE, 184–193.

Marieu, V., P. Bonneton, D. L. Foster and F. Ardhuin (2008): Modelling of vortex ripple morphodynamics, *J. Geophys. Res.* **113**, C09007, doi: 10:1029/2007JC004659.

Martin, C. S. (1970): Effect of a porous sand bed on incipient sediment motion, *Water Resour. Res.* **6**(4), 1162–1174.

Masselink, G. and Hughes, M. G. (2003): *Introduction to Coastal Processes and Geomorphology*: Edward Arnold, 354 pp.

Maxey, M. R. (1990): On the advection of spherical and non-spherical particles in a non-uniform flow, *Philos. Trans. R Soc. Lond.* **333**, 289–307.

Maxey, M. R. and S. Corrsin (1986): Gravitational settling of aerosol particles in randomly orientated circular flow fields, *J. Atmospherical Sci.* **43**, 1112–1134.

Murray, S. P. (1970): Settling velocity and vertical diffusion of particles in turbulent water, *J. Geophys. Res.* **75**(9), 1647–1654.

Nakato, T., F. A. Locher, J. R. Glover, and J. F. Kennedy (1977): Wave entrainment of sediment from rippled beds, *Proc. ASCE*, **03**(WW1), 100–183.

Nezu, I. and H. Nakagawa (1993): Turbulence in open-channel flows, IAHR Monograph, Balkema, 281 pp.

Nielsen, P. (1979): *Some basic concepts of wave sediment transport*, PhD Thesis, Series Paper 20, ISVA, Tech University of Denmark, 160 pp.

Nielsen, P. (1983): Entrainment and distribution of different sand sizes under water waves, *J. Sed. Pet.* **53**(2), 423–428.

Nielsen, P. (1984): Field measurements of time averaged suspended sediment concentrations, *Coastal Engrg.* **8**, 51–72.

Nielsen, P. (1988): Three simple models of wave sediment transport, *Coastal Engrg.* **12**, 43–62.

Nielsen, P. (1991): Combined convection and diffusion: A new framework for suspended sediment modelling, *Proc. Coastal sediments'*91, ASCE, 418–431.

Nielsen, P. (1992): *Coastal bottom boundary layers and sediment transport*: World Scientific, Singapore, 324 pp.

Nielsen, P. (1993): Turbulence effects of the settling of suspended particles, *J. Sed. Pet.* **63**(5), 835–838.

Nielsen, P. (1998): Coastal groundwater dynamics, *Proc. Coastal Dynamics* '97, Plymouth, ASCE, New York, 546–555.

Nielsen, P. (2002): Shear stress and sediment transport calculations for swash zone modeling, *Coastal Engrg.* **45**(1), 53–60.

Nielsen, P. (2006): Sheet flow sediment transport under waves with acceleration skewness and boundary layer streaming, *Coastal Engrg.* **53**(9), 749–758.

Nielsen P. (2009): *Coastal and Estuarine Processes*: World Scientific, 341 pp., ISBN 978-981-283-711-0 or 978-981-283-712-7(pbk).

Nielsen, P. and You Z-J. (1996): Eulerian mean velocities under non-breaking waves over horizontal bottoms, *Proceedings of the 25th International Conference on Coastal Engineering*, Orlando, ASCE, 4066–4078.

Nielsen, P., S. Robert, B. Moeller-Christiansen and P. Oliva (2001): Infiltration effects on sediment mobility under waves, *Coastal Engrg.* **42**(2), 105–114.

Nielsen, P., K. U. van der Wal and L. Gillan (2002): Vertical fluxes of sediment in oscillatory sheet-flow, *Coastal Engrg.* **45**(1), 61–68.

Nielsen P. and D. P. Callaghan (2003): Shear stress and sediment transport calculations for sheet flow under waves, *Coastal Engrg.* **47**(3), 347–354.

Nielsen, P. and I. A. L. Teakle (2004): Finite mixing length modelling of the diffusion of fluid momentum and particles in turbulence, *Phys. Fluids* **16**(7), 2342–2348.

Nielsen, P. and P. A. Guard (2011): Vertical scale and shear stresses in wave boundary layers over movable beds, *Proceedings of the 32nd International Conference on Coastal Engineering*, Shanghai China 2010. Retrievable from http://journals.tdl.org/ICCE/.

O'Donoghue, T., M. Li, J. Malarkey, S. Pan, A. G. Davies and B. A. O'Connor (2004): Numerical and experimental study of wave generated sheet flow, *Proceedings of the 29th International Conference on Coastal Engineering*, Lisbon, World Scientific, 1690–1702.

Open University (1999): *Waves, Tides and Shallow-water Processes*, 2nd edn.: Butterworth-Heineman, 227 pp.

Ribberink J. S. and Z. Chen (1993): Sediment transport of fine sand under asymmetric oscillatory flow, Delft Hydrualics Report H840.20, part VII.

Ribberink, J. S. and A. A. Al-Salem (1994): Sediment transport in oscillatory boundary layers in cases of rippled beds and sheet flow, *J. Geophys. Res.* **99**(C6), 12707–12727.

Ribberink, J. S, I. Katapodi, K. A. H. Ramadan, R. Koelewijn and S. Longo (1994): Sediment transport under non-linear waves and currents, *Proceedings* of the 24th*International Conference on Coastal Engineering*, Kobe ASCE, 2527–2541.

Ribberink, J. S, C. M. Dohmen-Janssen, D. M. Hanes, S. R. McLean and C. Vincent (2000): Near bed sand transport mechanisms under waves, *Proceedings of the 27th International Conference on Coastal Engineering*, Sydney, ASCE, New York, 3263–3276.

Richardson, J. F. and W. N. Zaki (1954): Sedimentation and fluidization, Part 1, *Trans. Inst. Chem. Engrg.* **32**, 35–53.

Riedel, H. P. (1972): Direct measurement of bed shear stress under waves, PhD Thesis, Queens University, Kingston Ontario, 142 pp.

Rouse, H. (1937): Modern conceptions of the mechanics of turbulence, *Trans. ASCE*, **102**, 463–543.

Rubin, D. M. and R. M. Hunter (1987): Bedform alignment in directionally varying flows, *Science* **237**, 276–278.

Rutherford, J. C. (1994): *River mixing*: John Wiley, Chichester, USA, 347 pp.

Sato, S. (1986): Oscillatory boundary layer flow and sand movement over ripples, PhD thesis, University of Tokyo.

Sato, S, M. B. Kabiling and H. Suzuki (1992): Prediction of near bottom velocity history by a non-linear dispersive wave model, *Coastal Engrg. Japan* **35**(1), 67–82.

Sato, S. (1996): Numerical simulation of 1993 southwest Hokkaido earthquake tsunami around Okushiri Island, *J. Waterway, Port and Coastal Engrg.* ASCE, **122**(5), 209–215.

Sawamoto, M. and T. Yamashita (1986): Sediment transport rate due to wave action, *J. Hydrosci. Hydraul. Engrg.* **4**(1), 1–15.

She, K, L. Trim and D. Pope (2005): Fall velocities of natural sediment particles: as simple mathematical presentation of the fall velocity law, *J. Hydraul. Res.* **43**(2), 189–195.

Sherman, D. J. and E. J. Farrell (2008): Aerodynamic roughness lengths over movable beds: Comparison of wind tunnel and field data, *J. Geophys. Res.* **113**, F02S08, doi:10.1029/2007/JF000784.

Shields, A. (1936): Anwendung der Ähnlichkeitsmechanik und Turbulenzforschung auf die.g.eschiebe-bewegung, *Mitt Preuss Versuchsanstalt für Wasserbau und Schiffbau*, No 26, Berlin.

Sistermans, P. G. J. (2002): Graded sediment transport by non-breaking waves and a current, PhD thesis, Delft University Tech., 205 pp.

Sleath J. F. A. (1970): Measurements close to the bed in a wave tank, *J. Fluid Mech.* **42**, 111–123.

Sleath, J. F. A. (1984): *Sea Bed Mechanics*: Wiley, 335 pp.

Sleath, J. F. A. (1987): Turbulent oscillatory flow over rough beds, *J. Fluid Mech.* **182**, 369–409.

Sleath, J. F. A. (1999): Conditions for plug formation in oscillatory flow, *Cont. Shelf Res.* **19**, 1643–1664.

Snyder, W. H. and J. L. Lumley (1971): Some measurements of particle velocity autocorrelation functions in a turbulent flow, *J. Fluid Mech.* **48**, 41–71.

Soulsby, R. L. (1997): *Dynamics of Marine Sands: A Manual for Practical Applications*: Thomas Telford, London.

Sumer, B. M., A. Kozakiewicz, J. Fredsoe and R. Deigaard (1999): Velocity and concentration profiles in sheet flow layer of mobile bed, *J. Hydraul. Engrg.* **122**(10), 549–558.

Teakle, I. A. L. and P. Nielsen (2004): Modelling suspended sediment profiles under waves: A finite mixing length theory, *Proceedings of the 29th International Conference on Coastal Engineering*, Lisbon, World Scientific, 1780–1792.

Teisson, C., M. Ockenden, P. Le Hir, C. Kranenburg and L. Hamm (1993): Cohesive sediment transport processes, *Coastal Engrg.* **21**, 126–162.

Testic, F. Y., S. I. Voropayev and H. J. S. Fernando (2005): Adjustment of sand ripples under changing water waves, *Phys. Fluids* **17**, 072104.

Terrile, E., A. J. H. M. Reniers, M. J. F. Stive, M. Tromp and J. Verhagen (2006): Incipient motion of coarse particles under regular shoaling waves, *Coastal Engrg.* **53**(1), 81–92.

Tomkins, M. R., P. Nielsen and M. G. Hughes (2003): Selective entrainment of sediment graded by size and density under waves, *J. Sedimentary Res.* **73**(6), 906–911.

Tooby, P. F., G. L. Wick and J. D. Isacs (1977): The motion of a small sphere in a rotating velocity field: A possible mechanism for suspending particles in turbulence, *J. Geophys. Res.* **82**(15C), 2096–2100.

Thorne, P. D., A. G. Davies and P. S. Bell (2009): Observation and analysis of sediment diffusivity profiles over sandy rippled beds under waves, *J. Geophys. Res.* **114**, C02023, doi:10.1029/2008JC0044944.

Traikovsky, P., D. E. Hay, J. D. Irish and J. F. Lynch (1999): Geometry, migration and evolution of wave orbital ripples at LEO-15, *J. Geophys. Res.* **104**(C1), 1505–1524.

Traikovsky, P. (2007): Observations of wave orbital scale ripples and a non-equilibrium time dependent model, *J. Geophys. Res.* **112**, C06026.

Trowbridge J. and O. S. Madsen (1984): Turbulent wave boundary layers: 1 Model formulation and first order solution, *J. Geophys. Res.* **89**(C5), 7989–7997.

Van Doorn, Th. (1982): experimenteel onderzoek naar het snelheidsveld in de turbulente bodemgrenslaagin een oscillernde stroming in een golftunnel, Delft Hydraulics Laboratory, Report M. 1562 1a.

Van Rijn, L. C. (1984a): Sediment pickup functions, *J. Hydraul. Engrg.* **110**(10), ASCE, 1494–1502.

Van Rijn, L. C. (1984b): Sediment transport Part II: Suspended load transport, *J. Hydraul. Engrg.* **110**(10), ASCE, 1613–1641.

Van Rijn, L. C. and F. J. Havinga (1995): Transport of fine sands by current and waves, Part II, *J. Waterway Port Coastal and Ocean Engrg.* **121**(2), 123–133.

van Rijn L. C. (2007): Unified view of sediment transport by currents and waves. I: Initiation of motion, bed roughness, and bed-load transport. *J. Hydraul. Engrg.* ASCE, **133**(6), 649–667.

van Rijn L. C. (2007): Unified view of sediment transport by currents and waves. II: Suspended transport. *J. Hydraul. Engrg.* ASCE, **133**(6), 668–689.

van Rijn L. C. (2007): Unified view of sediment transport by currents and waves. III: Graded beds, *J. Hydraul. Engrg.* ASCE, **133**(7), 761–775.

van Rijn L. C., Walstra D. J. R., van Ormondt M. (2007): Unified view of sediment transport by currents and waves, IV: Application of morphodynamic model, *J. Hydraul. Engrg.* ASCE, **133**(7), 776–793.

van Rijn L. C. (2009): Dune erosion due to storms, *Coastal Engrg.* **56**(4), 441–457.

Villaret, C. and G. Perrier (1992): Transport of fine sand by combined waves and current: An experimental study, Technical Note, Electricite de France, 81 pp.

Watanabe, A. and S. Sato (2004): A sheet flow transport rate formula asymmetric forward leaning waves and currents, *Proceedings of the 29th International Conference on Coastal Engineering*, Lisbon, World Scientific, 1703–1714.

Winterwerp, J. C. (1998): A simple model for turbulence induced flocculation of cohesive sediments, *J. Hydraul. Res.* **36**(3), 309–326.

Winterwerp, J. C. and W. G. M. van Kesteren (2004): *Introduction to the Physics of Cohesive Sediment Dynamics in the Marine Environment, Developments in Sedimentology*, Vol. **56:** Elsevier, 466 pp, ISBN 0444515534.

Zala Flores, N. and J. F. A. Sleath (1998): Mobile layer in oscillatory sheet flow, *J. Geophys. Res.* **103**(C6), 783–793.

Chapter 5

Morphology of Erosional and Accretionary Coasts

Paul D. Komar
Professor Emeritus
Oregon State University, USA

Tsuguo Sunamura
Professor Emeritus
Osaka University, Japan

1. Introduction

Some coasts are predominantly erosional, where waves break directly against rocky shores or where eroding cliffs are located immediately landward from a beach. Other coasts are dominated by accretion, where the long-term accumulation of sand or gravel in the beach has prevailed over periodic episodes of erosion during extreme storms. This distinction in coastal type is in large part controlled by the tectonic setting, recognized in the classification developed by Inman and Nordstrom (1971) that places the coast within the framework of global plate tectonics. Their classification recognizes that on tectonically active convergent margins, Japan and the Northwest coast of the United States (US) being prime examples, vertical land movements and mountain building have tended to produce erosional rocky coasts. In contrast, coasts on passive or divergent margins like the east coast of the US, have received large quantities of sediments from rivers so that extensive beaches have developed. Of course, wherever one finds a beach, no matter the tectonic setting, one is locally dealing with an accretional environment.

Somewhat different processes operate on these two types of coasts, giving rise to distinctive morphologies and different types of problems in the form of impacts on people living within the coastal zone. On the erosional convergent-margin coast, the problem generally is one of waves undercutting the rocks or sea cliff upon which houses have been constructed. At times, erosion of the sea cliff induces the occurrence of a landslide that leads to major losses of property. Problems also exist on accretionary coasts, mainly during extreme storms that episodically cut back the beach and wash away homes, or where a storm surge washes over a barrier island.

Research has led to a better understanding of the processes, important for both erosional and accretionary coasts. The objective of this chapter is to summarize our understanding of those processes, and how they yield the morphologies of these distinctive coastal types.

2. Sea Cliff Erosion — Processes and Morphology

2.1 *Episodic and localized nature of cliff recession*

The multiple factors that control sea-cliff recession and the relationships between the factors are diagrammed in Fig. 5.1. Waves erode the toe of a cliff, which increases the overall slope angle and stress, augmenting cliff instability. At the culmination mass movement occurs, which results in cliff recession and major morphological change. The debris supplied from mass movement usually forms talus at the cliff base. The talus deposit, which is transported in time alongshore and/or offshore by the action of waves and currents, constitutes a part of the beach sediment that otherwise has been derived from nearby rivers, eroding beaches and/or cliffs. Beach morphology and nearshore bottom topography control the sizes of waves that reach the toe of the cliff, while the water level influences the elevation of maximal wave attack.

Cliff recession distances have been documented using a variety of techniques (e.g., Sunamura (1992, p. 88)), of which the most common is to measure distances on sequential aerial photographs and historic maps. Average recession rates are determined by dividing the measured distances by the length of the time interval.

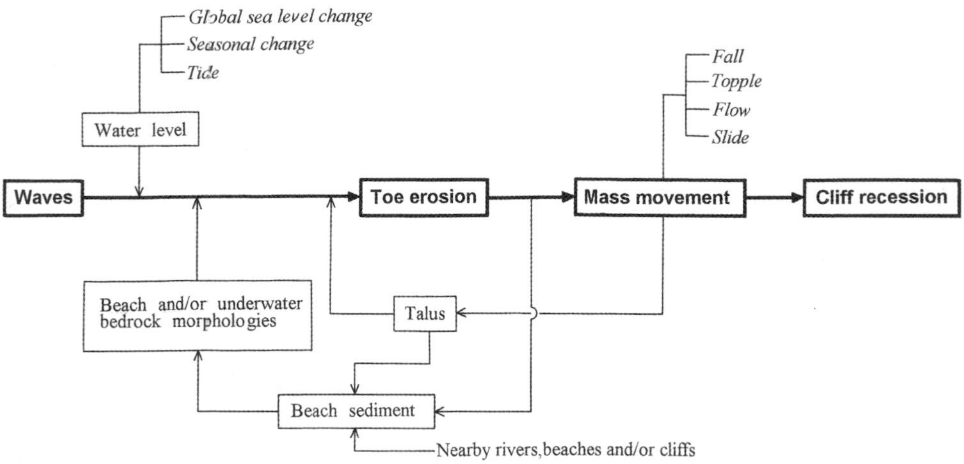

Fig. 5.1 The interaction of factors that affect the toe erosion of sea cliffs [modified from Sunamura (1983)].

Long-term average rates summarized on a lithological basis are on the order of: 10^{-3} m/year for granitic rocks; 10^{-3} to 10^{-2} m/year for limestone; 10^{-2} m/year for flysch and shale; 10^{-1} to 10^{0} m/year for chalk and Tertiary sedimentary rocks; 10^{0} to 10^{1} m/year for Quaternary deposits; and 10^{1} m/year for unconsolidated volcanic ejecta (Sunamura, 1983): Thus, lithology and cohesiveness of the cliff material greatly control the rates of cliff recession. It should be noted, however, that toe erosion by waves originally causes cliff recession. The importance of waves as a destructive agent emerges when considering an erosional event during a storm. Such a short-term event makes the long-term average rate highly unreliable.

On the north shore of Long Island, New York, a hurricane on 14 September 1944 cut back a bluff composed of glacial deposits by a horizontal distance of over 12 m in a single day (Davies *et al.*, 1972): This retreat provides a striking contrast to the 80 year average which is 0.5 m/year for this coast (Bokuniewiez and Tanski, 1980): Similar examples can be found for locations along the California coast (Griggs and Savoy, 1985): During storms in January 1983, waves removed about 14 m of the bluff top in one section of Miocene mudstone-siltstone cliffs at Santa Cruz, which had receded at an average rate of 0.2 m/year from 1931 to 1982. Near Capitola, 10 km east of Santa Cruz, a sea cliff having the same geology was cut back nearly up to 1.5 m to 3 m over night by a storm, while the long-term average recession rate including this site is only 0.3 m/year. The dotted, stepped line in Fig. 5.2 shows the cumulative erosion distances at the toe of a cliff cut into glacial deposits along the Ohio shore of Lake Erie, based on 5 years of accurate measurements by Carter and Guy (1988): Yearly-based average erosion rates vary considerably, ranging from 0 m/year for 1980 to 2.5 m/year for 1976. It is found that a 5 year average, 0.8 m/year, masks these fluctuations resulting from the individual erosion events represented by the stepped line.

The upper most curve in Fig. 5.3 gives recession rates during 16 years (1966 to 1982) along a 300 m stretch of till cliff at Easington, Holderness, UK, while the remaining diagrams plot the patterns of erosion for four sequential sub-periods (Richards and Lorriman, 1987): Aside from a considerable variation in space-averaged erosion rates among the four sub-periods, a highly localized recession pattern is seen in this coastal stretch, especially at Sites A and B (Fig. 5.3) which are only 40 m apart. Site A suffered little erosion during the first two sub-periods, 1966–1970 and 1970–1974, but receded dramatically during the subsequent period, 1974–1979, with an average rate of 4 m/year. On the other hand, Site B was cut back at 4 m/year during the first period (1966–1970), with no marked recession in the following periods. In the end, both sites had approximately the same recession rate when averaged over the full term of 16 years (top diagram): Figure 5.3 shows that not only at the two sites, but also in the remaining portion of the coast, the degree of variability in erosion rates becomes smaller as the time scale is longer, as suggested first by Cambers (1976).

The erosion-rate variability can be examined by applying the variation coefficient σ/R, in which R is the space-averaged value of erosion rates and σ is the standard

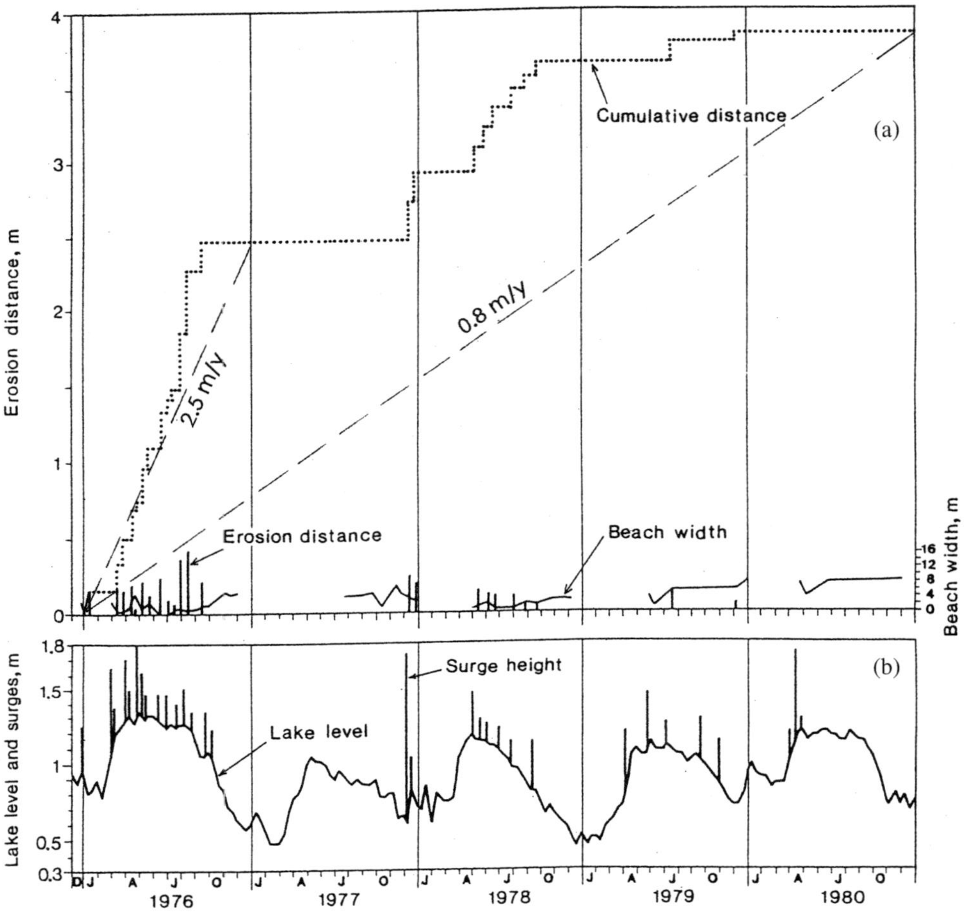

Fig. 5.2 (a) Erosion events affecting a bluff cut into glacial deposits at the Helen Drive site on the Ohio shore of Lake Erie, and variations in beach width at the bluff toe. (b) Lake-level changes and storm-surge heights, factors affecting the bluff erosion. Original data in (a) and (b) are from Carter and Guy (1988): The lengths of vertical lines in (a) and (b) indicate the magnitudes of erosion events and storm surges. The dotted line in (a) represents the cumulative erosion distance [from Sunamura (1992)].

deviation of R. The ratio σ/R decreases as the variability decreases. Figure 5.4 is a plot of σ/R values against the length of the averaging interval, using data from Easington (Fig. 5.3) together with data from two receding cliffs (Byobugaura and Okuma) composed of Tertiary sedimentary rocks on the Pacific coast of Japan (Sunamura, 1973): The value of σ/R decreases with increasing time for the three cliffs, although the σ/R-decreasing trend is different for the three sites. In other words, the spatial variation in erosion rates becomes smaller as the interval of time involved in the averaging increases. This suggests that a spot left uneroded will recede sooner or later.

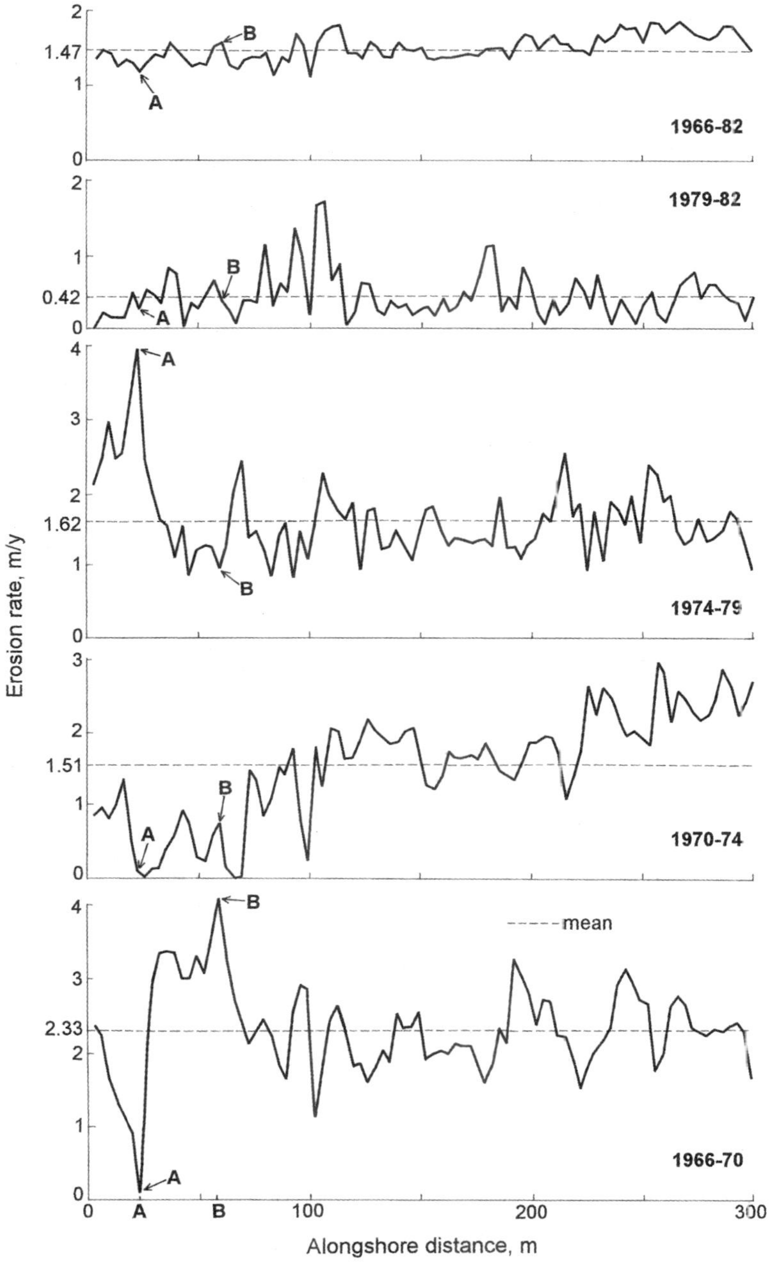

Fig. 5.3 Alongshore variations for different time intervals of recession rates along the top of a till cliff at Easington, Holderness, on the east coast of England [modified from Richards and Lorriman (1987)].

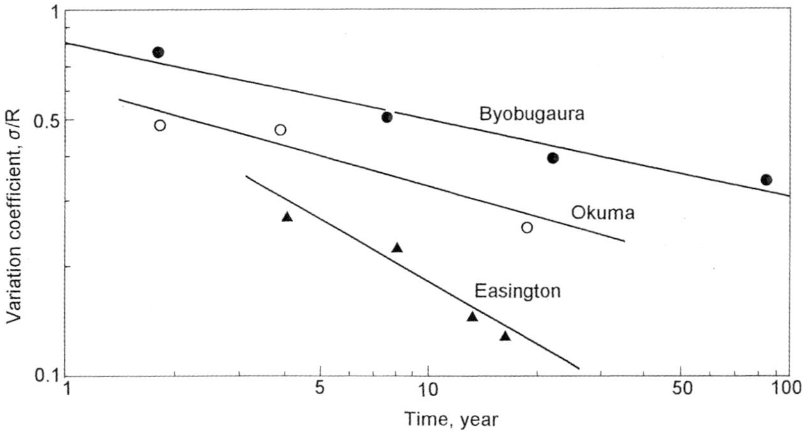

Fig. 5.4 The variation coefficient of the recession rate, σ/R, plotted against the length of the averaging interval for three sea-cliff sites [from Sunamura (2004)].

2.2 *Mechanisms of toe erosion*

The assailing force of waves reaching the toe of a cliff exerts two kinds of action on the cliff face — hydraulic and mechanical. The hydraulic force consists of compression, tension and shearing. On some occasions it also includes an explosive force that may be produced when air is trapped within a joint or fault-associated opening in a cliff. Yamamoto *et al.* (1990) were the first to measure wave pressures inside a vertical, rectangular crevice having a wedge-shaped cross-section, formed in a vertical cliff contained within a wave flume. Pressure gauges mounted on the side wall and ceiling of the crevice recorded considerably higher values than measured pressures acting on the cliff face outside the crevice. It was found that the pressure increase within the crevice depended on its size and shape, and on the incident wave height [cited in Sunamura (1994, Fig. 11)]. The rise in pressure exerts an excess outward force on the side wall of the opening, as well as on its roof. This may facilitate the removal of blocks bound by joints and faults, a process known as "quarrying" or "plucking". The process is closely associated with the propagation of wave pressure into cracks and fissures (Wolters and Müller, 2004). The mechanical action arises when waves are armed with beach sediment, and consists of abrasive and impact forces. These various assailing forces occur almost simultaneously when waves strike the cliff face. No studies have been conducted to individually measure one of these forces in the field because of the complexity of the phenomena and the difficulty of obtaining measurements. Recently, the first attempt to measure impact forces of waves armed with pebbles was made by Williams and Roberts (1995) on the macrotidal coast of South Wales. They used a specially designed instrument to provide measurements of the momentum of pebbles. Even in the controlled conditions of the laboratory, few attempts have been made to measure wave forces which are important in cliff erosion.

The resisting force of rocks forming a cliff is determined principally by their mechanical strength, which is closely associated with lithology (rock types); this can be evaluated in terms of compressive, tensile and shear strengths or wear resistance. These strengths have been tested in cliff-erosion studies [for details, see Sunamura (1992, p. 79)]. The resistance of cliff material against the assailing force of waves is considered to be different according to the type of action involved in the force. The most suitable parameter for expressing the rock resisting force has not been fully investigated, but compressive strength has been widely used as a surrogate for resistance (Sunamura, 1992, p. 65 and 79). The next problem is how to relate the strength reduction due to discontinuities such as cracks, joints, faults and bedding planes to the overall resistance of the cliff. In his study of shore platforms, Tsujimoto (1987) attempted to evaluate this effect by relating the compressive strength to V_f/V_ℓ, where V_f and V_ℓ are the sonic velocities measured, respectively, in the field on a rock mass having discontinuities and in the laboratory with an intact rock specimen. Compressive strength converted from Schmidt hammer readings (references cited in Goudie, 2006) has also been used to express rock resistance to wave attack (Budetta *et al.* 2000; Stephenson and Kirk, 2000). If the cliff material is prone to weathering, strength reduction is brought about by (1) deterioration of cementing material between minerals, and (2) development and extension of cracks and joints, both induced by alternate wet-dry weathering (e.g., Emery, 1960; Carter and Guy, 1988), frost action (e.g., Carter and Guy, 1988), salt crystallization (e.g., Johannessen *et al.*, 1982; Mottershead, 1989 and 1994; Takahashi *et al.*, 1994), and the combined action of salt and frost weathering (Robinson and Jerwood, 1987). Mottershead (1994) attempted to evaluate weathering-induced deterioration through strength testings on Lower Cretaceous sandstone used in sea walls at Weston-Super-Mare, Avon, UK. The results indicated that the strength of weathered (honeycombed) rock is approximately 80% that of fresh rock. Sunamura (1996) indicated that weathering-induced reduction in rock strength can be expressed in terms of an exponential decay function of time. To access the strength of weathered part of rocks, non-destructive testing devices such as a needle-type penetrometer (e.g. Suzuki and Hachinohe, 1995; Hachinohe *et al.*, 1999) and the Equotip hardness tester (e.g. Aoki and Matsukura, 2007; Viles *et al.*, 2011) have been used in geomorphological studies. Chemical weathering is also responsible for strength reduction for some rock types such as limestone (e.g., Trudgill, 1976). On calcareous rocky coasts, biological agencies may produce two contrasting effects on mechanical strength [for a review, see Trudgill (1985, p. 136–155) and Trenhaile (1987, p. 64–82)]. In some places rock-boring or rock-snapping organisms reduce the strength due to their many burrows, whereas in other locations encrusting organisms give rise to an increase in strength. Healy (1968) measured compressive strengths of sandstone and siltstone riddled by boring mollusks, and found that these rocks have only 20%–25% of the strength of the solid (intact) rocks.

Erosion occurs when the assailing force of waves is greater than the resisting force of cliffs. A wave-flume experiment using a model cliff composed of a mixture of cement, sand and water (Sunamura, 1977) indicated that a fundamental relationship governing cliff erosion may be expressed by

$$\frac{dX}{dt} = k\left[C + \ln\left(\frac{\rho g H}{S_c}\right)\right], \tag{5.1}$$

where X is the erosion distance, t is time, H is the wave height in front of the cliff, S_c is the compressive strength of the cliff material, ρ is the density of water, g is the acceleration due to gravity, C is a dimensionless constant, and k is a constant with dimensions LT^{-1}. In this relationship the assailing force is represented by $\rho g H$, and the resisting force by S_c. A critical wave height, H_c, necessary to cause toe erosion, can be derived by setting $\frac{dx}{dt} = 0$ in the above equation, yielding

$$H_c = \frac{S_c e^{-c}}{\rho g}. \tag{5.2}$$

Application of Eqs. (5.1) and (5.2) was made to Byobugaura and Okuma cliffs, Japan, both being found in severely eroding coasts in micro-tidal environments, with little alongshore variation in wave regime and rock strength (Sunamura, 1982a). The Byobugaura cliff is 10–60 m high and 9 km long, is composed of Pliocene mudstone, and has no beach development at its base due to prevailing longshore currents. The Okuma cliff, composed of interbedded Pliocene mudstone and sandstone layers, 30–35 m high and 1 km long, is fronted by a narrow, sand beach. The former has a compressive strength of $14.9\,\mathrm{kg/cm^2}$, and the latter $26.4\,\mathrm{kg/cm^2}$. Using data of cliff strength, wave climate and erosion rates over two time intervals, the constants k and C were determined and H_c was calculated. The result showed that the larger but rarer storm-wave events are solely responsible for cliff erosion at both sites, the smaller but more frequent waves having produced no erosion. Since only those waves exceeding a threshold value [Eq. (5.2)] can cause erosion, the short-term average erosion rate should be related to the frequency of occurrence of waves higher than H_c.

As compared with Byobugaura, the Okuma cliff with larger S_c showed smaller H_c. One would have anticipated that with the larger cliff strength, Okuma would have had a larger critical wave height needed to initiate erosion, but the opposite result was found. This was explained by considering the difference in the assailing force of waves between the two sites. Calculations showed that for the same wave height, the assailing force acting on the Okuma cliff with a beach was 1.8 times greater than on the Byobugaura cliff where there is no beach, suggesting that beach sand at the foot of the cliff acts as an abrasive to intensify the assailing force. Such an abrasive effect has been measured directly by Robinson (1977) using a micro-erosion meter; erosion rates of Upper Lias shale cliffs at Whitby on the northeast Yorkshire coast of England were found to be 15 to 18.5 times higher where there is a beach.

In order to examine cliff erosion processes associated with beach sediment, Sunamura (1982b) performed a wave flume experiment by systematically changing the amount of beach sand and the size of the beach at the foot of a steep, model cliff. No erosion took place when there was no beach, indicating the ineffectiveness of hydraulic action alone. Erosion increased with increasing size of the beach, and it attained a maximum when the beach top abutting the cliff was located at about the still-water level. With a further increase in beach size, the erosion rate dropped and finally became null when the beach top was placed at the wave runup limit. Erosion took place when swash mobilized the beach material at the cliff-beach junction, and produced a vortex rotating around a horizontal axis. Sand grains entrapped within this vortex scraped against the cliff material, working as an abrasive tool.

Along cliffed coasts composed of glacial deposits, locations with a lower beach elevation have a higher erosion rate; this is indicated by the work of McGreal (1979) on the Kilkeel coast in Northern Ireland, Pringle's (1985) investigation on the Holderness coast in England, and the study of Carter and Guy (1988) on the Lake Erie shore of Ohio. Furthermore, locations with a smaller volume of beach material have a higher erosion rate, as suggested by Jones and Williams (1991) on the west coast of Wales. These findings indicate that the beach works as a wave-energy dissipater, and no significant mechanical action induced by the beach sediment is likely to be involved in the erosion, results that are quite different from the field and laboratory findings described above. Hydraulic action alone is sufficient to produce erosion in very soft cliff materials such as glacial deposits. Efficacy of mechanical action becomes more notable where the cliff material is harder. Thus, the relationship between the erosion rate and beach elevation varies with the strength of the cliff material (Sunamura, 1992, Fig. 6.19).

2.3 *Factors controlling toe erosion*

Toe erosion is obviously dependent on the wave-energy level in the offshore. However, other factors that control erosion at the cliff base include: (1) the water level; (2) the beach fronting a cliff or nearshore bottom topography; and (3) the presence or absence of talus. The water-level factor affects the energy level of the waves rushing at a cliff, and includes in order of decreasing periodicity in their fluctuations: (1) global eustatic sea-level change; (2) seasonal variations in sea levels; and (3) tides including storm surges. When considering sea-cliff erosion in tectonically active areas such as the circum-Pacific, crustal movement is of importance in addition to eustatic sea-level change; the combined effect is the relative sea-level change as measured on tide gauges. The influence of relative sea-level change on cliff erosion has been documented along the coast of Oregon in the Pacific Northwest of the U.S. (Komar and Shih, 1993). A marked spatial variability in cliff erosion has been produced on this 500 km long coast having a north–south orientation. The tectonic movement is associated with the subduction of oceanic plates beneath the continental North American plate. An earthquake-induced subsidence occurred in the year 1700, and has been followed by a gradual aseismic uplift. Along the southern half of the Oregon coast and near the Oregon-Washington border in the north, this tectonic uplift has exceeded the global eustatic rise in sea level during the past century, resulting in a drop in relative sea-level that has largely eliminated sea-cliff erosion, although evidence remains for the large-scale erosion that followed the abrupt subsidence during the subduction earthquake 300 years ago. Along the north-central Oregon coast, the present rise in eustatic sea-level exceeds the tectonic uplift, and the relative sea-level rise continues to be a factor in sea-cliff erosion, although the erosion rates are highly variable due to site-specific factors (Komar *et al.* 2007).

The seasonal occurrence of variations in monthly mean sea-levels is due mainly to changes in water temperatures and densities, with additional contributions due to the geostrophic effects of offshore currents (Komar and Enfield, 1987). At most locations, mean water levels are higher during the summer due to the warmer water. Along the west coast of North America, monthly mean sea levels have been increased by up to 20 cm during El Niño events, with the 1982–1983 and 1996–1998 events having been most dramatic (Flick and Cayan, 1985; Flick, 1998; Komar, 1986). On enclosed water bodies such as the Great Lakes, seasonal changes in water levels are experienced due to fluctuations in precipitation and evaporation that govern the inflow versus the outflow from the lakes.

The tidal or lake-level variations are frequently disturbed by meteorological factors, resulting in episodic water-level changes known as storm surges. The most important phenomenon that may accelerate sea-cliff erosion is an abnormal rise in water level caused by strong onshore winds and an abrupt reduction in atmospheric pressure. Figure 5.2(b) shows water-level fluctuations on Lake Erie, where storm surges are superimposed on seasonal variations (Carter and Guy, 1988). A comparison with the measured erosion distances (Fig. 5.2(a)), paying attention to the gaps in the beach-width curve representing intervals of debris and/or ice cover at the bluff toe when no erosion could take place, reveals a good correspondence between bluff erosion and extreme water levels. A remarkable water-level rise may occur on tidal coasts when storm surges are superimposed on high tides. The well-known 1953 storm surge on the North Sea caused dramatic erosion of glacial-sediment cliffs, an event that was well documented by Williams (1960, p. 82–87) at Covehithe, Suffolk, England: a 9 m high cliff was cut back 9 m during about two hours at the peak of the surge with a height of 2.3 m yielding a total sea-level rise of 3.5 m above MSL.

Beaches front most eroding cliffs. Where the cliff material is so soft that only the hydraulic action of waves is sufficient to cause erosion, critical for its initiation is whether waves running up the beach are able to reach the cliff toe. This ability is controlled not only by wave characteristics in deep water and the beach morphology (beach width and slope), but also by the water-level factors discussed above. The relationship between the width of the fronting beach and the rate of erosion of sea cliffs was examined by Everts (1991), selecting five sites along the 90 km coast of the Oceanside littoral cell north of San Diego, California, where the beach is backed by cliffs composed of Tertiary sandstones. At each site the erosion rate decreased as the beach became wider; a marked decrease occurred when the beach width exceeded 20 m, and erosion ceased altogether when it attained a width of 60 m.

The reduction or removal of the fronting beach can lead to a dramatic increase in sea-cliff erosion. This was seen at Big Lagoon, north of Eureka on the northern California coast (Savoy and Rust, 1985). The long-term recession rate of a sandy cliff backing the beach was 0.6 m/year, but during the 1982–1983 El Niño the cliff was cut back by more than 9 m during one winter when a combination of storm waves and high tides removed the beach.

The role of the morphology and behavior of the fronting beach is further illustrated by sea-cliff erosion into Pleistocene sandstones along the central Oregon coast (Komar and Shih, 1991, 1993; Shih and Komar, 1994). The beaches developed along the 24 km length of the Lincoln City littoral cell range from dissipative to nearly reflective, depending on the grain size of the beach. The coarser-grained reflective beaches are steeper sloped and respond more quickly to winter storms with larger changes in beach-profile levels than the fine-grained, gently-sloping dissipative beaches. As a result, the reflective beach is a weaker buffer against wave attack, and the sea cliff is more susceptible to erosion compared with areas where the cliff is fronted by a dissipative beach. A further decrease in buffer protection on the reflective beach is brought about by the more pronounced development of embayments eroded by rip currents, allowing for easy landward penetration of storm waves to the toe of the sea cliff.

When a beach acts as a buffer between the ocean and a sea cliff, elevated water levels due to a combination of high tides and storm-wave runup are needed before the water can reach the base of the cliff to produce erosion. Thornton *et al.* (1987), Shih *et al.* (1994), Ruggiero *et al.* (1996) and (Ruggiero *et al.*, 2001) have developed quantitative, process-based models that account for the water-elevation factors that are important to cliff and foredune erosion: the mean sea-level, the predicted tidal stage, processes such as storm surge that increase the measured tide, and wave setup and runup that are dependent on the wave height and period. Thornton *et al.* applied their model to analyses of the frequency of wave attack and resulting dune retreat along Monterey Bay, California. The analyses of Shih *et al.* and Ruggiero *et al.* are applied to sea-cliff erosion along the coast of Oregon, where the changing beach morphology also plays a role, particularly the formation of rip-current embayments that locally lower the berm. Cliff erosion is highly episodic, and occurs at a specific site only under the right combination involving the presence of a rip-current embayment to reduce the buffer protection of the beach, followed by a severe storm that reaches the coast during high spring tides.

The width of the fronting beach also depends on the overall budget of sediments, being influenced by human activities such as harbor works, dam construction on rivers, and beach mining or nourishment (Komar, 1996). A reduction in beach levels caused by human activities can produce unintentional results of rapid sea-cliff erosion. A classic example is the erosion of Hallsands in South Devon, England, where a fishing village was located on a low cliff composed of schist. During the five years from 1897 to 1902, approximately half a million cubic meters of shingle were removed from the beach fronting the cliff, to be used in construction of a dockyard at Plymouth. As a result, the entire beach was lowered by as much as 3.6 m, and this removal resulted in rapid cliff erosion, leading to the ruin of the fishing village (Robinson, 1961; Hails, 1975). Accelerated sea-cliff erosion induced by neighboring harbor works has occurred in Half Moon Bay on the central California coast (Lajoie and Mathieson, 1985). The construction of a breakwater between 1956 and 1960 in the northern half of this 10 km log-spiral shaped shoreline resulted in the depletion of beach sand in front of the cliff to the south of the breakwater. This caused a remarkable increase in cliff erosion following breakwater construction, from 0.1 m/year before construction when the wide fronting beach offered protection to the cliff, to 2 m/year of cliff retreat after construction resulted in the loss of the beach. These examples illustrate the degree of buffering protection offered by a fronting beach to sea-cliffs, and the resulting cliff erosion when that protection is reduced; additional studies dealing with cliff erosion intensified by anthropogenic factors are reviewed by Sunamura (1992, p. 218–225).

The offshore bottom topography and its effect on wave refraction can also control the assailing force of waves acting on the cliff face. Robinson (1980) examined the long-term variation in erosion rates of glacial sand cliffs at Dunwich in Suffolk, England. The average rates were: 1.6 m/year from 1589 to 1753, 0.85 m/year from 1753 to 1824, 1.5 m/year from 1824 to 1884, 1.15 m/year from 1884 to 1925, but then dropping to only 0.15 m/year from 1925 to 1977. This drop in the erosion rate during the past recent 50 years can be attributed to the reduction in wave energy reaching the coast. During previous centuries, wave refraction over the offshore Sizewell Bank focused the wave energy on the shoreward Dunwich site, but the northward growth of the Bank during this century now produces a divergence in wave refraction and a reduction in the assailing force of the waves.

Wave-induced erosion of cohesive nearshore bottom has been extensively studied along the Canadian shore of the Great Lakes (Coakley *et al.*, 1986; Davidson-Arnott, 1986; Kamphuis, 1987; Davidson-Arnott and Ollerhead, 1995; Davidson-Arnott and Langham, 2000), where rapidly receding, glacial-till cliffs are exposed with a narrow beach at their foot, and nearshore bottom profiles also cut into till are covered with a sparse veneer of sand. These studies have been conducted with a view that nearshore lowering will reduce the dissipation of wave energy reaching the cliff toe, which in turn will facilitate cliff erosion. According to measurements by Davidson-Arnott and Ollerhead (1995) of the lowering rates with a micro-erosion meter off an eroding bluff west of Port Dalhousie on the south shore of Lake Ontario, the downcutting attained 3 to 4 cm during the period of May to October 1992, and it was found that the lowering of the till bed is controlled greatly by the surficial sediment, which accelerates or decelerates erosion depending on the thickness of the sediment cover. A series of laboratory experiments using a wave flume, in which a model till profile was made of intact samples taken from the field, indicated that such sediment control dominates the erosional processes (Bishop *et al.*, 1992; Skafel and Bishop, 1994), and also showed that the degree of downcutting depends on the type of wave breaking — plunging breakers are more effective than spilling breakers (Skafel, 1995). The possibility of strength reduction with time (i.e., softening of the underwater, exposed till surface) has recently been reported from the field (Davidson-Arnott and Ollerhead, 1995), and also from the laboratory (Skafel and Bishop, 1994). This factor, which acts to promote downcutting, would be important in erosional processes on the till shore profile. Further research is needed to elucidate the mechanism for softening.

Talus can protect the foot of the cliff from wave attack until it is removed by waves and currents; thus, the presence of talus is a controlling factor for toe erosion. How long talus resides at the cliff base depends on variables such as: (1) the dimension of the talus, (2) the composition of the debris, and (3) the wave and current conditions. According to Griggs and Johnson (1979, Fig. 12), talus composed of a considerable amount of debris, including large blocks of sandstone and siltstone, falling from a 25 m high Tertiary cliff at Santa Cruz, California, resided only a few years at the foot of the cliff. In the Lincoln City littoral cell on the mid-Oregon coast, the life time of talus at the base of a cliff fronted by a dissipative beach is much longer than where a steeper reflective beach is found (Komar and Shih, 1993). The talus remains for several years to decades when protected by a dissipative beach because of the less-active, low-intensity infragravity dominated swash, resulting in a slow recession of the cliff; with a more-reflective beach dominated by swash with a strong incident-wave component, the talus is rapidly removed yielding higher cliff-recession rates.

2.4 *Slope instabilities and mass movement*

Cliff undercutting by waves results in slope instabilities, eventually leading to intermittent mass movement (e.g., Brunsden and Lee, 2004). Such movement can be destructive to coastal properties. For example, the initial movement of the Tertiary mudstone cliff in the Jump-Off Joe area on the Oregon coast resulted in the loss of a dozen homes during the 1940s (Sayer and Komar, 1988). In 1982 approval was given to construct a condominium on this unstable site, and movement once again destroyed the structure before it could be inhabited.

A variety of terms are used for mass movement phenomena. "Landslide" is a general term, which may be further classified into four types of mass movement: falls, topples, flows and slides. These types depend mainly on lithological factors such as geological structures, stratigraphic features, and geotechnical properties. Hybrid types exist between two or more of these failure modes.

Falls denote movement of a mass that travels most of the distance through air as a freely-falling body. They are subdivided into rock falls, debris falls, and earth falls, according to the type of cliff material. Cliff retreat associated with rock falls has been reported from the Welsh bold coasts composed of densely-jointed rocks (Williams and Davies, 1987; Jones and Williams, 1991).

Topples differ from falls in the fact that little free-fall movement takes place because rotation of a block around a fixed hinge dominates the motion. Topples are common on precipitous, sometimes overhanging, cliffs composed of vertically-jointed hard rocks like those found on the Liassic coasts of South Wales (Davies *et al.*, 1991, 1998; Williams *et al.*, 1993).

Flows move with increasing velocity toward the surface of the moving body; no block movement is present due to differential shearing within the body. Flows occurring on some coasts cut into soft clayey materials with fluidized potential are called mudflows, and are frequently initiated by mudslides. The sliding mass may disintegrate during its movement — the flow component increases, resulting in the generation of mudflows in the latter phase of the event. However, the distinction between mudslides and mudflows is not easy in the field. These two failure modes contribute significantly to the recession of cliffs in soft materials such as clay (e.g., Hutchinson, 1970, 1973), mudstone (e.g., Grainger and Kalaugher, 1987), and till (e.g., Bryan and Prince, 1980).

Slides are shearing displacements occurring on a distinct slip surface, and the sliding mass exhibits block movement. There are two major movements: planar slides (also called "translational slides") and rotational slides ("slumps"). The former has an almost linear sliding surface, whereas the latter is along a circular plane. Planar sliding with a high-angle failure plane may be found on till bluffs (e.g., McGreal, 1979; Jones and Williams, 1991), chalk cliffs (Hutchinson, 1972; Middlemiss, 1983), and Tertiary mudstone cliffs (e.g., Sunamura, 1992, p. 111–114). Rotational slides can be categorized into "shallow-seated" and "deep-seated" slides, depending on the depth of the sliding surface. Shallow-seated slides are observed on till cliffs in Northern Ireland (McGreal, 1979), and along the Canadian shore of the Great Lakes (e.g., Quigley *et al.*, 1977; Edil and Vallejo, 1980). Deep-seated slides frequently occur on the London Clay cliffs in the UK (e.g., Hutchinson, 1973, 1986).

In addition to basal erosion by waves, the most common controlling factors for the occurrence of mass movement are rainfall and groundwater. Rainfall results in direct impacts on the cliff face, inducing gully erosion and slumping (e.g., Kuhn and Shepard, 1980, 1983; Bird and Rosengren, 1986), and frequently affects phreatic conditions. The latter includes: (1) groundwater level (e.g., McGreal, 1979; Edil and Vallejo, 1980), (2) seepage (e.g., Buckler and Winters, 1983), and (3) porewater pressure (e.g., McGreal and Craig, 1977; Allison and Brunsden, 1990). Of these variables, porewater pressure has been related by Allison and Brunsden (1990) to the pattern of mudslide movement that has occurred on the Isle of Purbeck, England, based on continuous monitoring of the failure processes. Frost (or freeze-thaw) action is also common in the mid- to high-latitude regions, which promotes downslope movement of cliff-face material (e.g., Hutchinson, 1986; Jones and Williams, 1991). In tectonically active areas, earthquakes sometimes trigger mass movement (e.g., Bird, 1994; Griggs, 1994). Vegetation factors such as the growth of tree roots within joints may induce cliff failure (Griggs and Johnson, 1979; Bird, 1994).

Most mass movement phenomena have occurred during the action of storm waves which facilitates cliff undercutting, or during and immediately after a long spell of wet weather which in many instances causes a rise in groundwater level, promoting a reduction in shear strength of the cliff material. Other cliff failures, however, have occurred suddenly and unexpectedly during normal weather conditions, as seen from an event at Scarborough on the east coast of England (Anon., 1994). The event began in the morning of 4 June 1993, with shallow-seated rotational slides on a 60 m high till cliff, where the toe had been protected by a sea wall. The sliding lasted for several days, leading to the gradual collapse of a four-star hotel.

Whatever the cause, mass movement acts to render the cliff face profile more stable — a more gently sloping profile develops. Simultaneously, debris masses are supplied to the base of the cliff. Waves remove the debris, undercutting the base so that the overall cliff profile once again becomes steep and unstable, and mass movement ensues. As seen in this recurrence, a cyclic change between steep and gentle profiles occurs during the cliff-recession process. This indicates that similar cliff profiles will recur with a certain time interval. A possible recurrence period for the parallel profile is estimated at 10 years for the Taito-misake cliff cut into Pleistocene mudstone on the Pacific coast of Japan (Sunamura, 1983), and at 30 years period for the London Clay cliff at Warden Point on the Isle of Sheppey, England (Hutchinson, 1986). Investigations of such site-specific values, preferably on a physical basis including cliff-recession processes, are necessary for the long-term coastal-zone management of rocky shores.

3. Accretionary Coastal Landforms

The beach is the fundamental accretionary landform on the coast. It may be part of a larger system such as a spit or barrier island, but is best considered on its own in order to focus on the processes of waves and currents, sediment transport, and on its resulting morphology that also depends on the range of grain sizes supplied by the sediment sources. Many beaches are essentially two-dimensional, showing little variation in the longshore, while others are highly three-dimensional with rhythmic bars or series of cusps and embayments extending along the shore.

This examination of accretionary beaches will progress from the simplest to the more complex. Accordingly, the review begins with sand beaches having profiles that are essentially two-dimensional, with the focus being on how the waves determine the slope of the beach face, the elevation of the berm, and form offshore bars. The next stage of complexity is the fully three-dimensional sand beach, where a variety of morphological forms include beach cusps, crescentic bars, and series of embayments formed by rip currents. Important at this stage of beach complexity is the morphodynamics model of Wright and Short (1983), which classifies the sand beaches according to the sequence of morphological forms, a sequence that depends on the wave conditions. This examination of accretionary coastal systems ends with a consideration of beaches that are mixtures of wide ranges of sediment grain sizes, beaches that contain significant proportions of both sand and a coarse-grain fraction, pebbles and cobbles. Such mixed beaches tend to be more complex in that they can have a large variety of grain sorting patterns, and also because the processes that control their morphologies are still poorly understood.

3.1 *The two-dimensional profiles of sand beaches: Form and variation*

In their simplest forms, the profiles of sand beaches consist of a series of components toward the offshore: a nearly-horizontal dry berm, the beach face that is acted upon by the swash of waves, and a series of troughs and bars within the surf zone dominated by wave breaking. The morphologies of these components depend on the beach sediment grain size, on the wave conditions, and on the range of tides.

A number of investigators have examined the factors that control the slope of the beach face, and have attempted to develop semi-empirical relationships to predict its inclination. The slope is governed by the asymmetry in the intensity of the wave runup versus the backwash, and the resulting asymmetrical cross-shore sediment transport. Due to the frictional drag on the runup and perhaps some water percolation into the beach, the return backwash tends to be weaker than the shoreward uprush. This asymmetry moves sediment onshore until a slope is built up over which gravity supports the backwash and enhances offshore sediment transport. When the same amount of sediment is transported seaward as is moved landward, the beach-face slope becomes effectively constant and is in a state of dynamic equilibrium. The equilibrium slope also depends in part on the quantity of water lost through percolation, which begins to be important for coarse-sand beaches, and even more so for gravel and cobble beaches.

Field studies have empirically demonstrated that beaches composed of coarse particles have steeper slopes (Bascom, 1951; Wiegel, 1964); this has also been shown in laboratory wave-tank experiments (Bagnold, 1940; Rector, 1954). These studies further demonstrated that the slope depends on the wave conditions, either on the wave steepness or on the wave energy. The predictive relationships combine these sediment and wave parameters. For example, Dalrymple and Thompson (1976) employed the dimensionless settling velocity, H_∞/w_sT, where w_s is the mean settling velocity of the beach sediment, H_∞ is the deep-water wave height, and T is the wave period. The use of this ratio is based on physical arguments involving the cross-shore movement of suspended sediments by waves. Figure 5.5 from Dalrymple and Thompson combines data from several laboratory studies, showing a general decrease in beach slope with increasing H_∞/w_sT. Sunamura (1984) analyzed both laboratory and field data to obtain somewhat different empirical relationships. The field data are scattered, but can be reasonably described by the equation

$$S_0 = \tan\alpha = \frac{0.12}{(H_b/g^{1/2}D^{1/2}T)^{1/2}}, \tag{5.3a}$$

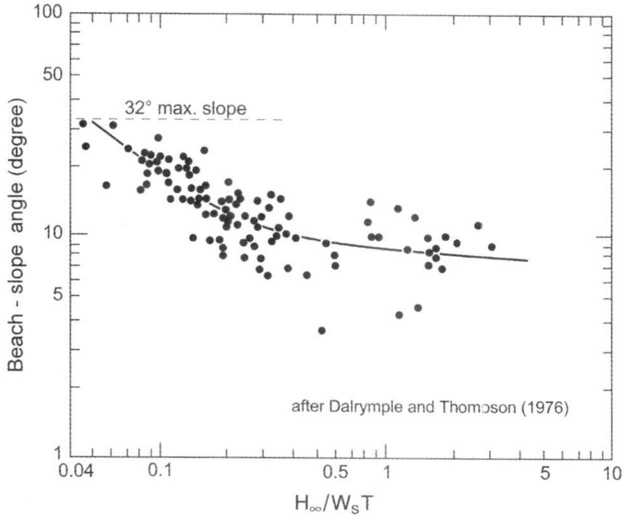

Fig. 5.5 A compilation of laboratory data to relate the beach-face slope to H_∞/w_sT where w_s is the settling velocity of the beach sand [modified from Dalrymple and Thompson (1976)].

where H_b is the wave-breaker height and D is the mean diameter of the beach sediment. The slopes of laboratory beaches are on average greater than those of natural beaches for the same sediment size and wave parameters; the reason for this difference has not been established, but possibly is due to the general absence of infragravity energy on the laboratory beaches, water motions having periods that are greater than the incident waves, the presence of which on natural beaches would act to reduce the beach slope. Using $L_\infty = (g/2\pi)T^2$ for the deep-water wavelength, Eq. (5.3a) from Sunamura (1984) can be converted to

$$S_0 = 0.12(2\pi)^{0.25}\left(\frac{D}{H_b}\right)^{0.25}\left(\frac{H_b}{L_\infty}\right)^{-0.25}, \tag{5.3b}$$

showing an inverse dependence on the wave steepness as found by other investigators, and also depending on the ratio of the sediment grain size to the wave height.

The general dependencies of beach slopes on sediment sizes and wave conditions are clear from the trends of the field and laboratory data. However, there has not been full agreement as to the choice of which dimensionless ratios are most appropriate for establishing predictive relationships or accounting for the physical processes that control the beach slope. If the slope is established mainly by the cross-shore transport of suspended sediments as envisioned in the analysis of Dalrymple and Thompson (1976), then a ratio containing the grain settling velocity w_s would be appropriate; on the other hand, if water percolation into and out of the beach face is important, then the grain diameter (and range of sizes) would be more relevant. In order to resolve such questions, it is now more important to focus on the processes that control the beach slope than on further empirical analyses.

The elevation and morphology of the berm is governed by the level of wave runup on the beach face, and by the cycle of the tides. Takeda and Sunamura (1982) [see Sunamura (1989)] used the Hunt equation for the runup height as a function of the wave conditions, obtaining a relationship for the berm elevation B_h:

$$B_h = 0.125 H_b^{5/8} (gT^2)^{3/8}. \tag{5.4}$$

The proportionality coefficient is based on laboratory and field measurements, the latter having been obtained along the Pacific coast of Japan where the mean tidal range is approximately 1 m (the berm height was measured relative to mean sea level). Sediment sizes ranged 0.2 mm to 1.3 mm without showing differences in berm elevations.

Most beach profiles contain one or two longshore bars, while a few reveal the presence of three or more (multiple bars). In laboratory experiments starting with a uniform beach slope, an offshore bar generally develops in the position of wave breaking with the water depth at the bar position being approximately equal to the breaker height. Experiments have demonstrated that multiple bars can form if large waves are followed by small waves; the large waves build a bar offshore in deeper water, while the subsequent low waves pass over that outer bar and break within the inshore where they form a second bar. Tidal variations can also produce multiple bars, with high- and low-tide bars being formed, although the high-tide bar may become partially flattened when exposed at low tide. Dolan and Dean (1985) have argued that multiple bars, specifically those in Chesapeake Bay where some 4 to 17 bars are found, can be formed by repeated wave breaking such that each bar corresponds to a zone of concentrated breaking. This certainly is the pattern once the bars have formed, but it is questionable whether dozens of bars can develop from an initially flat bed simply by repeated breaking. One difficulty is that with a spectrum of waves present, waves break at all water depths when crossing a smooth, uniform slope (i.e., the surf zone is saturated). Alternative hypotheses for the formation of multiple bars include their association with standing waves produced by wave reflection from the beach (Carter *et al.*, 1973; Lau and Travis, 1973; Davies and Heathershaw, 1984; Heathershaw and Davies, 1985; O'Hare and Davies, 1990), with standing or progressive edge waves (Bowen and Inman, 1971; Aagaard, 1991; Howd *et al.*, 1991), with the transformations of the incident waves (Boczar-Karakiewicz and Davidson-Arnott, 1987), or with breaker-induced vortices (Zhang and Sunamura, 1994).

There is evidence both pro and con for each of these hypotheses for the formation of bars, with no consensus having emerged. The break-point hypothesis is valid at least for some bars, and is most easily demonstrated with monochromatic waves in laboratory channels or with very narrow spectra of regular swell waves on natural beaches. There is a problem in application of the break-point hypothesis with random waves having a large range of heights and periods, as bars are observed to form within the wave-saturated portion of the surf zone where waves are breaking at all depths (Sallenger and Howd, 1989; Holman and Sallenger, 1993). The hypotheses involving bar formation due to standing waves formed by wave reflection, by sediment movements associated with progressive or standing edge waves, by the existence of higher harmonics in shoaling waves, or due to breaker-induced vortices interacting with cross-shore currents all appear theoretically reasonable and have received some support from laboratory and/or field observations. It is possible that depending on the circumstances, these various hypotheses could at different times account for the formation of bars. In a review of the research that has been undertaken at the Field Research Facility, Duck, North Carolina, Holman and Sallenger (1993) concluded that bar generation at that site is not simply a result of breakpoint generation or of sediment transport toward the nodes of a single standing infragravity wave, but instead results from a hybrid of mechanisms. They note that determining the respective contributions of the different mechanisms will require the implementation of complex numerical models combined with extensive programs of field-data collection.

As summarized here, the component parts of the two-dimensional profiles of sand beaches depend on the wave conditions, and their combined responses during storms versus low-wave conditions account for the observed variations in profiles on many beaches as diagrammed schematically in Fig. 5.6. Such shifts in beach profiles were first documented by Shepard (1950) and Bascom (1953) on the west coast of the US where storm waves dominate the winter and lower-energy waves occur throughout the summer. Based on this marked seasonality, Shepard referred to the two profile types as *summer profiles* versus *winter profiles*. More recently, Larson *et al.* (1988) used the terms *bar profile* and *berm profile*, eliminating any seasonal implication and reflecting the usual presence of bars versus an accretional berm in the two profile types. Whichever terms are used, important is the cycle itself, which shows a tendency for the beach to change its morphology in response to the intensity of the wave conditions. With low-energy waves, the profile is typically characterized by a wide berm, and by a smooth offshore profile with little or no development of bars except perhaps in relatively deep water where inactive bars

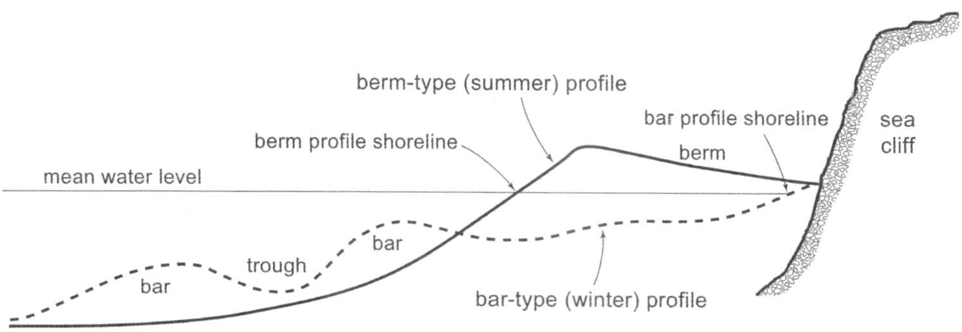

Fig. 5.6 The bar-type profile which forms during a storm, versus the berm-type profile that occurs under lower-energy wave conditions. The cycle is often seasonal, as noted [from Komar (1998)].

may be found (Fig. 5.6). In contrast, under higher wave conditions, the berm is cut back by the intensified wave swash, and the sediment is carried offshore to form one or more bars.

In that the component parts of the beach profile are found to depend on sediment and wave parameters, it can be expected that the change in total form from a berm profile to a bar profile will have a similar dependence. The earliest attempts to predict the form of the profile focused on the wave steepness H_∞/L_∞. For example, Johnson (1949) found in wave-channel experiments that with a wave steepness greater than 0.03 an offshore bar always forms (bar profile), whereas if the steepness is less than 0.025 an offshore bar is never formed (berm profile). Subsequent laboratory studies found other critical values for the wave steepness. In particular, experimenting with higher and longer-period waves in a large channel, Saville (1957) obtained a critical wave steepness of 0.0064, much lower than found in studies utilizing small-scale wave channels. It was concluded that the change in form from a berm profile to a bar profile depends on the wave energy level or wave height as well as on the wave steepness, and also on the sediment grain size or settling velocity. The more recent analyses, therefore, have used these parameters as well as the wave steepness. For example, Dean (1973) developed a model for the shift from a berm to a bar profile based on a consideration of the trajectory of suspended sand particles during their fall to the bottom, acted upon at the same time by the horizontal water-particle velocities of the waves. If the grain fall requires a short time relative to the wave period, then the particle will be acted upon predominantly by onshore velocities. On the other hand, if the fall velocity is low, then the grains will tend to shift offshore. Based on this model, Dean (1973) established that the dimensionless ratio w_s/gT is important in governing the critical wave steepness. Using laboratory data from various sources, he obtained the relationship

$$\mathit{critical}\,\frac{H_\infty}{L_\infty} = \frac{1.7\pi w_s}{gT}, \tag{5.5}$$

for the critical wave steepness. For those data sets the relationship correctly predicted whether the transport is onshore or offshore in 87.5% of the cases, that is, whether there is a berm profile or a bar profile. Allen (1985) applied this relationship to beaches on Sandy Hook, New Jersey, and found that with a proportionality coefficient of 2.0 and using the breaker height rather than the deep-water wave height, the model correctly predicted 98% of the erosion events, the shift to a bar profile. On the other hand, it predicted only 45% of shifts to an accretionary berm profile; Allen attributed this poorer prediction of the accretionary phase to the considerable time lag that can exist on natural beaches for profile shifts in that direction. Kriebel *et al.* (1986) carried out a number of additional laboratory experiments, and found that using large-scale tests yielded proportionality coefficients between 4 and 5, rather than the 1.7 value found by Dean (1973) in Eq. (5.5). Other combinations of parameters have been proposed. Kraus and Larson (1988) and Larson *et al.* (1988) compared the critical wave steepness for the transition from bar profiles to berm profiles with the dimensionless number H_∞/w_sT, utilizing data from large wave tanks, and obtained

$$critical \frac{H_\infty}{L_\infty} = 0.00070 \left(\frac{H_\infty}{w_sT}\right)^3. \tag{5.6}$$

Kraus *et al.* (1991) extended the comparison to field data collected from a variety of coasts, and found that Eq. (5.6) accounts for the transition of the profile if H_∞ is taken to be the mean wave height, whereas if it is taken as the significant wave height then the coefficient becomes 0.00027. As discussed by Dalrymple (1992), the relationship of Eq. (5.6) is counter intuitive since for a fixed value of H_∞/w_sT, increasing the wave steepness leads to a shift from a bar (storm) profile to a berm profile, opposite to the expected dependence and trend established by other studies. Dalrymple demonstrated that this results from a direct relationship between the wave steepness H_∞/L_∞ and H_∞/w_sT, and suggested that the relationship be rearranged into a single dimensionless number

$$\frac{gH_\infty^2}{w_s^3T} = \frac{2\pi}{0.0007} \approx 9.0 \times 10^3, \tag{5.7}$$

where bar-type profiles occur at higher values of this ratio.

Over the years there has been a number of studies in a variety of coastal settings that have qualitatively demonstrated that profiles do respond to changing wave conditions as illustrated schematically in Fig. 5.6. Two of the earliest investigations were those of Shepard (1950) and Bascom (1953), both on the California coast, respectively at La Jolla and Carmel. As well as demonstrating the shift in profile form due to seasonal variations in wave-energy levels, the contrasting sediment grain sizes of the beaches at those two locations showed that the coarser-sand beach (Carmel) experienced much greater changes; on the La Jolla beach the cut-and-fill cycle changed the elevation of the berm by on the order of 1 m to 1.5 m, while at Carmel the change was 2 m to 3 m. This may have resulted in part from differences in the wave-energy levels at the two locations, but a more significant factor is clearly the contrasting beach sediment grain sizes. The general observation is that, at least within the range of sand beaches, coarser-sand beaches experience greater elevation changes over the seasonal cycle, and also in response to individual storms. This has been shown further by beach-profile series on the Oregon coast where the beaches have the same wave climate (Shih and Komar, 1994). That study also illustrates that due to the large vertical changes of the coarse-sand beach profiles, and also due to their rapid cut back by storm waves, coastal developments backing the coarse-sand beaches are more susceptible to episodic erosion than are properties backing finer-sand beaches. A factor in this response is that the coarser-grained beaches are steeper and the surf zone is narrower, so with the same wave input from the offshore, the wave energy is more concentrated on the coarser-sand beach. Another factor is the difference in the wave swash characteristics, being dominated by the incident waves on the steep beach, while infragravity swash motions are more important on the fine-grained beaches which tend to filter out the incident-wave energy.

While a beach may erode rapidly during the one or two days of a storm, its subsequent recovery can extend over weeks to years. This has been shown, for example, by the study of Thom and Hall (1991) on the coast of Australia, based on monthly beach profiles collected for 16 years (1972–1988). The study was centered on a 5.5 km crescent-shaped pocket beach, which largely ensures the conservation of the total volume of sand so that any long-term trends of erosion or accretion do not result from losses or additions of sand to the system. Figure 5.7 shows time series of sand volumes on the beach profiles above a low-tide datum, expressed as the volume of sand per meter longshore length. Each series is clearly divided into two accretion dominated periods (ADP) and a single erosion dominated period (EDP). The EDP interval was initiated by a major storm in 1974 that eroded the beach, while a second storm in 1978 cut back the beach further. Subsequent to the 1978 storm, the beach volumes recovered, with the initial rate of volume increase ($0.269\,m^3/m/day$) having been the most rapid and with subsequent rates having progressively declined. It is seen in Fig. 5.7 that this accretion dominated period extended over several years, with most of the volume recovery having been completed within five years. Thom and Hall (1991) showed that much of the volume gain, especially since 1981, was in the growth of a foredune at the back of the beach. With time, the foredune shifted landward as well as growing upward. The foredunes contained some 10% of the total volume of subaerial sand above the datum, which can serve as a reserve to be returned to the beach during another major storm, thereby providing a buffer to the coast and backshore properties during subsequent erosion events.

With research investigations of sand beaches now spanning more than 50 years, we have developed a fairly good understanding of the processes of the waves and currents that affect their morphologies. At the same time, we have considerably improved our capability to predict the rates of sand movement on beaches under the combined action of waves with superimposed currents. The coupling of these research products has in recent years led to the development of numerical models that simulate the variations in the two-dimensional beach profiles in response to the changing waves, tides and currents. There are now a number of such models, formulated at various coastal laboratories throughout the world, that while differing in their details still have the same basic components, and for the most part yield similar results in applications. An example of these models is that developed in England by Southgate and Nairn (1993) and Nairn and Southgate (1993). Their model includes analyses of wave transformations in the offshore, wave breaking and decay within the surf zone, wave setup effects on the mean water levels in the presence of waves, the generation of undertow which can be important to cross-shore sediment movements, and the transport of the suspended and bedload sediments are independently evaluated. The models developed by several investigators differ

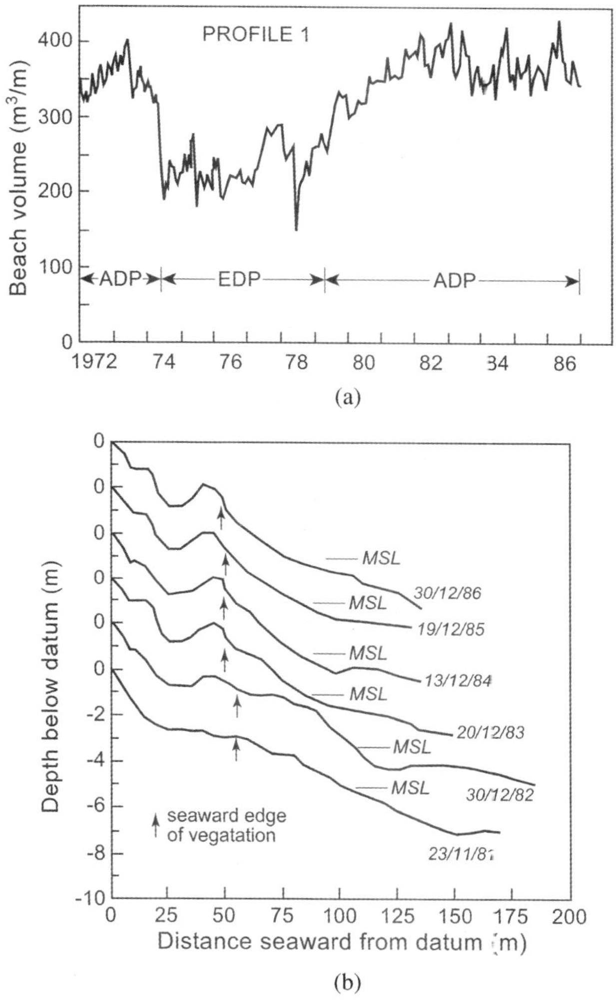

Fig. 5.7 (a) Beach volume changes for a profile time series obtained at Moruya, a pocket beach on the New South Wales coast of Australia. Two distinct accretion-dominated periods (ADP) are separated by an erosion-dominated period (EDP) that occurred during a major storm. (b) A series of profiles obtained during the accretion-dominated period since 1981, showing the accretion of the beach and the formation of a foredune. [from Thom and Hall (1991)].

primarily in how they evaluate these processes and in the computational procedures used in their models. Figure 5.8 shows the results of the model compared with the evolution of a beach profile in a large-scale wave channel, the two examples contrasting the results using uniform waves of constant height and period versus random waves. In both cases the model has been reasonably successful in duplicating the details of the measured profiles and their evolution with time from an initially uniform slope, including the formation of a single offshore bar with uniform waves, and multiple bars with random waves; Southgate (1991) provides additional

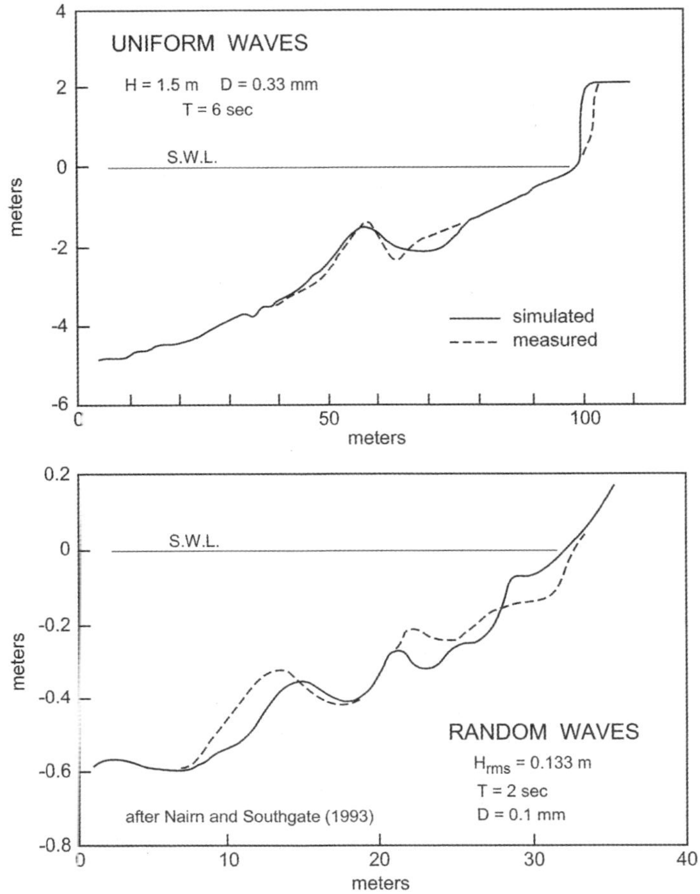

Fig. 5.8 Results of the profile morphology predicted by the numerical model of Nairn and Southgate (1993) compared with measured beach profiles in a large wave channel, using uniform and random waves [modified from Nairn and Southgate (1993)].

comparisons and sensitivity tests of their model components. Based on such results, Nairn and Southgate (1993) concluded that the processes related to the incident waves can be predicted to a level of accuracy that is sufficient to permit a confident estimate of the evolution of the beach-profile morphology, but the description of the long-period (infragravity) structure and its influence on the sediment transport is still inadequate.

Similar models have been developed in Japan by Dibanjnia *et al.* (1994) and Watanabe and Shiba (1994) for the cross-shore sediment transport and accompanying profile changes. Their model predictions again have been compared with laboratory data obtained in large wave channels, and show good agreement for conditions where the waves break either by plunging or spilling.

Refinements in the numerical beach profile models are still needed. Their results are best in the erosional phase of profile change, but generally underestimate the initially rapid offshore sediment transport with the post-storm return of low waves, leading again to the conclusion that the models inadequately account for wave reflection from the beach and infragravity long-period wave motions. In spite of still having shortcomings, numerical beach profile models have seen a significantly increased use in applications, for example in predicting the modifications of profiles in beach nourishment projects.

A more extensive review of the research into the variations of profiles on sand beaches, and the development of numerical models to simulate those variations can be found in Komar (1998, Chapter 7).

3.2 *Three-dimensional sand beaches*

Sand beaches commonly have morphologies that are three-dimensional, including offshore bars that are crescentic or trend obliquely toward the offshore, bar forms that can have a longshore rhythmicity. This three dimensionality can extend to the exposed part of the beach, which often consists of alternating cusps and embayments having scales of a few meters (beach cusps) to hundreds of meters (giant cusps).

Crescentic bars, Fig. 5.9, are remarkable in the regularity of their lunate shapes and uniform spacings along the shore. There may be an associated series of cusps on the beach if the crescentic bars in effect attach to the shore. The range in lengths of the resulting longshore rhythmicity is difficult to establish, since for many reported occurrences it is not possible to determine conclusively whether crescentic bars or some other form of rhythmic topography is being described. Crescentic bars appear to range from about 100 m to 2,000 m in length, with a predominance of 200 m to 500 m (Komar, 1998). The regularity of crescentic bars has inspired a

Fig. 5.9 Well-developed crescentic bars within a pocket beach near Cape Kalaa, Algeria, on the Mediterranean Sea [from Clos-Arceduc (1962)].

number of suggestions as to their origin. The mechanism proposed by Bowen and Inman (1971) provides a reasonable explanation, a hypothesis that is supported by wave-basin experiments and field comparisons. The hypothesis involves the velocity field associated with edge waves present on a sloping beach, particularly with the net drift velocities. Beach sediment moves about under the currents of the edge waves, until the sand reaches zones where the velocity is below the threshold of motion, where it then accumulates. It was shown conceptually by Bowen and Inman (1971) that with this model of sediment movement by edge waves, sand would accumulate into the form of crescentic bars; Holman and Bowen (1982) provide a more quantitative analysis in having generated simulations of crescentic-bar formation in numerical computer models, using Bagnold-type sediment transport equations. The similarity of the theoretical forms of crescentic bars to the observed forms is remarkable, and provides strong evidence in support of the edge-wave hypothesis for their formation.

The field examination by Bowen and Inman (1971) of their hypothesis for the generation of crescentic bars centered on the example shown in Fig. 5.9 of bars on the Mediterranean Sea coast of Algeria. Their wavelength is approximately 500 m, so the edge-wave length needed to account for their formation would have had to been about 1,000 m. The morphology of the bars suggests a mode 2 edge wave, and on this basis Bowen and Inman (1971) calculated that the required edge-wave period for bar formation would have had to been on the order of 40 s to 50 s, that is, within the infragravity range. At the time of their study in the early 1970s, the existence of such long-period (infragravity) edge waves had not been documented conclusively in the nearshore, and this led to objections to this edge-wave origin. However, there have been number of field experiments (e.g., Huntley *et al.*, 1981; Oltman-Shay and Guza, 1987) in subsequent years that have documented the existence of long-period infragravity edge waves containing considerable energy at periods equal to and in excess of the 40 s to 50 s needed to account for crescentic bars as proposed by Bowen and Inman (1971).

Another large-scale rhythmic shoreline form of importance are embayments eroded into the berm by rip currents. In most instances rip currents erode sand from the berm and transport it offshore, forming embayments at the rip positions and leaving cusps midway between (Komar, 1971). Rip currents, and hence the embayments and cusps, typically have spacings that range from tens to hundreds of meters. The shoreline consisting of embayments and cusps is only the surface expression of the underwater topography that is molded by the water circulation and waves. The seaward flowing rips tend to erode channels, the result being a segmented offshore bar, often with each bar segment attached by an underwater shoal to the shoreline cusp. This form of rhythmic topography with embayments cut by rips may be important to property erosion, in that the embayments remove most of the buffer protection offered by the beach. Investigations of erosion along the Oregon coast have found that the development of rip-current embayments is the primary factor in the loss of property located in foredunes (Komar and Rea, 1976;

Komar, 1983) and atop sea cliffs backing beaches (Komar and McDougal, 1988; Shih and Komar, 1994).

The importance of crescentic bars and rip-current embayments in controlling the three-dimensional morphology of the beach has led to their inclusion in the morphodynamics classification formulated by Wright and Short (1983), depicted in Fig. 5.10. The end-member beach types in this classification are represented by *Dissipative* and *Reflective* beaches. Dissipative beaches are characteristically low in slope such that the waves break well offshore and continue to loose energy as bores while they cross the wide surf zone — the beach is dissipative in the sense of being highly effective in decreasing the energy of the incident waves. It also was noted that Dissipative beaches can be dominated by long-period infragravity water motions, some of which may be in the form of edge waves. In the opposite extreme, Reflective beaches are steep and the incident waves break close to shore and wash directly up the beach face. Accordingly, Reflective beaches are dominated by the incident waves, and if edge waves are present they most probably are subharmonic edge waves where the period is twice that of the incident waves (Guza and Davis, 1974), the edge-wave mode that may be responsible for the formation of beach cusps (Guza and Inman, 1975). Between these extremes are several Intermediate beach categories, beach types that are more three-dimensional with the presence of crescentic bars, rip-current embayments and beach cusps. Starting at the top of the diagram in Fig. 5.10, the Dissipative state (a) is the condition already described, while the Intermediate Longshore Bar-Trough state (b) differs in there being more relief in the bar and trough elevations and the beach face is steeper than on the Dissipative beach. In response to this morphology, most of the wave dissipation occurs by breaking over the bar. The waves reform within the deeper trough, and break for a second time on the beach face. The steep beach face can be locally reflective, so beach cusps often develop. The Rhythmic Bar and Beach state (c) is similar to (b), but the distinctive feature is the development of crescentic bars and the accompanying large-scale cusps along the shoreline. The Transverse Bar and Rip topography, state (d), consists of welded oblique bars and intervening troughs occupied by strong rip currents. The Ridge-and-Runnel/Low-Tide Terrace state (e) is mainly characterized by the presence of a swash bar that is migrating shoreward, confining a narrow and deep longshore trough at the base of the beach face. Weak and irregularly spaced rip currents may be present, in places cutting through the bar. The final stage (f) is the fully Reflective beach condition, where the sole three-dimensional element is in the form of beach cusps on an otherwise steep beach.

The occurrence of these various beach states was found by Wright and Short (1983) to depend on the beach slope and wave parameters according to the dimensionless ratio:

$$\varepsilon = \frac{H_b \omega^2}{2gS^2} = \frac{2\pi^2 H_b}{gT^2 S^2}, \tag{5.8a}$$

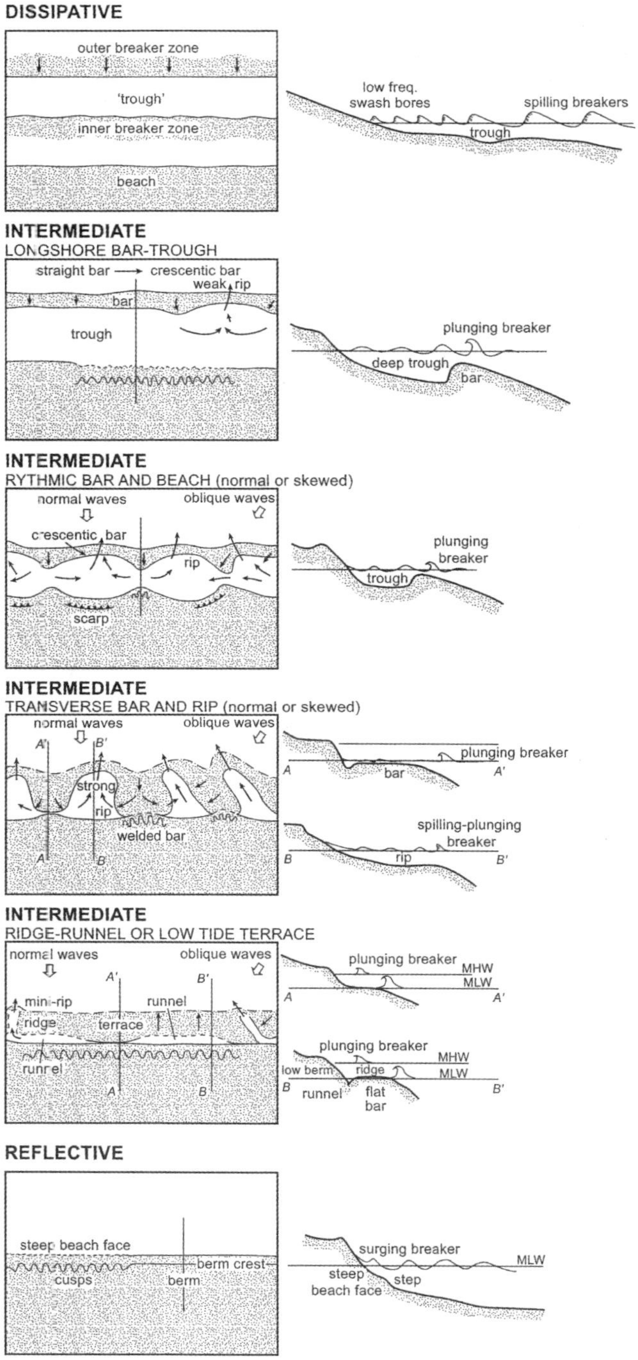

Fig. 5.10 The morphodynamic classification of beaches. The plan and profile configurations of six morphodynamic beach states are recognized, depending on the overall beach slope and wave conditions [modified from Wright and Short (1983)].

where $\omega = 2\pi/T$ is the radian frequency, and S is the beach slope. This ratio can be rearranged into the form of an Iribarren number, using $L_\infty = gT^2/2\pi$ for the deep-water wavelength, yielding

$$\xi_b = \frac{S}{(H_b/L_\infty)^{1/2}} = \left(\frac{\pi}{\varepsilon}\right)^{1/2}. \tag{5.8b}$$

The Dissipative extreme was found, by Wright and Short, to occur when ε ranges from 30 to 100, corresponding to very low values of the Iribarren number ($\xi_b = 0.2$ to 0.3). At the opposite extreme, the Reflective beach condition occurred when $\varepsilon < 1$ or approximately $\xi_b > 2$. The four intermediate stages are more difficult to assess in terms of these dimensionless ratios since their values can range widely depending on the local conditions of the bottom slope and wave parameters, permitting, for example, the outer surf zone to be dissipative while the steep beach face of the inner surf zone is more reflective.

When the wave conditions change the beach state within the sequence shown in Fig. 5.10 shifts. With a storm, ε generally increases while ξ_b decreases, and this acts to force the beach morphodynamics from Reflective towards Dissipative; when viewed as a change in the beach profile, this is seen to effectively be a shift from a berm-type profile to a bar-type profile as categorized earlier in Fig. 5.6, where attention was limited to the two-dimensional profile. The full sequence as outlined in Fig. 5.10 is more complex, especially with the return of lower wave conditions following the storm. During the storm the beach may shift quickly to the state (a) fully-Dissipative condition having a bar-type profile, but with lower wave conditions after the storm the beach may successively pass through states (b), (c), etc., characterized by an increasing three dimensionality having various cuspate or rhythmic shoreline features (crescentic bars, rip-current embayments, and welded bars). Characterizing the morphology by a simple two-dimensional beach as in Fig. 5.6 is therefore an oversimplification. Short (1979) has stressed the importance of these successive beach states with increasing or decreasing wave heights and energies, and this is further discussed by Wright and Short (1983). They point out that most beaches will not typically range through all morphological states diagrammed in Fig. 5.9, but will be constrained within limits that depend on the grain size of the beach sediment (which largely controls its average slope), and the total range in the wave climate. There will be a variability in wave climate about some mode, and this will define the range in morphodynamics about some corresponding mode. Thus, a fine-grained sand beach that experiences a high-energy wave climate will be dominated by states (a) through (c), that is, toward the Dissipative end within the spectrum of beach types. In contrast, a coarse-grained ocean beach or a sandy beach within a lake will be observed to cycle through states (d), (e) and (f), depending on the wave conditions.

The modality of the beach morphodynamics will control the overall variability of beach changes. Beaches that are at one of the extremes, either Dissipative or

Reflective, tend to show the least variability in their three-dimensional morphology or in a simple set of beach profiles. In contrast, beaches that have a central modality within the Intermediate beach states show a greater range in morphologies, and developments backing such beaches tend to be more susceptible to erosion problems.

The morphodynamic classification of Wright and Short (1983) was based on observations made along the coast of Australia, but has been shown to apply to other coasts, in some cases with minor modifications. Based on two years of daily observations of bar patterns and wave conditions at the Field Research Facility in Duck, North Carolina, Lippmann and Holman (1990) developed a classification that differs to only a small degree from the morphodynamics classification of Wright and Short (1983). Two categories were actually never observed at Duck during the two-year study of Lippmann and Holman, the extreme end members, the fully Dissipative and fully Reflective beach morphologies. The daily documentation of the bar conditions permitted assessments of the frequencies of occurrence of the various bar forms and their residence times. The most stable form is equivalent to the Ridge-and-Runnel/Low Tide Terrace Intermediate category of Wright and Short (1983), Fig. 5.10, which for the Duck location is the most reflective condition achieved by the beach, the morphology that is maintained during prolonged periods of low-wave conditions.

3.3 *Mixed sand and gravel beaches*

Thus far this review has focused on the "simple" beaches composed entirely of sand. But many beaches contain a significant coarse-grained component, gravel (shingle) and cobbles, and sometimes even boulders. Beaches containing both sand and these coarse grains are generally more complex than beaches composed only of sand, in both their morphologies due to the variety of grain sorting patterns and because the mixing of sand and gravel in the deposit affects the porosity and permeability ("conductivity") of the beach, which in turn alters the swash dynamics, sediment transport, beach slopes, and so on. Unfortunately, such "mixed" beaches also have not been studied to nearly the same extent as sand beaches. In a recent review of such beaches, Mason and Coates (2001) concluded: "Beaches containing a mixture of both sand and gravel have aroused only sporadic interest." The result is that much remains unknown about this type of beach so the prediction of their morphological responses to changing wave conditions is still uncertain.

A classification of the range of beach types based on their proportions of sand versus gravel and cobbles, and how these contrasting sizes are hydraulically sorted, has been developed by Jennings and Shulmeister (2002) based on analyses of 42 gravel beaches found on the South Island of New Zealand. Their classification is depicted in Fig. 5.11, with the addition of pure sand beaches so as to represent the full spectrum from pure coarse-grained beaches to sand beaches, with two

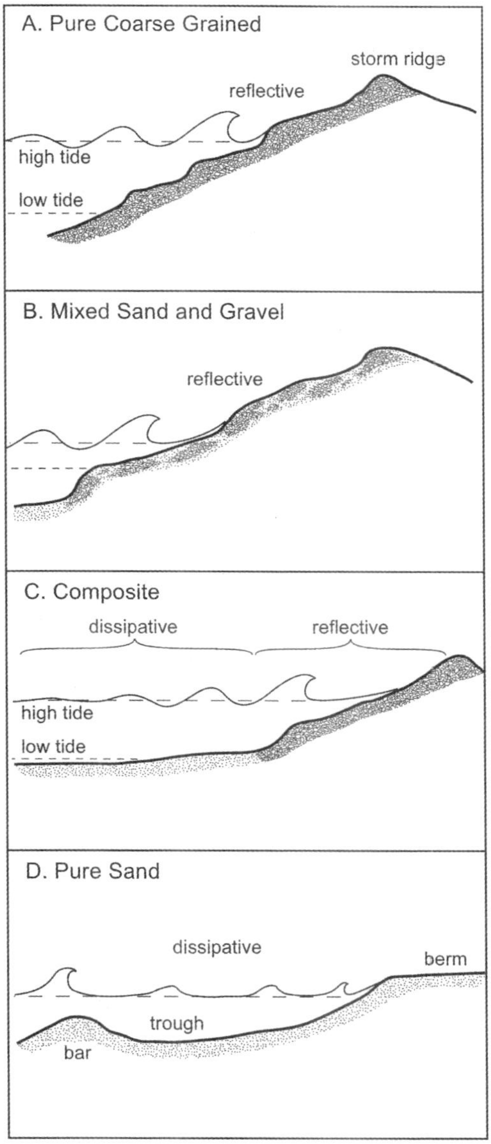

Fig. 5.11 The classification of beaches based on their proportions of coarse sediments (gravel and cobbles) versus sand, with the resulting differences in their morphologies [extended from Jennings and Shulmeister (2002)].

intermediate categories. The four categories in this classification become:

(a) *pure coarse-grained beaches*
Those composed of particle sizes ranging from pebbles to cobbles and boulders, with minimal sand;

(b) *mixed sand-and-gravel beaches*
Those consisting of high proportions of both coarse particles and sand, with there being an intimate mixing of the two size fractions in the beach deposit;
(c) *composite beaches*
Those having a higher proportion of sand which has been sorted by the waves and nearshore currents, so the beach consists of an upper foreshore or backshore ridge composed of gravel and cobbles, and a lower foreshore of sand, generally with a distinct boundary between them; and
(d) *pure sand beaches*
Beaches consisting almost entirely of sand, and if coarse particles are present the quantity is insignificant so it does not appreciably affect the morphology and dynamics of the beach.

Even a "pure" coarse-grained beach typically does contain some sand in the voids between the gravel and cobbles, but with an insufficient proportion to affect the porosity and permeability of the deposit as a whole, which is important in that with enough sand it can alter the runup of the wave swash on the beach and the resulting mobility of the gravel particles. This critical proportion of sand is not well established, but can be taken as approximately 5% to 10% of the deposit as a whole. The distinction between beach categories (b) and (c) is dependent on both the proportions of sand versus the coarse-grained component, and on the ability of the waves and currents to sort and separate those respective sizes. Generally, the classic form of a mixed sand-and-gravel beach is where there is still a relatively small portion of sand, sufficiently small that the volume can mainly reside within the voids between the coarser gravel and cobble clasts. In this sense the "mixed" beach is an intimate mixture of the full range of grain sizes. There may be a degree of sorting and separation of some of the sand so it is locally concentrated on the beach face or is carried by the waves to the immediate offshore, leaving a beach that can still be classified as mixed sand-and-gravel but with a sub-tidal offshore sand deposit. With increasing quantities of sand provided by the sediment sources to the coast, the volume of this offshore sand accumulation increases to the extent that it is exposed at times of low tide, at which point it would be classified as a category (c) composite beach, since there is a fronting sand beach at least during low tides. With a still greater quantity of sand, a well established sand beach can be formed in front of the gravel ridge, with the sand beach visible at all tidal stages; the waves might only reach the gravel ridge during storms that erode back and inundate the fronting sand beach.

Pure coarse-grained beaches that consist entirely of gravel and cobbles tend to always remain Reflective due to their persistent steep profile slopes. As shown by Wright and Short (1983), this imparts a degree of stability to the beach by virtue of the large sizes of the particles and perhaps also because a significant portion of the wave energy is reflected. The stability of mixed sand-and-gravel beaches is uncertain and often unpredictable due to the added proportions of sand which can fill the voids between the gravel particles, reducing the extent of percolation of the wave swash runup so it retains more of its energy, resulting in the cut back of the beach profile during storms. Composite beaches are interesting in that if the fronting sand deposit is sufficient, it in effect provides a Dissipative sand beach backed by a Reflective coarse-grained ridge, the two most stable end members in the morphodynamics classification of Wright and Short (1983). Because of this relative stability of pure coarse-grained beaches, some mixed beaches, and particularly composite beaches that have both Dissipative and Reflective elements, it has been recognized that constructing a comparatively small ridge of gravel and cobbles at the back of a sand beach can provide the same degree of protection to shore-front properties as does a large volume of sand added in a beach nourishment project, and in some cases can even substitute for a hard structure such as a riprap revetment or seawall. Such shore-protection structures having the forms of natural coarse-grained beaches are termed "cobble berms" or "dynamic revetments", the latter term recognizing that they are expected to be dynamic, in contrast to riprap "static" revetments (Ahrens, 1990). Amongst other advantages, the construction of dynamic revetments has been acceptable in sensitive areas such as parks, where riprap revetments or seawalls are unacceptable [e.g., Allan and Komar (2002, 2004)].

The processes acting on mixed sand-and-gravel beaches, such as wave swash energies and runup elevations, are made complex by the presence of the sand which to varying degrees fills the voids between the gravel and cobble particles. The importance of this sand content on otherwise coarse-grained beaches is illustrated indirectly by its effects on the beach slopes. Figure 5.12 is a graph of the foreshore slopes

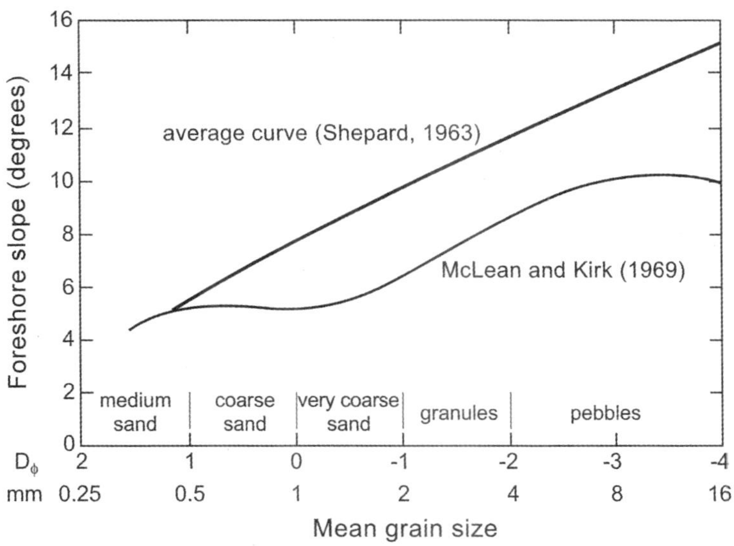

Fig. 5.12 The beach-face slope versus the grain size on coarse-grained beaches, according to the data of Shepard (1963) for California beaches and McLean and Kirk (1969) for the mixed beaches of New Zealand [from Komar (1998)].

of beaches ranging in mean grain sizes from medium sand through pebbles, based on the tabulation of Shepard (1963) for relatively well sorted beaches in southern California, and McLean and Kirk (1969) for mixed sand-and-gravel beaches in New Zealand. The California beaches show a uniform increase in slope, depending on the mean sediment grain size. The curve from the study of McLean and Kirk (1969) for mixed sand-and-gravel beaches shows a much different pattern. Due to the overall poorer sorting with the addition of sand to the gravel, these mixed beaches have lower slopes than those measured by Shepard (1963) even for the same mean grain size. The wavy nature of the McLean and Kirk (1969) curve is due to the nature of the sources of sediments to the New Zealand beaches, sources that yield beaches which consist of pebbles with diameters of 4 mm to 16 mm or of sand having a median diameter of 0.5 mm, or mixtures of the two. When the individual modes of pebbles or sand occur alone, the sediment sorting is good and the resulting beach is steeper; when the modes are mixed, the sorting is poor so the water percolation from the swash into the beach is reduced and the beach slope is lowered. Accordingly, the curve of McLean and Kirk (1969) in Fig. 5.12 rises to higher slopes for sands of 0.5 mm diameter and for pebbles, but with the slopes reduced for intermediate mean grain sizes where the grain sorting and permeability are both low.

The importance of swash infiltration into the beach and how it determines the beach slope has been more directly analyzed by Masseling and Ling Li (2001) in a series of numerical models. It was found that swash infiltration increases the onshore asymmetry in the swash flow, thereby enhancing onshore sediment transport and resulting in relatively steep beach-face gradients. However, this accretionary effect of swash infiltration is only evident when the rate of infiltration is sufficiently large, that is, when the total infiltration volume over a wave cycle exceeds about 2% of the uprush volume. This threshold is attained when the beach sediment grain size is coarser than about 1.5 mm, implying that the correlation between the beach-face slope and sediment size found for sandy beaches is not due to enhanced swash asymmetry caused by infiltration. For gravel beaches, however, swash infiltration is concluded to be the dominant factor in controlling the beach-face gradient, with increased sediment size and permeability being responsible for the steeper beach slopes.

While the responses of sand beaches to storms have been thoroughly studied, our understanding of the comparable morphologic responses of coarse-grained beaches to changing wave energies is still rudimentary. Furthermore, different responses have been found in the several studies that have been undertaken. Some found that similar to sand beaches, high-energy waves erode back the beach face and berm of gravel beaches, transporting the gravel toward the offshore where it is deposited at the base of the gravel deposit, but does not form a bar as is generally the case for sand beaches [e.g., Sherman (1991)]. While storm waves may result in the offshore transport of some of the gravel, it has also been observed on many coarse-grained beaches that a portion of the gravel is swept landward by the intense wave swash and is deposited at the top of the beach face or is carried over the top of the gravel ridge, resulting in its landward migration. This response has been found for example by Carter and Orford (1993) of the coast of Ireland, and by Everts *et al.* (2002) and Allan and Komar (2002, 2004) on the US west coast. These US west coast studies both involved composite beaches having a gravel and cobble ridge at the back of a wide sand beach. At the time of a storm, the sand of the fronting beach is moved to offshore bars, reducing its slope and making it more dissipative; at the same time, on the gravel ridge at the landward side of the sand beach, the storm waves produce a net landward movement of the coarse particles so that both the crest elevation of the gravel deposit and its slope increase; while the fronting sand beach had become more dissipative in response to the storm, the gravel ridge had become more reflective. This difference in responses was undoubtedly caused by their contrasting permeabilities and how that determines the balance between the intensity of the swash runup of the waves versus the backwash.

In should be apparent from even this brief summary that "mixed" beaches of sand and gravel deserve more than "sporadic interest" (Mason and Coates, 2001). With our presently having only a rudimentary understanding of their processes and morphologic responses, it is difficult to satisfactorily manage coasts having this type of beach, or to make sound assessments in engineering applications.

4. Summary and Conclusions

The objective of this chapter has been to review the status of our understanding of the processes that are important to the erosion of rocky coasts with the formation of sea cliffs, and the processes that govern the morphologies of accretionary beaches. Important in both cases is the wave energy, and the associated forces of impact against cliffs or the runup elevations of the swash on the sloping beach. Laboratory experiments have demonstrated the responses of cliffs under the attack of waves, governed also by the material composition of the cliff and the availability of sediments on a fronting beach that can serve as an abrasion agent. The processes of cliff erosion in the field are extremely complex due to the highly variable rock compositions and presence of fabric such as joints and bedding. Sea-cliff erosion also tends to be episodic, occurring only when extreme high tides combine with storm wave runup, and when the level of the fronting beach has been lowered by the presence of a rip current that locally cuts away an embayment, a combination of factors that allows the water to reach the base of the cliff.

The morphologies of accretionary beaches also reflect the wave conditions, and the sediment grain sizes composing the beach. The components of the beach profile — its slope, the elevation of the berm, the number of offshore bars, etc. — depend on wave and sediment parameters. These dependencies govern the change from a berm-type (summer) profile to a bar-type (winter) profile with varying wave conditions, a seasonal cycle that is observed on many beaches. Some beaches are simple in being essentially two-dimensional, so the variations are reflected in the components of the profile, while other beaches are more complex in being three-dimensional, including longshore variations due to the presence of crescentic bars, rip-current embayments and beach cusps. Advanced morphodynamic classifications of beaches account for this three dimensionality, and how it changes with the wave conditions and sediment characteristics. A still greater complexity in accretionary beaches is found when they are composed of both sand and a coarse-grained component, gravel and cobbles. The mixing of the markedly different grain sizes can have a significant effect on the processes, especially those that depend on the permeability of the beach, resulting in a considerable variability of sorting patterns of the different grain sizes. In this range of accretionary beach types, these "mixed" beaches have received the least research attention.

References

Aagaard, T. (1991): Multiple-bar morphodynamics and its relation to low-frequency edge waves, *J. Coastal Res.* **7**, 801–813.

Ahrens, J. P. (1990): Dynamic revetments, *Proc 22nd Int. Conf. on Coastal Engrg.* ASCE, 1837–1850.

Allan, J. C. and P. D. Komar (2002): A dynamic revetment and artificial dune for shore protection, *Solving Coastal Conundrums, Proc. 28th Int. Conf. on Coastal Engrg.*, ASCE, 2044–2056.

Allan, J. C. and P. D. Komar (2004): Environmentally compatible cobble berm and artificial dune for shore protection, *Shore & Beach*, **72**, 9–18.

Allen, J. R. (1985): Field evaluation of beach profile response to wave steepness as predicted by the Dean model, *Coastal Engrg.* **9**, 71–80.

Allison, R. J. and D. Brunsden (1990): Some mudslide movement patterns, *Earth Surface Processes Landforms*, **15**, 296–311.

Anon. (1994): The Scarborough landslide, *Quart. J. Engrg. Geology* **27**, 3–6.

Aoki, H. and Y. Matsukura (2007): A new technique for non-destructive field measurement of rock-surface strength: an application of the Equotip hardness tester to weathering studies, *Earth Surface Processes and Landforms*, **32**, 1759–1769.

Bagnold, R. A. (1940): Beach formation by waves: Some model experiments in a wave tank, *J. Inst. Civil Engrs.* **15**, 26–52.

Bascom, W. H. (1951): The relationship between sand size and beach face slope, *Trans. Amer. Geophys. Union*, **32**, 866–874.

Bascom, W. H. (1953): Characteristics of natural beaches, *Proc. 4th Int. Conf. on Coastal Engrg.*, ASCE, 163–180.

Bird, E. C. F. (1994): Cliff hazards and coastal management, *J. Coastal Res.* Special Issue, **12**, 299–309.

Bird, E. C. F. and N. J. Rosengren (1986): Changes in cliff morphology at black rock point 1973–1986, *Victoria Naturalist*, **103**, 106–113.

Bishop, C., M. Skafel and R. Nairn (1992): Cohesive profile erosion by waves, *Proc. 23rd Int. Conf. Coastal Engrg.* ASCE, 2976–2989.

Boczar-Karakiewicz, B. and R. G. D. Davidson-Arnott (1987): Nearshore bar formation by non-linear wave processes — A comparison of model results and field data, *Marine Geology* **77**, 286–304.

Bokuniewicz, H. and J. Tanski (1980): Managing localized erosion of coastal bluffs, *Proc. Coastal Zone* '80, ASCE, **3**, 1883–1898.

Bowen, A. J. and D. L. Inman (1971): Edge waves and crescentic bars, *J. Geophys. Res.* **76**(36) 8662–8671.

Brunsden, D. and E. M. Lee (2004): Behaviour of coastal landslide systems: An interdisciplinary view, *Zeit. Geomorphologie N. F.*, Suppl. Bd., **134**, 1–112.

Bryan, R. B. and A. G. Price (1980): Recession of the Scarborough bluffs, Ontario, Canada, *Zeit. Geomorphologie N. F.*, Suppl. Bd., **34**, 48–62.

Buckler, W. R. and H. A. Winters (1983): Lake Michigan bluff recession, *Ann. Assoc. Amer. Geographers*, **73**, 89–110.

Budetta, P., G. Galietta and A. Santo (2000): A methodology for the study of the relation between coastal cliff erosion and the mechanical strength of soil and rock masses, *Engrg. Geology* **56**, 243–256.

Cambers, G. (1976): Temporal scales in coastal erosion systems, *Trans. Inst. British Geographers*, New Ser., **1**, 246–256.

Carter, C. H. and D. E. Guy, Jr. (1988): Coastal erosion: processes, timing and magnitudes at the bluff toe, *Marine Geology* **84**, 1–17.

Carter, R. W. G. and J. D. Orford (1993): The morphodynamics of coarse clastic beaches and barriers: A shore and long term perspective, *J. Coastal Res.***15**, 158–179.

Carter, T. G., P. L.-F. Liu and C. C. Mei (1973): Mass transport by waves and offshore sand bedforms, *J. Waterways, Harbors, & Coastal Engrg. Div.* ASCE, WW2, 165–184.

Clos-Arceduc, A. (1962): Etude sur les vuesaeriennes, des alluvionslittoralesd'allureperiodique, cordons littorauxetfostons, *Bull. Soc. Fr. Photogramm.* **4**, 13–21.

Coakley, J. P., N. A. Rukavina and A. Zeman (1986): Wave-induced subaqueous erosion of cohesive tills: preliminary results, *Proc. Symp. Cohesive Shores*, National Res. Council, Canada, 120–136.

Dalrymple, R. A. (1992): Prediction of storm/normal beach profiles, *J. Waterway, Port, Coastal & Ocean Engrg.* ASCE, **118**, 193–200.

Dalrymple, R. A. and W. W. Thompson (1976): Study of equilibrium beach profiles, *Proc. 15th Int. Conf. on Coastal Engrg.*, ASCE, 1276–1296.

Davidson-Arnott, R. G. D. (1986): Rates of erosion of till in the nearshore zone, *Earth Surface Processes Landforms* **11**, 53–58.

Davidson-Arnott, R. G. D. and J. Ollerhead (1995): Nearshore erosion on a cohesive shoreline, *Marine Geology* **122**, 349–365.

Davidson-Arnott, R. G. D. and D. R. L. Langhan (2000): The effects of softening on nearshore erosion of a cohesive shoreline, *Marine Geology* **166**, 145–162.

Davies, A. G. and A. D. Heathershaw (1984): Surface wave propagation over sinusoidally varying wave topography, *J. Fluid Mechanics* **144**, 419–443.

Davies, D. S., E. W. Axelrod and J. S. O'Conner (1972): *Erosion of the north shore of Long Island*, Technical Report, Marine Sciences Research Center, State University of New York, Stony Brook, 18, 101 pp.

Davies, P., A. T. Williams and P. Bomboe (1991): Numerical modelling of Lower Lias rock failures in the coastal cliffs of South Wales, *Proc. Coastal Sediments* '91, ASCE, 1599–1612.

Davies, P., A. T. Williams and P. Bomboe (1998): Numerical analysis of coastal cliff failure along the Pembrokeshire Coast National Park, Wales, UK, *Earth Surface Processes Landforms*, **23**, 1123–1134.

Dean, R. G. (1973): Heuristic models of sand transport in the surf zone, *Conference on Engineering Dynamics in the Surf Zone*, Sydney, Australia, 208–214.

Dibanjnia, M., T. Shimizu and A. Watanabe (1994): Profile change of a sheet flow dominated beach, *Proc. 24th Int. Conf. on Coastal Engrg.* ASCE, 1946–1960.

Dolan, T. J. and R. G. Dean (1985): Multiple longshore sand bars in the upper Chesapeake Bay, *Estuarine Coastal Shelf Science*, **21**, 726–743.

Edil, T. B. and L. E. Vallejo (1980): Mechanics of coastal landslides and the influence of slope parameters, *Engrg. Geology* **16**, 83–96.

Emery, K. O. (1960):*The Sea off Southern California*: John Wiley & Sons, New York.

Everts, C. H. (1991): Seacliff retreat and coarse sediment yields in southern California, *Proc. Coastal Sediments* '91, ASCE, 1586–1598.

Everts, C. H., C. D. Eldon and J. Moore (2002): Performance of cobble berms in Southern California, *Shore & Beach*, **70**, 5–14.

Flick, R. E. (1998): Comparison of California tides, storm surges, and mean sea level during the El Niño winters of 1982–1983 and 1996–1998, *Shore & Beach*, **66**, 6–11.

Flick, R. E. and D. R. Cayan (1985): Extreme sea levels on the coast of California, *Proc.* 19th *Coastal Engrg.Conf.* ASCE, 886–898.

Grainger, P. and P. G. Kalaugher (1987): Intermittent surging movements of a coastal landslide, *Earth Surface Processes Landforms*, **12**, 596–603.

Griggs, G. (1994): California's coastal hazard, *J. Coastal Res.* Special Issue **12**, 1–15.

Griggs, G. and R. E. Johnson (1979): Coastal erosion, Santa Cruz County, *California Geology*, April, 66–76.

Griggs, G. and L. Savoy (1985): Sea cliff erosion, In: *Living with the California Coast* (eds. G. Griggs and L. Savoy), pp. 29–34: Duke University Press, Durham, North Carolina.

Goudie, A. S. (2006): The Schemidt Hammer in geomorphological research, *Progress in Physical Geography* **30**, 703–718.

Guza, R. T. and R. Davis (1974): Excitation of edge waves by waves incident on a beach, *J. Geophys. Res.* **79**, 1285–1291.

Guza, R. T. and D. L. Inman (1975): Edge waves and beach cusps, *J. Geophys. Res.* **80**(21), 2996–3012.

Hachinohe, S., N. Hiraki and T. Suzuki (1999): Rates of weathering and temporal changes in strength of bedrock of marine terraces in Boso Peninsula, Japan, *Engrg. Geology* **55**, 29–43.

Hails, J. R. (1975): Sediment distribution and Quaternary history, *J. Geological Society of London*, **131**, 19–35.

Healy, T. R. (1968): Bioerosion on shore platforms in the Waitemata Formation, Auckland, *Earth Science J.* **2**, 26–37.

Heathershaw, A. D. and A. G. Davies (1985): Resonant wave reflection by transverse bedforms and its relation to beaches and offshore bars, *Marine Geology* **62**, 321–338.

Holman, R. A. and A. J. Bowen (1982): Bars, bumps and holes: Models for the generation of complex beach topography, *J. Geophys. Res.* **87**, 456–468.

Holman, R. A. and A. H. Sallenger (1993): Sand bar generation: A discussion of the Duck experiment series, *J. Coastal Res.* Special Issue No. **15**, ed. A. D. Short, 76–92.

Howd, P. A., A. J. Bowen, R. A. Holman and J. Oltoman-Shay (1991): Infragravity waves, longshore currents, and linear sand bar formation, *Proc. Coastal Sediments* '91, ASCE, 72–84.

Huntley, D. A., R. T. Guza and D. B. Thronton (1981): Field observations of surf beat, 1. Progressive edge waves, *J. Geophys. Res.* **83**, 1913–1920.

Hutchinson, J. N. (1970): A coastal mudflow on the London Clay cliffs at Beltinge, north Kent, *Geotechnique* **20**, 412–438.

Hutchinson, J. N. (1972): Field and laboratory studies of a fall in Upper Chalk cliffs at Joss Bay, Isle of Thanet, In: *Stress-Strain Behaviour of Soils* (R. H. G. Parry, ed.), pp. 692–706. *Proc. Roscoe Memorial Symposium,* G. T. Foulis, Henley-on-Thames, Oxfordshire.

Hutchinson, J. N. (1973): The response of London Clay cliffs to differing rates of toe erosion, *Geologia Applicata e Idrogeologia*, **8**, 221–239.

Hutchinson, J. N. (1986): Cliffs and shores in cohesive materials: geotechnical and engineering geological aspects, *Proceedings of the Symposium on Cohesive Shores*, National Research Council, Canada, 1–44.

Inman, D. L. and C. E. Nordstrom (1971): On the tectonic and morphological classification of coasts, *J. Geology* **79**, 1–21.

Jennings, R. and J. Shulmeister (2002): A field based classification scheme for gravel beaches, *Marine Geology* **186**, 211–228.

Johannessen, C. L., J. J. Feiereisen and A. N. Wells (1982): Weathering of ocean cliffs by salt expansion in a mid-latitude coastal environment, *Shore & Beach*, **50**, 26–34.

Johnson, J. W. (1949): Scale effects in hydraulic models involving wave motion, *Trans. Amer. Geophys. Union*, **30**, 516–525.

Jones, D. G. and A. T. Williams (1991): Statistical analysis of factors influencing cliff erosion along a section of the West Wales coast, UK, *Earth Surface Processes Landforms*, **16**, 95–111.

Kamphuis, J. W. (1987): Recession rate of glacial till bluffs, *J. Waterway Port Coastal Ocean Engrg.* **113**, 60–73.

Komar, P. D. (1971): Nearshore cell circulation and the formation of giant cusps, *Geological Society of America Bulletin*, **82**, 2643–2650.

Komar, P. D. (1983): The erosion of Siletz Spit, Oregon, in: *Handbook of Coastal Processes and Erosion* (P. D. Komar, ed.), pp. 65–76: CRC Press, Boca Raton, Florida.

Komar, P. D. (1986): The 1982–1983 El Niño and erosion on the coast of Oregon, *Shore & Beach* **54**, 3–12.

Komar, P. D. (1998): *Beach Processes and Sedimentation*, 2nd edn.: Prentice-Hall, Upper Saddle River, New Jersey.

Komar, P. D. and C. C. Rea (1976): Erosion of Siletz Spit, Oregon, *Shore & Beach* **44**, 9–15.

Komar, P. D. and D. B. Enfield (1987): Short-term sea-level changes and coastal erosion, in: *Sea-level Fluctuations and Coastal Evolution* (D. Nummedal *et al.*, eds.) Special Publication No. **41**, pp. 16–27, Soc. Econ. Paleontologists and Mineralogists, Tulsa, Oklahoma.

Komar, P. D. and W. G. McDougal (1988): Coastal erosion and engineering structures: The Oregon experience, *J. Coastal Res.* **4**, 76–92.

Komar, P. D. and S.-M. Shih (1991): Sea-cliff erosion along the Oregon coast, *Proc. Coastal Sediments* '91, ASCE, 1558–1570.

Komar, P. D. and S.-M. Shih (1993): Cliff erosion along the Oregon coast: A tectonic-sea level imprint plus local controls by beach processes, *J. Coastal Res.* **9**, 746–765.

Komar, P. D., J. C. Allan and P. Ruggiero (2011): Sea level variations along the U.S. Pacific Northwest Coast: Testonic and climate controls, *J. Coastal Res.* **27**, 808–823.

Kraus, N. C. and M. Larson (1988): *Beach profile change measured in the tank for large waves*, 1956–1957 and 1962, Technical Report CERC-88-6, US Army Corps of Engineers.

Kraus, N. C., M. Larson and D. L. Kriebel (1991): Evaluation of beach erosion and accretion predictors, *Coastal Sediments* '91, ASCE, 572–587.

Kriebel, D. L., W. R. Dally and R. G. Dean (1986): Undistorted Froude scale model for surf zone sediment transport, *Proc.* 20th *Coastal Engrg. Conf.*, ASCE, 1296–1310.

Kuhn, G. G. and F. P. Shepard (1980): Coastal erosion in San Diego, California, *Proc. Coastal Zone* '80, ASCE, **3**, 1899–1918.

Kuhn, G. G. and F. P. Shepard (1983): Beach processes and sea cliff erosion in San Diego County, California, In: *Handbook of Coastal Processes and Erosion* (P. D. Komar, ed.), pp. 266–264: CRC Press, Boca Raton, Florida.

Lajoie, K. R. and S. A. Mathieson (1985): San Francisco to AÒo Nuevo, in: *Living with the California Coast* (G. Griggs and L. Savoy, eds.), pp. 140–177: Duke University Press, Durham, North Carolina.

Larson, M., N. C. Kraus and T. Sunamura (1988): Beach profile change: Morphology, transport rate, and numerical simulation, *Proc. 21st Coastal Engrg. Conf.* ASCE, 1295–1309.

Lau, J. and B. Travis (1973): Slowly varying Stokes waves and submarine longshore bars, *J. Geophys. Res.* **78**(21), 4489–4497.

Lippmann, T. C. and R. A. Holman (1990): The spatial and temporal variability of sand bar morphology, *J. Geophys.Res.* **95**(C7), 575–590.

Mason, T. and T. T. Coates (2001): Sediment transport processes on mixed beaches: A review for shoreline management, *J. Coastal Res.* **17**, 643–657.

Masseling, G. and L. Li (2001): The role of swash infiltration in determining the beachface gradient: A numerical study, *Marine Geology* **176**, 139–156.

McGreal, W. S. (1979): Factors promoting coastal slope instability in southeast County Down, N. Ireland, *Zeit. Geomorphologie, N.F.* **23**, 76–90.

McGreal, W. S. and D. Craig (1977): Mass-movement activity: An illustration of differing responses to groundwater conditions from two sites in northern Ireland, *Irish Geography* **10**, 28–35.

McLean, R. F. and R. M. Kirk (1969): Relationship between grain size, size-sorting, and foreshore slope on mixed sand-shingle beaches, *New Zealand J. Geology and Geophysics* **12**, 138–155.

Middlemiss, F. A. (1983): Instability of chalk cliffs between the South Foreland and Kingstone, Kent, in relation to geological structure, *Proc. Geologists' Association*, **94**, 115–122.

Mottershead, D. N. (1989): Rates and patterns of bedrock denudation by coastal salt spray weathering: A seven-year record, *Earth Surface Processes Landforms*, **14**, 383–398.

Mottershead, D. N. (1994): Spatial variations in intensity of alveolar weathering of a dated sandstone structure in a coastal environment, Westonn-super-Mare, UK, in: *Rock Weathering and Landform Evolution* (D. A. Robinson and R. B. G. Williams, eds.), pp. 151–174: John Wiley & Sons, Chichester.

Nairn, R. B. and H. N. Southgate (1993): Deterministic profile modelling of nearshore processes. Part 2. Sediment transport and beach profile development, *Coastal Engrg.* **19**, 57–96.

O'Hare, T. J. and A. G. Davies (1990): A laboratory study of sand bar evolution, *J. Coastal Res.* **6**, 531–544.

Oltman-Shay, J. and R. T. Guza (1987): Infragravity edge wave observations on two California beaches, *J. Phys. Oceanogr.* **17**, 644–663.

Pringle, A. W. (1985): Holderness coast erosion and the significance of ords, *Earth Surface Processes Landforms*, **10**, 106–124.

Quigley, R. M., P. J. Gelinas, W. T. Bou and R. W. Packer (1977): Cyclic erosion-instability relationships: Lake Erie northshore bluffs, *Canadian Geotechnical J.* **14**, 310–323.

Rector, R. L. (1954): *Laboratory study of the equilibrium profiles of beaches*, US Army Corps of Engineers, Beach Erosion Board, Tech. Memo. No. 41, 38 pp.

Richards, K. S. and N. R. Lorriman (1987): Basal erosion and mass movement, in: *Slope Stability* (M. G. Anderson and K. S. Richards, eds.), pp. 331–357: John Wiley & Sons, Chichester.

Robinson, A. H. W. (1961): The hydrography of Start Bay and its relationship to beach changes at Hallsands, *Geographical J.* **127**, 63–77.

Robinson, A. H. W. (1980): Erosion and accretion along part of the Suffolk coast of East Anglia, England, *Marine Geology* **37**, 133–146.

Robinson, D. A. and L. C. Jerwood (1987): Sub-aerial weathering of chalk shore platforms during harsh winters in southeast England, *Marine Geology* **77**, 1–14.

Robinson, L. A. (1977): Marine erosive processes at the cliff foot, *Marine Geology* **23**, 256–271.

Ruggiero, P., P. D Komar, W. G. McDougal and R. A. Beach (1996): Extreme water levels, wave runup and coastal erosion, *Proc.* 25th *Int. Conf. on Coastal Engrg.*, ASCE, 2793–2805.

Ruggiero, P., P. D. Komar, W. G. McDougal, J. J. Marra and R. A. Beach (2001): Wave runup, extreme water levels and the erosion of properties backing beaches, *J. Coastal Res.* **17**(2), 407–419.

Sallenger, A. H. and P. A. Howd (1989): Nearshore bars and the break-point hypothesis, *Coastal Engrg.* **12**, 301–313.

Saville, T. (1957): Scale effects in two dimensional beach studies, *Int. Association of Hydraulic Res.* Lisbon, Portugal, **A3**, 1–8.

Savoy, L. and D. Rust (1985): The northern California coast: Oregon border to Shelter Cove, in: *Living with the California Coast* (G. Griggs and L. Savoy, eds.), pp. 81–105: Duke University Press, Durham, North Carolina.

Sayre, W. O. and P. D. Komar (1988): The Jump-Off Joe Landslide at Newport, Oregon: History of Erosion, Development and Destruction, *Shore & Beach* **56**, 15–22.

Shepard, F. P. (1950): *Beach cycles in Southern California,* US Army Corps of Engineers, Beach Erosion Board, Tech. Memo. No. 20, 26 pp.

Shepard, F. P. (1963): *Submarine Geology*: Harper & Row, New York.

Sherman, D. J. (1991): Gravel beaches, *National Geographic Res. Exploration*, **7**, 442–452.

Shih, S.-M. and P. D. Komar (1994): Sediments, beach morphology and sea cliff erosion within an Oregon coast littoral cell, *J. Coastal Res.* **10**, 144–157.

Shih, S.-M., P. D. Komar, K. Tillotson, W. G. McDougal and P. Ruggiero (1994): Wave run-up and sea-cliff erosion, *Proc. 24th Coastal Engrg. Conf.* ASCE, 2170–2184.

Short, A. (1979): Three-dimensional beach stage model, *J. Geology* **87**, 553–571.

Skafel, M. G. (1995): Laboratory measurement of nearshore velocities and erosion of cohesive sediment (till) shorelines, *Coastal Engrg.* **24**, 343–349.

Skafel, M. G. and C. T. Bishop (1994): Flume experiments on the erosion of till shores by waves, *Coastal Engrg.* **23**, 329–348.

Southgate, H. N. (1991): Beach profile modelling: Flume data comparions and sensitivity tests, *Coastal Sediments* '91, ASCE, 1829–1841.

Southgate, H. N. and R. B. Nairn (1993): Deterministic profile modelling of nearshore processes. Part 1. Waves and currents, *Coastal Engrg.* **19**: 27–56.

Stephenson, W. J. and R. M. Kirk (2000): Development of shore platforms on Kaikoura Peninsula, South Island, New Zealand. Part One: The role of waves, *Geomorphology*, **32**, 21–41.

Sunamura, T. (1973): Coastal cliff erosion due to waves — field investigations and laboratory experiments, *J. Faculty Engrg.* University of Tokyo, **32**, 1–86.

Sunamura, T. (1977): A relationship between wave-induced cliff erosion and erosive force of waves, *J. Geology* **85**, 613–618.

Sunamura, T. (1982a): A predictive model for wave-induced cliff erosion, with application to Pacific coasts of Japan, *J. Geology* **90**, 166–178.

Sunamura, T. (1982b): A wave tank experiment on the erosional mechanism at a cliff base, *Earth Surface Processes Landforms*, **7**, 333–343.

Sunamura, T. (1983): Processes of sea cliff and platform erosion, in: *Handbook of Coastal Processes and Erosion* (P. D. Komar, ed.), pp. 233–265: CRC Press, Boca Raton, Florida.

Sunamura, T. (1984): Quantitative predictions of beach-face slopes, *Geological Soc. Amer. Bull.* **95**, 242–245.

Sunamura, T. (1989): Sandy beach geomorphology elucidated by laboratory modeling, in: *Applications in Coastal Modeling* (V. C. Lakhan and A. S. Trenhail, eds.), Elsevier Oceanography Series **49**, pp.159–213: Elsevier, Amsterdam.

Sunamura, T. (1992): *Geomorphology of Rocky Coasts*: John Wiley & Sons, Chichester.

Sunamura, T. (1994): Rock control in coastal geomorphic processes, *Trans. Japanese Geomorphological Union*, **15**, 253–272.

Sunamura, T. (1996): A physical model for the rate of coastal tafoni development, *J. Geology* **104**, 741–748.

Sunamura, T. (2004): Cliffs, erosion rates, in: *Encyclopedia of Coastal Science* (M. L. Schwartz, ed.), pp. 240–241: Springer, Dordrecht.

Suzuki, T. and S. Hachinohe (1995): Weathering rates of bedrock forming marine terraces in Boso Peninsula, Japan, *Trans. Japanese Geomorphological Union*, **16**, 93–113.

Takahashi, K., T. Suzuki and Y. Matsukura (1994): Erosion rates of sandstone used for a masonry bridge pier in the coastal spray zone, in: *Rock Weathering and Landform Evolution* (D. A. Robinson and R. B. G. Williams, Eds.), pp. 175–192: John Wiley & Sons, Chichester.

Takeda, I. and T. Sunamura (1982): Formation and height of berms, *Trans. Japanese Geomorphological Union*, **3**, 145–157 (in Japanese with English abstract).

Thornton, E. B., T. McGee, S. P. Tucker and D. M. Burych (1987): Predicting erosion on the recessive Monterey Bay shoreline, *Proc. Coastal Sediments* 87, ASCE, 1809–1824.

Thom, B. G. and W. Hall (1991): Behaviour of beach profiles during accretion and erosion dominated periods, *Earth Surface Processes and Landforms* **16**, 113–127.

Trenhaile, A. S. (1987): *The Geomorphology of Rock Coasts*: Oxford University Press, Oxford.

Trudgill, S. T. (1976): The marine erosion of limestone on Aldabra Atoll, Indian Ocean, *Zeit. Geomorphologie N.F.*, Suppl. Bd., **26**, 164–200.

Trudgill, S. T. (1985): *Limestone Geomorphology*: Longman, London.

Tsujimoto, H. (1987): Dynamic conditions for shore platform initiation, *Science Report, Inst. Geosci.* University of Tsukuba, **8A**, 45–93.

Viles, H., A. Goudie, S. Grab and J. Lalley (2011): The use of the Schmidt Hammer and Equotip rock hardness assessment in geomorphology and heritage science: A comparative analysis, *Earth Surface Processes and Landforms*, **36**, 320–333.

Watanabe, A. and K. Shiba (1994): Numerical modeling of beach profile change under sheet-flow condition using nonlinear wave theory, *Proc. 24th Coastal Engrg. Conf.*, ASCE, 2785–2798.

Wiegel, R. L. (1964): *Oceanographical Engineering*: Prentice-Hall, Englewood Cliffs, New Jersey.

Williams, A. T. and P. Davies (1987): Rates and mechanics of coastal cliff erosion in Lower Lias rocks, *Proc. Coastal Sediments* '87, ASCE, 1855–1870.

Williams, A. T. and G. T. Roberts (1995): The measurement of pebble impacts and wave action on shore platforms and beaches: The swash force transducer (swashometer), *Marine Geology* **129**, 136–143.

Williams, A. T., P. Davies and P. Bomboe (1993): Geometrical simulation studies of coastal cliff failures in Liassic strata, south Wales, UK, *Earth Surface Processes Landforms*, **18**, 703–720.

Williams, W. W. (1960): *Coastal Changes*: Routledge and Kegan Paul, London.

Wolters, G. and G. Müller (2004): The propagation of wave impact induced pressure into cracks and fissures, in: *Coastal Chalk Cliff Instability* (R. N. Mortimore and A. Duperret, eds.), Engineering Geology Special Publications, Vol. 20, pp. 121–130. Geological Society of London, London.

Wright, L. D. and A. D. Short (1983): Morphodynamics of beaches and surf zones in Australia, in: *Handbook of Coastal Processes and Erosion* (P. D. Komar, ed.), pp. 35–64: CRC Press, Boca Raton, Florida.

Yamamoto, K., T. Sunamura and T. Uda (1990): Experimental study on pressure rise in rock fissure due to wave action, *Proc. Coastal Engrg.*, JSCE, **37**, 314–318 (in Japanese).

Zhang, D. P. and T. Sunamura (1994): Multiple bar formation by breaker-induced vortices: A laboratory approach, *Proc. 24th Int. Coastal Engrg. Conf.*, ASCE, 2856–2870.

Chapter 6

Numerical Modeling of Nearshore Wave Transformation

Hitoshi Nishimura
Professor Emeritus
The University of Tsukuba, Japan

Masahiko Isobe
(Vice) President
Kochi University of Technology, Japan

1. Introduction

On solving coastal engineering problems, one of the most fundamental technologies often required is the prediction of wave transformation in the nearshore area. The wave transformation includes shoaling, refraction, diffraction, reflection, transmission, and dissipation due to bottom friction and breaking. Although analytical solutions are available for some simple situations, numerical or physical modeling is needed to predict the complex, actual wave fields. Notable progresses in the numerical modeling technology have been enabling their application to more general situations with higher accuracy. This chapter describes various model equations provided for numerical prediction of wave transformation.

Among many factors of wave transformation, refraction and diffraction are fundamental in wave field modeling. Wave shoaling is automatically included in refraction analysis and reflection can be treated as the boundary condition in diffraction analysis. Wave transmission and dissipation are normally taken into account through addition of an energy dissipation term in wave equations. Model equations are, therefore, categorized into those for refraction, diffraction, and combined refraction and diffraction.

As waves propagate in an area with nonuniform depth, local variation in wave celerity causes distortion of crestlines and consequently wave direction. Wave refraction thus produced allows geometrical analyses since it is solely a kinematic phenomenon. Local variation in wave ray spacing yields convergence or divergence of wave energy, consequently resulting in wave amplitude variation. This is a dynamic aspect of wave refraction. The water depth variation along a wave orthogonal causes wave shoaling at the same time. Model equations for wave refraction analysis will be presented in Sec. 2.

On the other hand, wave diffraction is basically a process of wave energy dispersion to smooth out the lateral gradient of the amplitude. Detailed discussion on this effect will appear in Sec. 3. In Sec. 4, numerical models of combined refraction and diffraction will be presented. During the last quarter century, numerous progresses have been achieved in this category, including mild-slope equation, nonlinear shallow water equations, Boussinesq equations and nonlinear mild-slope equations. The models have been derived and improved to take into consideration the effects of nonlinearity and dispersivity as well as irregularity, current, dissipation, permeability and so forth. Thus advanced simulation technology will become a more useful tool for better control of the nearshore wave field in the near future.

Selection of a suitable numerical model is a problem that coastal engineers often encounter. For this purpose, theoretical applicabilities of various numerical models are summarized in Sec. 5 on the basis of the thorough reviews in preceding sections.

2. Wave Refraction

2.1 *Law of wave refraction*

In a fixed system of Cartesian coordinates $\mathbf{x} = (x, y)$ on the still water surface, another local orthogonal coordinates (ξ, η) is defined along a crestline and a wave orthogonal as shown in Fig. 6.1. A simple geometrical estimation of the crestline movement leads to the Snell's law describing variation in the wave direction:

$$\frac{\partial \alpha}{\partial \eta} = \frac{1}{C} \frac{\partial C}{\partial \xi}, \tag{1}$$

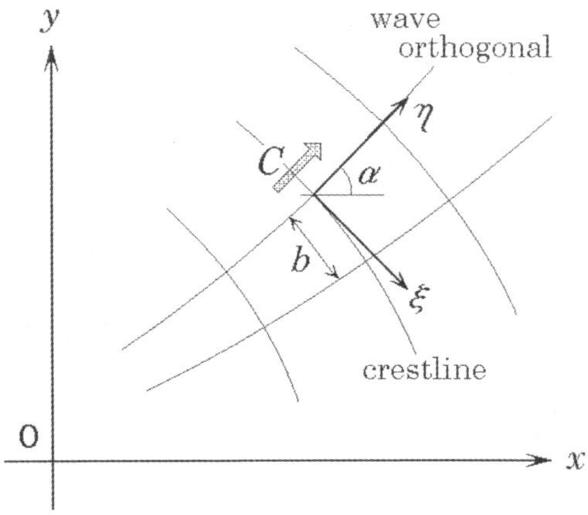

Fig. 6.1 Definition sketch of local coordinate system.

where α is the angle that the wave orthogonal makes against the ξ-axis, and C is the local wave celerity.

When a wave propagates from one location P_1 to another location P_2, it always follows a path that minimizes the traveling time. This well-known Fermat's principle is formulated as

$$\int_{P_1}^{P_2} \frac{d\eta}{C} = \min . \tag{2}$$

The above variational problem is ascribed to the corresponding Euler equation, which also proves to be equivalent to Eq. (1).

Variation in the wave orthogonal spacing b is determined through further geometrical analysis as follows:

$$\frac{1}{b}\frac{\partial b}{\partial \eta} = -\frac{\partial \alpha}{\partial \xi} \quad \text{or} \quad \frac{1}{b}\frac{\partial^2 b}{\partial \eta^2} = -\frac{1}{C}\frac{\partial^2 C}{\partial \xi^2} . \tag{3}$$

Conservation of wave energy between the wave orthogonals is described as

$$\frac{\partial}{\partial \eta}(bEC_g) = 0 \tag{4}$$

where $E = \rho g a^2/2$ is the wave energy per unit suface area, ρ is the water density, g is the gravitational acceleration, and a is the wave amplitude. Variation in the group velocity C_g brings the effect of wave shoaling into the above expression.

Let us express the wave-induced surface elevation as

$$\zeta = a \cdot e^{i\varphi} , \tag{5}$$

where a is the wave amplitude function. If the phase function φ is assumed in a simple form of

$$\varphi = \mathbf{k} \cdot \mathbf{x} - \sigma t , \tag{6}$$

(σ: the angular frequency, t: the time), irrotationality of the wave number vector $\mathrm{k} = \nabla\varphi = (k\cos\alpha, k\sin\alpha)$ is written as follows (Noda *et al.*, 1974):

$$\nabla \times \mathbf{k} = 0 \quad \text{or} \quad \frac{\partial(k \sin\alpha)}{\partial x} - \frac{\partial(k\cos\alpha)}{\partial y} = 0 . \tag{7}$$

The above expression easily proves to be equivalent to Eq. (1) through simple coordinate conversion, and is more suitable in a framework of orthogonal grids. The law of energy conservation is similarly rewritten in the following form for estimating wave amplitude:

$$\nabla \cdot (E\mathbf{C_g}) = 0 , \tag{8}$$

where $\mathbf{C}_g = (C_g \cos\alpha,\ C_g \sin\alpha)$ is the group velocity vector.

2.2 *Karlsson's energy conservation equation*

Karlsson (1969) proposed the following energy conservation equation for analyzing the refraction of waves with a directional spectrum:

$$\nabla \cdot (\varepsilon \, \mathbf{C}_{\mathrm{g}}) + \frac{\partial(\varepsilon \, C_{g\theta})}{\partial \theta} = 0, \tag{9}$$

where

$$C_{g\theta} = \frac{C_g}{C}\left(\frac{\partial C}{\partial x}\sin\,\alpha + \frac{\partial C}{\partial y}\cos\,\alpha\right), \tag{10}$$

in which the new independent variable θ represents the wave direction angle and ε is the wave energy density, $\varepsilon(\sigma,\theta)d\sigma d\theta$ being the wave energy per unit surface area distributed in a small frequency range from σ to $\sigma+d\sigma$ and in a small direction range from θ to $\theta+d\theta$. This equation essentially describes wave energy transport due to wave refraction in the 3D field of (x, y, θ) and implicitly includes the refraction law mentioned in the previous section. Refraction of waves in each frequency band can be separately treated here since energy transfer between the frequency bands is not considered in the equation.

In numerical application of the equation, it should be noticed that discretization in the wave direction can yield significant nonphysical dispersion of wave energy over direction ranges in addition to real spatial dispersion.

2.3 *Effect of current*

Let us consider waves with an absolute angular frequency σ_0 propagating in a basin where a current with horizontally nonuniform velocity $\mathbf{U}$ coexists. The local wave number k and absolute wave celerity C are related to each other as

$$k \cdot C = \sigma_0 \tag{11}$$

while the relative wave motion observed in a frame moving with the current has the following intrinsic angular frequency σ:

$$\sigma = \mathbf{k} \cdot (\mathbf{C} - \mathbf{U}) = \sigma_0 - \mathbf{k} \cdot \mathbf{U}. \tag{12}$$

For this wave-current coexistent field to be simple superposition of conservative waves and uniform current, the following dispersion relation must hold:

$$\sigma^2 = gk \tanh kh. \tag{13}$$

These two equations are combined to determine the local wave number k and intrinsic angular frequency σ.

In such a field of stationary waves with varying intrinsic frequency, the wave action E/σ is conserved as:

$$\nabla \cdot \left(\frac{E}{\sigma}\mathbf{C}_g\right) = 0, \tag{14}$$

which describes the wave amplitude variation. More rigorous theoretical background of these expressions was presented by Mei (1983) for waves coexistent with a slowly varying current.

3. Wave Diffraction

3.1 *Helmholtz equation*

For a horizontal bed, separation of variables technique is applied to solve the governing Laplace equation under the given bottom boundary condition. First, the velocity potential φ is determined as

$$\varphi = Z(z)\,\psi(x,t), \tag{15}$$

where

$$Z(z) = \frac{\cosh k(h+z)}{\cosh kh} \tag{16}$$

and k is the wave number.

On assuming a simple sinusoidal oscillation in time, further separation of variables for ψ and the complex water surface elevation ζ:

$$\psi = \hat{\psi}(\mathbf{x})\,e^{-i\sigma t} \tag{17}$$

$$\zeta = \hat{\zeta}(\mathbf{x})e^{-i\sigma t} \tag{18}$$

and the dynamic surface boundary condition yields

$$\hat{\psi} = -\frac{ig}{\sigma}\hat{\zeta}(\mathbf{x}), \tag{19}$$

where σ is the angular frequency. For such an oscillation, the kinematic surface boundary condition requires the dispersion relation in the form of Eq. (13).

Substituting the expressions (15) through (17) into the Laplace equation, we obtain the Helmholtz equation with respect to $\hat{\zeta}$:

$$\nabla^2\hat{\zeta} + k^2\hat{\zeta} = 0. \tag{20}$$

3.2 *Analytical solutions*

The Helmholtz equation (20) is analytically solved for sinusoidal waves around simply arranged straight breakwaters. For uniform waves obliquely incident at an angle α to a semi-infinite breakwater, Penny and Price (1952) derived the following solution based on Sommerfeld's theory on the diffraction of light:

$$\zeta/a = I(\lambda^+)\exp[ikr\cos(\alpha+\theta)] + I(\lambda^-)\exp[ikr\cos(\alpha-\theta)], \tag{21}$$

where $(r,\, \theta)$ is the polar coordinate system defined as in Fig. 6.2, a is the amplitude of incident waves, and by using Fresnel integrals $C(\lambda)$ and $S(\lambda)$,

$$I(\lambda) = \frac{1+i}{2}\int_{-\infty}^{\lambda}\exp\left(-i\frac{\pi\lambda^2}{2}\right)\mathrm{d}\lambda$$

$$= \frac{1}{2}[1 + C(\lambda) + S(\lambda)] - \frac{i}{2}[C(\lambda) - S(\lambda)],$$

$$\lambda\pm = -\sqrt{\frac{4kr}{\pi}}\sin\frac{\alpha\pm\theta}{2}. \tag{22}$$

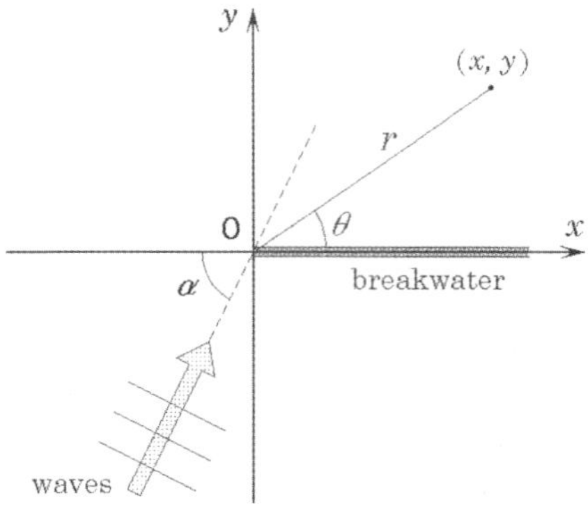

Fig. 6.2 Definition sketch of polar coordinate system.

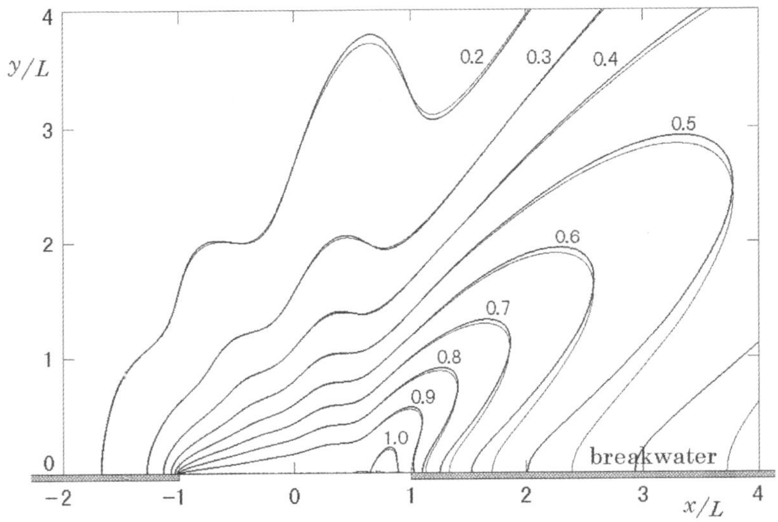

Fig. 6.3 Relative amplitudes of diffracted waves behind breakwaters (Penny and Price solution).

They also proposed a method to compose approximate solutions for various arrangement of multiple breakwaters by superposing the above solutions. So-called diffraction diagrams depicting such solutions had long been used for practical purposes.

Another typical problem is the diffraction of uniformly incident waves passing through a gap of a straight breakwater. A rigorous solution for this case is given in the form of Mathieu function series (McLachlan, 1964). Figure 6.3 shows an example of diffraction diagrams representing the approximate and rigorous solutions for a condition that the gap width is two wavelengths and the angle of wave incidence is 30°. In the figure, the local wave amplitudes are normalized with the incident wave amplitude and the spatial coordinates with the wavelength L. Accuracy of

Penny and Price's approximate solution is rather acceptable except for the close lee of right-side breakwater, although the boundary condition of no wave transmission across the breakwaters is not fully satisfied.

3.3 *Green-function method*

Plan configuration of coastal area is often complicated especially in the presence of artificial structures. In such cases, numerical approaches are required to analyze local wave behavior, in which diffraction and multiple reflection of waves become important. In the numerical calculation of direct integration type based on 2D discretization of equations and variables, the effect of wave diffraction tends to be underestimated, and multiple reflection of waves often raises a difficulty in maintaining stability of the computation. Another difficult problem lies in the treatment of computational boundaries allowing free transmission of waves. The Green-function method, a kind of boundary element method, provides a useful tool for the present purpose, avoiding the difficulties mentioned above.

In the following sections, discussions are focused on a basin with uniform depth, for which a simple analytical expression of Green function is available. For these years, however, efforts are being paid for extended application of the method to arbitrary bathymetry, where local Green functions are to be numerically calculated from the mild slope equation (Nishimura and Kwak, 1998).

(1) Outline of the method

Consider a semi-infinite plane $y \geq 0$ with a perfectly reflective boundary on the x-axis. If a wave source with intensity Γ is located at a point $\mathrm{x}_0 = (x_0, 0)$ on the boundary, the field of semi-circular waves emitting from the source is given as

$$\hat{\zeta} = \Gamma \cdot H_0^{(1)}(k|\mathbf{x} - \mathbf{x}_0|), \tag{23}$$

where $H_0^{(1)}$ denotes the zeroth-order Hankel function of the first kind. This expression satisfies the Helmholtz equation (20) as well as the boundary condition

$$\frac{\partial \hat{\zeta}}{\partial y} = 0 \quad \text{on} \quad y = 0. \tag{24}$$

Next, a basin surrounded by physical and/or imaginary boundaries Ω is considered. A wave field produced by distributed wave source $\gamma(\xi)$ along the boundary may be obtained through linear superposition of unit solution (23) as:

$$\hat{\zeta}\,(\mathbf{x}) = \int_{\Omega} \gamma(\xi) \cdot H_0^{(1)}(k|\mathbf{x} - \mathbf{x}_\xi|)\, d\xi, \tag{25}$$

where ξ is the coordinate taken along the boundary and $\mathbf{x}_\xi$ is the corresponding position vector. Solution is completed by determining the source distribution $\gamma(\xi)$ so that resulting wave field may satisfy all the boundary conditions imposed.

In numerical treatment, the total boundary Ω is divided into small segments $\Omega_i (i = 1.2, \ldots, N)$ and the effect of each segment is represented by the wave source intensity Γ_i assumed on it. Then,

$$\hat{\zeta}(\mathbf{x}) = \sum_{i=1}^{N} \Gamma_i \cdot H_0^{(1)}(k|\mathbf{x} - \mathbf{x}_i|), \tag{26}$$

where x_i is the position vector of each wave source.

On applying the present method to a basin of complicated plan shape, it is recommended to divide the total computational region into several concave polygonal subregions for separate treatment. This procedure gives explicit physical meaning to local wave sources on each boundary, and reduces the sizes of matrices handled. A common imaginary boundary thereby appears between two neighboring subregions, where matching conditions are imposed for continuation of solutions. The computational procedure described above is first applied to the subregion of wave incidence, and then to successive subregions. Such series of computations are to be repeated until all the boundary conditions are satisfied.

(2) Physical boundary

Let us again look into the problem of semi-infinite plane $y \geq 0$ with a solid boundary along the x-axis. For this particular case, as is well known, the wave field produced by a unit wave Γ located at an arbitrary point $\mathbf{x}_* = (x_*, y_*)$ in the plane is given by the following Green function:

$$G(\mathbf{x}; \mathbf{x}_*) = \frac{i}{4}[H_0^{(1)}(k|\mathbf{x} - \mathbf{x}_*|) + H_0^{(1)}(k|\mathbf{x} - \bar{\mathbf{x}}_*|)], \tag{27}$$

where $\bar{\mathbf{x}}_* = (x_*, -y_*)$ denotes the mirror image point of x_* with respect to x-axis.

Since the incident complex wave amplitude, $\hat{\zeta}_I$, satisfies the Helmholtz equation, application of Green's theorem to G and $\hat{\zeta}_I$ yields

$$\zeta_I(x, 0) = -\frac{i}{2}\int_{-\infty}^{\infty} \frac{\partial\hat{\zeta}_I(x_0, y_0)}{\partial y_0}|_{y_0=0} \cdot H_0^{(1)}(k|x - x_0|)\, dx_0. \tag{28}$$

This relationship between the complex wave amplitude $\hat{\zeta}_I$ and its normal gradient to the boundary holds generally for any composite waves incident to the boundary as proved on the basis of linear superposition of component wave fields.

Similarly, the following expression can be deduced for the wave field $\hat{\zeta}_R$ produced by a source distribution along the boundary:

$$\hat{\zeta}_R(\mathbf{x}) = -\frac{i}{2}\int_{-\infty}^{\infty} \frac{\partial\hat{\zeta}_R(x_0, y_0)}{\partial y_0}|_{y_0=0}\, H_0^{(1)}(k|\mathbf{x} - \mathbf{x}_0|)\, dx_0, \tag{29}$$

where $\mathbf{x}_0 = (x_0, 0)$. Comparison of the above equation with Eq. (25) leads to a simple expression of source intensity, that is,

$$\gamma(x_0) = -\frac{i}{2}\left.\frac{\partial\hat{\zeta}_R(x_0, y_0)}{\partial y_0}\right|_{y_0=0}. \tag{30}$$

These incident and reflected waves are superposed in front of a boundary to produce a composite wave field $\hat{\zeta}_I + \hat{\zeta}_R$. Perfect reflection of waves which causes zero surface gradient at the boundary in the normal direction to the boundary:

$$\frac{\partial}{\partial y}(\hat{\zeta}_I + \hat{\zeta}_R) = 0 \quad \text{on} \quad y = 0, \tag{31}$$

directly determines the source intensity distribution along the same boundary to produce reflected waves as follows:

$$\gamma_R(x_0) = \frac{i}{2} \left. \frac{\partial \hat{\zeta}_I(x_0, y_0)}{\partial y_0} \right|_{y_0=0} . \tag{32}$$

The simplest assumption for description of partial reflection may be

$$\gamma_R(x_0) = \frac{iK_\mathrm{R}}{2} \left. \frac{\partial \hat{\zeta}_I(x_0, y_0)}{\partial y_0} \right|_{y_0=0} , \tag{33}$$

where K_R is the reflection coefficient. It is noted as an extreme case that the simplest expression of $\gamma_R(x_0) = 0$ becomes a very accurate condition for an absorbing boundary allowing free transmission of waves.

(3) Imaginary boundary

Rational treatment of continuation condition along imaginary boundaries is vitally important to maintain the accuracy of total analysis.

Consider an imaginary boundary Ω_{12} separating subregions D_1 and D_2. Two alignments of wave sources γ_{R1} and γ_{R2} are to be assumed along such a boundary on its both sides for the subregions D_1 and D_2, respectively. In each subregion D_j ($j = 1, 2$), waves $\hat{\zeta}_{Ij}$ emitting from the other boundaries are incident to the boundary Ω_{12}, and waves $\hat{\zeta}_{Rj}$ emitting from the distributed source γ_{Rj} are superposed on it to produce an overall wave field $\hat{\zeta}_{Ij} + \hat{\zeta}_{Rj}$. These two wave fields must coincide along the common boundary Ω_{12} in terms of both the wave amplitude and their gradient, that is,

$$\hat{\zeta}_{I1} + \hat{\zeta}_{R1} = \hat{\zeta}_{I2} + \hat{\zeta}_{R2} \quad \text{on} \quad \eta = 0 \tag{34}$$

$$\frac{\partial}{\partial \eta}(\hat{\zeta}_{I1} + \hat{\zeta}_{R1}) = \frac{\partial}{\partial \eta}(\hat{\zeta}_{I2} + \hat{\zeta}_{R2}) \quad \text{on} \quad \eta = 0, \tag{35}$$

where η-axis is taken normal to the boundary from D_1-side to D_2-side with its origin on the boundary, and

$$\hat{\zeta}_{Rj}(\mathbf{x}) = \int_{\Omega_{12}} \gamma_{Rj}(\xi) \cdot H_0^{(1)}(k|\mathbf{x} - \mathbf{x}_\xi|)\, d\xi, \tag{36}$$

$$\gamma_{R1} = -\frac{i}{2} \frac{\partial \hat{\zeta}_{R1}}{\partial \eta} \Big|_{\eta=0} , \quad \gamma_{R2} = \frac{i}{2} \frac{\partial \hat{\zeta}_{R2}}{\partial \eta} \Big|_{\eta=0} . \tag{37}$$

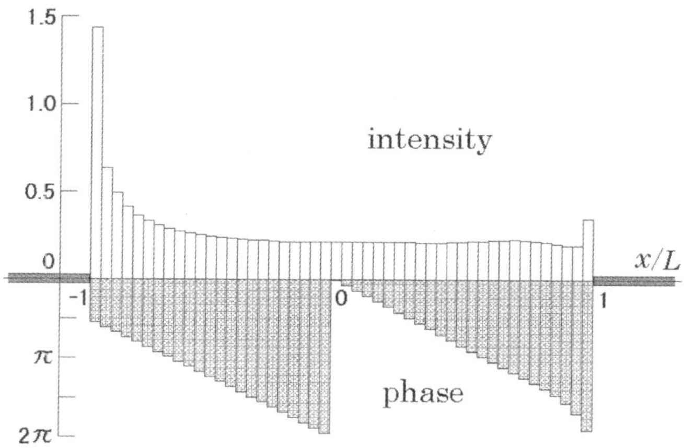

Fig. 6.4 Source intensity distribution for diffracted waves.

From Eqs. (35) and (37),

$$\gamma_{R1} + \gamma_{R2} = \frac{i}{2}\frac{\partial}{\partial \eta}(\hat{\zeta}_{I1} - \hat{\zeta}_{I2})|_{\eta=0}. \tag{38}$$

Substituting Eqs. (36) and (38) into Eq. (34), we obtain a simple integral equation for γ_{R1} or γ_{R2}. In numerical solution of the integral equation, computation of inverse matrix is required only once for each boundary of this type.

(4) Trial computation

The Green-function method was applied to the same problem of a breakwater gap treated in Sec. 3.2. Wave fields on the front side (subregion D_1) and lee side (subregion D_2) of the breakwaters were determined so as to satisfy the matching conditions along the gap, which was divided into 48 boundary segments. In this simple case of semi-infinite subregion D_2, the corresponding wave source γ_{R2} can be regarded solely as that of transmitted waves.

As shown in Fig. 4, a nonuniform source intensity distribution represents the transmission of uniform incident waves. Figure 6.5 shows a calculated amplitude distribution of transmitted waves behind the breakwaters again in comparison with the corresponding rigorous solution. A surprisingly good agreement observed in the figure implies validity and usefulness of the present method.

4. Combined Refraction and Diffraction

4.1 *Mild-slope equation*

The assumptions which enable a simple mathematical formulation of wave transformation on a sloping bottom are smallness of the wave amplitude and mildness of the bottom slope. These two assumptions lead to the mild-slope equation (MSE) which

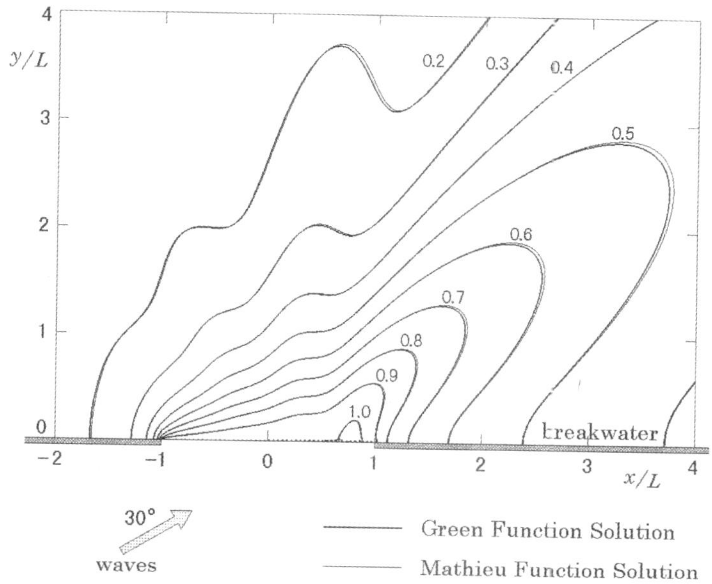

Fig. 6.5 Relative amplitudes of diffracted waves behind breakwaters (Green-function solution).

was first derived by Berkhoff (1972). The MSE including effects of both refraction and diffraction has widely been used in engineering practices.

(1) Original mild-slope equation

(a) Derivation. For a horizontal bed, the velocity potential φ is expressed by Eq. (15). The energy conservation yields

$$\int_{-h}^{0} \varphi \left(\nabla^2 \varphi + \frac{\partial^2 \varphi}{\partial z^2} \right) dz = 0, \tag{39}$$

where the upper limit of integral is taken at the still water level upon the assumption of small amplitude. By substituting Eq. (15) into Eq. (39), the following time-dependent form of the mild slope equation can be obtained (Smith and Sprinks, 1975), where the Laplace equation and boundary conditions were used and second- and higher-order terms with respect to the bottom slope neglected:

$$\nabla(CC_g \nabla \psi) + (k^2 CC_g - \sigma^2)\psi - \frac{\partial^2 \psi}{\partial t^2} = 0. \tag{40}$$

On assuming a sinusoidal oscillation given by Eq. (17), the mild slope equation is obtained as

$$\nabla(CC_g \nabla \hat{\psi}) + k^2 CC_g \hat{\psi} = 0. \tag{41}$$

In spite of the assumption which results in the vertical distribution of the velocity potential as given by Eq. (16), the MSE proves to be applicable up to a slope of 1/3 (Booij, 1983), covering most of practical situations. A finite series expression of the velocity potential was proposed to improve accuracy of the MSE (Massel, 1993).

(b) Physical interpretation. To give an idea of refraction and diffraction effects included in the MSE, the complex amplitude of the velocity potential is expressed in terms of the amplitude a and phase angle $\hat{\varphi}$ (Berkhoff *et al.*, 1982; Ebersole, 1985) as

$$\hat{\psi} = ae^{i\hat{\varphi}}. \tag{42}$$

Substitution of the above expression into the mild-slope equation, Eq. (41), leads to the following extended expressions of the eikonal equation and energy conservation for real and imaginary parts, respectively:

$$(\nabla\hat{\varphi})^2 = k^2 + \frac{\nabla(CC_g\nabla_a)}{CC_g a}, \tag{43}$$

$$\nabla(a^2 CC_g \nabla\hat{\varphi}) = 0. \tag{44}$$

If, for example, there appears a local maximum of a on a horizontal bed, the second term added in Eq. (43) takes a negative value, causing smaller substantial wave number $|\nabla\hat{\varphi}|$ smaller than k determined from the dispersion relation. The wave amplitude then tends to decrease as suggested by Eq. (43). This dispersion of wave energy is interpreted as an effect of diffraction.

(2) Variations of mild-slope equation

(a) Helmholtz equation: An alternative form of the MSE can be derived by substituting the following transformation:

$$\hat{\psi}'(\mathbf{x}) = \sqrt{CC_g}\hat{\psi}(\mathbf{x}), \tag{45}$$

into Eq. (41) and neglecting terms of the second-order with respect to the bottom slope (Radder, 1979):

$$\nabla^2\hat{\psi}' + k^2\hat{\psi}' = 0. \tag{46}$$

The above elliptic equation of Helmholtz type is often more convenient to deal with.

(b) Slowly varying current: Kirby (1984) derived a mild-slope equation for the presence of a slowly varying current:

$$\nabla(CC_g\nabla\phi) + (k^2CC_g - \sigma^2)\psi - (\nabla\mathbf{U})\frac{D\psi}{Dt} - \frac{D^2\psi}{Dt^2} = 0, \tag{47}$$

where $\mathbf{U}$ is the current velocity vector in the horizontal plane, the total diferential operator D/Dt denotes

$$\frac{D}{Dt} = \frac{\partial}{\partial t} + (\mathbf{U}\nabla) \tag{48}$$

and the intrinsic angular frequency $\sigma = \sigma_0 - \mathbf{kU}$ (σ_0: the absolute angular frequency) satisfies the dispersion relation, Eq. (13).

(c) Permeable bed: The MSE for a permeable bed was derived in a similar manner to the case of a fixed bed (Rojanakamthorn *et al.*, 1990).

(d) Energy dissipation: Dalrymple *et al.* (1984) introduced an energy dissipation term into the MSE as

$$\nabla(CC_g\nabla\hat{\psi}) + (k^2CC_g + i\sigma f_D)\hat{\psi} = 0. \tag{49}$$

By substituting Eq. (42) into the above equation, the following equation is obtained from the imaginary part:

$$\nabla(a^2CC_g\nabla\hat{\varphi}) = -a^2 f_D, \tag{50}$$

from which f_D is regarded as an energy dissipation coefficient. Dalrymple *et al.* (1984) suggested various formulations of f_D for a porous bottom, viscous mud bottom, laminar bottom boundary layer, densely packed surface film and so on.

(3) Breaking transformation

An energy dissipation term is also employed to reproduce the effect of wave breaking (Isobe, 1987). Breaking point is first determined (Watanabe *et al.*, 1984) as

$$\gamma = \gamma_b, \tag{51}$$

where γ denotes the ratio between the water particle velocity at the still water level, u_c, and wave celerity C:

$$\gamma = \frac{u_c}{C}. \tag{52}$$

Now, the breaking criterion is given as

$$\gamma_b = 0.53 - 0.3\exp[-3\sqrt{h/L_o}] + 5\tan^{3/2}\beta\exp[-45(\sqrt{h/L_o} - 1)^2], \tag{53}$$

where $\tan\beta$ is the bottom slope and L_o the deepwater wavelength. The coefficient f_D is so fixed as to yield constant wave height to water depth ratio γ_s on a uniformly sloping beach and to vanish for γ smaller than the critical ratio γ_r.

$$f_D = \frac{5}{2}\sqrt{\frac{g}{h}}\sqrt{\frac{\gamma - \gamma_r}{\gamma_s - \gamma_r}}\tan\beta, \tag{54}$$

$$\gamma_s = 0.4(0.57 + 0.53\tan\beta), \tag{55}$$

$$\gamma_r = 0.135. \tag{56}$$

4.2 *Time-dependent mild-slope equations*

(1) Time-dependent mild-slope equations for regular waves

Since the time-dependent MSE (40) is of hyperbolic type, it can be split into two simultaneous first-order partial differential equations (Nishimura *et al.*, 1983; Copeland, 1985; Watanabe and Maruyama, 1986). As an example, the equations proposed by Watanabe and Maruyama (1986) are cited below.

First, ψ'' is defined as

$$\psi''(\mathbf{x}, t) = n\psi(\mathbf{x}, t), \tag{57}$$

where $n = C_g/C$. For a sinusoidal oscillation expressed by Eq. (17), the mild-slope equation (41) is alternatively written in a hyperbolic form as

$$\nabla\left(\frac{C^2}{n}\nabla\psi''\right) - \frac{1}{n}\frac{\partial^2\psi''}{\partial t^2} = 0. \tag{58}$$

Through definitions of

$$\zeta = -\frac{1}{gn}\frac{\partial\psi''}{\partial t} \tag{59}$$

and

$$\mathbf{Q} = \frac{C^2}{gn}\nabla\psi'', \tag{60}$$

a corresponding set of first-order equations is obtained as follows:

$$\frac{\partial\zeta}{\partial t} + \nabla\mathbf{Q} = 0, \tag{61}$$

$$\frac{\partial\mathbf{Q}}{\partial t} + \frac{C^2}{n}\nabla(n\zeta) = 0. \tag{62}$$

An energy dissipation term is to be added to Eq. (62) for analysis of breaking transformation:

$$\frac{\partial\mathbf{Q}}{\partial t} + \frac{C^2}{n}\nabla(n\zeta) + f_D\mathbf{Q} = 0. \tag{63}$$

The numerical model based on Eqs. (61) and (63) has been tested through application to wave field analysis including refraction, diffraction and breaking. Figure 6.6 shows a comparison between calculated and measured breaking lines and wave height distributions.

(2) Time-dependent mild-slope equations for random waves

Random waves consist of an infinite number of component waves with different frequencies. The sinusoidal oscillation with a representative angular frequency $\bar{\sigma}$ is removed from ψ' of random waves:

$$\psi'(\mathbf{x}, t) = \tilde{\psi}'(\mathbf{x}, t)e^{-i\bar{\sigma}t}. \tag{64}$$

Introduction of a Padé approximant to k^2:

$$k^2 = \frac{b_0 + b_1(\Delta\sigma) + b_2(\Delta\sigma)^2}{1 - a_1(\Delta\sigma)}, \tag{65}$$

leads to the following time-dependent mild-slope equation for random waves (Isobe, 1994):

$$\nabla^2\tilde{\psi}' - ia_1\nabla^2\left(\frac{\partial\tilde{\psi}'}{\partial t}\right) + b_0\tilde{\psi}' + ib_1\frac{\partial\tilde{\psi}'}{\partial t} - b_2\frac{\partial^2\tilde{\psi}'}{\partial t^2} = 0. \tag{66}$$

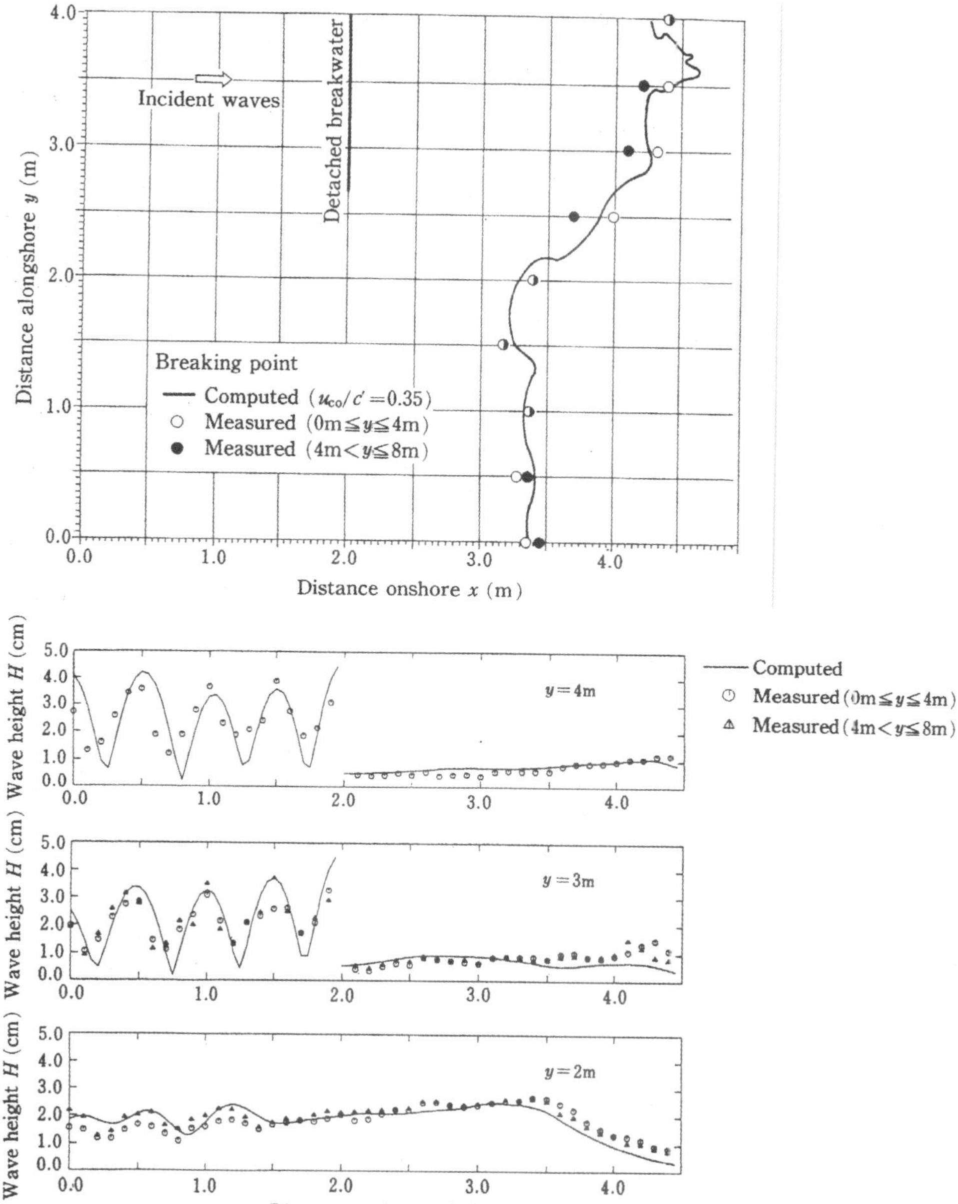

Fig. 6.6 Location of breaker line and cross-shore distribution of wave height (Watanabe and Maruyama, 1986).

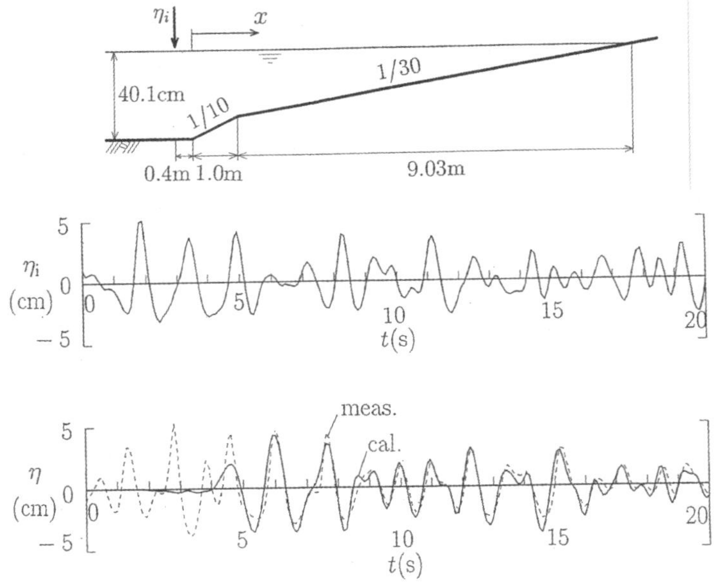

Fig. 6.7 Time history of water surface fluctuation of shoaling random waves; top: bottom configuration, middle: measured incident wave history used as input data for numerical calculation, bottom: calculated and measured wave histories at 4 m shoreward from measuring point of the above incident wave history (Isobe, 1994).

All the coefficients in the above equation are independent of the frequency. It is, therefore, used to calculate $\tilde{\psi}'$ composed of infinite number of waves with different frequencies, and to directly estimate transformation of random waves. Figure 6.7 shows a comparison of calculated and measured water surface fluctuations due to shoaling random waves.

4.3 *Parabolic equation*

Since the mild-slope equation is a partial differential equation of elliptic type, every part of the boundary affects the whole wave field of concern. This implies that dependent variables have to be determined simultaneously for all the grid points and the computation requires fairly much time. The computational labor is drastically saved through parabolic approximation allowing step by step solution from offshore to onshore boundary.

(1) Derivation of parabolic equation

An alternative form, Eq. (46), of the MSE can be rewritten as

$$\frac{\partial^2 \hat{\psi}'}{\partial x^2} + \frac{\partial^2 \hat{\psi}'}{\partial y^2} + k^2 \hat{\psi}' = 0. \tag{67}$$

If the wave direction is roughly confined in the x-direction, the phase change in the direction may approximately be separated as

$$\hat{\psi}' = \breve{\psi}' e^{i \int \mathbf{k} d\mathbf{x}}, \tag{68}$$

in which $\breve{\psi}'$ is assumed to be a weak function of x. In an expanded form of the x-derivative in Eq. (67):

$$\frac{\partial^2 \hat{\psi}'}{\partial x^2} = \left(\frac{\partial^2 \breve{\psi}'}{\partial x^2} + 2ik \frac{\partial \breve{\psi}'}{\partial x} + i \frac{\partial k}{\partial x} \breve{\psi}' - k^2 \breve{\psi}' \right) e^{i \int \mathbf{k} d\mathbf{x}}, \tag{69}$$

the first term in parentheses is consequently neglected to yield a parabolic equation:

$$2ik \frac{\partial \breve{\psi}'}{\partial x} + \frac{\partial^2 \breve{\psi}'}{\partial y^2} + i \frac{\partial k}{\partial x} \breve{\psi}' = 0. \tag{70}$$

Note that the above equation is valid only when waves propagate mostly in the x-direction.

(2) Variations of parabolic equation

Various parabolic approximations have been proposed to expand the range of application. An equation for obliquely propagating waves to the Cartesian coordinates was derived by Kirby, (1986a), whereas parabolic equations in the curvilinear coordinates by Lozano and Liu (1980), Isobe (1986) and others. Nonorthogonal coordinates were employed by Kirby (1988), and Liu and Boissevain (1988) to improve the applicability of parabolic equations.

A parabolic equation with energy dissipation term was proposed by Dalrymple *et al.* (1984). Isobe (1987) presented a parabolic equation model for combined refraction, diffraction and breaking of random waves. Liu and Tsay (1983) and Kirby (1986b) tried to take into account wave reflection ignored in the basic parabolic equations.

Applications of various parabolic equations to actual fields are also found in Dingemans *et al.* (1984) and Shimizu *et al.* (1992). Figure 6.8 shows an example of such application in which the spatial distributions of the significant wave height and mean direction are compared between calculation and measurement.

Parabolic equations for diffraction of weakly nonlinear Stokes-type waves were derived by Yue and Mei (1980) for uniform depth, and by Kirby and Dalrymple (1983) and Liu and Tsay (1984) for variable depth.

4.4 *Nonlinear shallow-water equations*

When the wavelength is extremely long in comparison with the water depth, the momentum equation in the vertical direction reduces to the hydrostatic pressure distribution:

$$p = \rho g(\zeta - z). \tag{71}$$

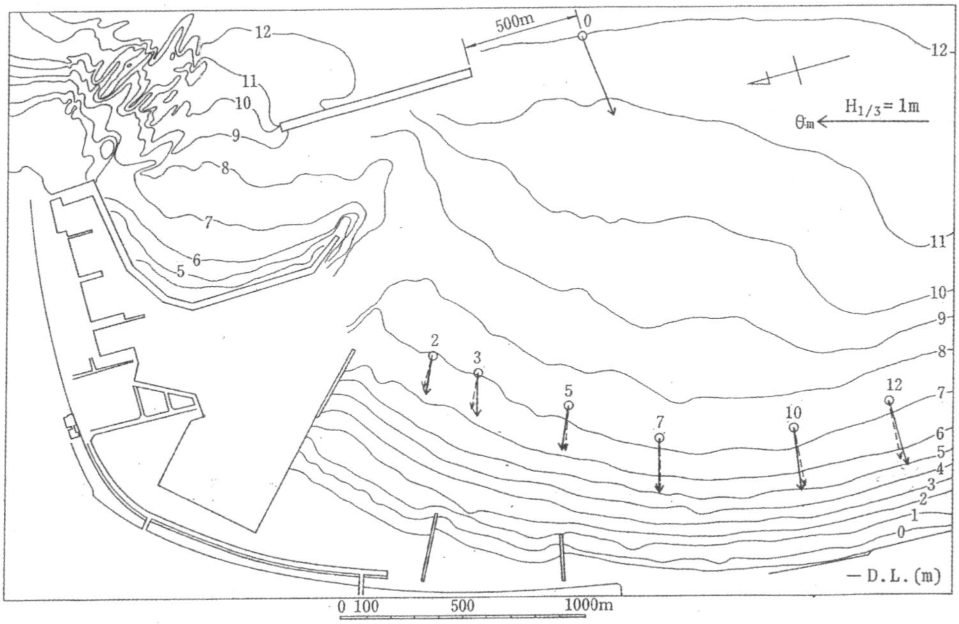

Fig. 6.8 Comparison of the spatial distributions of significant wave height (length of arrows) and mean direction (direction of arrows) between calculation by parabolic model (dash lines) and measurement (solid lines) (Isobe *et al.*, 1988).

The basic equations and boundary conditions are then simplified and nonlinear shallow-water equations are written (Stoker, 1967) as

$$\frac{\partial \zeta}{\partial t} + \nabla[(h+\zeta)\mathbf{u}] = 0, \tag{72}$$

$$\frac{\partial \mathbf{u}}{\partial t} + (\mathbf{u}\nabla)\mathbf{u} = -g\nabla\zeta \tag{73}$$

As can be proved by the characteristics method, the equations do not allow any waves with permanent form on a horizontal bed, since vertical acceleration has been ignored. The nonlinear shallow-water equations should, therefore be used when wave transformation due to other effects such as sloping bottom and/or energy dissipation are predominant (Carrier and Greenspan, 1958). An advantage of the above equations is that they automatically reproduce a discontinuity due to wave breaking without any aids of empirical relationship or constant (Keller *et al.*, 1960; Freeman and Le Méhauté, 1964). This is the primary reason why these equations are used to analyze the process of wave breaking and run-up (Hibberd and Peregrine, 1979; Ryrie, 1983).

Kobayashi and his co-workers added a bottom friction term to the momentum equation and extensively studied the wave transformation in the surf zone, including runup, set-up and reflection under various conditions of smooth and rough bottoms, gentle and steep slopes, and regular and irregular waves (Kobayashi *et al.*, 1989;

Kobayashi and Wurjanto, 1990, 1992). Although discrepancies were found between calculation and measurement in wave shoaling and breaking point, the model reproduced wave transformation in very shallow water fairly well. The nonlinear shallow-water equations were also used by Asano (1994) and Kobayashi and Karjadi (1994) in analyses of swash oscillation due to obliquely incident waves, while those for a permeable bed was presented by Kobayashi and Wurjanto (1990).

4.5 *Boussinesq equations*

(1) Original Boussinesq equations

Boussinesq equations are to deal with weakly nonlinear and weakly dispersive shallow-water waves. Those for variable water depth were derived by Peregrine (1967).

Basic assumption here is $O(H/h) \sim O((h/L)^2)$ as in the cnoidal wave theory. By retaining the terms up to the orders of $(H/h)^2$ and $(H/h)\cdot(h/L)^2$, the following Boussinesq equation is obtained:

$$\frac{\partial\zeta}{\partial t} + \nabla[(h+\zeta)\bar{\mathbf{u}}] = 0, \tag{74}$$

$$\frac{\partial\bar{\mathbf{u}}}{\partial t} + (\bar{\mathbf{u}}\nabla)\bar{\mathbf{u}} + g\nabla\zeta = -\frac{h^2}{6}\frac{\partial}{\partial t}\nabla(\nabla\bar{\mathbf{u}}) + \frac{h}{2}\frac{\partial}{\partial t}\nabla[\nabla(h\bar{\mathbf{u}})], \tag{75}$$

where $\bar{\mathbf{u}}$ is the horizontal components of the depth-average particle velocity.

The Boussinesq equations have been verified by many researchers through application for wave transformation in shallow water (Abbott *et al.*, 1978; Madsen and Warren, 1984; Rygg, 1988; Freilich *et al.*, 1990; Elgar *et al.*, 1993). A numerical calculation method has also been developed on the basis of Fourier transform (Liu *et al.* 1985; Kirby, 1990).

(2) Various Boussinesq equations

(a) Modified Boussinesq equations: The above Boussinesq equations are applicable only for shallow water because of the underlying assumption. The water depth to deepwater wavelength ratio h/L_o has to be smaller than 0.22 to maintain the error within 5% in terms of the wave celerity. In order to improve the accuracy, the following modified momentum equation was derived as substitute for Eq. (75) (Madsen *et al.*, 1991; Madsen and Sørensen, 1992):

$$\begin{aligned}\frac{\partial\mathbf{Q}}{\partial t} + \nabla\left(\frac{|\mathbf{Q}\cdot\mathbf{Q}|}{(h+\zeta)}\right) + g(h+\zeta)\nabla\zeta = {} & \left(B+\frac{1}{3}\right)h^2\frac{\partial}{\partial t}\nabla(\nabla\mathbf{Q}) + Bgh^2\nabla(h\nabla^2\zeta) \\ & + Bgh^2(|\nabla\cdot\nabla\zeta|\nabla h) + \frac{h}{6}\nabla h\frac{\partial}{\partial t}(\nabla\mathbf{Q}) \\ & + \frac{h}{6}\frac{\partial}{\partial t}|\nabla\cdot\mathbf{Q}|(\nabla h),\end{aligned} \tag{76}$$

where $\mathbf{Q}$ denotes the 2D vector of the flow rate per unit width. Equation (76) with $B = 1/15$ can be used for the range of $h/L_o < 0.5$ within 5% error in the wave celerity (Madsen *et al.*, 1991).

Another set of modified Boussinesq equations was derived by taking the horizontal velocity at an arbitrary elevation as a dependent variable (Nwogu, 1993) and applied to the nonlinear random wave interaction (Nwogu, 1994). Chen and Liu (1995) proposed parabolic models through application of Fourier transform to the modified Boussinesq equations.

(b) Breaking dissipation: Models have been proposed to incorporate the effect of wave breaking in the Boussinesq equations (Karambas and Koutitas, 1992; Schäffer *et al.*, 1993; Kabiling and Sato, 1993). Figure 6.9 shows an example of calculated wave height distribution and nearshore current.

(c) Permeable bed: Equations for weakly nonlinear shallow water waves on a permeable bed can be derived upon similar assumptions to those for Boussinesq equations (Cruz *et al.*, 1997).

4.6 *Nonlinear mild-slope equations*

Nadaoka and Nakagawa (1993) and Nadaoka *et al.* (1994) first presented the equations for fully-nonlinear and fully-dispersive waves. Alternative formulations were also found in Isobe (1994) and in Nochino (1994).

Equations are derived by expanding the dependent variables such as the velocity, velocity potential or pressure into a series using a set of vertical distribution functions. The expression is substituted into the basic equations and they are integrated in the vertical direction. Since no assumption is made on the nonlinearity and dispersivity, the resultant equations are available even for analysis of strongly nonlinear transformation of waves. The number of terms in the series (usually two or three) is to be decided depending on the accuracy required.

Isobe (1994) expanded the velocity potential φ using vertical distribution functions Z_α selected in accordance with the problem of concern,

$$\varphi(\mathrm{x}, \mathrm{z}, t) = \sum_{\alpha=1}^{N} Z_\alpha(z; h(\mathrm{x})) f_\alpha(\mathrm{x}, t) \equiv Z_\alpha f_\alpha. \tag{77}$$

By applying the variational principle to the Lagrangian which is equivalent to the set of basic equations and boundary conditions, the following nonlinear mild-slope equations are obtained:

$$Z_\partial^\zeta \frac{\partial \zeta}{\partial t} + \nabla(A_{\alpha\beta} \nabla f_\beta) - B_{\alpha\beta} f_\beta + (C_{\beta\alpha} - C_{\alpha\beta})(\nabla f_\beta)(\nabla h)$$

$$+ \frac{\partial Z_\beta^\zeta}{\partial h} Z_\alpha^\zeta f_\beta (\nabla \zeta)(\nabla h) = 0, \tag{78}$$

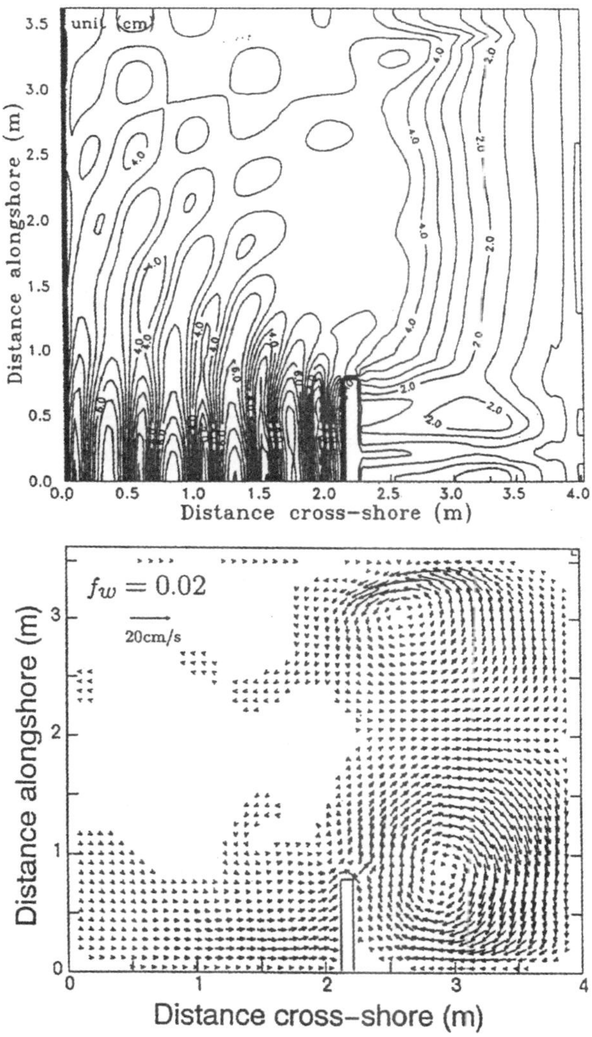

Fig. 6.9 Wave height distribution and nearshore current calculated by Boussinesq equations (Kabiling and Sato, 1993).

$$g\zeta + Z_\beta^\zeta \frac{\partial f_\beta}{\partial t} + \frac{1}{2} Z_\gamma^\zeta Z_\beta^\zeta (\nabla f_\gamma)(\nabla f_\beta) + \frac{1}{2} \frac{\alpha Z_\gamma^\zeta}{\partial z} \frac{\partial Z_\beta^\zeta}{\partial t} f_\gamma f_\beta$$

$$+ \frac{\partial Z_\gamma^\zeta}{\partial h} Z_\beta^\zeta f_\gamma (\nabla f_\beta)(\nabla h) = 0. \tag{79}$$

In the above equations, the unknowns are ζ and f_α ($\alpha = 1$ to N). Once the equations are solved numerically, the velocity potential is determined from Eq. (77). In case that the vertical distribution functions Z_α are employed from the theory for a horizontal bed, terms of second- or higher-orders with respect to the bottom slope are to be consistently neglected in Eqs. (78) and (79). Figure 6.10 compares calculated

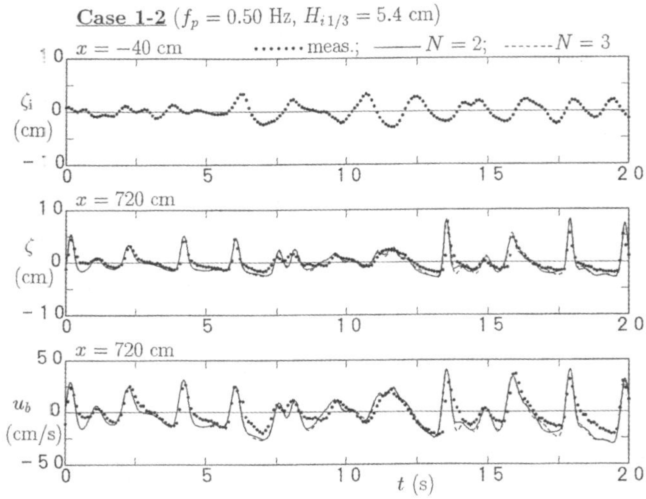

Fig. 6.10 Water surface fluctuation ζ and bottom velocty u_b of random waves on shoaling water; top: incident wave profile ζ_i (Isobe, 1994).

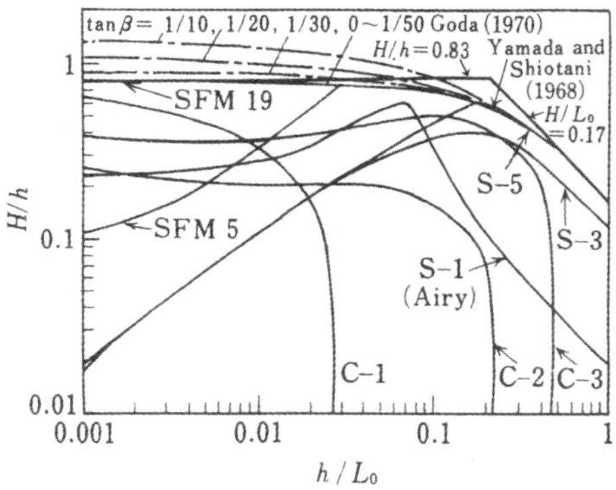

Fig. 6.11 Constant rms error lines for dynamic surface boundary condition; S: Stokes wave, C: cnoidal wave, SFM: stream function method.

and measured water surface fluctuation and bottom velocity on a sloping bed. Good agreements are found for both of them in spite of strongly nonlinear wave deformation prior to breaking.

5. Assumptions and Applicabilities of Wave Equations

Fundamental wave theories are for waves of permanent type which propagate on a uniform depth without deformation. Stokes and cnoidal wave theories represent this category. Both the theories approximately satisfy two nonlinear boundary conditions at the free surface, with accuracy depending on the order of solution. Figure 6.11

shows contours of 1% error in terms of the dynamic free surface boundary condition, where S denotes the Stokes wave theory, C the cnoidal wave theory, and SFM the stream function wave theory. Applicability of the Stokes wave theory is limited by the wave steepness H/L in deep water and by Ursell's parameter $U_r = HL^2/h^3$ in shallow water. As to the cnoidal wave theory, limiting conditions are the relative water depth h/L on deeper side and the wave height to water depth ratio H/h in shallow water.

Comparison of the assumptions underlying in the derivation of the mild-slope equation and Boussinesq equations suggest that they respectively coincide with the small amplitude wave theory (or first-order Stokes wave theory) and first-order cnoidal wave theory. The above discussion on accuracy may, therefore, be reasonably extended to each corresponding equations. The assumption of hydrostatic pressure distribution restricts the use of nonlinear shallow water equations to shallower depth than the Boussinesq equations.

Additional assumptions classifying wave equations are those on the bottom slope and directionality of local wave field. Table 6.1 summarizes the assumptions limiting application of each set of equations. It should be noted that some improved versions have been proposed such as the mild-slope equation allowing weak nonlinearity or the Boussinesq equations for more dispersive waves. These trials for improvement, however, are not necessarily successful for all the wave properties including the water surface profile, velocity and pressure.

Finally, availability of the equations is summarized in Table 6.2 for each factor of wave transformation, to be analyzed. Prior to the selection of suitable equations

Table 6.1 Assumptions behind various wave equations.

Wave equations	Bottom slope $L\|\nabla\|h/h$	Progressive $\alpha \exp^{i(kx-\sigma t)}$	Nonlinearity $H/L, H/h$	Relative depth $(h/L)^2$
Wave ray equation	first-order	yes	small[e]	arbitrary
Karlsson's equation	first-order	yes	small	arbitrary
Helmholtz equation	0	no	small	arbitrary
Takayama's method	0	no	small	arbitrary
MSE	first-order	no	small[f]	arbitrary
T-D MSE	first-order	no	small	arbitrary
Numerical wave equation	first-order[a]	no	small	arbitrary
MSE for random waves	first-order	no	small	arbitrary
Parabolic equation	first-order	yes[d]	small[f]	arbitrary
NL shallow water equations	arbitrary[b]	no	arbitrary	very small
Boussinesq water equations	arbitrary[b]	no	second-order	first-order
NL MSE	arbitrary[c]	no	arbitrary	arbitrary

[a]correction for shoaling coefficient required.
[b]shallow water assumed.
[c]dependent on vertical distribution functions.
[d]diffraction effect included.
[e]nonlinear wave celerity applicable.
[f]weakly nonlinear version available.

Table 6.2 Applicability of various wave equations.

Wave equations	Shoaling	Refraction	Diffraction	Reflection	Breaking model	water depth	w–c interaction	Irregularity	Non-linearity	Area L	Area M	Area S	Remarks
Wave ray equation	◎	◎	×			◎	○	○	△	○	○	○	nonlinear wave celerity available
Karlsson's equation	◎	◎	▽	△	○	◎	○	◎	×	○	○	○	
Helmholtz equation	○	○	◎	◎	×	◎	×	○	×			○	piecewisely uniform water depth
Takayama's method			◎	◎	×	◎	×	◎	×		○	○	uniform water depth
MSE	◎	◎	◎	◎	○	◎	○	○	△			○	
T-D MSE	◎	◎	◎	◎	○	◎	○		×			○	breaking model available, boundary conditions flexible
Numerical wave equation	○	◎	◎	◎		◎		◎	×			○	
MSE for random waves	◎	◎	◎	◎	○	◎		◎	×			○	

(*Continued*)

Table 6.2 (*Continued*)

Wave equations	Shoaling	Refraction	Diffraction	Reflection	Breaking model	water depth	w–c interaction	Irregularity	Non-linearity	Area L	Area M	Area S	Remarks
Parabolic equation	◎	◎	◎	Δ	○	◎	○	○	Δ		○	○	
NL shallow water equations	◎	◎	◎	◎	○	×		◎	◎			○	very shallow water
Boussinesq equations	◎	◎	◎	◎	○	○	○	◎	◎			○	improvement available for deep water
NL MSE	◎	◎	◎	◎		◎		◎	◎			○	

◎: applicable in original form.
○: applicable in improved form.
Δ: partially applicable in improved form.
∇: theoretically not applicable but practically acceptable.
blank: subject to further research.
×: not applicable.
area = L: from deep to shallow water.
M: around coastal structures.
S: within a harbor.

to a specific wave field analysis, one should well examine the significance of each factor and characteristics of wave transformation involved: refraction, diffraction, wave-current interaction, and dissipation as well as nonlinearity, disspersivity, and so forth. It is seen that the nonlinear mild-slope equations is applicable for the widest range of wave analyses covering most factors of wave transformation.

6. Concluding Remarks

Recent progress in the numerical modeling of nearshore wave field has enabled the treatment of combined refraction and diffraction problems even for nonlinear random waves. There still remain, however, some more factors to be included in future models: that is, dissipation due to wave breaking, strong wave-current interaction, undertow in the surf zone, swash oscillation and so forth. Existing linear theories may be enough to predict integral properties such as wave energy and radiation stress, while nonlinear wave profiles are essential for the estimation of wave overtopping, wave forces on structures, sediment transport and so forth. Three-dimensional or quasi-three-dimensional analysis is another direction of progress to be considered. Further development in this research field is expected for our better understanding and more useful modeling of coastal process.

References

Abbott, M. B., H. M. Petersen and O. Skovgaard (1978): On the numerical modelling of short waves in shallow water, *J. Hydraul. Res.* **16**, 173–203.

Asano, T. (1994): Swash motion due to obliquely incident waves, *Proc. 24th Int. Conf. on Coastal Engrg.* ASCE, 27–41.

Berkhoff, J. C. W. (1972): Computation of combined refraction-diffraction, *Proc. 13th Int. Conf. on Coastal Engrg.* ASCE, 471–490.

Berkhoff, J. C. W., N. Booij and A. C. Radder (1982): Verification of numerical wave propagation models for simple harmonic linear water waves, *Coastal Engrg.* **6**, 255–279.

Booij, N. (1983): A note on the accuracy of the mild-slope equation, *Coastal Engrg.* **6**, 255–279.

Carrier, G. F. and H. P. Greenspan (1958): Water waves of finite amplitude on a sloping beach, *J. Fluid Mech.* **4**, 97–109.

Chen, Y. and P. L.-F. Liu (1995): Modified Boussinesq equations, and associated parabolic models for water wave propagation, *J. Fluid Mech.* **288**, 351–381.

Copeland, G. J. M. (1985): A practical alternative to the mild-slope wave equation, *Coastal Engrg.* **9**, 125–149.

Cruz, E. C., M. Isobe and A. Watanabe (1997): Boussinesq equations for wave transformation on porous beds, *Coastal Engrg.* **30**, 125–156.

Dalrymple, R. A., J. T. Kirby and P. A. Hwang (1984): Wave diffraction due to areas of energy dissipation, *J. Waterway, Port, Coastal, and Ocean Engrg. ASCE* **110**, 67–79.

Dingemans, M. W., M. J. F. Stive, A. J. Kuik, A. C. Radder and N. Booij (1984): Field and laboratory verification of the wave propagation model CREDIZ, *Proc. 19th Int. Conf. on Coastal Engrg.* ASCE, 1178–1191.

Ebersole, B. A. (1985): Refraction-diffraction model for linear water waves, *J. Waterway, Port, Coastal and Ocean Engrg. ASCE* **111**, 939–953.

Elgar, S., R. T. Guza and M. H. Freilich (1993): Dispersion, nonlinearity and viscosity in shallow-water waves, *J. Waterway, Port, Coastal and Ocean Engrg. ASCE* **119**, 351–366.

Freeman, J. C. and B. M. Le Méhauté (1964): Wave breakers on a beach, and surges on a dry bed, *Proc. ASCE*, 90, No. HY2, 187–216.

Freilich, M. H., R. T. Guza and S. L. Elgar (1990): Observations of nonlinear effects in directional spectra of shoaling gravity waves, *J. Geophys. Res.* **95**(C6), 9645–9656.

Hibberd, S. and D. H. Peregrine (1979): Surf and run-up on a beach: A uniform bore, *J. Fluid Mech.* **95**, 323–345.

Isobe, M. (1986): A parabolic refraction-diffraction equation in the ray-front coordinate system, *Proc. 20th Int. Conf. on Coastal Engrg. ASCE* 306–317.

Isobe, M. (1987): A parabolic equation model for transformation of irregular waves due to refraction, diffraction and breaking, *Coastal Engrg. in Japan*, **30**, 34–47.

Isobe, M. (1994): Time-dependent mild-slope equations for random waves, *Proc. 24th Int. Conf. on Coastal Engrg. ASCE*, 285–299.

Kabiling, M. B. and S. Sato (1993): Two-dimensional nonlinear dispersive wave-current and three-dimensional beach deformation model, *Coastal Engrg. in Japan* **36**, 195–212.

Karambas, Th. V. and C. Koutitas (1992): A breaking wave propagation model based on the Boussinesq equations, *Coastal Engrg.* **18**, 1–19.

Karlsson, T. (1969): Refraction of continuous ocean wave spectra, *J. Waterways and Harbors Div., Proc. ASCE* **95**, 437–448.

Keller, H. B., D. A. Levine and G. B. Witham (1960): Motion of a bore over a sloping beach, *J. Fluid Mech.* **7**, 302–316.

Kirby, J. T. (1984): A note on linear surface wave-current interaction over slowly varying topography, *J. Geophys. Res.* **89**, No. Cl, 745–747.

Kirby, J. T. (1986a). Rational approximations in the parabolic equation method for water waves, *Coastal Engrg.* **10**, 355–378.

Kirby, J. T. (1986b). On the gradual reflection of weakly nonlinear Stokes waves in regions with varying topography, *J. Fluid Mech.* **162**, 187–209.

Kirby, J. T. (1988): Parabolic wave computations in non-orthogonal coordinate systems, *J. Waterway, Port, Coastal, and Ocean Engrg. ASCE* **114**, 673–685.

Kirby, J. T. (1990): Modelling shoaling directional wave spectra, *Proc.* 22nd *Int. Conf. on Coastal Engrg. ASCE* 109–122.

Kirby, J. T. and R. A. Dalrymple (1983): A parabolic equation for the combined refraction-diftraction of Stokes waves by mildly varying topography, *J. Fluid Mech.* **136**, 453–466.

Kirby, J. T. and R. A. Dalrymple (1984): Verification of a parabolic equation for propagation of weakly-non-linear waves, *Coastal Engrg.* **8**, 219–232.

Kobayashi, N. and E. A. Karjadi (1994): Swash dynamics under obliquely incident waves, *Proc.* 24th *Int. Conf. on Coastal Engrg. ASCE* 2155–2169.

Kobayashi, N. and A. Wurjanto (1990): Numerical model for waves on rough permeable slopes, *J. Coastal Res.* Special Issue, No. 7, 149–166.

Kobayashi, N. and A. Wurjanto (1992): Irregular wave setup and run-up on beaches, *J. Waterway, Port, Coastal, and Ocean Engrg. ASCE* **118**, 368–386.

Kobayashi, N., G. S. DeSilva and K. D. Watson (1989): Wave transformation and swash oscillation on gentle and steep slopes, *J. Geophys. Res.* **94**(C4), 951–966.

Liu, P. L.-F. and P. L. Boissevain (1988): Wave propagation between two breakwaters, *J. Waterway, Port, Coastal, and Ocean Engrg. ASCE* **114**, 237–247.

Liu, P. L.-F. and T.-K. Tsay (1983): On weak reflection of water waves, *J. Fluid Mech.* **131**, 59–71.

Liu, P. L.-F. and T. K. Tsay (1984): Refraction-diffraction model for weakly nonlinear water waves, *J. Fluid Mech.* **141**, 265–274.

Liu, P. L.-F., S. B. Yoon and J. T. Kirby (1985): Nonlinear refraction-diffraction of waves in shallow water, *J. Fluid Mech.* **153**, 185–201.

Lozano, C. and P. L.-F. Liu (1980): Refraction-diffraction model for linear surface water waves, *J. Fluid Mech.* **101**, 705–720.

Madsen, P. A. and O. R. Sørensen (1992): A new form of the Boussinesq equations with improved linear dispersion characteristics, Part 2. A slowly-varying bathymetry, *Coastal Engrg.* **18**, 183–204.

Madsen, P. A. and I. R. Warren (1984): Performance of a numerical short-wave model, *Coastal Engrg.* **8**, 73–93.

Madsen, P. A., R. Murray and O. R. Sørensen (1991): A new form of the Boussinesq equations with improved linear dispersion characteristics, *Coastal Engrg.* **15**(4), 371–388.

Massel, S. R. (1993): Extended refraction-diffraction equation for surface waves, *Coastal Engrg.* **19**, 97–126.

McLachlan, N. W. (1964): *Theory and Application of Mathieu Functions*: Dover Publication, p. 401.

Mei, C. C. (1983): *The Applied Dynamics of Ocean Surface Waves*: John Wiley & Sons, New York, 89–98.

Nadaolca, K. and Y. Nakagawa (1993): Derivation of fully-dispersive wave equations for irregular wave simulation and their fundamental characteristics, *Proc. Japan Soc. Civil Engrs.*, No. 467, 83–92 (in Japanese).

Nadaoka, K., S. Beji and Y. Nakagawa (1994): A fully-dispersive nonlinear wave model and its numerical solution, *Proc. 24th Int. Conf. on Coastal Engrg. ASCE* 427–441.

Nishimura, H. and M. S. Kwak (1998): Green-function analysis of a wave field over arbitrary bathymetry, *Proc. 26th Int. Conf. on Coastal Engrg. ASCE* 551–576.

Nishimura, H., K. Maruyama and H. Hiraguchi (1983): Wave analysis by direct numerical integration, *Proc. 30th Japanese Conf. on Coastal Engrg. JSCE* 123–127.

Nochino, M. (1994): Fully-nonlinear coupled vibration equations for irregular water waves and their basic characteristics, *Proc. Coastal Engrg.*, JSCE, **41**, 16–20 (in Japanese).

Noda, E. K., C. J. Sonu, V. C. Rupert and J. I. Collins (1974): Nearshore Circulations under Sea Breeze Conditions and Wave-Current Interactions in the Surf Zone, Tetra Tech Rep. No. TC-149-4, p. 216.

Nwogu, O. (1993): Alternative form of Boussinesq equation for nearshore wave propagation, *J. Waterway, Port, Coastal, and Ocean Engrg. ASCE* **119**, 618–638.

Nwogu, O. (1994): Nonlinear evolusion of directional wave spectra in shallow water, *Proc.* 24th *Int. Conf. on Coastal Engrg. ASCE* 467–481.

Penny, W. and T. Price (1952): The diffraction theory of sea waves and shelter afforded by breakwater, *Philos. Trans. R Soc. Lond. Ser. A* **224**, 236–253.

Peregrine, D. H. (1967): Long waves on a beach, *J. Fluid Mech.* **27**, 815–827.

Radder, A. C. (1979): On the parabolic equation method for water-wave propagation, *J. Fluid Mech.* **95**, 159–176.

Rojanakamthorn, S., M. Isobe and A. Watanabe (1990): A mathematical model of wave transformation over a submerged breakwater, *Coastal Engrg. in Japan* **32**, 209–234.

Rygg, O. B. (1988): Nonlinear refraction-diffraction of surface waves in intermediate and shallow water, *Coastal Engrg.* **12**, 191–211.

Ryrie, S. C. (1983): Longshore motion generated on beaches by obliquely incident bores, *J. Fluid Mech.* **129**, 193–212.

Schäffer, H. A., P. A. Madsen and R. Deigaard (1993): A Boussinesq model for waves breaking in shallow water, *Coastal Engrg.* **20**, 185–202.

Shimizu, T., A. Ukai and M. Isobe (1992): Field verification of numerical models for calculation of nearshore wave field, *Proc. 23rd Int. Conf. on Coastal Engrg. ASCE* 590–603.

Smith, R. and T. Sprinks (1975): Scattering of surface waves by a conical island, *J. Fluid Mech.* **72**, 373–384.

Sørensen, O. R., H. A. Schäffer, P. A. Madsen and R. Deigaard (1994): Wave breaking and induced nearshore circulations, *Proc. 24th Int. Conf. on Coastal Engrg. ASCE* 2583–2594.

Stoker, J. J. (1967): *Water Waves*: Interscience, New York, pp. 567.

Watanabe, A. and K. Maruyama (1986): Numerical modeling of nearshore wave field under combined refraction, diffraction and breaking, *Coastal Engrg. in Japan* **29**, 19–39.

Watanabe, A., T. Hara and H. Horikawa (1984): Study on breaking condition for compound wave trains, *Coastal Engrg. in Japan* **27**, 71–82.

Yue, D. K. P. and C. C. Mei (1980): Forward diffraction of Stokes waves by a thin wedge, *J. Fluid Mech.* **99**, 33–52.

Chapter 7

Physical Modeling — Renewed Importance

J. William Kamphuis
Professor Emeritus
Queen's University, Canada

1. Introduction

In a period of exponential growth in computing capability, the use of an apparently archaic tool such as physical modeling needs to be justified. Consider two attributes of physical models:

(a) *Qualitatively*, physical models are normally close simulations of the prototype.
(b) Viewing and studying a physical model's operation *adds* to the modeler's physical understanding of the problem.

Numerical models, on the other hand:

(c) can give *spurious* solutions that have little similarity to the prototype, if not formulated correctly.
(d) do not add to the modeler's understanding; they essentially *reflect* the modeler's input.[a]

Finally:

(e) Physical models were traditionally used to visualize a problem and to obtain some qualitative indications of possible solutions (see Points (a) and (b)). Today they are expected to provide more quantitative results. To translate the qualitative impressions from physical models into quantitative results is a difficult

[a]This statement refers specifically to the modeler who constructed the model. Established numerical models with built-in, long-term experience, based on the combined understanding of experts involved in developing that model can of course add to the understanding of a user.

task. It requires a thorough understanding of scaling, modeling technology, scale effects[b] and laboratory effects.[c]

Points (c) and (d) indicate that a problem must be clearly understood, before a numerical model produces valid solutions. Equations governing the processes, numerical methods, transfer functions and calibration coefficients must all be known from the outset. Points (a) and (b), however, indicate that physical model tests will produce qualitative answers, even if the problem is not fully understood. Thus physical modeling will continue to be an important tool to solve complex problems for which not all the equations and coefficients are known. Such is the case for sediment transport by waves and currents. For more tractable problems (such as fluid flow models with relatively simple boundary conditions) numerical modeling is used. Points (a) to (d) describe the distinct roles of the two types of models. If a problem is not clearly understood, the visual and qualitative feedback from a physical model will be very valuable to further understanding. A poorly formulated numerical model, on the other hand, will not be useful in such a case and an almost correct numerical model will be misleading.

Point (e) highlights the main difficulty with physical modeling. It is its major source of criticism and has been the reason for the decline in popularity of physical models after 1970, as will be seen later. The above five points will be referred back to in the remainder of this chapter.

2. Historical Development of Physical Modeling

The historical information was drawn mainly from Freeman (1929), Rouse and Ince (1957), Garbrecht (1987) and Levi (1995) and hence reflects only the history of physical modeling in Europe and America. An example will first be given of an historical physical experiment.

Aristotle (384–322 BP) firmly believed that knowledge must advance through observation. Yet, after his death, his writings were transformed into the ultimate authority on everything. One of Aristotle's technological theories was that a vacuum cannot exist. There was of course no proof! This would mean that a siphon can operate over any difference in elevation. Giovanni–Battista–Baliani set in motion a series of events in 1630, that ended the almost 2000 year reign of this Aristotelian misconception. In a letter to Galileo Galilei, he calls on actual physical experience gained while diving in the sea. He postulates that as divers experience an increase in pressure with depth, resulting from the height and density of the sea water above them,

[b]Scale effect results when some parameter (such as viscosity or roughness) cannot be scaled correctly.

[c]Input conditions, such as boundary conditions, are not modeled correctly (Examples: many years of wave climate are simulated by one or a few combinations of wave height, period and direction; directional waves are simulated with a long-crested wave generator).

we are all subjected to a pressure resulting from the height and density of air above us. He goes on to postulate that, as a result, there must exist an absolute zero pressure and hence a vacuum must form in a siphon, once zero pressure is reached. Baliani estimated atmospheric pressure to be equivalent to about 25 m of water. Next, Baliani, who is a public official in Genoa, confidently builds a water supply line, which includes a 25 mm copper pipe siphon over a 21 m high hill. When the siphon does not work, he again writes to Galileo, who replies in typical professorial fashion on 6 August 1630:

> "I truly regret that you did not ask my opinion about the siphon's performance BEFORE you incurred the expenses, for I could have saved you the costs by pointing out the impossibility of the matter."

The letter continues by pointing out that a column of water breaks, under suction when it exceeds 12 m in length, or perhaps even less. We now have two wildly different estimates of atmospheric pressure. Interestingly enough, one of the great philosopher-scientists of the day, René Descartes, continues to follow Aristotle and disputes the whole idea of a vacuum. He states that the pump material or leakage is the cause of Galileo's limit to the length of a water column under suction.

The matter is eventually settled in 1644 by an ingenious physical experiment performed by Viviani at the request of Evangelista Torricelli. Since a 12 m vertical length of tubing is rather prodigious, Torricelli suggests using mercury and Viviani in effect builds a number of mercury barometers, from which atmospheric pressure is correctly deduced. Although Torricelli *et al.* did not perceive the value of their new tool in weather forecasting, they demonstrated atmospheric pressure and the possibility that a vacuum can exist. They used a physical (mercury/air) model to put to rest the long-standing theorem ascribed to Aristotle.

Let us now trace the development of physical modeling of hydraulic phenomena, as we know it today. We begin with the hydraulic machinery tests of John Smeaton (1724–1792). In the introduction of his gold-medal winning paper of 1759 to the British Royal Society, he states:

> "What I have to communicate on this subject was originally deduced from experiments made on working models, which I look upon as the best means of obtaining the outlines in mechanical inquiries. But in this case it is very necessary to distinguish the circumstances in which a model differs from a machine in large; otherwise a model is more apt to lead us from the truth rather than towards it. Hence the common observation, that a thing may do very well in a model that will not answer in large. And, indeed, though the utmost circumspection be used in this way, the best structure of machines cannot be fully ascertained, but by making trials with them, when made of their proper size. It is for this reason that though the models referred to and the greater part of the following experiments, were made in the years 1752 and 1753, yet I deferred offering them to the Society, until I had an opportunity of putting the deduction made therefrom in real practice, in a variety of

> cases, and for various purposes; so as to be able to assure the Society that I have found them to answer."

This statement reflects both Points (a) and (e). It correctly sums up the dilemma of obtaining quantitative results from physical model tests.

After Smeaton, Ferdinand Reech (1805–1880) was instrumental in developing hydraulic modeling further and in fact he (not Froude) was first to state that the velocity scale in a model should equal the square root of the length scale. In 1872, William Froude (1810–1879) obtained funding for a 76 m long ship model towing tank at Torquay. Mobile bed river models were initiated by Louis Jerome Fargue (1827–1910) with a model of improvements to the Garonne at Bordeaux in 1875. This was followed by Osborne Reynolds (1842–1912), who built a mobile bed model of the Mersea Estuary in 1885 and first related the time scale of a model to the square root of the length scale. Vernon-Harcourt (1839–1907) subsequently built a model of the Seine and in his report he sums up his efforts to obtain quantitative results:

> "If I succeed in demonstrating with the model that the originally existing conditions can be reproduced typically; and if, moreover, by placing regulating works in the model, the same changes can be reproduced that were brought about by the training works actually built, then I am sure that I can take the third and most important step, namely, of investigating, with every promise of success, the probable effect of the projects that have been proposed. . . ."

This is Vernon-Harcourt's attempt to deal with Point (e). The implication is also that one need not understand the flow processes completely in order to determine quantitative results from a hydraulic model, reinforcing Points (a) and (b).

By 1900, the torch of physical modeling was passed into Germany. Hubert Engels began the laboratory in Dresden in 1891; it boasted a complete new rivers laboratory by 1913. Theodor Rehbock founded a facility in Karlsruhe in 1901, which by 1921 had become very large and its work was very well known. A river laboratory was opened in Berlin in 1903 and hydraulic machinery facilities were opened in Berlin (1903), Brunn (1910) and Munich (1920). Work was also started in neighboring countries. Laboratories were opened in Leningrad (1907) and Toulouse (1908); Manchester re-activated Reynolds' laboratory in 1908 and new laboratories were set up in Padua (1910), Vienna (1912), Stockholm (1917), Delft (1927) and Zurich (1928).

The following 40 years were truly the halcyon days of the large modeling laboratories. The experience and knowledge gained from earlier studies were applied to ever greater and more complex problems and experience gained with each model ensured that the quality of the quantitative results improved with time.

3. The Decline Of Physical Modeling

However, something went wrong. Gradually, it was discovered that there was a limit to what physical models can do. By 1970, modeling experts expressed concern that the models looked right, but needed too much subjective interpretation to quantify their qualitative results. In spite of using the most appropriate scaling relationships, the scale effects and laboratory effects in many models were large. Attempts to decrease scale effects by building larger models had also increased the costs of physical models. All this came at the time when the meteoric rise in the use of the digital computer was beginning and the answer was obvious. Since scale effects did not plague the computer solutions, they would obviously have the potential to give better answers. In time, the use of computer models looked even more attractive as computational costs decreased rapidly. Hence, since 1970, the use of physical models has declined and numerical models are used ever more often. To continue the discussion, it is necessary to introduce some definitions at this point. Kamphuis (1991, 1996, 1996a, 2000, 2000a) distinguishes between two classes of physical models: Design and Process Models.

The Design Model simulates actual complex prototype situations in order to provide specific information that can be used in design or in retrospective study of failures. Models to determine the effects of proposed construction such as dams and navigation channels on estuarine flows and salinity, models of accretion and erosion near harbor entrances, outfall design models simulating effluent plumes, perhaps in stratified tidal flow and models of breakwater stability are all examples of design models. These models are built to simulate a certain prototype with its specific geometry and boundary conditions.

The Process Model studies a particular physical process such as how stratification affects sedimentation in an estuary, how bedform ripples are related to vortices and how the vortices move sediment up into the water column or how wind waves cause mixing and influence dispersion. The process to be modeled and the boundary conditions are specified in a relatively abstract manner and these, along with the scales can therefore be defined to minimize laboratory and scale effects.

Design models may be further subdivided into Long-Term and Short-Term Design Models. Long-term studies refer to the classical hydraulic models that simulate response to fluid flow and wave climates that span months or years. Such models also normally cover extensive prototype distances (kms). Models that study the effects of construction in estuaries and models of shore morphology near harbors are examples of long-term design studies. Short-term models, on the other hand, simulate a prototype situation that takes place in hours or days, for example, one particular storm or series of storms. Thus, an outfall design study and a breakwater stability study are short-term design models. Usually these short-term design models also cover smaller prototype sizes. (10–100 m).

It is clear that the long-term design model suffers most from scale and laboratory effects. Since a large prototype must be modeled for specific long-term boundary conditions, in a modeling facility of limited size, the modeler is left with few options but to build the best possible model, calibrate it and live with all the scale and laboratory effects. As a result, theoretical scaling relationships are of mainly academic interest and are only relevant in the broadest sense for planning such models. Technically, such models should be only useful to give qualitative results. Le Méhauté (1990) echoes Points (a) and (b) when he says that, in spite of all the shortcomings, scale models remain the "best analog computers." The value of qualitative results must not be underestimated. They describe and often re-define the problem and give many indications toward possible solutions. On the other hand, in the light of Point (e), any quantitative results derived from long-term design models must normally be attributed to the understanding, experience and ingenuity of the modeler.

But, the trend to require more accurate and quantitative answers to more precise question (see Point (e), above) dictates that the model results be trusted without putting undue faith in the modeler. Because of the scaling and laboratory limitations, hydraulic modeling has therefore moved away from the long-term design model toward the short-term design or process models. This trend will continue.

Historically, however, the most common type of physical model has been the long-term design model. Even now, such models may still be a good shortcut to results when answers are needed and the intermediate steps (equations and interactions of various details) are not clearly understood. But even the best of such models amount to almost the same technology described by Vernon-Harcourt over 100 years ago. Thus the long-term design model has been the prime candidate for replacement by numerical models and in fact, it has virtually been replaced. For example, 2-Dimensional, or sometimes layered (Quasi 3-Dimensional) numerical models have essentially replaced physical estuary models. Similarly, morphology near harbors is now normally modeled by a combination of 1-Dimensional shoreline change models and 2-Dimensional or Quasi 3-Dimensional flow models that incorporate sediment transport. The scale effects and laboratory effects in long-term coastal and estuary models are large enough, such that numerical models can provide better answers in most cases. It is the rapid decrease in the use of such long-term design models that has resulted in the decline of the world's prominent coastal laboratories.

4. Physical Models that Work

The short-term design models offer many more options than the long-term design models. The short-term input conditions are better defined and simpler, which means that laboratory effects are usually smaller. The process model is really an abstraction of prototype conditions and therefore, it offers even more alternatives.

It can be set up specifically to minimize laboratory and scale effects and thus the process model will yield the most reliable quantitative results. Since the prototype sections tested in short-term design models and process models are usually smaller than for a long-term design model, smaller model scales (larger models simulating smaller prototype sections) become economically possible.

The scales for a long-term design model are typically 100–300. For a short-term design model scales are typically 30–50 and for a process model they are even smaller. Using a larger modeling facility will make even further improvement. Hence, large wave flumes are used to reduce scale and laboratory effects. Example are the CRIEPI flume in Japan (Kajima *et al.*, 1982), the Delta Flume in the Netherlands (Smith *et al.*, 1994), the Großer Wellenkanal in Germany (Grüne and Führböter, 1975) and the Oregon Large Wave Flume in the US (McKee Smith, 1994). Scales of the order of 3–10 are possible in such large facilities. Wave tunnels, such as at Queen's University (Brebner and Riedel, 1973) and Delft (Ribberink and Al-Salem, 1990) are also used to reduce scales in process models.

Theoretical scaling relationships become much more relevant when such small scales are used and the details of scaling and the laws of similitude must be clearly understood by the modeler, for such models to be successful. Also, since it not necessary to simulate long prototype durations, in process models and short-term design models the cost of a model does not depend on the time scales. This means model test runs are relatively short and can readily be repeated. To define scale effects, it becomes possible to test scale series.[d]

The superior quantitative results of short-term design models and process models have resulted in a trend away from long-term design modeling. It means that proper understanding of the details of model scaling is becoming even more important and relevant. One indication of this is the relatively recent publication of four large volumes which deal primarily with scaling of physical models — Martins (1989), Shen (1990), Hughes (1993) and Chakrabarti (1994).

Along with the move toward more precise modeling comes the tendency to reduce laboratory effects as much as possible by improving laboratory equipment and methodology. For a long-term design model, the many simplifying assumptions meant that the actual model input conditions could be quite crude. Since the input conditions can be quite accurately specified, it is reasonable that more care is taken in the model to simulate the input conditions correctly. This has resulted in rapid improvements to modeling equipment. For example, in only a few years, wave generation has gone from paddle-generated regular waves, through long crested irregular waves to directional seas. It is now possible to suppress unwanted long wave activity and Dalrymple (1989) already talks about "designer" waves in which the computer program driving the directional wave generator takes into account the reflections off the walls of the basin to produce a carefully prescribed wave, exactly at the structure to be tested.

[d]A series of models of the same prototype but at different scales.

Clearly the model of the future uses a very large facility to model prototypes of relatively limited size to minimize scale effect, and uses sophisticated, computer controlled boundary conditions (waves, currents, water levels, water quality, wind) to minimize laboratory effect. The small scales now attainable only in large wave flumes and oscillating tunnels need also to be achieved in wave basins. Such large-area models might be called a "Level 2000 Models" (in view of the new millennium). However, such sophisticated models will only give reliable quantitative results if scaling is properly understood and applied. Because of the renewed importance of understanding scaling and scale effects, these will be discussed below.

5. Scaling and Scale Effects

A full explanation of scaling cannot be given here and the reader will have to refer to other publications for more detail. Also, the non-technical reader may safely skip Sec. 5, without loss of a general understanding of physical modeling.

Model scales may be derived either from equations or by dimensional analysis. Both methods have their strengths and limitations and hence both should be used. Neither method can completely describe a physical model since a model simulates the prototype better than either equations or dimensional analysis. In fact, that is the reason physical models are used. On the one hand, physical models can simulate non-linear and complex processes and boundary conditions that are not clearly understood and therefore cannot yet be expressed by the equations or a numerical model. On the other hand, physical models also yield detailed results, while dimensional analysis only describes a process in general form.

There are many types of coastal models such as models of wave agitation in harbours, wave interaction with structures or spreading of contaminants by currents and waves. Extensive examples may be found in the literature, such as in Sharp (1981). Kamphuis (1996, 1996a) describes the scaling of the coastal mobile bed sediment transport and morphology model which is perhaps the most difficult of all physical hydraulic models. The scaling of such a model will be briefly reviewed here.

5.1 *Scaling from equations*

Because the fluid flow of such a model can be described by equations, we will first use equations to derive the fluid flow scales. All the equations that apply to the problem must be collected and they must be correctly formulated and dimensionally consistent. Each equation will highlight certain aspects and can be used to derive particular scales. The assumption is that each equation applies to the model as well as to the prototype. Dividing the equation for the prototype by the same equation for the model means that scale relationships may be derived by simply substituting scales for the variables in the equations. For example:

$$\Gamma = M\frac{dV}{dt} \Rightarrow \Gamma_p = M_p\frac{dV_p}{dt_p}; \quad \Gamma_m = M_m\frac{dV_m}{dt_m} \Rightarrow n_\Gamma = \frac{\Gamma_p}{\Gamma_m} = n_M\frac{n_M}{n_t}. \tag{1}$$

Here Γ is force, M is mass, V is velocity and t is time. The subscripts p and m refer to prototype and model and n denotes scale of the particular variable (prototype/model value). The model scales may be derived directly from the equations as is done in Kamphuis (1996, 1996a) or in another way as some authors prefer to first make the equations dimensionless, for example, Langhaar (1951), Shen (1990) and Hughes (1993).

Fluid flow in a coastal morphology model is described by the continuity equation and the Navier Stokes equations. From these equations, it is seen that all distance scales must be equal and all the velocity scales must be equal. Dividing an inertia term by a gravity term yields:

$$\frac{{n_u}^2}{n_g n_l} = n_F = 1. \tag{2}$$

Here u is a velocity, g is gravitational acceleration, l is a typical length and F is the Froude number. The design condition stated in Eq. (2) is that the Froude number must be the same in model and prototype.[e] Dividing a viscosity term into an inertia term yields the Reynolds number scaling:

$$\frac{n_u n_l}{n_v} = n_R = 1, \tag{3}$$

where ν is the kinematic viscosity of the fluid. Usually, g and ν are roughly the same for model and prototype and hence $n_g = n_\nu = 1$. Equation (2) then becomes $n_u = (n_l)^{1/2}$, while Eq. (3) yields $n_u = 1/n_l$. Unless $n_u = n_l = 1$, which occurs only in the prototype, Eqs. 2 and 3 cannot both be satisfied and, using physical reasoning, it must be decided if gravity or viscosity is the parameter of lesser importance, so that either Eq. (2) or Eq. (3) can be ignored. For the usual hydraulic models, gravity effects are considered to be the more important; thus the Froude number will be represented correctly (Eq. (2)) while the incorrect representation of the viscosity terms will cause the model not to behave exactly like the prototype. The difference between the model and the prototype behavior, resulting from not being able to adhere to certain scaling criteria is called Scale Effect and it is imperative that the modeler understand the consequences each time an improper scaling is introduced. For example, not satisfying viscosity requirements may not be serious as long as the flow in the model is always rough-turbulent, but for coastal models in which the orbital velocity passes through zero twice every wave cycle, not adhering to Eq. (3) could have some serious implications.

[e] In Sec. 5.2, it will be seen that, in fact, all dimensionless ratios should be the same in the model and prototype.

To learn more about the waves in a coastal model, wave action can be incorporated with the depth integrated version of the Navier Stokes equations (Phillips, 1977). Further insight into wave kinematics may be achieved by using additional equations such as the small amplitude expressions for wave dispersion and orbital velocity, and an equation for bottom friction. Usually, there are enough inaccuracies in the shear stress and the radiation stress that n_ρ can be considered equal to 1, even when sea water is modeled by fresh water. This results in one possible scaling combination based on modeling the Froude criterion correctly:

$$n_x = n_y = n_z = n_{(p/\rho g)} = n_d = n_k = n_\eta = n_H = n_L = n_\tau = n$$
$$n_u = n_v = n_w = n_U = n_V = n_{u'} = n_{v'} = n_{w'} = n_{u_w} = n_{v_w} = n_t = n_T = \sqrt{n},$$
$$n_S = n^2 \tag{4}$$

where x, y are horizontal distances, normally defined as shore perpendicular and shore parallel direction respectively, z is vertical distance, p is pressure, d is depth, k is bottom roughness, η is wave setup, H is wave height, L is wave length, τ is shear stress, and the unsubscripted n represents the general model scale. Further, u, v and w are velocities in the$x$, y and z directions, U and V are the depth-averaged velocities in the x and y directions, u', v' and w' are turbulent velocity fluctuations, u_w and v_w are the wave generated velocities, T is wave period and S is the radiation stress of the waves. Examples of other equations and their conversion into scaling relationships may be found in Hughes (1993).

The selection of the relevant equations is where the real ingenuity and thinking of the modeler become important. Which equations should be used? Which are the most important? Which equations are not necessary or redundant? Note that if an equation is not included in the analysis, some important relationships will be completely missed.

The model scale (n) is normally determined from the ratio of a required prototype length and the available modeling space. For example, if in a long-term design model a prototype distance of 10 km must be modeled in 50 m of a modeling basin, n would be 200. Similarly for a process model 250 m of prototype may be modeled in the same 50 m of modeling space resulting in n = 5. Again, if a 5 m high wave must be modeled in a facility that cannot produce wave heights greater than 0.2 m, n must be less than 25, etc.

Often, particularly for long-term design models, it is not possible to use the same horizontal and vertical scales and the model must be geometrically distorted. If $n = 200$, as in the above example, and this same scale were used as the vertical scale, all model depths, wave heights, bedforms, etc would become very small. In such cases, it is usual to choose the vertical scale to be smaller than the horizontal scales. A depth scale of 30 to 50 is usual. This would result in a geometric model distortion of 4 to 7.

Denoting the time scales for the currents by subscript c we can now write for the fluid flow in a geometrically distorted coastal model:

$$\begin{aligned}
&n_z = n_{(p/\rho g)} = n_d = n_k = n_\eta = n_H = n_L = n_\tau = n \\
&n_x = n_y = Nn \\
&n_u = n_v = n_U = n_V = n_w = n_{u'} = n_{v'} = n_{w'} = n_{u_w} = n_{v_w} = n_t = n_T = \sqrt{n}, \\
&n_{tc} = n_{Tc} = N\sqrt{n} \\
&n_S = n^2 \qquad (5)
\end{aligned}$$

where n without subscript denotes the vertical scale and N is the geometric distortion. Geometric distortion of the model introduces further major scale effect. All slopes, cross-sections, velocity and pressure gradients, the viscosity terms and the turbulence terms are distorted. Thus geometric model distortion should be avoided. This is normally not possible for long-term design models, but is possible for all other models. Also, because the number of waves per unit horizontal distance is now incorrect, wave diffraction is no longer correct and hence a wave model involving wave diffraction must be undistorted.

5.2 *Scaling with dimensional analysis*

Dimensional analysis is a powerful tool that is particularly useful if the problem is not sufficiently understood, so that the applicable equations are not well known. It assumes that all fundamental equations are dimensionally homogeneous and that all physical quantities may be expressed as a power product of a number of independent basic entities. Details of the method have been discussed in many references such as: Hughes (1993), Ivicsics (1980), Kamphuis (1975, 1985 and 1991), Langhaar (1951) and Yalin (1971).

A dimensional expression in which an dependent variable A is a function of K independent variables (a_i)

$$A = f(a_1, a_2, a_3, a_4, \ldots, a_{K-1}, a_K), \qquad (6)$$

may be reduced to a dimensionless expression with $(K - r)$ dimensionless independent variables (X_i)

$$\Pi_A = \phi_A(X_1, X_2, X_3, \ldots, X_{k-r}), \qquad (7)$$

where Π_A is the dimensionless version of A, r is the number of base units pertinent to the problem (usually three: mass, $[M]$, length $[L]$, and time $[T]$). Because dimensional analysis uses functional relationships such as Eqs. (6) and (7), it says nothing about the actual relationships, only that their dimensions must be correct.

For the fluid phase of a coastal model, one possible relationship would be:

$$A = f(H, T, d, k, \rho, \mu, g, x, y, z, t). \qquad (8)$$

Here A is any dependent variable, for example wave orbital velocity, wave length, etc., and μ is the dynamic viscosity of the fluid. The list of independent variables must include all the variables needed to define the dependent variable and should exclude all variables that are not independent or are irrelevant. This is a rather daunting assignment, particularly for the "uninitiated".

A large number dimensionless expressions could represent Eq. (8) and Yalin (1971) and Kamphuis (1975, 1985 and 1991) follow an organized approach to determine a first reasonable set of dimensionless variables. For a problem with three basic dimensions ($[M]$, $[L]$ and $[T]$), it is necessary to find three independent variables and to divide each other independent variable by a power product of these three "repeaters". It would be natural to pick the three variables that are most difficult to vary experimentally as the repeaters, but it is a good rule of thumb to pick two variables that are difficult to vary and one length. The initially derived dimensionless ratios can be re-combined at any time to form new dimensionless ratios and the trick is to arrive at those ratios that best describe the problem at hand.

Choosing ρ, g and d as repeating variables, one possible dimensionless expression for Eq. 8 is:

$$\Pi_A = \phi\left(\frac{H}{d}, T\sqrt{\frac{g}{d}}, \frac{k}{d}, \frac{\mu}{\rho d\sqrt{gd}}, \frac{x}{d}, \frac{y}{d}, \frac{z}{d}, t\sqrt{\frac{g}{d}}\right). \tag{9}$$

The number of independent variables has decreased by three. Are these the correct ratios? Here the detailed knowledge by the modeler about the physics of the problem becomes important. For example, the first ratio $X_1 = H/d$ would suffice if a process near breaking is described, but in deep water, (H/gT^2) would be better. Since all the variables are dimensionless, (H/gT^2) could be obtained by replacing the first dimensionless ratio $X_1 = (H/d)$ by $X_1/X_2^2 = (H/d)/(gT^2/d) = (H/gT^2)$. It is also usually better to relate the horizontal distances x and y to a typical horizontal length. For waves this should obviously be wave length. Since wave length is not independent from the other parameters in Eq. (8) (it is completely defined by H, T and d), it was not included there. If the dependent variable A in Eq. (8) were chosen to be L, then Eq. (9) would yield $\Pi_L = L/d$. Multiplying power products of this ratio into some of the dimensionless ratios in the right-hand side of Eq. (9) and replacing X_8 by $X_8X_2^{-1}$ results in a much more useful function for deeper water:

$$\Pi_A = \phi_A\left(\frac{H}{L}, \frac{L}{gT^2}, \frac{k}{d}, \frac{\mu}{\rho d\sqrt{gd}}, \frac{x}{L}, \frac{y}{L}, \frac{z}{d}, \frac{t}{T}\right). \tag{10}$$

Throughout this manipulation, the number of independent dimensionless variables has remained the same; the new dimensionless products always replace only one of the original products. But the parameters to be used depend very much on the problem at hand. For example, with x as the cross-shore direction and y as the alongshore direction, then X_5, X_6 and X_7 could be re-combined to form $m = z/x$ (beach slope), $\alpha = x/y$ (incident wave angle) and z/d (relative depth). Replacing X_4 by $X_4^{-1}X_1^{1/2}$ to yield a more common Reynolds number and choosing u_w as the dependent variable, Eq. (10) becomes:

$$\frac{u_w}{\sqrt{gd}} = \phi_{u_w}\left(\frac{H}{L}, \frac{L}{gT^2}, \frac{k}{d}, \frac{\sqrt{gH}d}{v}, m, \alpha, \frac{z}{d}, \frac{t}{T}\right). \tag{11}$$

Dimensional analysis is clearly a very flexible method, but for the dimensionless relationships to be most useful the modeler needs to have a thorough understanding of what exactly needs to be modeled, otherwise the resulting relationships will contain less meaningful dimensionless parameters.

Model scaling from dimensions assumes that the functions ϕ apply both to the prototype and the model. Since the actual detailed shape of the 8-Dimensional ϕ in Eq. (11) is not known, all the independent dimensionless variables in the model must individually be made the same as in the prototype to make $\phi_m = \phi_p$. In other words, the scales for each of the dimensionless variables must equal one. For example, for the first ratio on the right-hand side of Eq. (11):

$$\left(\frac{H}{L}\right)_p = \left(\frac{H}{L}\right)_m \quad \Rightarrow \frac{H_p}{H_m} = \frac{L_p}{L_m} \quad \Rightarrow n_H = n_L \quad \Rightarrow n_{H/L} = 1. \tag{12}$$

The left-hand side of Eq. (11) is a Froude number and the fourth term on the right-hand side is a Reynolds number. As with the scaling from equations discussed earlier, in the usual situation where $n_\rho = n_\mu = n_\nu = n_g \approx 1$, it is not possible to satisfy both the Reynolds and Froude number criteria. Finally, introducing additional dependent variables into Eq. (8), results in additional dimensionless relationships and we would eventually arrive at exactly the same scales as in Eq. (4) or in Eq. (5) for a distorted model.

Usually the bottom roughness, k, cannot be scaled down correctly by the model scale, n. For a fixed bed model, the bed would need to be made impossibly smooth. For a mobile bed model, the bed form size does not scale down by a factor n and scaling the sediment itself down by n would cause the sediment to go into suspension for all prototype sediments finer than gravel. This means bottom roughness is almost always distorted, bringing with it a new set of scale effects.

5.3 *Scaling sediment transport*

To determine the scales required for the mobile bed portion of a coastal model, the sediment transport may be assumed to be a function of shear stress on the bottom. The following dimensional relationship may then be written:

$$A = f(\rho, \mu, g, D, v_*, \rho_s, a_B). \tag{13}$$

Here D is grain size, v_* is shear velocity $\{= (\tau/\rho)^{1/2}\}$, τ_o is bottom shear stress, ρ_s is sediment density and a_B is wave orbital amplitude at the bottom. Eq. (11) may be reformulated as:

$$\Pi_A = \phi_A\left(\frac{v_* D}{v}, \frac{v_*^2}{\Delta g D}, \frac{\rho_s}{\rho}, \frac{a_B}{D}\right) = \phi_A(R_*, M_*, \rho_*, G_*), \tag{14}$$

where $\Delta = (\rho_s - \rho)/\rho$, R_* is grain size Reynolds number, M_* is mobility number, ρ_* is relative density of the sediment and G_* the geometric link between the waves and the sediment. The first and second ratios form the axes of the Shields diagram and the second ratio could be expressed as excess mobility over a critical value for large grain sizes. Shear stress or shear velocity can be related to wave and current action using theoretical or empirical equations. Although all four ratios in Eq. (14) must be the same in model and prototype, the best any scale model can do is to simulate M_*, ρ_* and G_* correctly. Because R_* is then not modeled correctly, a Reynolds number scale effect will be present (model Reynolds numbers will be much smaller than required). Usually such a model is not even possible, because the grain size would need to be scaled down by the model scale and this makes the model material size too small. Detailed discussions on shear stress and the various scale effects may be found in Kamphuis (1975, 1985 and 1991).

One common approach is to satisfy both the R_* and the M_* criteria by using lightweight sediment. Kamphuis (1975, 1985 and 1991) shows that, for a coastal model, this introduces a number of serious scale effects. Hence, sand should be used in a coastal model.

For model particle size, in a sand model, the fourth ratio in Eq. (14) states $n_D = n$. Noda (1972) and Kriebel *et al.* (1986) both find $n_D \approx (n)^{1/2}$. Both criteria result in impossibly small model grain size. It is normally best simply to use the smallest sand grain size that will not go into suspension in the model. This is of order of 0.1 mm and it determines n_D. For example, using $D_M = 0.1\,\text{mm}$ for a prototype beach sand with $D_p = 0.25\,\text{mm}$ results in $n_D = 2.5$. In a conventional model, the model scale might be $n = 50$ and $n_\Delta \approx 1$.

Often model scaling using equations is combined with the use of dimensional analysis. Introducing the equations for friction factor and wave shear stress over a flat bottom from Kamphuis (1975) into Eq. (14) results in:

$$n_{v_*} = n^{1/8} n_D^{3/8} = 2.3; \quad n_{R_*} = n^{1/8} n_D^{11/8} = 5.7$$
$$n_{M_*} = n^{1/4} n_D^{-1/4} n_\Delta^{-1} = 2.2; \quad n_{\rho_*} = 1; \quad n_{G_*} = n n_D^{-1} = 20. \tag{15}$$

Because sediment transport processes are highly dependent on mobility number it is good that the model mobility number is only slightly smaller than required. There will, however, be scale effects, particularly because of the improper scaling of G_*, which causes the grain size to be much too large in relation to the driving forces of the waves. Many prototypes and almost all sand models exhibit bedforms, which complicate the scaling further.

Note that in a large, Level 2000 model basin, n might be as low as 5, in which case $n_{v_*} = 1.72, n_{R_*} = 4.3, n_{M_*} = 1.18, n_{\rho_*} = 1$ and $n_{G_*} = 2.0$, which represents a large improvement, particularly in n_{M_*} and n_{G_*}.

The above discussion was concerned with underwater sediment transport which was assumed to be caused by wave-generated shear stress. That would be the case when the model simulates sediment pickup and erosion. When the model must simulate suspended load and deposition of sediment, the suspending mechanisms and the fall velocity ω become important. There are several theories about how material becomes suspended and Fredsoe and Deigaard (1992) present a good summary. Scales can be derived from the equations for turbulent eddy viscosity, mixing length and the parameters in the various turbulence models. The net result is that:

$$n_{v_*} = n^{1/2}. \tag{16}$$

This is convenient and as expected we can now set:

$$n_{v_*} = n_\omega = n_\mu = n_v = n^{1/2}, \tag{17}$$

where ω is the fall velocity scale. However, Eq. (15) shows that Eq. (17) is only possible when $n_D = n$, for zero roughness distortion. Usually there is roughness distortion in a model and since $N_D = n_D/n < 1, n_{v_*} < (n)^{1/2}$. Clearly the v_* scale will be less than the other velocity scales if roughness distortion is present, that is, v_* in the model will be too large. We can either make:

$$n_\omega = n_{v_*} = n^{1/2} N_D^{3/8} \quad or \quad n_\omega = n_u = n_v = n^{1/2}. \tag{18}$$

The first condition provides for correct settling in the vertical direction (without transport by currents); the second condition simulates settling in the presence of currents and an exaggerated suspension mechanism. The required shear stress scale for modeling sedimentation is normally different from that required for an erosion model. Usually both need to be represented, for example, onshore erosion is deposited offshore, and hence, a choice must be made if the sediment transport model is primarily an erosion or sedimentation model.

Until now the discussion was concerned with sediment motion in the relatively gentle environment offshore of the breaking zone. In the breaking zone, bottom shear is not the prime moving force, but wave energy dissipation rate and wave momentum (radiation stress) cause the sediment transport. The following example function may be derived from Eq. (8) after replacing k with D and ρ_s:

$$\Pi_A = \phi_A \left(\frac{H_b}{d_b}, \frac{H_b}{L_b}, m, \alpha_b, \frac{t}{T}, \frac{\sqrt{gH_b}D}{v}, \frac{H_b}{\Delta g D}, \frac{\rho_s}{\rho}, \frac{H_b}{D} \right). \tag{19}$$

The last four ratios in this equation are other forms of R_*, M_*, ρ_* and G_*. To continue the earlier example, for $n = 50, n_\Delta \approx 1$ and $n_D = 2.5$:

$$\begin{aligned} n_{R*} = n^{1/2} n_D = 17.7; \quad n_{M*} = n n_D^{-1} n_\Delta^{-1} = 21; \quad n_{\rho*} = 1; \\ n_{G*} = n n_D^{-1} = 20, \end{aligned} \tag{20}$$

which indicates serious scale effects, particularly because both M_* and G_* in the model are an order too small. Considerable improvement may be brought about by a larger, Level 2000 model in which case ($n_{R_*} = 5.6$, $n_{M_*} = 2$, $n_{\rho_*} = 1$ and $n_{G_*} = 2$).

Ideally, particularly if the model involves structures and wave diffraction, the model should be undistorted but mobile bed models have what could be termed an unavoidable natural distortion. The beach in a model is formed by the waves and currents and will take on a profile that is a function of these conditions and the composition of the beach. For example, a sand model beach is usually steeper than its sandy beach prototype. This natural distortion may be defined as:

$$N_n = \frac{n_x}{n_z} = \frac{\chi_P}{\chi_M}\frac{Z_M}{Z_P} = \frac{m_M}{m_P} = n_m^{-1}. \tag{21}$$

To determine N_n it is really necessary to perform preliminary model tests. From model and prototype beach slopes, it is then possible to determine N_n. For model planning purposes, natural distortion may be computed approximately from known equations for beach profiles (Kamphuis, 1996). Other recipes may also be found in the literature to serve as a first approximation. They are based on eroding beach or dune model results in which the eroded material is deposited offshore. Hughes (1993) summarizes these test results to find:

$$N_n = \left(\frac{n}{n_\omega^2}\right)^\theta \qquad 0.25 \le \theta \le 0.50. \tag{22}$$

6. Implications for Physical Modeling

For a wave model with sediment transport and erosion close to shore and deposition further offshore, Eq. (4) or Eq. (5), Eq. (9), Eq. (14), Eq. (18), Eq. (19) and Eq. (21) must be considered simultaneously. It is not difficult to see that all of the scaling criteria developed above cannot be satisfied, even for a relatively simple coastal model. On the other hand, all the above equations are helpful for understanding the processes involved and for determining how to build models.

Points (a) and (b) continue to pique our interest in physical modeling. Although the qualitative images provided by a physical model need not correspond directly to prototype reality, at least they are based on a degree of physical similarity so that many of the complex processes, their interactions and the complicated boundary conditions are all reasonably modeled. The physical modeler is, of course, cognizant of Point (e) and knows that a model study cannot be quantitatively successful unless the modeler understands the shortcomings of the model and works around them as best possible. Only a modeler who understands which parameters are important and what are the scale effects resulting from ignoring some of the scaling laws, a modeler who goes through detailed reasoning similar to Sec. 5 of this chapter will be successful in obtaining good quantitative answers — results which neither numerical modeling nor field experimentation can provide. Le Méhauté (1990) says:

> "A return to experimental techniques, which may have declined during the last two decades because of the successful development of mathematical modeling appears not only desirable but inevitable. The need for experimentation will actually be dictated by (the gaps in) mathematical modeling."

Clearly, the future for physical models lies in reducing the scale and laboratory effects and the cost. To reduce scale effects, future physical models will inevitably tend toward process model abstractions of the problem rather than design models of the problem itself. Process model results do not provide direct solutions to practical problems, however; they must be seen as abstract building blocks that can be used as steps toward a solution of a practical problem. Usually a number of such building blocks are required to solve a practical problem. This leads to a concept that was called Composite Modeling in Kamphuis (1995, 1996, 1996a, 2000, 2000a). It combines a physical modeling phase, analysis of the model results and a computation phase as shown, for example in Fig. 7.1.

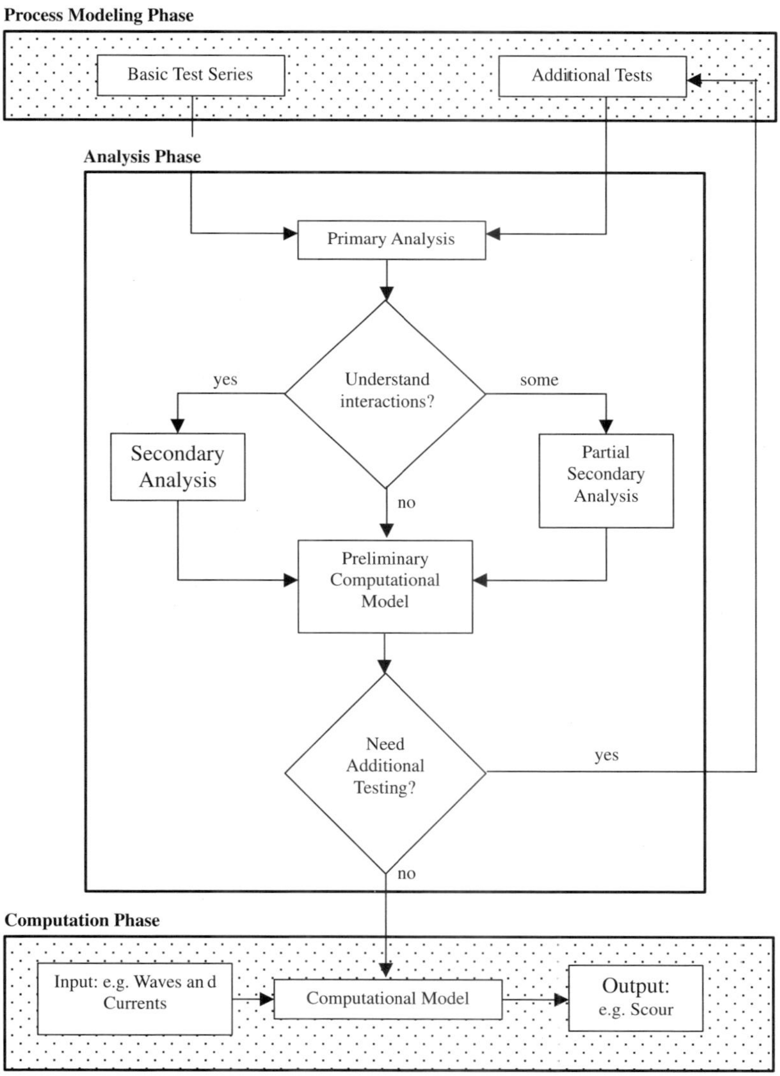

Fig. 7.1 Composite modeling (from Kamphuis, 2000).

The physical modeling phase consists of performing a number of relatively simple process models. These models will all be similar but for different input conditions. This means the experience gained with each individual model test is immediately applied in the subsequent model tests, which makes the physical modeling process very efficient.

The computational model that links the process model results could be a complete numerical model, in which case the physical process models simply provide appropriate coefficients and transfer functions. It is more likely, however, particularly for engineering studies that need to provide useful answers within a limited time and budget for prototypes with complicated boundary conditions, that a complete numerical model is not possible. The computational model will then normally be a relatively simple statistical summation of a number of carefully determined process modeling results. It could of course also be a simple mathematical model using empirical relationships derived from process model tests. One example of the latter is a 1-D shoreline change model, which uses sediment transport equations derived from physical model results and from field observations. For a composite model to be successful, the non-linear interactions between the various building block units need to be carefully estimated from the process model results. This means many measurements must be taken in time, over the duration of the model run, as the results for each process model develop.

With respect to cost: the physical models components are simple and repetitive. They do not require experienced technicians, nor long calibrations or long testing periods. The results are continuously improved as the tests proceed. The testing of various prototype scenarios takes place in the inexpensive computational phase.

7. Example

An example of process model studies and composite modeling with waves, currents and sediment transport will now be given. Circular artificial islands were used for oil exploration in the Canadian Beaufort Sea in the 1980s. Since there was no rock or even gravel available for scour protection, these islands were built entirely of sand. During the winter (October–June) the ice surrounding the island provides an excellent shore protection against erosion by the waves and currents. However, during the open water season (July, August and September) the large volume of sand (6-8 Million m^3) itself forms the only real "protection" against erosion of the centre of the island where the buildings and the drilling rig are located. To determine the very complex, three-dimensional erosion and deposition of sediment under waves and currents, 52 process model tests were run for the test conditions shown in Table 7.1 (Kamphuis and Kooistra, 1986; Kamphuis and Larsen, 1987). The four different model scales and three model grain sizes provide scale series for investigation of scale effects. The physical model layout is shown in Fig. 7.2.

Table 7.1 Process model tests of artificial islands summary of test conditions.

Model Scales — n	200	100	75	50
Model Particle Sizes — D_{50}(mm)	0.56	0.18	0.11	
Prototype Wave Heights(m)	6.5	4.75	3.0	
Prototype Wave Periods(sec)	8.0	10.0		
Wave Types	Regular	Irregular		

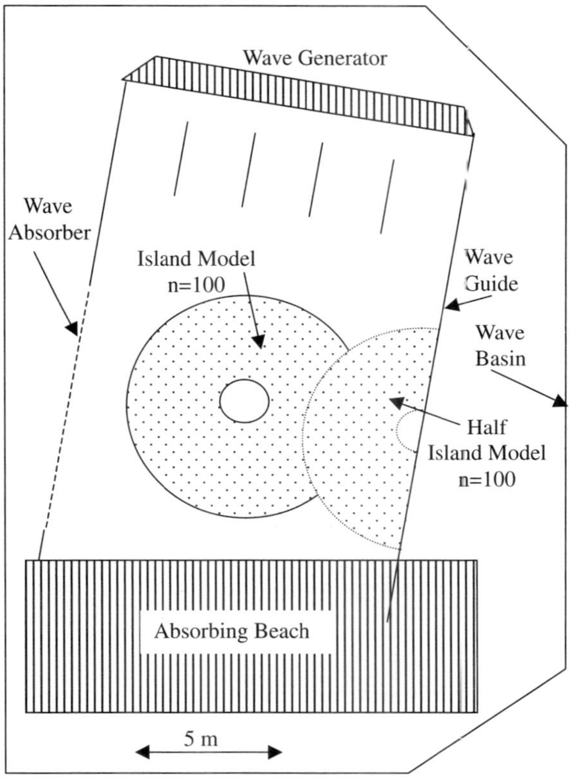

Fig. 7.2 Layout of physical model (from Kamphuis, 2000).

Complete models were run at scales $n = 200$ and $n = 100$; half models were run at $n = 100$, $n = 75$ and $n = 50$.

The primary output of the process models was volume erosion rate as a function of time for various input wave conditions. These volume erosion rates were combined through simple computer calculations to form a composite model to predict the erosion of such islands for various complex wave climate scenarios. The example shown in Fig. 7.3 compares the composite model results with measurements for one actual prototype island that sustained severe (and rapid) erosion.

The process models described in Fig. 7.2 and Table 7.1, which were run in a relatively small facility with an irregular, long-crested wave generator. Clearly

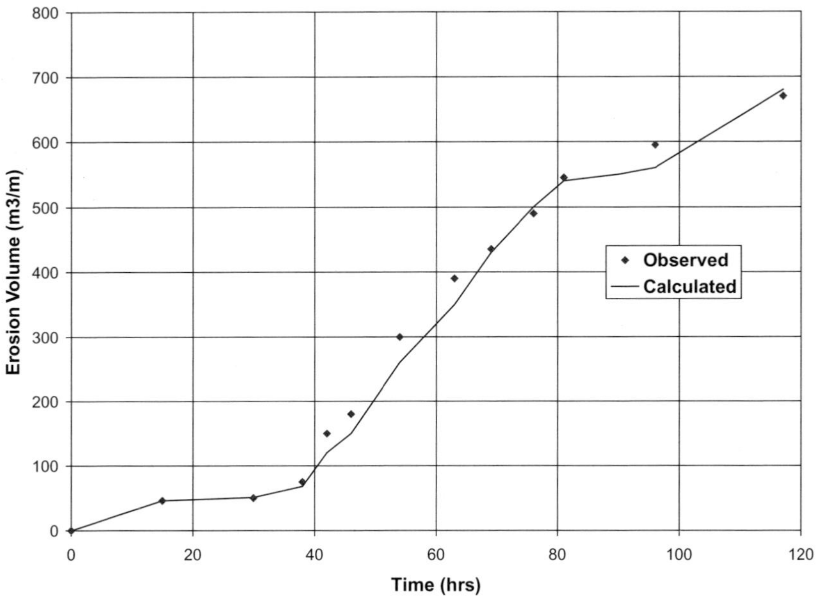

Fig. 7.3 Result from composite model (from Kamphuis, 2000).

the results could be improved substantially if they were to be done in a large, sophisticated "Level 2000" modeling facility with directional waves. That is the obvious direction in which physical modeling needs to proceed.

8. Conclusions

The following conclusions may now be drawn:

(a) Points (a) to (d) show that there will continue to be a need for physical modeling of coastal problems.
(b) The corollary is that numerical modeling and field research, even if used together will not be able to solve all coastal problems on their own.
(c) The decline in interest in physical modeling was the result of disenchantment of the profession with the cost and the scale and laboratory effects of long-term design models.
(d) The future of physical modeling lies in better interpretation of the model results. This will require reductions in scale effects, laboratory effects and costs.
(e) This means the trend to use short-term design models and process models will continue. The models will be scaled more correctly. They will also be more abstract and less of a direct simulation of a prototype problem.
(f) Reducing scale and laboratory effects will require the development of new, larger laboratory facilities with more sophisticated boundary controls ("Level 2000"

facilities). In parallel with this, a thorough understanding of scaling techniques and scale effects is needed to make use of these superior tools.

(g) Cost of physical modeling can be limited by using the concept of composite modeling. Many relatively simple process modeling runs are conducted and these are integrated with simple computer programs. Low cost is achieved since all the process models are simple and relatively similar, and all the what-if scenarios are calculated in the inexpensive computational phase.

The following general conclusions may also be drawn with respect to physical modeling of coastal areas:

(h) Numerical models will not be able to supply the required accuracy of results by themselves. There will be an increased need for physical modeling to supply more accurate information about the processes to be modeled numerically.
(i) The prototype can provide some data, but requirements for close experimental observation, manipulation of the boundary conditions and test repeatability will require extensive use of properly designed physical models.
(j) Engineering problems will need good answers to questions involving complex physical environments and boundary conditions, within a limited budget and time frame. This will result in an increased need for physical modeling of processes. Such model results can then be placed in many different scenarios with simple computer programs, at relatively low cost.

In future, there must be greater integration of field research, physical modeling and numerical modeling. Physical modeling will play a pivotal role between field observations, which are usually spot measurements of uncontrollable physical events, and the more-or-less full understanding of the problem needed for a numerical model. The integration will in part consist of the use of Level 2000 facilities. These models will be so large as to approximate field results that can be controlled, closely observed and repeated. At the same time these models are steered by sophisticated computer programs, which resemble or are based on existing numerical modeling techniques. Field observations, Level 2000 physical modeling and numerical modeling will thus be combined to take full advantage of their strengths.

References

Brebner, A and P. H. Riedel (1973): A new oscillating water tunnel, *J. Hydr. Res.* IAHR, **11**, 107–121.

Chakrabarti, S. K. (1994): Offshore structure modeling, *Advanced Series on Ocean Engrg.* **9**, 470 p.

Dalrymple, R. A. (1989): Physical modeling of littoral processes, in: *Recent Advances in Physical Modeling* (R. Martins, ed.): Kluwer Academic Publishers, pp. 567–588.

Fredsoe, J. and R. Deigaard (1992): Mechanics of Coastal sediment transport, *Advanced Series on Ocean Engrg.* **3**, 367 p.

Freeman, J. R., ed. (1929): *Hydraulic Laboratory Practice*, ASME, New York.

Garbrecht, G., ed. (1987): *Hydraulics and Hydraulic Research, a Historical Review*: Balkema Press, Rotterdam.

Grüne, J. and A. Führböter (1975): A large wave-channel for full-scale modeling of wave dynamics in the surf zone, *Proc. Int. Symp. on Modeling Techniques*, ASCE, San Francisco, 82–100.

Hughes, S. (1993): *Physical Models and Laboratory Techniques in Coastal Engineering*, Advanced Series on Ocean Engineering Vol. 7: World Scientific Publishing, Singapore, 567 p.

Ivicsics, L. (1980): *Hydraulic Models*: Water Resources Publications, Fort Collins, 310 p.

Kajima, R., T. Shimizu, K. Maruyama and S. Saito (1982): Experiments on beach profile change with a large wave flume, *Proc.* 18th *Int. Conf. on Coastal Eng.*, ASCE, 1385–1404.

Kamphuis, J. W. (1975): *The Coastal Mobile Bed Model*, Civil Eng. Rep. No 75, Queen's University, Kingston, Canada, 114 p.

Kamphuis, J. W. (1985): On understanding scale effect in coastal mobile bed models, in: *Physical Modeling in Coastal Engrg.* (R. Dalrymple, ed.), Balkema, pp. 141–162.

Kamphuis, J. W. and Kooistra, J. (1986): On volume erosion and scale effects for hydraulic models of large conical islands, *Proc. Symp. on Scale Effects in Modeling Sediment Transport Phenomena,* IAHR, Toronto, pp. 10–23.

Kamphuis, J. W. and Larson, M. R. (1987): A mobile bed scale model series under attack by constant, regular and irregular waves, *Proc. Coastal Sediments* '87, ASCE, New Orleans, 154–174.

Kamphuis, J. W. (1987): *The Minuk Sacrificial Beach Island Model Tests*, J. W. Kamphuis, Associates Limited report prepared for ESSO Resources Canada, 202 p.

Kamphuis, J. W. (1991): Physical modeling, In: *Handbook of Coastal and Ocean Engrg.* (J. Herbich, ed.), Gulf Publishing, **2, Chapter 21** pp. 1049–1066.

Kamphuis, J. W. (1995): Composite modeling — an old tool in a new context, *Proc. IAHR Congress, London*, **2**, 230–235.

Kamphuis, J. W. (1996): Physical modeling of coastal processes, in: *Advances in Coastal and Ocean Engrg.* (P. L.-F. Liu, ed.): World Scientific Publishing, Singapore, **2**.

Kamphuis, J. W. (1996a): Coastal hydraulic models, *ASCE Short Course, Orlando*, 45 p.

Kamphuis, J. W. (2000): *Introduction to Coastal Engrg. and Management*: World Scientific Publishing, Singapore, 437 p.

Kamphuis, J. W. (2000a): Designing with models, Keynote Address, *Proc.* 27th *ICCE*, Sydney, 19–32.

Kriebel, D. L., W. R. Dally and R. G. Dean (1986): Undistorted Froude model for surf zone sediment transport, *Proc.* 20th *Int. Conf. on Coastal Eng.* ASCE, 1296–1310.

Langhaar, H. L. (1951): *Dimensional Analysis and Theory of Models*: John Wiley and Sons, 165 p.

LeMehaute, B. (1990): Similitude, *Ocean Engrg. Science*, *The Sea*, Vol. 9, John Wiley and Sons, 955–980.

Levi, E. (1995): *The Science of Water — The Foundation of Modern Hydraulics*, Translation from El Agua Segun La Ciencia by D. A. Medina, ASCE Press, New York, 649 p.

Martins, R., ed. (1989): *Recent Advances in Physical Modeling*: Kluwer Academic Publishers, 620 p.

McKee Smith, J. (1994): Undertow at Supertank, *Proc. Coastal Dynamics* '94, ASCE, Barcelona, 220–232.

Noda, E. K (1972): Equilibrium beach profile scale model relationship, *J. Waterways, Harbors and Coastal Eng.*, ASCE, **98**, 511–528.

Phillips, O. M. (1977): *The Dynamics of the Upper Ocean*: Cambridge University Press, Cambridge, 261 p.

Ribberink, J. S. and A. Al -Salem (1990): Bedforms, sediment concentrations and sediment transport in simulated wave conditions, *Proc.* 22nd *Int. Conf. on Coastal Eng.*, ASCE, 2318–2331.

Rouse, H and S. Ince (1957): *History of Hydraulics*, Iowa Inst. of Hyd. Res., Re-published by Dover Press, (1963) New York.

Sharp, J. J. (1981): *Hydraulic Modeling*, Butterworths, 240 p.

Shen, H. W., ed. (1990): *Movable Bed Physical Models*, Kluwer Academic Publishers.

Smith, G. M, J. W. W. Seijffert and J. W. van der Meer (1994): Erosion and overtopping of a grass dike — Large scale model experiments, *Proc.* 24th *Int. Conf. on Coastal Eng.*, ASCE, 2639–2652.

Yalin, M. S. (1971): *Theory of Hydraulic Models*, MacMillan Press, 266 p.

Notation

a_i independent variable,
a_B wave orbital amplitude at the bottom,
A dependent variable,
b (subscript) at breaking,
B (subscript) along the bottom,
c (subscript) current,
d depth,
D grain size,
f function,
F Froude number,
g gravitational acceleration,
G_* geometric link between the waves and the sediment,
H wave height,
k bottom roughness,
K number of independent variables,
l typical length,
L wave length,
m beach slope,
m (subscript) model,
M mass,
M_* mobility number,
n general model scale (prototype/model value), vertical scale in a distorted model,(when subscripted) model scale for the subscripted variable,
N geometric distortion of the model,
N_n natural distortion of a sediment model,
p pressure,
p (subscript) prototype,
r number of base units,

R	Reynolds number,
R_*	grain size Reynolds number,
S	radiation stress of the waves,
t	time,
T	wave period,
u	velocity in the x direction,
u'	turbulent velocity fluctuation in the x direction,
U	depth-averaged velocity in the x direction,
v	velocity in the y direction,
v'	turbulent velocity fluctuation in the y direction,
v_*	shear velocity $\{= (\tau_o/\rho\ )^{1/2}\}$,
V	velocity, depth-averaged velocity in the y direction,
w	velocity in the z direction, (subscript) wave,
w'	turbulent velocity fluctuation in the z direction,
x	horizontal distance, normally shore perpendicular,
X_i	dimensionless independent variable,
y	horizontal distance, normally shore parallel
z	vertical distance,
α	angle of wave incidence,
Δ	$(\rho_{s-}\rho)/\rho$,
ϕ	dimensionless function,
Γ	force,
η	wave setup,
μ	dynamic viscosity of the fluid,
ν	kinematic viscosity of the fluid,
Π_A	Dimensionless version of A,
θ	exponent,
ρ	fluid density,
ρ_s	sediment density,
ρ_*	relative density of the sediment $(= \rho_s/\rho)$
τ	shear stress,
τ_o	bottom shear stress,
ω	fall velocity.

Summary

This chapter discusses physical modeling as applied to coastal problems. Intuitive reasons for using physical models are first given. Physical models usually add to the understanding of a problem, while numerical models reflect the input by the modeler. Next, physical modeling is placed in an historical perspective. The decline in physical modeling is shown to be the result of disenchantment with large, costly long-term design models that need much interpretation. In the next section, physical models that actually work are discussed. Section 5 presents the techniques of scaling. The entire chapter is written in such a format that the non-technical reader can safely skip this relatively theoretical section. Future implications for physical modeling are discussed next. It is clear that the future belongs to short-term design models and process models. These can be simulated at smaller scales (larger models simulating smaller prototype sections). It is also evident that future models will be larger, with more sophisticated controls (Level 2000 models). The more abstract model results will need to be integrated with computations, resulting in composite models. Finally, physical modeling in the future will form a vital link between numerical modeling (which can only reflect the modeler's knowledge) and field research (in which experimentation is costly, difficult to control and difficult to repeat).

Keywords

History; laboratory effects; modeling; numerical modeling; physical modeling; scaling; scale effects.

Chapter 8

Field Observations and Data Management

Gary P. Mocke
Maritime Director
WorleyParsons RSA, South Africa

Francois Smit
Lead Coastal Engineer
WorleyParsons RSA, South Africa

1. Introduction

Field observation has long been a key element of coastal engineering. As Leonardo da Vinci aptly put it some 500 years ago:

> *"Although nature commences with reason and ends in experience it is necessary for us to do the opposite, that is, to commence with experience and from this to proceed to investigate the reason."*

Most of this book deals with the "discovery of the reason". In this chapter an overview is given of what is involved with the planning, execution, dissemination and management of field observations. Some perspective is also given regarding expected future trends in coastal observational research.

Due to the complexity of coastal processes, which may vary significantly even for adjacent locations, some level of empirical analysis based on locally measured or observed site conditions is standard coastal engineering practice. Field observations can therefore be considered an integral part of modern coastal engineering. With the advent of more robust and reliable field observation equipment over the last two decades it has become possible to make *in situ* measurements of processes which could previously barely be measured under controlled laboratory conditions. Similarly, the advent of rapid and cheap communications and the worldwide adoption of the internet have lead to the increasing use of real-time transmittal of captured data through technologies such as GPRS, microwave links, wireless networking and satellite transmission. However, this does not mean that field observation is a recent addition to the arsenal of tools of the coastal engineer.

One of the first "reports" of a process level coastal observation exercise is that by Leonardo da Vinci (1452 to 1519), who observed (Reti, 1974):

> *"The wave of the sea that strikes the oblique shore breaks on it with* impetus. *After the break, the impetus of the wave is consumed as it runs along the shore which it has struck. And receding, it strikes the base of the following wave and slows it down. As the crest of the wave advances more rapidly than its base, the crest is left without support and for this reason it falls upon the aforesaid base, which after the percussion enters underneath the foam and abandons it on the shore. This foam recedes against the base of the following wave."*

Recent coastal hydrodynamics advances described elsewhere in this volume have succeeded in theoretically modeling the observation of Da Vinci some 500 years after he provided the following explanation of why one would want to do field studies:

> *"I will treat of such a project. But first of all I shall make a few experiments and then demonstrate why bodies are forced to act in this manner. This is the method that one has to pursue in the investigation of phenomena of nature. It is true that nature begins by reasoning and ends by experience; but nevertheless, we must take the opposite route: as I have said, we must begin with experiment and try through it to discover the reason."*

This chapter will attempt to show how one would go about "experimenting" in the coastal zone, to thus gain an understanding of what actually happens there. Data of this nature aid the broadening of coastal engineering knowledge in a dual manner, namely qualitatively through the understanding of how things happen, and quantitatively as input to the development and verification of theoretical models.

The focus of this chapter will be primarily on field observation techniques and approaches aimed at understanding near shore coastal processes from the process level detail to the more macro-scale level of resultant coastal changes. The chapter deals first with planning issues, within the context of temporal and spatial scales of change, and then discusses measurement techniques and instruments, followed by an overview of issues related to project execution. This includes a summary of some major coastal process measurement initiatives, long term coastal monitoring programs and associated data reduction and management issues. It ends with a brief overview of perspectives for the future.

2. Objectives and Planning

The effective measurement of coastal processes such as waves, water levels, currents and sediment transport requires a well planned field observation system. The planning needs to incorporate a clear definition of project objectives, which could be directed anywhere from routine long-term monitoring through to highly precise and detailed process measurements. Project objectives should be evident from a clear definition of the scientific and engineering questions that need to be addressed. Special precautions should be taken to ensure full data coverage for highly critical measurements, possibly through the placement of co-located backup instruments to provide redundancy, or by regular monitoring of performance through a real-time data communications shore link. Planning should ensure that data is delivered promptly and is quality checked before being added to a data base. Early analysis of at least part of the dataset will help to ensure data quality and relevance, facilitating timely improvements to monitoring programs when necessary. Finally, an over-riding factor in field measurement campaigns is often that of resource availability. This may relate to available expertise and instrumentation; however, the most frequent limiting factor is funding levels.

Coastal field studies are generally performed for the purpose of scientific research, numerical model validation or the pre- and post-construction monitoring of the impact of coastal structures or other shore development initiatives (i.e. beach nourishment). Figure 8.1 relates these types of studies to temporal and spatial scales of coastal processes. Project monitoring frequently only requires fairly low-level observations such as periodic measurements of beach profiles and near shore bathymetry in the project area and the interpretation of remotely sensed data, such as the analysis of shorelines extracted from satellite and/or aerial imagery. The quantification of conditions forcing the observed and likely future topographic responses, however, requires the measurement of underlying wave, water level, current and even sediment transport processes. If the measurements are being used for sediment-flux determinations the bathymetric surveys could be seen as an "indirect" observation technique whereas *in situ* sediment flux measurements would provide "direct" observations.

Besides providing important quantitative insights on processes at the study site scientific research studies contribute to a more general non-site specific understanding of coastal phenomena. The data emanating from such measurements further serves an important function in the validation of mathematical models of coastal processes. Recent years have seen the development of increasingly sophisticated

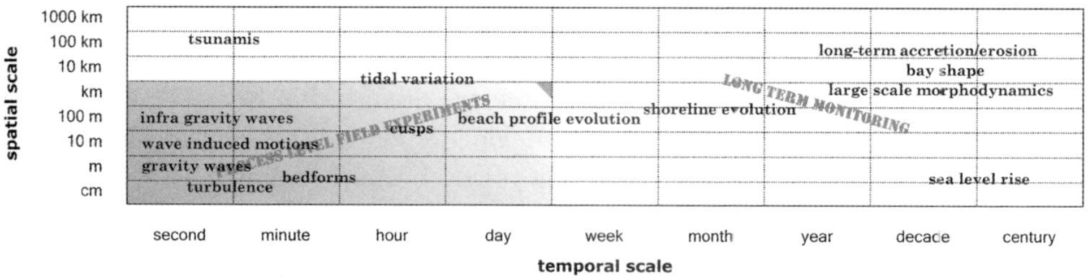

Fig. 8.1 Temporal and spatial scales of coastal processes.

process based mathematical models, many of which have only been validated by theoretical and laboratory studies. Improved instrumentation and measurement techniques, as well as large-scale field exercises providing synoptic measurements of inter-related processes, have however provided important insights into the development and verification of predictive models (see De Vriend *et al.*, 1988, Mangor, 2004 and Kamphuis, 2010).

3. Measurement Techniques

3.1 *Introduction*

Technologies for measuring coastal processes such as waves, water levels, winds, currents, sediment transport and bed evolution are continuously subject to improvement. This is particularly true of highly dynamic environments such as the surf zone where rapid changes and wave breaker related effects such as aeration requires specialized instrumentation and deployment techniques.

Horikawa (1988), Hemsley *et al.* (1990), Morang *et al.* (1997a,b); special editions of the Coastal Engineering Journal on instrumentation (1998) and video imaging (2007); and Pandian *et al.* (2007) may be referred to for comprehensive reviews regarding the development of near-shore measurement techniques. An overview of some of the more commonly used available methods and devices is presented below, with attention given to the measurement of waves, water levels, currents, sediment transport and coastal topography. Other aspects the coastal engineer may get involved with in the field, such as the measurement of physical and biological water quality parameters, are not discussed here.

3.2 *Waves and water levels*

3.2.1 *Wave gauges*

As illustrated in Figure 8.1, waves transcend a range of spatial and temporal scales. Of most relevance to littoral processes are gravity waves, for which a number of measurement techniques have been developed, including staffs, capacitance, resistance and induction gauges, bottom-mounted pressure gauges, surface buoys measuring and integrating water surface accelerations, high frequency acoustic Doppler current profilers (ADCPs) with wave measurement capability, as well as remote sampling techniques such as radar and satellite altimetry (discussed in Sec. 3.7).

These days the most often deployed gauges for coastal engineering studies are either surface (e.g., Waverider® buoys) or bottom mounted (e.g. ADCP or pressure gauge) instruments. These instrument types are briefly described here. Wave gauges can further be broadly classified according to whether they record directional characteristics or not. Directional measurements provide information on the wave energy directional distribution, essential for wave transformation and sediment transport computations. Directional wave measurements may be made from an array of non-directional gauges, or otherwise by gauges specifically designed to provide wave directional information.

Surface Wave Data Buoys

Arguably the most reliable, robust and easily deployed surface wave gauges have historically been the Waverider non-directional and directional instruments manufactured by Datawell in the Netherlands. Others in wide use include the TRI-AXYS buoys developed by AXYS Technologies Inc. Wave buoys have seen successful deployment throughout the world and those from the principal suppliers have reputations for reliability and durability. The instruments calculate wave parameters through integration of measured vertical water surface acceleration. Data transmission between the buoy and onshore receiver stations are typically through radio or satellite transmission, with options such as GPRS also available. These instruments can effectively provide a continuous update of wave conditions as well as data quality. Satellite tracking also assists in retrieving lost instruments. Recent developments include the addition of transducers on these instruments, providing the capability to measure currents using acoustic Doppler technology. The incorporation of solar panels on the buoys allows for increased deployment durations.

The durability of the instruments is illustrated by the experience narrated by retired coastal engineer J. Michael Hemsley, who was involved in a US Army Corps of Engineers, Coastal Engineering Research Center (now called the Coastal and Hydraulics Laboratory) project named "Monitoring Completed Coastal Projects" where they were monitoring the effectiveness of the dolosse restored jetties at Manasquan Inlet, New Jersey, in the late 1980s (pers. comment, 2011):

> *"Once long ago — probably nearly 30 years now — a Corps of Engineers non-directional Waverider was lost in a storm off Manasquan, New Jersey in the USA. It was long gone before any search could be mounted, and it was back in the day when data were reported via line-of-site radio links. Months later, a friend at the Corps' North Atlantic Division called to report that an American flag freighter had seen the buoy in the North Atlantic and reported it to the US Coast Guard, who called the Corps of Engineers office because the buoy was marked as Corps property. Unfortunately, they did not attempt to retrieve it.*
>
> *Much later, a call came from a US Air Force employee in the Canary Islands. Seems a fisherman had found it and turned it over to the local harbormaster, who called the Air Force base to report it. After determining that it was certainly our buoy, the Air Force flew it to Atlanta, Georgia, USA, and we had it shipped from there to our laboratory in Vicksburg, Mississippi, USA.*
>
> *After a good cleaning and battery change, the buoy was tested using the classic Datawell rubber cord bouncing system, and it was put back into service. It had not been hurt during its long Atlantic trip — a testament to the durability of the Datawell Waverider buoy."*

Pressure Gauges

Pressure gauges incorporate strain or piezo-resistive gauges to measure sub-surface pressure. The gauges include temperature sensors and can often optionally accommodate conductivity and/or turbidity sensors. By incorporating an electromagnetic current meter, the instruments, often referred to as *puv* (*p*-pressure, $u + v$ velocity) gauges, can determine wave direction from the orbital velocity signal. The Valeport range of recorders comprises well-known examples of such instruments.

Pressure gauges may be bottom mounted in shallow water. In deep water deployments, they may need to be suspended high enough in the water column to detect the full spectrum of orbital motions, bearing in mind that higher frequency wave orbital motions do not penetrate the water column as deeply as those associated with lower frequencies.

Although compact and relatively inexpensive, *puv* gauges share the disadvantages of pressure sensors with regard to maintenance requirements and data transmission restrictions due to bottom mounting, with cable communication generally used for real-time output. Data processing and storage is therefore often done on board these instruments by employing linear wave theory. Raw data is of course stored as well for possible future analysis using custom post-processing routines. A relative advantage over buoy type devices, however, is the capacity for extracting long period tidal water levels and current data. The current data, which may also include the vertical velocity component, provides important supplementary information for sediment studies.

ADCPs with Wave Measurement Capability

ADCPs, discussed further in Sec. 3.4, were principally developed to measure current velocity profiles through the water column through measuring the Doppler shift of sound. Subsequently the Iterative Maximum Likelihood Method (IMLM) was employed to allow the measurement of wave conditions using a bottom mounted ADCP directed upward to the water surface (Terray *et al.*, 1999). ADCPs typically use a Janus configuration of acoustic beams (beams typically oriented at a 20° to 30° angle to the vertical), which can be used as an array of sensors to determine wave directionality. An ADCP equipped with a pressure sensor can provide three independent estimates (orbital velocity, surface echo ranging and pressure) of the non-directional wave height energy spectrum, thereby providing a confidence check on data quality (as for pressure gauges, the effectiveness of the pressure sensor in resolving the wave spectrum is limited by the deployment depth of the instrument). The IMLM allows the independent resolution of the wave field in all (arbitrarily sliced) directions, which means that separate swells from different directions can be resolved, even when of similar frequencies. Primers on the measurements of waves using different manufacturers ADCPs provides detailed background information and is available online (see e.g. Teledyne RDI, Nortek and Sontek, 2001).

Horizontally looking ADCPs using narrow beam configurations have also been developed that allow for the measurement of directional waves (in addition to surface currents) across a horizontal range that can extend for a couple of hundred metres from the instrument.

In very shallow water and in the surf zone, acoustic Doppler velocimeters (ADVs), having very small sampling volumes, can be used to make directional wave measurements using *puv* principles.

Instrument Placement

As discussed by Morang *et al.* (1997a), there are no firm guidelines on placing wave gauges, which is usually a function of site conditions, resource and scheduling constraints, and historical data availability. In the event of process-level field experiments or specific coastal project design/monitoring, it is recommended that one or more instruments be placed in close vicinity to the area of interest. If the scope exists for longer deployments or where regional data is required, deep-water placements covering seasonal variations are recommended. The optimal arrangement is simultaneous shallow and deep-water deployments, thus providing transfer functions for transforming more readily available offshore data to the site location.

When deploying ADCPs, the intended depth of deployment and expected wave conditions to be measured are important parameters to take into consideration when deciding on which type of instrument to use. Higher frequency transponders are necessary for shallower waters (e.g. a 1200 kHz system is suitable for near-shore waters between 5 m and 14 m water depth, whilst a 150 kHz system would be more suitable for water depths around 100 m). Trade-offs between minimum resolvable wave heights and periods, depths of deployment and transponder frequency are inevitable. Equipment manufacturers such as Teledyne RDI, Nortek and Sontek all provide performance specifications of instruments and guidance on equipment choice and setup.

Buoys are also used as Lagrangian measurement instruments, being allowed to drift with the tides and currents. The Queensland Environmental Protection Agency in Australia for example uses this approach to measure wave conditions in cyclones. As the occurrence and path a cyclone will follow are unknown until it actually forms, the measurement of wave conditions close to the center of a cyclone are unlikely to be measured by fixed instruments. The Queensland EPA makes use of a Bell 412 helicopter to drop a Waverider® buoy named BOB (Baby Ocean Buoy) offshore when a cyclone is heading towards the Queensland coast. The instrument then records wave conditions as it moves and is usually recovered after a period of 10 to 14 days (Queensland Government, 2013).

3.3 *Water levels*

Site specific process or project studies require continuous water level measurements in the vicinity of the study site. Such measurements are particularly important in the surf zone, where wave transformations result in both set-down and set-up of the mean water level. Changing water levels over long periods, such as those associated with sea level rise, are of significant importance for coastal zone management. The types of water level gauges in common use are radar, pressure, acoustic, stilling well floats and tidal poles. The latter is often used as a sanity check of digital recordings by other gauge types). The more recently developed radar gauges have advantages in being relatively cheap, do not need a stilling well installed, record at high frequencies and are easily maintained as they are deployed above the water surface and so do not come into contact with seawater and can be readily accessed.

Where gauges are not installed at a site, high water elevations may also be estimated from debris deposits on the beach, whereas inundation levels during extreme events can often be determined from water marks on elevated features.

Since 1933, the Permanent Service for Mean Sea Level (PSMSL) at the National Oceanography Centre in the UK has been responsible for the collection publication, analysis and interpretation of sea level data from the global network of tide gauges. As of 2010, the database of the PSMSL contained over 60,000 station-years of monthly and annual mean values of sea level from about 2,050 tide gauge stations around the world received from almost 200 national authorities (Richards and Holgate, 2011). On average, approximately 2,000 station-years of data are entered into the database each year.

In the USA the University of Hawai'i Sea Level Center (UHSLC) houses data from a global network of tide gauges as part of the Global Sea Level Observing System (GLOSS). The UHSLC maintains the Joint Archive for Sea Level (JASL) which acquires hourly data on an annual basis from across the globe. More than 100 agencies from 90 countries have contributed to JASL since 1987.

3.4 *Currents*

3.4.1 *Lagrangian measurements*

The measurement of current speed and direction covers a wide spectrum of technologies, cost and resources. Dye, drogues, floats and other tracers of Lagrangian drift are relatively simple methods often used in coastal sediment and water quality studies. Although usually not able to provide a completely faithful representation of water movement, these relatively inexpensive methods provide a good representation of flow patterns if deployed optimally. The first-level insights they provide can also be useful in developing a sampling strategy for more comprehensive follow-up field studies.

3.4.2 *Eulerian measurements*

Various levels of sophistication exist for point source Eulerian current measurements, extending from dueled impeller (direct reading) meters, through electromagnetic meters to state-of-the-art ADCPs.

Direct reading current meters: Although subject to fouling and mechanical problems, and slow response to changes in current direction, direct reading instruments (impellers) are relatively cheap and easy to deploy.

Electromagnetic current meters: These instruments display reasonable response and accuracy, they are however, relatively expensive to purchase and maintain, and require careful instrument calibration. The maximum speed that can be measured by commercially available electromagnetic current meters is of the order of 5 m/s, with a maximum velocity fluctuation response time of about 0.1 s. This limits the application of the meter in measuring an important feature of the surf zone, namely the high frequency turbulent motions.

ADCPs: The advent of ADCPs constituted an important breakthrough in current measurements. Initially used as profilers on ships, acoustic Doppler techniques have provided precise point and profile measurements useful for wave, boundary layer and turbulence measurements.

ADCPs contain transducers that transmit and receive sound pulses ("pings"). The sound waves get scattered by natural occurring particles in the water. By determining the Doppler shift of the sound pulses reflected back to the instrument, the velocity of the current can be determined. In order to determine the 3D current, three transducers are required. Instruments like the Teledyne RDI Workhorse contain four transducers in a Janus configuration (see Fig. 8.2), providing a level of redundancy and error estimating. The Janus configuration helps in rejecting errors in horizontal velocity due to tilting of the ADCP.

One of the principal advantages of ADCPs compared to other current gauges is that they measure currents in bins through the water column, allowing the velocity profile to be determined. Using other type of gauges would require a string of instruments to be suspended through the water column. The ADCP bins are evenly spaced, i.e. all bins are of the same size. The ADCP measures the average speed through the bin, whereas a string of current meters would each measure the current at a discrete point in space.

The velocity profiles provided by ADCP instruments, achieved without strings of multiple point source meters, provide important information on flow structure and fluxes of water motion. Velocity profiles provided by ADCP instruments have proven to be particularly useful in tidal inlet and estuarine studies, where circulation and exchange processes need to be quantified both laterally and vertically. One such technique has been the horizontal mounting of profilers on jetties or pipelines for "sidelooking" across inlets or navigation channels. Lane *et al.* (1997) discuss the use and limitations of ADCPs for quantifying net fluxes of water and suspended particulate matter across a macro-tidal estuary.

ADCPs can also be deployed on a moving vessel in so-called bottom tracking mode, whereby the instrument sends out bottom tracking pulses to determine the altitude above and the velocity over the bottom, in addition to the standard acoustic pulses used to determine current velocity. In this manner the ADCP can be used to survey circulation patterns, something that is not possible using a fixed current gauge.

An ADCP has no moving parts, so maintenance of the instrument is relatively straightforward and instruments are known to have seen years and even decades

Fig. 8.2 An ADCP being deployed within a lightweight frame off a vessel in the Gulf off Qatar. The head of the ADCP showing the four transducers is clearly visible. Courtesy E. Heynes.

of service with no significant problems. Biofouling of the transducer heads are one problem that occurs often. Whilst anti-fouling coating can be applied, simple measures such as the use of petroleum jelly mixed with chili powder have been proven successful to prevent biofouling on shorter deployment durations.

ADVs: Measuring either at a single point or overmultiple depth cells or "bins", ADVs measure velocities from the Doppler shift in the backscattered acoustic energy caused by moving suspended particles in the water column. As these measurements are made at a position displaced away from the acoustic meters, minimal disturbance of the local flow regime ensues.

Gliders: Multi-sensor gliders have found application in oceanography in recent years and are primarily deployed in deep water. Coastal gliders have also been developed that can operate in water depths as shallow as 4 m (in the case of the

Teledyne Webb Research Slocum Electric Glider). These instruments can operate autonomously for long periods of time and transmit captured information on currents, as well as other parameters such as temperature and salinity when surfacing. They can be used to assist in determining regional current patterns (Bouffard, *et al.*, 2010), important for developing an understanding of nearshore circulation as well as upwelling and downwelling events (Pattiaratchi, 2010). It is foreseeable that a fleet of coastal gliders could be deployed as part of large-scale coastal field measurement campaigns in the near future, providing invaluable information on offshore conditions used to better understand the observed near-shore conditions and serve as boundary conditions to numerical models describing these near-shore processes.

3.4.3 *Turbulence*

Wave breaker related turbulence is a fundamentally important surf zone process (Mocke, 2001), significantly influencing sediment transport and undertow current behavior. Although some important advances have been made in measuring breaker turbulence in controlled laboratory conditions (Govender *et al.*, 2004), the quantification of turbulence levels and characteristics in the field (and in particular the high energy surf zone) remains a key challenge for near-shore measurements.

A number of investigators have measured turbulence levels under non-breaking conditions, with particular attention given to the wave boundary layer. Only a limited number of turbulence measurements have however been made in the prototype surf zone. Although some attention has been given to extracting high-frequency turbulence quantities from acoustic sensors (Voulgaris and Trowbridge, 1998), their application in the surf zone is constrained by the highlevels of breaker related aeration. A Laser Doppler Velocimeter (LDV) was deployed by Agrawal and Aubrey (1992) in a low energy wave environment, whilst George *et al.* (1994) used hot film sensors to measure turbulence levels under moderate wave conditions. During the SANDYDUCK '97 experiment (see Sec. 4.2.3) ADV's mounted on a low level frame in 4–5 m water depth were used to measure three-dimensional velocities within 1 m of the bottom (Fredericks *et al.*, 2001). Whilst the instruments were outside the surf zone for long periods, three events were captured where strong winds and large waves were present and the instruments were in the outer part of the surf zone. Trowbridge and Elgar (2001) describe the results of the analysis of these measurements. Bryan *et al.* (2003) similarly used ADVs to estimate turbulent dissipation rates in the surf zone. Feddersen (2012) describes surf zone field measurements of turbulent energy dissipation in 2006 at Huntington Beach, California, where a cross-shore array of seven frames, each fitted with a buried pressure sensor and downward looking ADV, were deployed.

3.5 *Sediment transport*

Whether considered either as a rate or a concentration, sediment transport remains a very difficult quantity to measure. Typical transport rate measurement approaches include the use of sand tracers, ultrasonic flux meters and traps. Concentrations are generally measured using optical or acoustic sensors as well as simply by water column sampling. Although tracers provide an average transport rate distribution over a geographical area (see Black *et al.*, 2007 for a review), other methods generally only present single point measurements. Given the complex and highly variable interactions between sediment and near shore hydrodynamics, any characterization of sediment transport rates for an area would strictly require a synoptic array of measurement points. Such profiles are conventionally achieved by stacking point measurement devices in vertical or horizontal arrays.

Horikawa (1988) and White (1998) provide comprehensive reviews of most of the commonly used instrumentation and measurement techniques, including the development of many refined instruments for measuring sediment concentration flow fields. An example is the optical backscatter sensor (OBS) (Downey *et al.*, 1981), which has also been miniaturized (MOBS) (Hanes and Huntley, 1986) and increased in resolution in the form of the Fiber Optic Back scatter Sensor (FOBS). With the capacity for measuring suspended sediment concentrations at multiple positions in the lower 50 cm of the water column (Beach *et al.*, 1992) the FOBS provides important insights on sediment transport and suspension dynamics in the wave-current bottom boundary layer. The OBS type sensors are relatively robust and responsive to large fluctuations in concentration, thus making them suitable for surf zone deployments. Signal response is however highly sensitive to sediment grain size, requiring a careful calibration in mixed grain size environments such as mud-sand mixtures and near the seabed in the wave boundary layer. There is also the possibility of instrument proximity related bed disturbance as well as bubble contamination in highly aerated regions such as the surf zone (Smith and Mocke, 2002). These and other restrictions related to OBS application are discussed in White (1998), who suggests a number of caveats for use of such instruments.

Advances in acoustic theory and sonar design have demonstrated the possibility of resolving profiles of sand size particles to an accuracy of $\pm 10\%$ and $\pm 25\%$ for concentration and particle size respectively with spatial and temporal resolutions better than 0.01 m and 1 s, respectively (Taylor *et al.*, 1997). Shown in Fig. 8.3(a) and 8.3(b) are measurements of suspended sediment concentration (Williams *et al.*, 1998) made in an experimental wave flume using a 4 MHz acoustic backscatter sensor (ABS) for regular waves and irregular waves respectively. The progressive increase of suspended sediment concentrations due to a "wave pumping" effect is evident for wave grouping under the irregular wave case.

Although not originally developed for the purpose, ADCPs have been used to derive estimates of suspended sediment concentration by recording the relative intensity of the acoustic backscatter signal (Holdaway *et al.*, 1998; Roland, 2006). These ADCP instruments furthermore have profiling ranges significantly greater than the 1–2 m typical of existing ABS systems. Stanton (1996) discusses the development of the Coherent Acoustic Sediment Profiler (CASP) to measure all three

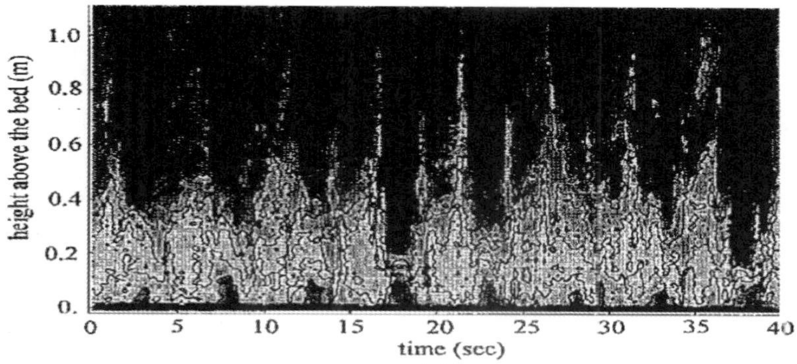

(a)

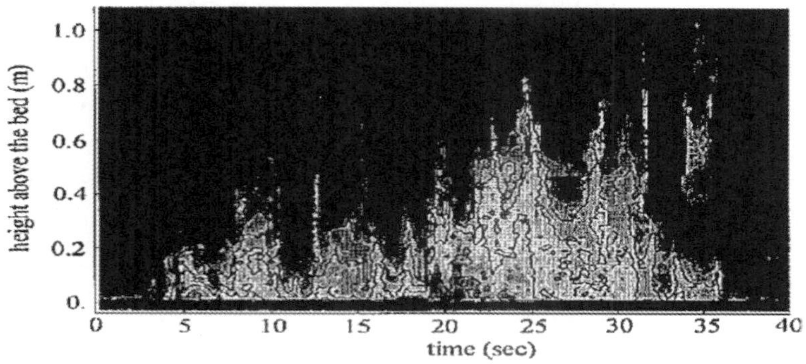

(b)

Fig. 8.3 Suspended sediment concentrations measured by a 4 MHz ABS (Williams *et al.*, 1998) for (a) regular waves ($H = 0.8$ m, $T = 5$ s) and (b) irregular waves (Hs = 0.82 m, TP = 4.9 s).

velocity components as well as the sediment concentration by means of four intersecting acoustic beams. Zedel *et al.* (1996) describe the development of a high-resolution multi-beam coherent Doppler profiler capable of measuring velocity and concentration profiles as well as 3-dimensional bathymetry over horizontal scales of up to 10 m. Using three 1 MHz ultrasonic transducers, Shen and Lemmin (1997) describe a method for simultaneously measuring the instantaneous concentration and corresponding 20 velocity profiles. Given their capacity for simultaneously measuring concentrations and velocities, and hence sediment transport fluxes, through the entire water column, acoustic profiling sensors therefore constitute a powerful new technology for improved near-shore process measurements.

As for optical sensors, ABS is highly sensitive to grain properties such as size fractions. Although, some attention (Hay and Sheng, 1992) has been directed at trying to acoustically detect grain size, much work remains for deciphering multi-grain size environments.

The preceding discussion has focused on the capacity for new instrumentation techniques to measure suspended sediment concentrations and transport. However, as discussed by White (1998), considerable obstacles remain with regards to the measurement of sediment concentrations and transport at or very near the bed. As no reliable electronic sensors are presently available, continued reliance is made to conventional sediment tracer and trap methods.

Similarly, ADVs can provide information on instantaneous suspended sediment concentrations at a single point. As discussed by Chanson (2008), the calibration of ADVs for suspended sediment concentration measurements is critical.

3.5.1 *Cohesive sediments*

Still in its infancy is the application of ABS to cohesive sediment environments. This arises from the complex inter-relationship between flow size, structure and concentration and the turbulent character of the flow which together produce a more complex pattern of back scatter than can be accounted for with present scattering models. As discussed by Eisma *et al.* (1996) amongst others in a special issue of the Journal of Sea Research, floc characteristics are best measured by *in situ* photographic or video imaging techniques.

In addition to suspended sediment characteristics, it is important to be able to quantify the susceptibility of cohesive sediments to deposition and erosion to/from the bed. Widdows *et al.* (1998) describe the use of *in situ* fluxes to evaluate deposition and erosion thresholds.

3.5.2 *Aeolian transport*

As comprehensively reviewed by Horikawa (1988), the principal procedures for performing field observations of wind-blown sand involve anemometers and/or sandtraps. It is important to distinguish between the different modes/phases of Aeolian transport, i.e. creep, saltation and suspension. Useful devices for measuring saltation are piezo-electric sensors and saltiphones. It is furthermore important to measure meteorological factors as well as the general properties of the blown sand. Anemometers, which may range in sophistication from propeller up to ultrasonic or hot-wire, measure the wind speed as an external force producing sand movement. Sandtraps, which may be aligned either vertically or horizontally, measure the weight or volume of the blown sand and can be configured as either total quantity or distribution type.

3.6 *Coastal topography/bathymetry and sediment characteristics*

3.6.1 *Introduction*

Surveys of sediment characteristics, beach profiles and near shore bathymetric configurations and changes are an important element of field investigations. As comprehensively discussed in CIRIA (1996), the information gained in this manner is vital for coastal engineers in the design of coastal projects as well as for evaluating the effects of coastal structures on adjacent shorelines. Historical data on beach profile and shoreline changes also provides critical input for establishing development setback lines and other coastal zone management initiatives. The measurement of topographical changes and associated volumetric analysis furthermore quantifies the performance of breakwater protection and beach nourishment projects. Sediment characteristics determined from beach and near-shore material sampling provide insights on seasonal and/or long-term changes in beach composition, sediment sources and the appropriateness of beach nourishment material. The information is further more essential input for coastal models. Appropriate measures for near-shore sediment sampling and analysis are presented in CIRIA (1996).

As sediment and topography changes evolve in response to fluid and sediment motion, they further provide important insights with regard to the constituent wind, wave, current and sediment processes in the measurement area. An example is the information on sediment transport pathways derived from sampled particle size and sorting characteristics (Gao *et al.*, 1994). Unfortunately, such measurements are typically carried out at infrequent intervals that far exceed the temporal scales of most of the forcing processes, resulting in, at best, an integrated interpretation of underlying processes. Imaging techniques for coastal monitoring, discussed further in Sec. 3.7.6 can provide the near continuous information required for process based interpretation.

Horikawa (1988), Morang *et al.* (1997b) and Gorman *et al.* (1998), amongst others, provide comprehensive reviews on technologies and interpretation methods for carrying out topographic surveys. Topographic/bathymetric survey techniques can be broadly categorized into four methods: (1) direct leveling, (2) acoustic sounding, (3) image analysis and (4) topographic and hydrographic lidar. The first two of these methods are discussed below; whilst the latter two are discussed in Sec. 3.7 on remote sensing.

3.6.2 *Direct leveling*

In the low-lying Netherlands, where dikes provide protection from flooding, the measurement of coastal foreshores has been undertaken since the 1800's using rowboats, sightlines and lead lines (Figure 8.4). Coastal profiles were measured to determine whether steepening of the foreshore took place, which would indicate a risk of a dike slide flow or "dijkval". Data was handwritten in ledgers, forming one of the earliest established coastal databases (Figure 8.5).

These days, topographic profiles above the waterline are easily undertaken using DGPS equipment. The surveying of beach profiles from the back of the dune to the low water line (and even deeper if wading) is standard practice for most coastal monitoring programmes.

For extending the profiles to shallow nearshore waters, amphibious vehicles such as the FRF's CRAB are deployed. Besides amphibious vehicles (Birkemeier and Mason, 1984; Dally *et al.*, 1994), helicopters have been used as platform from which to undertake direct levelling. The USACE Portland district developed the Helicopter

Fig. 8.4 Execution of a sounding near groin 5a along the coast of Delfland (between The Hague and Hook of Holland) on 17 May 1979. Courtesy Archive Hoogheemraadschap van Delfland, Netherlands, reference number 2083.

Borne Near-shore Survey System in 1960 (Craig and Team, 1985). The technique has been shown to be efficient and accurate (Pollock, 1994) and to have potential (Coppoolse *et al.*, 1992) for application in high energy surf zones and near coastal structures where water based survey methods may be hazardous.

3.6.3 *Acoustic sounding*

Acoustic depth sounders mounted on motorized vessels allow the rapid surveying of large areas of the seafloor. As the echo sounder simply measures the depth of the underlying water column, reference survey charts require both horizontal and vertical position fixing of the instrument location. The horizontal position can be determined by conventional land survey methods or Global Positioning Systems (GPS). Vertical fixing, however, needs to build in corrections for vessel draft and movements due to tidal, wave and surge actions. This is a particular problem in the near-shore zone, where the dynamic response of the vessel to water waves may be poorly understood.

Multi-beam echo sounders and sector-scanning sonar and acoustic ripple profilers (Bell *et al.*, 1998) provide measurements of higher resolution and frequency. Combined with heave-roll-pitch sensors, precise positioning through differential GPS and the computerized integration of navigation sensors, data management

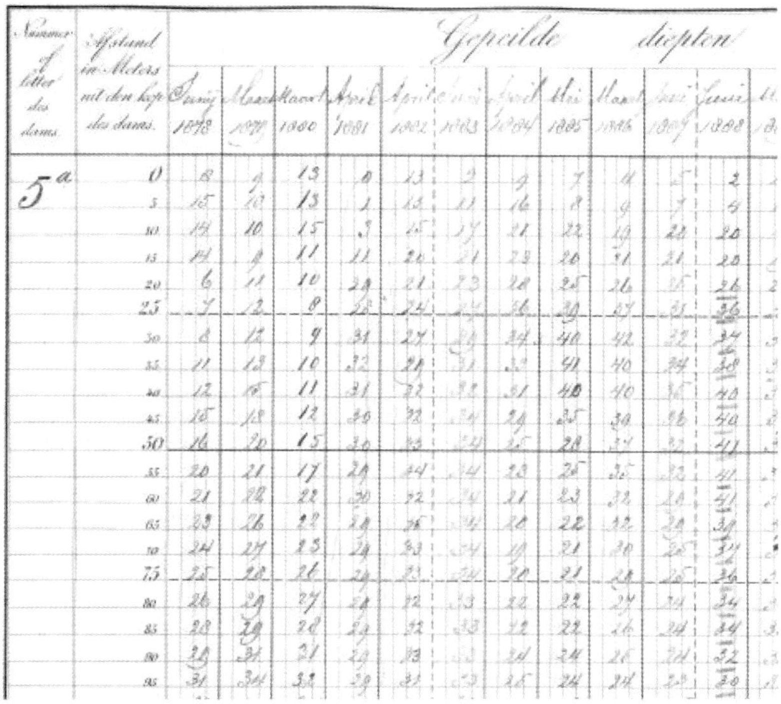

Fig. 8.5 Record of sounding of the depths in front of groyne 5a along the coast of Delfland (between The Hague and Hook of Holland, the Netherlands) from the period 1878–1897. The depth is in m below low water; the second column is the distance in m from the head of the groyne in the seaward direction. Courtesy Archive Hoogheemraadschap van Delfland, Netherlands, reference number 7830b.

and display, such instruments provide accurate and detailed bathymetry coverage. The data processing and management requirements of such systems are, however, considerable, requiring specialist hardware and software. Side-scan sonar is an invaluable tool for characterizing bed features as well as evaluating the condition of breakwaters and underwater structures. The basis and range of application of these instruments are further discussed in Morang *et al.* (1997b).

For the study of near-shore coastal processes the bathymetry in shallow waters needs to be measured. This is difficult to do using conventional survey vessels, as, in addition to vessel draft and instrument deployment depth limitations, there are inherent dangers in operating such vessels in the surf zone, around tidal inlets and coastal structures. To overcome these issues, personal watercraft such as jetskis fitted with echo sounders and GPS have been developed by numerous researchers, engineers and surveyors (see e.g. Dugan *et al.*, 2001; McMahan, 2001 and Lippmann and Smith, 2009). A similar approach was adopted by Mead (2000), who used a portable, floating survey system dubbed Horatio to survey remote surfing reefs in the Pacific, often by towing the system behind an inflatable kayak.

3.7 *Remote sensing*

Aerial photography and satellite and video imaging are powerful tools for field measurement and monitoring. An important impetus for the development of remote sensing techniques occurred with the Second World War, where hostilities prevented on site measurements. Lundahl (1948) describes some of the most interesting early developments for determining near-shore bathymetry at the time using aerial photography. An example is the application of parallax methods to relatively clear water reef areas, providing average near shore depths that differed by less than 15% from actual values.

3.7.1 *Aerial photographs*

Aerial photographs are frequently the only historical source of information on the geomorphology of a coastal area. Where stereographic pairs having overlap of more than 60% are available, detailed topographic maps can be generated using photogrammetry. As described by Sonu (1964) this interpretation can be extended into measuring underwater topography by multi-spectral and multi-temporal water transparency and sea bottom photogrammetry methods. Both of these approaches require relatively clearwater, and would therefore not be applicable in turbid regions such as the surf-zone. The inverse modeling of water depths from surface wave velocities and refraction can be used in such environments, provided there is minimal violation of wave stationarity, such as would occur over rapidly changing water depths. The interpretation of aerial photographs must include corrections for distortion effects related to camera tilt, variable scales, relief displacement and radial lens distortion.

3.7.2 *Satellite imagery*

Satellite images cover wide areas and all weather and seasonal conditions. As such, they are useful for quantifying large-scale coastal evolutionary and turbidity phenomena. Gorman *et al.* (1998) discuss the availability and applicability of satellite imagery, illustrating their use for detecting thermal plumes and suspended sediment runoff into the near-shore. Older imagery, such as the French Commercial Systeme Pour L'Observation deIaTerre (SPOT) and NASA's LANDSAT series of satellites, typically have image pixel resolutions in excess of 10 m. SPOT provides two identical high resolution visible (HRV) imagery systems, in 10 m resolution panchromatic mode. More recent satellites, including commercial ones such as Ikonos, Quickbird and GeoEye, provide sub-meter resolution panchromatic imagery useful in delineating shorelines, coastal assets and other features.

The Moderate Resolution Imaging Spectroradiometer (MODIS) instrument aboard NASA's Terra and Aqua satellites acquires data in 36 spectral bands. MODIS data includes sea surface temperature and ocean colour data and has been used to monitor the extent of dredged plumes in coastal waters (Kutser *et al.*, 2007).

3.7.3 *Lidar*

Airborne Lidar (Light Detection and Ranging) is used extensively in onshore topography and near-shore bathymetry mapping. Lidar is an optical remote sensing technology that measures the distance from the instrument to the land and/or sea bed by rapidly firing a laser beam. Topographic systems use a single near-infrared laser, whilst hydrographic Lidar systems make use of two lasers, one infrared and the other blue-green. The blue-green laser penetrates the water and reflects of the bottom, whilst the infrared reflects off the water surface. The difference in return time of the reflected beams allows the water depth to be determined. The beams are swept rapidly across the travel direction, with sounding rates typically around 900 Hz and sounding densities ranging from $2\,\mathrm{m} \times 2\,\mathrm{m}$ upwards, generating vast amounts of data (millions of soundings per hour). Coupled with a range of other sensors, including tide gauges, DGPS, and heave, roll and pitch sensors, the seabed and onshore beach topography can be accurately determined. Initial processing is usually done onboard, but extensive post-processing following sorties is required to produce survey datasets comparable in accuracy to single or multi-beam echo sounding surveys (IHO Order 1).

The principal advantage of hydrographic Lidar is that large areas that would take months or even years to survey by conventional means such as multi-beam echo sounding can be covered rapidly. Coverage rates of $20\,\mathrm{km}^2/\mathrm{h}$ are typical. Lidar effectively provides dense (2–5 m grid resolution) survey coverage in shallow waters where multi-beam echo sounding is challenging. The high density of points allows for objects such as underwater obstructions to be detected.

A disadvantage of hydrographic Lidar is penetration of the beam through the water surface, where scattering, reflection and absorption of the laser light serve to limit the intensity of the reflected return signal. Turbid waters and energetic surf zones can therefore be specifically problematic, but this can often be managed by waiting for suitable conditions before flying a sortie. The rule of thumb is that water depths up to three times the Secchi depth can be surveyed (Estep *et al.*, 1994).

The cost of mobilization associated with airborne Lidar surveying is usually one of the biggest factors preventing application. However, when surveying vast, clear, shallow water areas, the time and overall cost saving of deploying Lidar compared to echo sounding by boat makes these mobilization economically feasible.

The US Army Corps of Engineers (USACE) started using a Lidar system known as SHOALS (Scanning Hydrographic Operational Airborne Lidar Survey) operationally in 1994 under its National Coastal Mapping Program (NCMP). The system is mounted on an aircraft or helicopter to survey coastal areas and inlets (Estep *et al.*, 1994; Lillycrop *et al.*, 1996). Since 2004, the NCMP is executed by the USACE Joint Airborne Lidar Bathymetry Technical Center of Expertise (JALBTCX) using the replacement to the original SHOALS called the Compact Hydrographic Airborne Rapid Total Survey (CHARTS) system, which utilizes a Optech SHOALS-1000T sensor comprising a 1,000 pulse-per-second hydrographic laser and a 9,000 pulse-per-second topographic laser.

Lidar data and imagery collected along the whole of the US coastline by JALBTCX and other organizations are managed by NOAA's Coastal Services Center and are accessible through their Digital Coast portal (see Sec. 5.3)

Lidar Survey Example, Dubai: With the advent of major offshore developments such as the World project and the Palm Jumeirah and Palm Jebel Ali developments along the coast of Dubai in the United Arab Emirates, the need for an updated survey of the seabed off the Dubai coast became clear in 2003. Hydrographic Lidar was identified as a particularly suitable technique for rapidly surveying the shallow waters off the Dubai coastline, with the first survey carried out in 2004 (Sinclair and Smit, 2005).

Under normal conditions Dubai coastal waters are clear and ideal for laser surveying. The survey generally achieved excellent results, although ongoing dredging and reclamation impacted on water clarity, which reduced or prevented Lidar coverage in some areas. These gaps were all due to excessive turbidity prevalent in these areas at the time of capture. Off Jebel Ali the survey extended far enough offshore to capture the shipping channel in its entirety (Smit *et al.*, 2007).

The Lidar data set has been used in numerous coastal engineering and coastal management studies in Dubai, showing itself to be invaluable in providing accurate, detailed bathymetry data for use in the numerical modelling of coastal processes such as wave transformation, near-shore circulation, sediment transport and water quality.

3.7.4 *Synthetic aperture radar and radar altimetry*

Mounted on an aircraft, the scanning radar altimeter (SRA) system (Hwang and Walsh, 1997) can scan wave topography in a $\pm 22^\circ$ swath over a distance of 10 km in less than 20 min. The relatively high resolution mapping, having vertical resolution accuracies within 0.2 m, may be used to identify breaking patches and nonlinear properties such as wave asymmetry.

Radar altimeters on satellites have provided the means to measure the ocean's wave climate on a scale impossible to achieve with a buoy network. Altimeters calibrated against wave gauge measurements, are used to measure significant wave heights across the globe. Seasat, launched in 1978, provided the first picture of the global wave climate (Chelton *et al.*, 1981). Topex/Poseidon, Jason-1, Jason-2 (Ocean Surface Topography Mission) and Envisat have all provided valuable information on the global wave climate, including seasonal variability and extremes (Challenor *et al.*, 2006).

3.7.5 *High frequency radar*

HF Radar is used to remotely measure surface currents and waves. The coastal ocean dynamics application radar (CODAR) system is a portable land-based system developed at NOAA's Wave Propagation Laboratory. Two stations looking at the same patch of sea are needed to resolve current vectors as the system measures only the component of currents moving towards or away from it. The systems can cover ranges extending to 200 km for current measurements and 100 km for wave heights, with resolution in the kilometers range. This limits the application of these systems to open expanses of water. By decreasing the observation range to below 20 km, resolution on the order of 200 m to 500 m can typically be obtained.

3.7.6 *Digital videoimagery*

Sonu (1969) and Horikawa and Sasaki (1972) have utilized motor driven cameras to map near-shore wave and current patterns. Hotta and Mizuguchi (1980) operated an array of synchronized movie cameras to film the temporal evolution of sea surface elevation with respect to reference poles, thus providing wave time series. It has however been the more recent arrival of video and digital processing techniques which have made image analysis an attractive option for the measurement and monitoring of coastal processes.

Alport *et al.* (1998) used rectified time exposure video images of the central beaches of Durban, South Africa, to quantitatively depict shoreline and near-shore bar features in the vicinity of a coastal groyne system. Applied by the US Army Corps of Engineers in its Duck, N.C. field experiments, the technique has been applied to measuring sand bar morphodynamics (Lippmann and Holman, 1989) and wave run up characteristics (Holland *et al.*, 1991; Walton, 1993; Salmon *et al.*, 2007) as well as inter-tidal beach profiles (Aarninkhof *et al.*, 1997) and shoreline cuspate features. A number of fully autonomous video imaging stations have been installed at selected coastal locations worldwide since 1992 as part of the so-called ARGUS system (Holman, 1994). Data collected by this system is stored on a central database at the Coastal Imaging Laboratory of Oregan State University in the US, from where it is accessible to interested parties via the internet. Stereo imaging using multiple video cameras has provided measurements of foreshore topography, with vertical accuracy in the order of between 1 cm and 3 cm being achievable (Holland and Holman, 1997).

Aircraft-mounted video systems have further been found to provide rapid, inexpensive and accurate measurements of near-shore sand bars over several hundred kilometres of coast (Worley *et al.*, 1997). Rodriguez *et al.* (1997) describe the development of digital image processing techniques for measuring surf zone circulation and mixing. A special issue of the Coastal Engineering Journal (2007) provides an update on the state-of-the art in video imaging and discusses future trends.

4. Execution

4.1 *Introduction*

This section deals with the execution phase of field measurements, principally through presenting an overview of some of the major experimental observation programs executed to obtain process level data and insights.

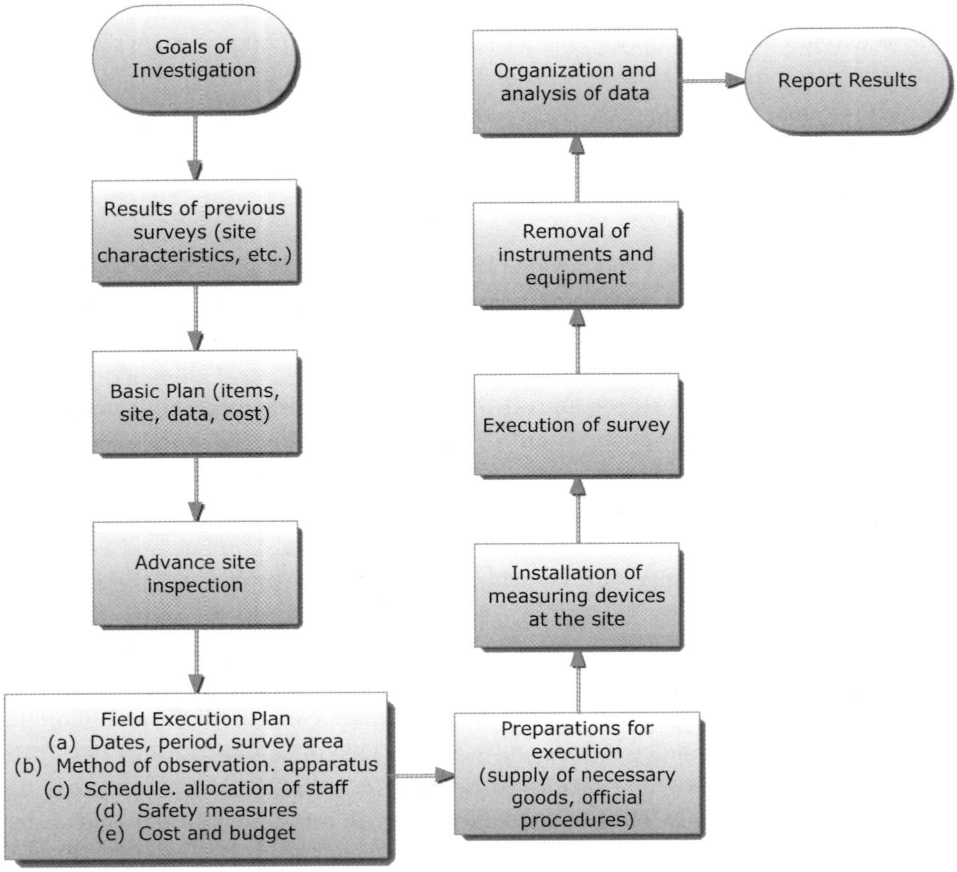

Fig. 8.6 Flow chart of a general plan for performing coastal engineering field studies.

However good the planning for a field observation program can be, and not withstanding the accuracy of the instruments, the ultimate success of the program is measured by the quality of data captured. Figure 8.6, adapted from Horikawa (1988), depicts the various stages in a field measurement program in flowchart form. Following a clear definition of project objectives and the basic plan, the next major steps are the drawing up of a detailed measurement execution plan, the installation of measuring devices and finally the analysis and presentation of data. Given the effort and cost associated with mobilizing a field measurement campaign, it is important to build in a measure of redundancy and backup with regard to hardware and manpower resources. This is especially critical in the event of scheduling constraints related to project "fast tracking" or the need for measurements over specific seasons or events.

4.2 *Large-scale field experiments*

4.2.1 *Introduction*

A great number of field experiments have been carried out worldwide. In a review of available field data on coastal and transport and morphodynamics Wallace (1993) lists over 150 field experiments, 24 of which are reviewed in detail. A further review (Arcilla *et al.*, 1994) is contained in the Proceedings of Coastal Dynamics '94, a specialist international conference on the role of large scale experiments in coastal research. These reviews, as well as a validation exercise of two morphodynamic models against field data (Wallace and Chesher, 1994), included assessments of the suitability of available data for assessing predictive model performance. The review of Wallace (1993) concluded that then-available experimental data sets generally did not satisfy basic suitability requirements, namely: (1) continuous measurement of boundary conditions such as directional wave spectra, (2) accurate and frequent bathymetry surveys, and (3) rapid analysis presentation and availability of data.

Outlined below are some of the most significant large scale coastal field experiments carried out over the past 30 years.

4.2.2 *NERC, Japan*

Comprising six series of experiments conducted at two Japanese east coast beaches over the period 1978–1984, these measurements include detailed wave, current and sediment transport observations (Horikawa and Hattori, 1987). This was the first major field experiment where a multiple synchronized camera system was deployed, in this case for the measurement of wave transformations, near-shore currents, and sediment transport, the latter two types of measurements being done through tracer methods. Simultaneous measurements of sediment concentrations and particle velocities using an ultrasonic flux meter, as well as sand trap measurements (Kraus *et al.*, 1981) providing the suspended sediment transport rate.

4.2.3 *Duck, USA (DUCK94, SUPERDUCK, DELILAH, SANDYDUCK97)*

In the early 1980's the Coastal Engineering Research Center (CERC) of the US Army Corps of Engineers initiated a series of field experiments at its Field Research Facility (FRF) at Duck, North Carolina. The earliest field measurements were primarily directed at obtaining beach profile measurements (Birkemeier and Mason, 1984), incorporating the use of the FRF's unique CRAB amphibious buggy. Somewhat more extensive exercises in 1985 (Kraus *et al.*, 1989; Bodge and Kraus, 1991) and SUPERDUCK in 1986 (Birkemeier *et al.*, 1989) incorporated sediment traps for the determination of long-shore transport rates. Both these latter experiments incorporated hydrodynamics and bathymetric measurements and experienced storm events over the measurement period.

The 1990 DELILAH [(Duck Experiment on Low-frequency and Incident-band Long-shore and Across-shore Hydrodynamics)] experiment [ ] initiated a series of increasingly complex, multi investigator field measurement exercises. The emphasis of DELILAH was surf-zone hydrodynamics in the presence of a changing barred bathymetry (Birkemeier, 1991; Thornton *et al.*, 1995), providing important insights into the three-dimensional structure of near-shore currents.

Following the success of DELILAH and canvassing of the coastal research community, the following focus topics were established as measurement priorities: small and medium scale sediment transport and morphology, waveshoaling, breaking and near-shore circulation, swash processes including sediment motion.

The DUCK '94 experiment (Birkemeier and Thornton, 1994) was conceived as a pilot experiment for Sandy Duck '97, the world's largest coastal field experiment carried out to date. Including continuing DUCK '94 investigations, the Sandy Duck experiment incorporated more than 25 participating organizations and experiments involving more than 100 scientists, students and technicians.

Emphasizing the quantification of sediment dynamics in addition to underlying hydrodynamics, the experiment involved a wide variety of sophisticated sensors that included a variety of acoustic probes, fiber-optic back scatter sensors (FOBS), side-scan sonars, acoustic timers and radar systems. These *in situ* deployments were complemented by video observation cameras and the Sensor Insertion System, a specially adapted crane mounted on the FRF research pier which allowed measurements during severe weather conditions (retired in 2011).

Sandy Duck '97 was an ambitious six-week measurement exercise focused on sediment transport that started on 22 September 1997. Consensus was established during planning and preparation for the experiment that three focus topics were essential to improving understanding of surf zone sediment transport:

- small and medium scale sediment transport and morphology;
- wave shoaling, wave breaking, and near-shore circulation; and
- swash zone processes, including sediment motion

Thirty basic studies were undertaken, with over 250 researchers and engineers from 26 organizations involved. Unlike the case for Duck '94, where a single cross-shore line of altimeter frames were deployed, multiple lines were set out to measure near-shore dynamics and bed level changes in both cross-shore and long-shore directions (Birkemeijer *et al.*, 1996).

4.2.4 *C^2S^2 and C-Coast, Canada*

The Canadian Coastal Sediment Study (C^2S^2), conducted in 1983 at Ponte Sapin, was directed at gaining an improved understanding of how to measure and predict near-shore sediment transport. Hydrodynamics, suspended sediment concentrations and bed levels were measured at a relatively exposed site having a sand veneer over a sand stone ledge (Fleming *et al.*, 1984, 1986). As also reported in Huntley (1982), the study found that then-available techniques were inadequate for accurately measuring sediment transport in the surf zone.

The Canadian Coastal Sediment Transport Programme (C-COAST), conducted at three contrasting coastal sites over the period 1987–1989, evolved out of C^2S^2 and measured sediment transport and the concurrent response of the bottom boundary under combined waves and currents (Greenwood *et al.*, 1990). These experiments provided very detailed suspended sediment transport measurements, in part measured by the then newly developed Remote Acoustic Sediment Transport (RASTRAN) system. Further insights were gained regarding bottom boundary layer response, including the large influence of bed forms on vertical gradients in suspended sediment concentrations.

4.2.5 *Walker Bay, South Africa*

The CAESAR3 (Swart, 1988) and CAESAR4 (Coppoolse *et al.*, 1992) field experiments were conducted in 1987 and 1990 respectively at the exposed Walker Bay test site of the South African Council for Scientific and Industrial Research (CSIR). Investigating the ratio of measured wave height to water depth (H/h) across the surf zone Nelson and Gonsalves (1992) quantified the significance of infra-gravity driven water depth modulations. Directed primarily at quantifying sediment transport rates in a highly energetic surf zone, the CAESAR4 exercise included a helicopter deployed mobile frame instrument for wave, current and suspended sediment concentration measurements. The helicopter was also used for rapid bathymetric coverage measurements.

4.2.6 *Egmond and NOURTEC, the Netherlands*

Carried out on a straight, sandy beach with a double long-shore bar system, the Egmond field experiments (Derks and Stive, 1984) conducted between 1981 and 1983 provided useful data for wave modelling in the surf-zone (Battjes and Stive, 1984, van Heteren and Stive, 1984). Sediment transport measurements were however not very successful, and only limited use could be made of morphological measurements (De Vriend *et al.*, 1988). Following a 2 Mm^3 shore face nourishment at Terschelling in 1993 the principal forcing conditions (waves, water levels, currents) and bathymetric response was monitored as part of the NOURTEC (Innovative nourishments techniques evaluation) project between 1993 and 1996 (Spanhoff *et al.*, 1997). Besides the monitoring program detailed process-level measurements were made over four campaigns of about six weeks each (Hoekstra *et al.*, 1997). This experiment thus provided a unique juxta-position of process and monitoring level information.

4.2.7 *EC MAST-III*

A series of experiments was conducted under the European Commission's Marine Science and Technology Research Programme (MAST-III). These included:

- COAST3D — aimed at improving understanding of the physics of coastal sand transport and morphodynamics, gathering validation data of sand transport and morphology suitable for testing numerical models of coastal processes, testing numerical models for predicting coastal sand transport and morphodynamics against data, delivering validated modelling tools and methodologies in a form suitable for coastal zone management (Soulsby, 1998).
- INDIA — Inlet Nearshore Dynamics Initiative: Algarve which aimed to obtain an improved understanding of interactive processes in and adjacent to a small tidal inlet (O'Connor *et al.*, 1998; Williams, 1998).
- SAND PIT — Sand transport and morphology of offshore sand mining pits/areas, which aimed to develop reliable techniques to predict the morphological behaviour of large-scale offshore sand pits.

4.2.8 *Sand engine*

The Sand Engine (de Zandmotor) is a Dutch pilot project aimed at fortifying the coast of Zuid-Holland. A 20 million cubic metres sand reclamation in the form of a hooked peninsula was applied at Ter Heijde, extending 1 km into the sea and 2 km alongshore. The reshaping of the reclamation through natural forcing is expected to lead to a wider, safer beach, with the hope being that the up to now five-yearly replenishment of the beaches along the coast will be unnecessary for at least 20 years. The project is closely monitored and can be followed online (www.dezandmotor.nl).

4.2.9 *Other coastal experiments*

Several other coastal experiments not described above have also been carried out, including the Holderness experiments (1993–1996) aimed at understanding coastal erosion (Prandle *et al.*, 1996) and the Near-shore Canyon Experiment (NCEX) (2003 to 2007) in La Jolla, California aimed at determining the effects of submarine canyons on waves, currents and subsequently near-shore morphology (Guza, 2007).

4.3 *Coastal monitoring programmes*

4.3.1 *Introduction*

Numerous coastal monitoring worldwide programmes are run by government departments and regional and local authorities worldwide, who have all recognized the inherent value of such programmes. These programmes are invariably located in developed nations, where funding for such endeavours are more readily available. Some examples of active coastal monitoring programmes are discussed below.

In addition to these large scale monitoring programmes the use of voluntary observers as a low cost alternative has also been attempted. Examples include the Coastal Observation Programme–Engineering (COPE) of the Queensland Beach Authority (Robinson, 1977), the Littoral Environment Observation (LEO) Programme of the USACE CRTC (Schneider, 1987) and the Continuous Low-level Environmental Observations (CLEO) programme in South Africa (Badenhorst and Swart, 1989). The success of these programmes in terms of accuracy of observations is however questionable, with Theron (1993) for example finding significant over- and under estimations of observed wave heights and periods in a comparison of CLEO data with field measurements.

4.3.2 *Field research facility*

The US Army Corps of Engineers Field Research Facility near Duck, North Carolina is an internationally recognized coastal observatory established in 1977. A 560 m long research pier extending to an approximate water depth of 7 m forms the centerpiece of the facility. A 43 m tall video mast provides support for radio antennae and video cameras. A number of unique vehicles are also housed at the facility, including the Coastal Research Amphibious Buggy (CRAB) which is used to measure sandy near-shore bathymetries, and the Lighter Amphibious Resupply Cargo (LARC-5) which is used to deploy instruments, support diving activities, tow sensors and collect data.

In addition to hosting several large-scale field experiments (described in Sec. 4.2.3), waves, winds, tides and currents are constantly recorded and real-time and archived data can be accessed online.

4.3.3 *Dutch coast — JARKUS*

The JARKUS measurements are annual surveys of beach profiles along the Dutch coast that have been undertaken since 1964. Profiles were measured from the back dune to approximately 800 m offshore until 1988, whereafter profiles extended to a depth of −8 m to −10 m (reached typically between 800 m and 1500 m from shore). A rich data set of profile change has been built up through the years, which is actively used to assist in managing the Dutch shoreline and nourishment regime (see e.g. Knoester, 1990 and Pot, 2011).

4.3.4 *Dubai coastal zone monitoring programme*

The Emirate of Dubai, part of the United Arab Emirates, a federation of seven emirates in the Arabian Peninsula, has been on a rapid development path over the last two decades. Rapid expansion of Dubai city, characterized by mega-developments along and off the Dubai coastline, required the establishment of a coastal monitoring programme by the Dubai Municipality to inform the coastal management process of the emirate. An initial programme was launched in 1997 and this was superseded by a more expansive programme involving continuous monitoring of waves, currents, water levels as well as regular beach topographic and near-shore hydrographic surveys, including hydrographic Lidar surveys (Smit and Mocke, 2007). The programme has delivered a valuable dataset of the changes associated with the rapid developments that took place along the Dubai coast during the first decade of the 21st century (see Mocke and Smit, 2008). Details on the still active programme, including measured data, can be accessed at www.dubaicoast.org.

4.3.5 *Other coastal monitoring programmes*

With authorities being more cognizant of coastal management and climate change issues than before, numerous coastal monitoring programmes are in existence these days. The Southwest Regional Coastal Monitoring Programme (http://www.channelcoast.org/southwest/) and the Coastal Observatory Liverpool Bay (http://cobs.pol.ac.uk/) are two such examples from the UK.

4.4 *Dealing with the unexpected (or when instruments go "walkabout")*

Despite the best planning and precautionary efforts the practicalities of carrying out measurements in the coastal zone means that things invariably will go wrong at some point, with instrument loss one of the more frequent occurrences. Engineers, oceanographers and researchers working in the coastal engineering field all have stories to tell of instruments lost and found, strange readings and even threats to personal safety.

Deploying instruments or undertaking sea bed leveling by helicopter is attractive, but inherently fraught with risk. The authors have personally witnessed personnel in the early 1990's deploying instrument frames whilst hanging from the skids of a helicopter hovering a mere meter or so above breaking waves (with patrolling sharks clearly visible beyond the breaker line) all in the interest of capturing dynamic surf zone conditions. This kind of intrepid but inherently risky behaviour is fortunately less likely to occur in today's more health and safety conscious environment, however, it is difficult to anticipate every possible eventuality when carrying out measurement campaigns in the field.

The loss of instruments due to mooring breaks, theft or vessel collisions are of constant concern. The Queensland EPA provides numerous case studies of buoys gone adrift (and often found again). The adventures of their floating Waverider BOB can be followed on their website (Queensland Government, 2013).

As field instruments may be subject to mechanical and electrical malfunction, quality-control procedures should be in place to ensure that measurements truly reflect site conditions. The likelihood of instrument malfunction also has implications for instrument deployment schedules, as an extended measurement period without instrument maintenance may result in significant data gaps. Gauges providing real-time telecommunications links to a shore station provide an early warning of any major malfunctions in the recording device. However, unless identified by specialized hardware or software, less evident malfunctions may only be detected through inter–comparisons with simultaneous measurements by co-located instruments. Although quality control procedures built into data analysis methods are useful in identifying suspect data, the user specified thresholds occasionally reject good records. Morang *et al.* (1997a) present an example where rapid increases in wave energy that reflected actual site conditions are rejected by a threshold specified for noise spike detection.

4.5 *Engineering project design monitoring*

Field measurements are a critical element in the design and performance assessment of coastal projects. In the design process for a new project (Thompson *et al.*, 1995; CIRIA, 1996), field observations are essential for obtaining a clear understanding of prevailing conditions at the project site. Together with historical observation, if such is available, such measurements contribute to the definition of project objectives and design and provide a basis for judging the success of a completed project.

Shown in Figure 8.7 is a typical instrumentation deployment for validating numerical models of wave transformation on an engineering design project.

The scale and complexity of a measurement programme is generally linked to the scale and importance of the coastal project. Generally, the most extensive field efforts are required at the outset of the project if site specific coastal processes need to be quantified. Although not necessarily at the "research level" of the

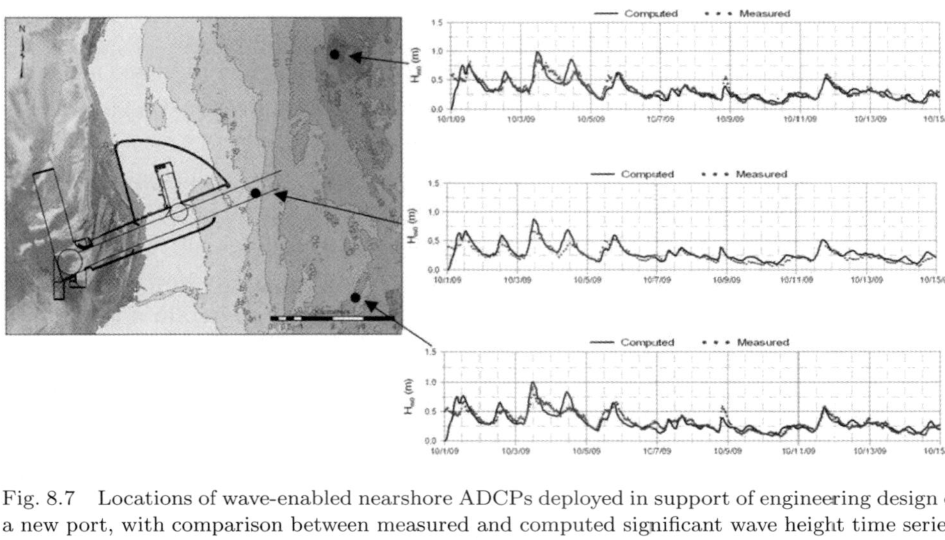

Fig. 8.7 Locations of wave-enabled nearshore ADCPs deployed in support of engineering design of a new port, with comparison between measured and computed significant wave height time series. [Mocke and Neal, (2011, June), *Overview of the Master Planning and Coastal Studies for the Design of a New Commercial Port and Naval Base in Qatar*. Powerpoint presentation presented at the 2nd Annual Marine and Coastal Engineering Middle East Conference, Doha, Qatar.]

field experiments described in Section 4.2, measurements need to provide sufficient information for developing predictive models of coastal response to the project. Post-construction monitoring is generally carried out over larger spatial and temporal scales, an example being the beach and shoreline evolution adjacent to a breakwater structure. Although topographic response can be monitored by relatively infrequent surveys, it may be necessary to measure underlying local wave and current processes at more micro-scale levels.

Quoting the statement by O'Brien (1984) that "There are few complete records of the design, construction, maintenance and demise of protective coastal structures", Hemsley (1990) discusses the development of a program for systematic monitoring of the performance of coastal projects. Established in 1981 by the US Army Corps of Engineers, the Monitoring Completed Coastal Projects (MCCP) Program collects data at more than 20 diverse sites around the United States. Verhagen (1996), CIRIA (1996) and Weggel (1995) discuss the monitoring of beach nourishment projects, with the latter author suggesting that nourishment is probably the least understood method of shore protection. Given that the nourishment sand is generally lost from the beach face over a period of several years, a monitoring of project performance is important to address common perceptions that nourishment is a futile exercise. Physical measurements should further be complemented by biological monitoring and cost benefit analyses in order to fully evaluate project performance (Weggel, 1995).

5. Data Reduction and Management

5.1 *General*

Data obtained from field measurements may take the form of digitized magnetic tapes, analog paper strip charts, signals, samples, photographic images, video-tapes, written numerical records and even memoranda of site observations. Electronic instruments measuring processes such as waves, water levels, currents and sediment concentrations require a system for manageable storage and processing of voltage signals, as further discussed below. Survey recordings are generally captured digitally for case of storage, quality control and analysis. Section 3.7.6 has already discussed the use of photographic and video images to obtain both qualitative and quantitative information on site conditions. The inherently digital form of video images is particularly advantageous for data storage, processing, and analysis purposes.

5.2 *Data processing and analysis*

Field measurement data may arrive in various spatial and temporal scales of resolution, calling for specialized processing and analysis procedures. An example is time series data, which requires analysis to separate short term fluctuations from any long term trends.

Voltage recordings such as those collected by wave gauges need to undergo a range of processing and analysis procedures before providing the wave statistical information required by coastal engineers and scientists. These procedures, which incorporate data transfer and conversion, spectral or parametric analysis, quality control and tabulation and plotting, are discussed in detailed in Horikawa (1988), Earle *et al.* (1995) and Morang *et al.* (1997a, 1997b). Some salient aspects are presented below.

With sampling frequencies of 2 Hz or even higher, wave, current and sediment concentration gauges produce far more data than can be handled by continuous collection. Data is therefore usually collected in "bursts" at predetermined intervals typically varying between 10 min and 6 h. Given that limitations exist on the amount of data that can be stored and analyzed, on an instrument collection procedures should be carefully designed to balance data requirements and resource constraints.

Data analysis procedures are highly dependent on the type of data. Given that wave recordings contain an assortment of energies and frequencies, spectral and/or statistical techniques are necessary for defining wave characteristics. The analysis procedures are further complicated where wave directional information is required (Horikawa, 1988). As emphasized by Morang *et al.* (1997a), wave data analysis is a complex procedure that requires a specialist understanding of wave mechanics. As discussed by Halpern (1980), the broad range of available current meters and related influences of mooring assembly, types of velocity sensors and recording schemes result in preferential measurement of different aspects of the spectrum of water motion. Furthermore, inappropriate modes of averaging velocities in time and/or space can be a key source of error in flow measurements (McCullough, 1980).

Gorman *et al.* (1998) present a comprehensive review of methods used to analyze maps, surveys and aerial photographs for shoreline and bathymetric changes. Procedures for analyzing such data need to take into account datum levels, grid and projection correction and other possible sources of error. Where possible, expected accuracy and error estimates should be properly documented, and added as maximum possible cumulative errors in the calculation of shoreline evolution rates and volumetric changes.

5.3 *Data management and transfer*

The manipulation, storage and documentation of bathymetric and other coastal spatial data sets have been greatly facilitated with the advent of Geographic Information Systems (GIS). Capable of storing data sets of variable scales, sources and attributes, GIS permits the processing and intercomparison of a broad range of coastal features (O'Regan, 1996, Ricketts, 1992, Jensen *et al.*, 1993). As such, GIS also enhances the collection, assessment, transfer and presentation of data.

In addition to having a variety of analytical statistical and modelling tools imbedded, there are increasing trends towards linking GIS with predictive models in the form of expert systems (O'Regan, 1996). An example is the linking of a wave transformation model to bathymetric data and a data bank of off shore wave conditions, including the construction of nested grids of variable bathymetric detail for computational purposes (CSIR, 1998).

The Arc Marine data model is a successful attempt to establish a common structure for spatial modelling and storage of marine related data (see for example Harris *et al.*, 2007).

6. Perspectives for the Future

This chapter has sketched the rapid advances in the art and science of field observation of coastal processes over recent decades. The question arises what still remains to be done? As in most other disciplines it is true that as we learn more through increasingly sophisticated studies, we realize how much there is yet to learn. It is then to be expected that coastal observational research will continue to evolve as, on the one hand, instrumentation becomes both more sophisticated and better able to deal with hostile near-shore coastal conditions and, on the other hand, theoretical insights expand. At the moment it would appear that the following key issues will help determine future trends:

6.1.1 *Validation of theoretical concepts*

With the advent of ever more powerful computing platforms, numerical modelling has become a key method in solving complex theoretical problems. It is increasingly becoming practice to design field observation programmes around key determinants in these theoretical frameworks. This approach has the benefit that it is very clearly focused on observing specific, frequently/mostly inter-related phenomena in such a manner that they can directly serve to validate the theoretical concepts being used as basis.

In this regard it should be noted that many high frequency responses in the surf zone, such as the very important breaking wave-induced turbulence, would require much more specific attention in future.

6.1.2 *Comprehensive, large-scale field experiments*

The execution of experiments aimed at validating theoretical concepts can frequently best be done through comprehensive programs encompassing multiple temporal and spatial scales of measurement. Sandy Duck 1997 is an example of such an approach, where researchers from around the world congregated to collaborate in one large experiment. A benefit of this approach is that expensive mobilization and demobilization costs are shared for a variety of semi-independent studies.

In addition, co-measured data on a variety of related and semi-related processes becomes available for all.

6.1.3 *Coherent field observation protocols*

Most reviews of field observation exercises conclude that the data are not accurate enough or that certain key variables had not been measured. It is suggested that protocols be derived to establish minimum requirements for both process related experiments and long term monitoring programmes.

6.1.4 *Open access central data repositories*

Increased information and data sharing is likely to result in reduced duplications of effort and the promotion of networking and collaborative scientific investigations. It would greatly facilitate accelerated learning if field data, at least at the process level, were to be made centrally available at some easily accessible website, for use by all. It would be necessary to sort out potential legal issues pertaining to copyright and ownership in some cases, but often a clear disclosure would be sufficient.

An example of such an open access system, is the Digital Coast portal administered by the Coastal Services Center (CSC) of the US National Ocean and Atmospheric Administration (NOAA) (http://www.csc.noaa.gov/digitalcoast/), which disseminates coastal data and derived products, data processing tools, training material and case studies. Content are contributed by numerous federal, state, county, non-governmental, private and academic entities. Data distributed through the portal include topographic and hydrographic data sets (including Lidar), recent and historical aerial imagery, land cover information, as well as benthic sampling data and coverage maps.

An analysis of the benefits and costs of the Digital Coast resource has counted the benefits as 3.5 times greater than the costs, with a return on investment of 254 percent (NOAA, 2012). It is estimated that in FY2011 alone, the download of data and tools by 142,000 users resulted in cost-reduction benefits that equal the entire historical investment in the resource. This is mainly because users spend less time looking for data, as data from a number of different sources are conveniently organized on the Digital Coast. In addition, users spend less time manipulating data as formatting is consistent and well-documented.

As stated in NOAA (2012):

> *"Many millions of dollars have been spent by government agencies and non-governmental organizations to develop the data, tools, training, and other products and services that are needed to efficiently and effectively manage the nation's coasts. The value of these resources, however, is greatly diminished by the fact that it is hard to find them, hard to determine their reliability, and hard to learn how to use them. Investments in the Digital Coast restore this lost value by identifying trustworthy products and services and making them easy to find, acquire, and use."*

6.1.5 *Field observations in developing countries*

As far as it is not already too late; the opportunity exists for developing countries to learn from properly documented and quantified coastal engineering successes and failures. Sources of funding in developing countries is generally a critical determinant of the level of detail that can be achieved in field studies. The challenge for coastal engineers is to use their collective pool of expertise and past experience to develop and implement cost effective coastal observation equipment and programmes in developing countries.

Acknowledgment

Special thanks are due to Dr. Nick Kraus and Dr. Jon Williams for constructive criticism made on earlier drafts of this chapter.

References

Aarninkhof, S. G., P. C. Janssen and N. G. Plant (1997): Quantitative estimations of bar dynamics from video images, Proceedings of the Coastal Dynamics '97, ASCE, Plymouth.

Arcilla, A. S., S. J. F. Marcel and N. C. Kraus, eds. (1994): Coastal Dynamics '94, ASCE.

Agrawal, Y. C. and D. G. Aubrey (1992): Velocity observations above a rippled bed using laser Doppler velocimetry, *J. Geophys. Res.* 97(C12), 20249–20259.

Alport, M., G. Mocke, K. Govender and A. Marais (1998): The measurement of near shore processes in the laboratory and the field using video imagery, Proceedings of the International Conference OCEANS '98. NICE, IEEE.

Badenhorst, P. and D. H. Swart (1989): CLEO beach observations and an application, *Proc. Coastal Zone* 89, 1688–1699.

Battjes, J. A. and M. J. F. Stive (1984): Calibration and verification of a dissipation model for random breaking waves, *Proc. 19th Int. Conf. on Coastal Engrg.* 1, 649–660.

Beach, R. A., R. W. Sternberg and R. Johnson (1992): A fiber optic sensor for monitoring suspended sediment, *Mar. Geol.* 103, 513–520.

Bell P. S., P. O. Thorne and J. J. Williams (1998): Acoustic measurements of sand ripple profile evaluation under controlled wave conditions, 4th European Conference on Underwater Acoustics, Rome, Italy, September 1998.

Birkemeier, W. A. and C. Mason (1984): The CRAB: A unique nearshore surveying vehicle, *J. Surveying Engrg.* 110, 1–7, ASCE.

Birkemeier, W. A. (1991): SAMPSON and DELILAH at the FRF, in the CERCular, Coastal Engineering Research Centre, US Army Corps of Engineers, Waterways Experiment Station, CERC-91-1, March 1991.

Birkemeier, W. A. and E. B. Thornton (1994): The DUCK94 Nearshore Field Experiment, Proc. Coastal Dynamics '94, ASCE, Barcelona.

Birkemeier, W. A., C. E. Long and K. K. Hathaway (1996): DELILAH, DUCK94 & Sandy-Duck: Three near shore field experiments, Proc. 25th Int. Conf. on Coastal Engrg., Orlando, Florida.

Birkemeier, W. A., C. F. Baron, M. A. Leffler, K. K. Hathaway, H. C. Miller and J. B. Strider (1989): SUPER DUCK near shore processes experiment data summary, Misc Paper CERC 89-16, CERC Field Research Facility, US Army Engrg. Waterways Expt. Station, Coastal Engineering Research Center, Vicksburg, MS, USA.

Black, K. S., S. Athey, P. Wilson and D. Evans (2007): The use of particle tracking in sediment transport: a review, In: Coastal and Shelf Sediment Transport, (P. S. Balson and M. B. Collins, eds.), Special Publications, 274, pp. 73–91, Geological Society of London.

Bodge, K. R. and N. C. Kraus (1991): Examination of long-shore transport rate magnitude, Coastal Sediments '91, Seattle, 1, 139–153, ASCE.

Bouffard, J., A. Pascual, S. Ruiz, F. Yannice and J. Tintoré (2010): Coastal and mesoscale dynamics characterization using altimetry and gliders: A case study in the Balearic Sea, *J. Geophys. Res.* 115, C10029, doi:10.1029/2009JC006087.

Bryan, K. R., K. P. Black and R. M. Gorman (2003): Spectral estimates of dissipation rate within and near the surf zone, *J. Phys. Oceanogr.* 33, 979–993.

Challenor, P., D. Woolf, C. Gommenginger, M. Srokosz, D. Cotton, D. Carter and N. Sykes (2006): Satellite altimetry: A revolution in understanding the wave climate, Ocean Surface Topography Science Team (OSTST) meeting, Venice. http://earth.esa.int/workshops/venice06/participants/453/paper_453_challenor.pdf

Chanson, H. (2008): Acoustic Doppler Velocimetry (ADV) in the Field and in Laboratory: Practical Experiences, International Meeting on Measurements and Hydraulics of Sewers (F. Lallarte and H. Chanson, eds.), Hydraulic Model Report No. CH70/08, Div. Civil Engrg., The University of Queensland, Brisbane, Australia, pp. 49–66.

Chelton, D. B., K. J. Hussey and M. E. Parke (1981): Global satellite measurements of water-vapor, wind speed and wave height, *Nature* 294(5841), 529–532.

CIRIA (1996): Beach Management Manual. Construction Industry Research and Information Association. Report 153.

Coastal Engrg. J. (1998): Special Issue on New Instrumentation in coastal engineering, edited by Garry L. Howell and Steven Hughes; *Coastal Engrg.*, 35(1&2).

Coastal Engrg. J. (2007): Special Issue on The CoastView Project: Developing coastal video monitoring systems in support of coastal zone management; *Coastal Engrg.*, 54(6&7).

Coppoolse, R. C., J. S. Schoonees, G. P. Mocke, and G. G. Smith (1992): Measurements of the hydrodynamics and sediment concentrations in the high energy surfzone at WalkerBay, Proc. 23rd Int. Conf. on Coastal Engrg., Venice, Italy.

Craig, R. and W. Team (1985): Surf zone and nearshore surveying with helicopter and a total station, Proc. US Army Corps of Engineers Surveying Conference, USAE Waterways Experiment Station, Vicksburg.

CSIR (1998): WAVEGIS-Technical Manual, Research Report ENV/S-198020B, Council for Scientific and Industrial Research, South Africa.

Dally, W. R., M. A. Johnson, and D. A. Osiecki (1994): Initial development of an amphibious ROV for use in big surf, *Marine Technical Soc. J.* 28(1), 3–10.

De Vriend, H., R. Reinalda, and H. Derks (1988): The Egmond field data and their use for the validation of mathematical models, SOGREAH seminar (A. DeGraauw and L. Hamm, eds.) (1988), pp. 134–151.

Derks, H. and M. J. F. Stive (1984): Field investigations in the TOW study programme for coastal sediment transport in the Netherlands, Proc. 19th Int. Conf. on Coastal Engrg., Houston, Part II, Chap. 123, 1830–1845, ASCE.

Downey, J. P., R. W. Sternberg and C. R. B. Lister (1981): New instrumentation for the investigation of sediment suspension processes in the shallow marine environment, in: C. A. Nittrouer (Ed.), Sedimentary Dynamics of Continental Shelves. *Mar. Geol.* 42, 19–34.

Dugan, J. P., W. D. Morris, K. C. Vierra, C. C. Piotrowski, G. J. Farruggia, and D. C. Campion (2001): Jetski-Based Nearshore Bathymetric and Current Survey System, *J. Coastal Res.* 17(4), 900–908.

Earle, M. D. M., D. Mcghee, and M. Tubman (1995): Field wave gaging program, wave data analysis standard, Instruction Report CERC-95-1, US Army Waterways Experiment Station, Vicksburg, Mississippi.

Eisma, D., A. J. Bale, M. P. Dearnaley, M. J. Fennessy, W. Van Lenssen, M. A. Maldiney, A. Pfeiffer, and J. T. Wells (1996): Inter comparison of in situ suspended matter (floc) size measurements, *J. Sea Res.* 36(1–2), 3–14.

Estep, L. L., W. J. Lillycrop, and L. E. Parson (1994): Estimation of maximum depth of penetration of a bathymetric lidar system using a Secchi disk database, *Marine Technology Soc. J.* 28(2), 31–36.

Fedderson, F. (2012): Observations of the surf-zone turbulent dissipation rate, *J. Phys. Oceanogr.* 42, 386–398.

Fleming, C. A., K. L. Philpott, and B. M. Pinchin (1984): Evaluation of coastal sediment transport estimation techniques, Phase 1: Implementation of longshore sediment transport models and calibration of wave hindcasting procedure, National Research Council of Canada, Canadian Coastal Sediment Study Report No. C2S2-10.

Fleming, C. A., B. M. Pinchin, and R. B. Nairn (1986): Evaluation of coastal sediment transport estimation techniques, Phase 2: Comparison with measured data, National Research Council of Canada, Canadian Coastal Sediment Study Report No. C2S2-19.

Fredericks, J. J., J. H. Trowbridge and G. Voulgaris (2001): Turbulence in the shallow nearshore environment during SandyDuck 97, Tech. Report WHOI-2001-02, Woods Hole Oceanographic Institution.

Gao, S., M. B. Collins, J. Lanckneus, G. De Moor and V. Van Lancker (1994): Grain size trends associated with net sediment transport patterns: an example of the Belgian Continental Shelf, *Mar. Geology*, 121, 171–185.

George, R., R. E. Flick and R. T. Guza (1994): Observations of turbulence in the surf zone, *J. Geophys Res.* 99(C1), 801–810.

Gorman, L., A. Morang and R. Larson (1998): Monitoring the coastal environment; Part IV: Mapping, shore line changes and bathymetric analysis, *J. Coastal Res.* 14(1), 61–92.

Govender, K., G. P. Mocke, Alport, M. (2004): Dissipation of isotropic turbulence and length scale measurements through the wave roller in laboratory spilling waves. *J. Geophys. Res. (Oceans),* 109(C8).

Greenwood, B., P. O. Osborne, A. J. Bowen, D. G. Hazen and A. E. Hay (1990a): Near shore sediment flux and bottom boundary dynamics: The Canadian Coastal Sediment Transport Programme (C-Coast), Proc. 22nd Int. Conf. Coastal Engrg., Delft, Ill, 2227–2240, ASCE.

Guza, R. T. (2007): Nearshore Canyon Experiment, Final Technical Report. Grant Number N00014-02-1-0415. Office of Naval Research, Virginia, USA.

Holdaway, G. P., P. D. Thorne, D. Flatt, S. E. Jones and D. Prandle (1999) Comparison between ADCP and transmissometer measurements of suspended sediment concentration, Continental Shelf Research, 19 (1999), 421–441.

Halpern, D. (1980): Moored current measurements in the upper ocean, In: *Air-Sea Interaction — Instruments and Methods* (F. Dobson, L. Hasse and R. Davis, eds.), New York: Plenum, pp. 127–140.

Harris, M. S., T. P. Tinker and E. E. Wright (2007): Online beach profile management and analysis system (PMAS), *J. Coastal Res.*, Special Issue, 50, 62–66.

Hanes, D. M. and D. A. Huntley (1986): Continuous measurements of suspended sand concentration in a wave dominated nearshore environment, *Cont. Shelf. Res.* 6(4), 585–596.

Hay, A. E. and J. Sheng (1992): Vertical profiles of suspended sand concentration and size from multifrequency acoustic backscatter, *J. Geophys. Res.* 97(C10), 15661–15667.

Hemsley, J. M. (1990): Monitoring completed coastal projects: Status of a program, *J. Coastal Res.* 6(2), 253–263.

Hemsley, J., D. O. Mcghee and W. M. Kucharski (1990): Near shore oceanographic measurements: Hints on how to make them, *J. Coastal Res.* 7(2), 301–315.

Hoekstra, P., K. T. Houman, A. Kroon, B. G. Ruessink, J. A. Roelvink and R. Spanhoff (1997): Morphological development of the Terschelling shoreface nourishment in response to hydrodynamic and sediment transport processes, Proc. 25th Int. Conf. on Coastal Engrg., ASCE, New York, 2897–2910.

Holland, K., R. Holman and A. Sallenger (1991): Estimation of overwash bore velocities using video technique, Coastal Sediments, ASCE, 489–497.

Holland, K. T. and R. A. Holman (1997): Video estimation of foreshore topography using trinocular stereo, *J. Coastal Res.* 13(1), 81–87.

Holman, R. A. (1994): The ARGUS program, Bulletin of the Coastal Imaging Lab., College of Oceanography, Oregon State University, Corvallis, OR 97331.

Horikawa, K., ed. (1988): *Nearshore dynamics and Coastal Processes: Theory, Measurement and Predictive Models*: Tokyo, Japan, University of Tokyo Press, 522 pp.

Horikawa, K. and T. Sasaki (1972): Field observations of near shore current system, Proc. 13th Int. Conf. on Coastal Engrg., ASCE, 635–652.

Horikawa, K. and M. Hattori (1987): Accomplishments of the Nearshore Environment Research Center Project, Proc. Coastal Sediments '87, ASCE, 756–771.

Hotta, S. and M. Mizuguchi (1980): A field study of waves in the surf zone, *Coastal Engrg. Japan* 23, 59–79.

Huntley, D. A. (1982): In-situ sediment monitoring techniques, a survey of the state of the art in the USA, National Research Council of Canada, Canadian Coastal Sediment Study Report No. C2521.

Hwang, P. A. and E. J. Walsh (1997): Remote sensing and coastal wave research, Proc. Coastal Dynamics '97, ASCE, Plymouth.

Jensen, J. R., D. J. Cowen, J. D. Atlhausen, S. Narumalani and O. Weatherbee (1993): An evaluation of the Coast Watch change detection protocol in South Carolina, Photogrammetric Engineering and Remote Sensing, 59(6), 1039–1046.

Kamphuis, J. W. (2010): *Introduction to Coastal Engineering and Management*, 2nd edn. Advanced Series on Ocean Engineering, Vol. 30: Singapore, World Scientific Publishing.

Knoester, D. (1990): De morfologie van de Hollandse Kustzone (analyse van het Jarkusbestand 1964–1986), Note GWAO-90.010, Rijkswaterstaat, the Netherlands.

Kraus, N., R. S. Farinato and K. Horikawa (1981): Field experiments on longshores and transport in the surf zone, *Coastal Engrg. Japan*, 24, 71–194.

Kraus, N. C., K. J. Gingerich and J. D. Rosati (1989): DUCK85 surfzone sand transport experiment, US Army Engineer Waterways Expt. Stn., Coastal Engineering Research Centre, Tech. Report CERC-89-5.

Kutser, T., L. Metsamaa, E. Vahtmae and R. Aps (2007): Operative monitoring of the extent of dredging plumes in coastal ecosystems using MODIS satellite imagery, *J. Coastal Res.* Special Issue 50.

Lane, A., D. Prandle, A. J. Harrison *et al.* (1997): Estuarine, Coastal and Shelf Science. 45/4, 433–451.

Lillycrop, W. J., L. E. Parson and J. L. Irish (1996): Development and operation of the SHOALS airborne Lidar hydrographic survey system, In: Laser Remote Sensing of Natural Waters: From Theory to Practice (Vl Feigelo & Yl Kopilevich, eds.), *Proc. SPIF*, 2964, pp. 26–37.

Lippmann, T. C. and G. M. Smith (2013): Shallow surveying in hazardous waters, Accessed 23 January 2013, http://www.thsoa.org/hy09/0514A_06.pdf.

Lippmann, T. and R. Holman (1989): Quantification of sand bar morphology: a video technique based on a wave dissipation, *J. Geophys. Res.* 94(C1), 995–1011.

Lundahl, A. C. (1948): Underwater depth determination by aerial photography, Symposium: Military Photographic Interpretation, Photogrammetric Engrg., XIV(4).

Mccullough, J. R. (1980): Survey of techniques for measuring currents near the ocean surface, In: Air-Sea Interaction-Instruments and Methods (F. L. Hasse and R. Davis, eds.), New York: Plenum, pp. 105–126.

McMahan, J. (2001): Hydrographic surveying from personal watercraft, *J. Surveying Engrg.*

Mangor, K. (2004): Shoreline Management Guidelines. DHI Water and Environment, Denmark.

Mead, S. T. (2000): Incorporating high-quality surfing breaks into multi-purpose reefs, PhD thesis, University of Waikato.

Mocke, G P. (2001): The structure and modelling of surf zone turbulence due to wave breaking. *J. Geophys. Res.* (Oceans), 106(C8), 17039–17057.

Mocke, G. P. and F. Smit (2008): Editorial and invited contribution: Challenges and impacts associated with coastal mega-reclamation developments, *Shore & Beach*, 76(4), American Shore & Beach Preservation Association (ASBPA).

Morang, A., R. Larson and L. Gorman (1997a): Monitoring the Coastal Environment; Part 1: Waves and Currents, *J. Coastal Res.* 13(1), 111–133.

Morant, A., R. Larson and L. Gorman (1997b): Monitoring the Coastal Environment; Part III: Geophys. and Res. Methods, *J. Coastal Res.* 13(4), 1064–1085.

National Oceanic and Atmospheric Administration (NOAA) (2012): Benefits and costs of the Digital Coast, Charleston, SC: NOAA Coastal Services Center.

Nortek. "New to Waves?", accessed 23 January 2013, <http://www.nortek-as.com/en/products/wave-systems/nortek-primers-getting-started-with-wave-1>.

O'Brien, M. P. (1984): Coastal Engineering, Proc. 19th[1] Int. Conf. on Coastal Engrg., ASCE, Houston, Texas.

O'Connor B. A., J. J. Williams, J. M. A. Dias, G. Voulgaris, M. A. Davidson, S. M. Arens, H. Howa and A. Sarmento (1998): INDIA: Inlet Dynamics Initiative: Algarve, Third Marine Science and Technology Conference Sponsored by The European Commission, Lisbon, Portugal.

O'Regan, P. R. (1996): The use of contemporary information technologies for coastal research and management — a review, *J. Coastal Res.* Charlottesville VA, 12(1), 192–204.

Pandian, P. K., O. Emmanuel, J. P. Ruscoe, J. C Side, R. E. Harris, S. A. Kerr and C. R. Bullen (2010): An overview of recent technologies on wave and current measurement in coastal and marine applications, *J. Oceanogr. Mar. Sci.* 1(1), 001–010.

Pattiaratchi, C (2010): Integrated Marine Observing System (IMOS) in Western Australia; Oceans Institute/AIMS Seminar, 10/03/2010–10/03/2010, Perth.

Pollock, C. A. (1994): Helicopter-borne nearshore Survey System, a valuable tool in difficult survey areas. *J. Coastal Res.* 11(3), 959–966.

Pot, R. (2011): System-description Noord-Holland Coast: A review of the nourishment strategy applied. M.Sc. Thesis, Delft University of Technology.

Prandle D., G. Ballard, A. Banaszek, P. Bell, D. Flatt, P. J. Hardcastle, A. Harrison, J. Humphery, G. Holdaway, A. Lane, R. Player, J. J. Williams and J. Wolf (1996): The Holderness coastal experiment '93–'96, POL Report, 44, 46 pp.

Queensland Government, 2013. Wave monitoring publications. [online] Available at:http://www.ehp.qld.gov.au/coastal/monitoring/waves/wave_monitoring_publications/index.html [Accessed on 27 January 2013].

Reti (1974): The Madrid Codices (of) Leonardo da Vinci, (tr. and ed. from the Spanish), New York, McGraw-Hill.

Richard, L. and S. Holgate (2011): Report by the Permanent Service for Mean Sea Level (PSMSL) for the Period 2007–2011 to the XXV General Assembly of the IUGG, Melbourne, Australia, June/July 2011.

Ricketts, P. J. (1992): Current approaches in geographic information systems for coastal management, *Marine Pollution Bulletin* 25(1–4), 82–87.

Robinson, D. A. and C. M. Jones (1977): Queensland Volunteer Coastal Observation Programme — Engineering (COPE), Third Australian Conference on Coastal and Ocean Engrg., Melbourne.

Rodriguez, A., E. Bahia, M. Diez, A. Sanchez-Arcilla, J. M. Redordo and M. Mestres (1997): Experimental study of mixing processes using images, Proc. Coastal Dynamics '97, ASCE, Plymouth.

Roland, P. J. (2006): Using acoustic backscatter to measure sediment flux in the surf zone, M.Sc. Thesis, Naval Postgraduate School, Monterey.

Salmon, S. A., K. R. Bryan and G. Coco (2007): The use of video systems to measure run-up on beaches, *J. Coatal Res.* Special Issue 50, 211–215.

Schneider, L. (1981): The Littoral Environment Observation (LEO) Data Collection Program. Coastal Engineering Technical Aid No. 81–5, USACE Coastal Engineering Research Center.

Shen, C. and U. Lemmin (1997): A two-dimensional acoustic sediment flux profiler, *Meas. Sci. Techno/.* 8, 880–884.

Sinclair, M. and F. Smit (2005): Dubai coastal zone Lidar survey, Proc. Arabian Coast 2005.

Smit, F., G. P. Mocke, R. Caprile and M. Sinclair (2007): Bathymetric Lidar data GIS applications for coastal zone management in Dubai, Coast GIS 2007.

Smit, F. and G. P. Mocke (2007): Monitoring and management of shoreline change along the dynamic Dubai coastline, In: Integrated Coastal Zone Management (ICZM) — The Global Challenge (R. Krishnamoorthy, ed.), Research Publishing Services, India.

Smith, G. G. and G. P. Mocke (2002): Interaction between breaking/broken waves and infragravity-scale phenomena to control sediment suspension transport in the surf zone, *Marine Geology* 187(2002), 329–345.

Sontek (2001): SonWave-PRO: Directional Wave Data Collection, Accessed 23 January 2013, http://www.sontek.com/pdf/technotes/SonWavePro-wp.pdf.

Sonu, C. J. (1964): Study of shore processes with aid of aerial photogrammetry, *Photogramm. Engrg.* 30(6), 932–941.

Sonu, C. J. (1969): Tethered balloon for study of coastal dynamics, *Amer. Soc. Photogramm.* Tech. Rep., 66, 91–103.

Soulsby, R. L. (1998): Coastal sediment transport: the COAST3D project, Proc. 26th Int. Conf. on Coastal Engrg., Copenhagen, Denmark.

Spanhoff, R., E. J. Biegel, J. Van Degraaf, and P. Hoekstra (1997): Shoreface nousrishment at Terschelling, the Netherlands: Feeder Berm or Breaker Berm?, Proc. Coastal Dynamics '97, ASCE, Plymouth.

Stanton, T. P. (1996): The Coherent Acoustic Sediment-flux Probe (CASP), USAE Waterways Experiment Station Contract Rep., CERC-96-1, 74 pp.

Swart, D. H. (1988): CAESAR-Coastal Area and Sediment Applied Research: Surf zone experiments in South Africa, (de Graauw and Hamm, eds.) (1988), 118–133.

Taylor, J., C. Vincent and J. Rees (1997): A "sound" basis for quantifying sediment movement, Feature article, Dredging and Port Construction (DPC) magazine, September 1997.

Teledyne RD Instruments (2013): WAVES PRIMER: Wave Measurements and the RDI ADCP Waves Array Technique, Accessed 23 January 2013, <http://www.rdinstruments.com/pdfs/waves_primer.pdf>.

Terray, E. A., W. M. Drennan and M. A. Donelan (1999): The vertical structure of shear and dissipation in the ocean surface layer, Woods Hole Oceanographic Institution.

Theron, A. K. (1993): Verification of CLEO Beach Parameters, CSIR Research Report 714.

Thompson, E. F., D. L. Ward, G. W. Domocrat, D. M. Pirier and J. G. Oliver (1995): General Design Process for Coastal Projects. *Shore and Beach.*

Thornton, E. B., C. M. Soares and T. P. Stanton (1995): Vertical profiles of long shore currents and bed shear stress, Proc. Coastal Dynamics '95, ASCE, Poland.

Trowbridge, J. and S. Elgar (2001): Turbulence measurements in the surf zone, *J. Phs. Oceanogr.* 31, 2403–2417.

Van Heteren, J. and M. J. F. Stive (1984): Wave kinematics and wave directionality in the surf zone, Proc. 19th Int. Conf. on Coastal Engrg., Houston, Part 1, Chapter 5.

Verhagen, H. J. (1996): Analysis of beach nourishment schemes, *J. Coastal Res.* Charlottesville VA: 12(1), 179–185.

Voulgaris, G. and J. Trowbridge (1998): Evaluation of the acoustic doppler veloumeter (ADV) for turbulence measurements, *J. Atmospheric and Oceanic Technology.*

Wallace, H. (1993): Coastal sand transport and morphodynamics: A review of field data, H. R. Wallingford Report SR 355 (Issue A).

Wallace, H. and T. J. Chesher (1994): Validation of coastal morphodynamic models with field data, Proc. Coastal Dynamics '94, ASCE, Barcelona.

Walton, T. (1993): Ocean City, Maryland, wave runup study, *J. Coastal Res.* 9(1), 1–10.

Weggel, J. R. (1995): A primer on monitoring beach nourishment projects, *Shore & Beach* 63(3), 20–24.

White, T. E. (1998): Status of measurement techniques for coastal sediment transport, *Coastal Engineering*, 35, 17–45.

Widdows, J., M. D. Brinsley, N. B. Bowley and C. D. Barrett (1998): A benthic annular flume for in situ measurement of suspension/feeding/biodeposition ration and erosion potential of inter-tidal cohesive sediments, Estuarine, Coastal & Shelf Science, 46, 27–38.

Williams, J. J. (1998): INDIA: Inlet Dynamics Initiative Algarve, Proc. OCEANS '98, IEEE, France.

Williams, J. J., P. S. Bell, L. E. Coates, P. J. Hardcastle, J. D. Humphery, S. P. Moores, P. O. Thorne and K. Trouw (1998): Evaluation of field equipment used in studies of sediment dynamics, POL Report 53, 45 pp.

Worley, C. R., T. C. Lippman, J. W. Haines and A. H. Sallenger (1997): An aerial video system for rapidly measuring sand bar morphology. Proc. Coastal Dynamics '97, ASCE, Plymouth.

Zedel, L., A. E. Hay, R. Cabrera and A. Lohrmann (1996): Performance of a single-beam pulse-to-pulse coherent Doppler profiler, *IEEE J. Ocean Engrg.* 21(3), 290–297.

Chapter 9

Beach Nourishment

Robert G. Dean
Professor Emeritus
University of Florida, USA

1. Abstract

Beach nourishment, the placement of large quantities of sand in the nearshore causes both physical and environmental changes. The physical changes and processes which are the main focus of this chapter include induced cross-shore and longshore sediment transport components, as their equilibration time-scales are important to the project performance. Both simple and detailed approaches exist to develop and evaluate alternate designs and to predict their performance characteristics. The simple approaches provide the interrelationships of the relevant parameters and reasonable estimates of project performance. Because the detailed approaches which require the use of numerical models are currently in an evolutionary state and some are proprietary, this chapter emphasizes on the more simple approaches.

The cross-shore and longshore project performance characteristics are treated separately. The cross-shore processes and performance are best represented by available equilibrium profile methodology. The equilibrium beach width resulting from placement of a particular sand size and volume density is shown to be very sensitive to the size of the nourishment sediment relative to that of the native sediment. The simple approach to plan-form evolution is through the PelnardConsidère differential equation and associated solutions. Many results of design interest can be established, including a simple relationship showing that the longevity of a beach nourishment project constructed on a long straight beach is proportional to the square of the project length and inversely proportional to the wave height to the 2.5 power. A brief review is provided for some of the numerical model characteristics available, to carry out detailed calculations of beach nourishment project performance.

2. Introduction

Beach nourishment comprises the placement of large quantities of good quality sediment to advance the shoreline seaward. In most cases, beach nourishment is placed where a background erosion has caused shoreline recession to a degree that remedial action is deemed warranted. Of the three alternatives to coping with a beach erosion problem: retreat, armoring and beach nourishment, beach nourishment is the only alternative which has the capability of maintaining the shoreline position and natural character of an eroding area. Beach nourishment is reasonably costly and usually can be justified only in areas where a significant upland economic base exists; examples include, Miami Beach, Florida and Ocean City, Maryland. However, with the increasing commercial, residential, recreational and environmental values and investments, and other interests along the shoreline, beach nourishment has become much more prevalent in the last decade and is expected to be applied more often in the foreseeable future.

Beach nourishment commenced in the United States (US) approximately five or six decades ago and was usually placed in reasonably contained beach segments. Other projects included those where the nourishment was a by-product of dredging for other purposes such as excavation for a harbor. The longevity of a nourishment project can be increased by structures including offshore (detached) breakwaters, terminal structures or groins. Although such structures have the potential of increasing the longevity of a beach nourishment project, in some locations the effects of structures on the adjacent shoreline are not as predictable as might be required in order to justify their incorporation into the project. Throughout this chapter we will see that sand quality (size) is an extremely important measure of the potential performance of a project both from a physical performance measure and also as an important environmental measure.

This chapter will describe the parameters usually required to justify a beach nourishment project, the basic mechanics of beach nourishment placement, and the subsequent evolution of the project. Examples of major projects will be presented along with a discussion of their performances. The important performance measures of project longevity and dry beach width will be examined in some detail and the role of structures will be considered. Relatively simple design and prediction methodologies will be examined and their appropriate uses discussed.

3. General Discussion

3.1 *Beach nourishment justification*

Usually there is a combination of parameters which will result in economic justification of a nourishment project. These parameters include the background erosion rate, the value of the upland assets to be protected and the value of the beach as a

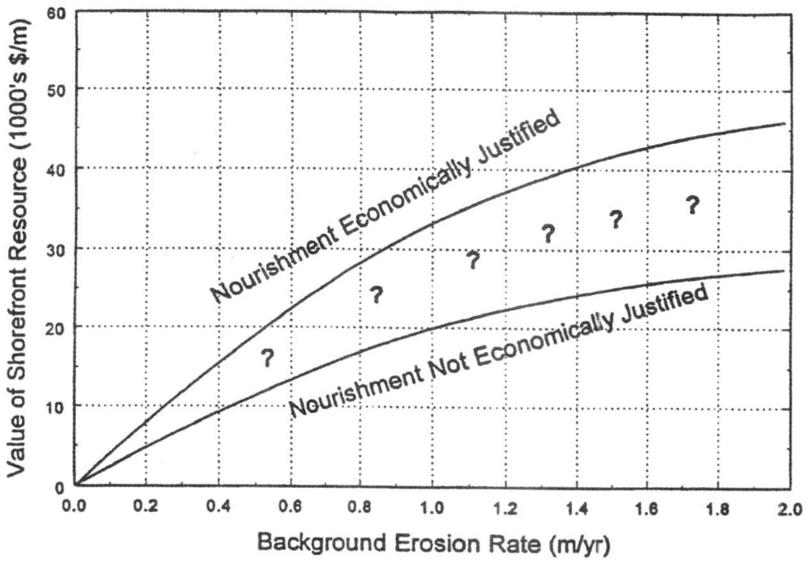

Fig. 9.1 Qualitative illustration of interrelationship between shorefront value and background erosion rate for nourishment justification.

recreational and/or environmental resource. In some rare cases, nourishment may be carried out to protect an historic site or simply to maintain a natural system.

Figure 9.1 illustrates, for the more normal situation requiring economic justification of the project, the qualitative relationship between the background erosion rate and the value per unit beach front length of the resource. In this figure the value of the resource includes the upland structures, and the recreational and environmental value of the beaches. As will be evident later, a significant parameter not evident in Fig. 9.1 is the length of the asset to be protected. Beach nourishment is most effective when carried out over substantial longshore distances.

3.2 *Beach nourishment placement*

In general, nourishment can occur through placement of the material on the beach or as an underwater mound, see Fig. 9.2. The advantages of placement on the beach are that the beneficial effects can be seen and serve for recreational purposes immediately; whereas, if the material is placed as an underwater mound the full benefits are not as noticeable nor as early. However, in some cases, if the quality of the material is considered marginal, offshore mound placement may allow the use of this material that otherwise would be deemed unsuitable for direct placement on the beach. Materials placed as an underwater mound have been carried out on a number of beach nourishment projects with varying degrees of success.

The possible benefits of placement as an underwater mound include the better quality (coarser) portion of this material migrating to shore and gradually

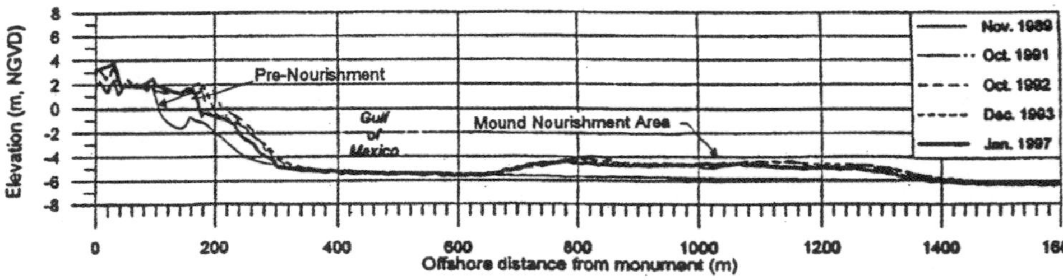

Fig. 9.2 Example of beach and offshore mound nourishment (Dean *et al.* 1997).

nourishing the beach or, if the mound is stable, providing wave energy reduction such that the beach landward of the mound will experience some stabilization against erosion.

Beach nourishment material is usually placed such that the beach is extended seaward and the initial underwater slope is substantially steeper than would be associated with an equilibrium beach profile (EBP). It is reasonable to establish as an upper limit of the designed beach nourishment project the natural berm height which represents the upper limit of wave uprush. Although this elevation depends on the tidal range and wave height, in many locations this elevation is on the order of 2 m to 3 m above mean sea level. One approach is to design the placed berm elevation slightly lower than is anticipated under natural conditions and to allow the natural uprush and overwash processes to increase the berm height to its natural elevation. Upon the placement of beach nourishment, the profile will commence to equilibrate and additionally the plan-form, which now represents a perturbation in the system, will commence to spread out laterally to the adjacent beaches, Fig. 9.3. The time-scales associated with these two processes are important project performance measures. Valuable references on beach nourishment have been prepared by the Dutch Rijkswaterstaat (1987), the National Research Council (1995) and Dean (2002).

3.3 *Important performance measures*

Two types of performance measures will be discussed: physical performance measures and environmental performance measures. The two relevant physical performance measures are primarily the project longevity and the dry beach width. The project longevity is a measure of the time difference between which successive nourishments are required to be conducted to maintain some desired minimum project conditions. Since the waves are the mobilizing factor which cause the spreading out of the nourishment, it will be shown that the project longevity is quite sensitive to the wave energy level in which the project is placed and that the alongshore length of the project is also very important to project longevity.

The dry beach width and the additional dry beach plan-form areas are also relevant measures of project performance. The nourishment grain size relative to the native is an important determinant in these two measures. Specifically, because

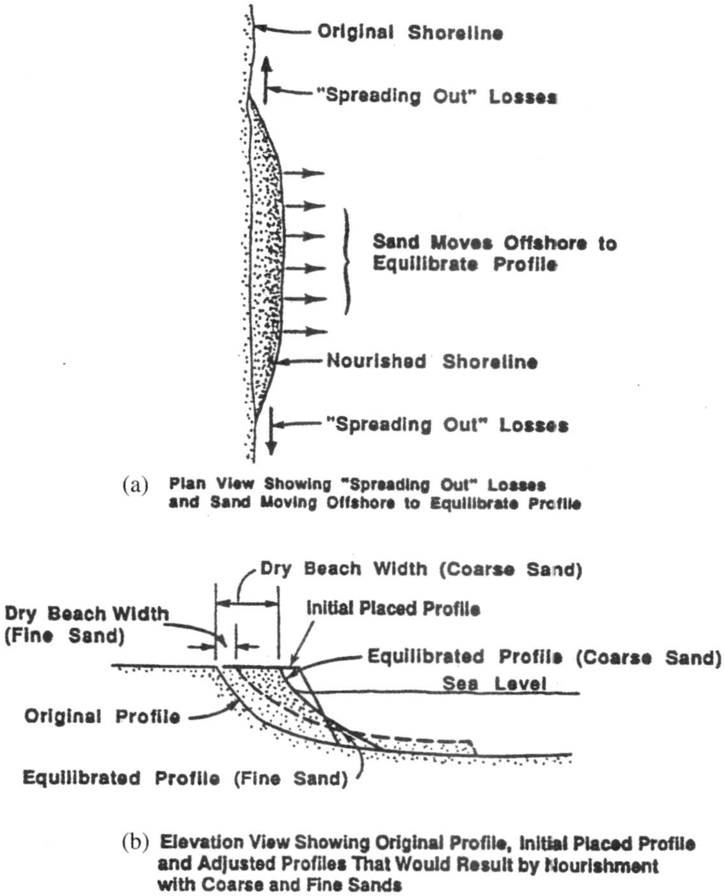

Fig. 9.3 Sand transport "Losses" and beach profiles associated with a beach nourishment project.

coarse sands are naturally associated with steeper slopes than fine sands, if the nourishment sand is coarser than the native, the additional dry beach width after equilibration will be substantially greater than if the nourishment sands were finer than the native. Thus, the dry beach width and the associated additional plan-form area are very sensitive to nourishment sediment size.

The primary environmental considerations are the nourishment sand characteristics relative to the native. The nourishment sand is usually more poorly sorted than the native sand and may thus contain a greater proportion offine material which can cause two adverse affects to the environment. The first is during placement, when silt and clay sized particles remain suspended and can reduce light penetration, interfere with respiration offish, and benthic flora and fauna including corals. The second possibility is through a partial weak cementation of sand grains together, resulting in a beach that is less suitable for sea turtle nesting. The latter can usually be improved by tilling the beach surface with normal agricultural equipment. Although beach nourishment with large quantities of compatible sands can smother some of the resident biota, it has been found that recolonization usually occurs within one or two years. This relatively rapid recovery is interpreted as due to the capability of the biota which reside in the nearshore area to cope with and recolonize after large naturally occurring changes. Beach nourishment can also be of substantial environmental benefit through the creation of wide beaches suitable for sea turtle and bird nesting.

3.4 *Role of stabilizing structures*

As noted previously, structures can be used to increase the longevity of the project either internally within a project or as terminal structures. In some cases, for example on barrier islands, projects are bounded on the two ends by fairly long jetties which serve as terminal structures. If the structure lengths exceed the active width of the beach, there is no possibility for good quality sand to be carried outside of the project limits and the sand can only be redistributed between the two terminal structures. In such a case, if the rate of longshore sediment transport is large it may be worthwhile to consider including intermediate structures within the project, for example as a field of groins.

3.5 *Examples of major projects*

Many major beach nourishment projects have been constructed, and in place, for sufficient durations to quantify their performances. Three such projects located in the US are discussed below:

(1) Miami Beach, Florida

The Miami Beach nourishment project was constructed in five phases over a five-year period from 1976 to 1981 and thus has been in place for approximately 30 years, at the time of writing this chapter, and has been monitored sufficiently to establish its performance. Most of this project is located on Miami Beach and is bounded by a large curved jetty at the north end of the island (at Baker's Haulover Cut) and a second large jetty at the southern end of the project at Government Cut, Fig. 9.4. This portion of the project extends over a 16 km length and comprised the initial placement of approximately seven million cubic meters of sand over the five-year construction period. The placement of this material resulted in a widening of the beach of approximately 100 m. The source of the sand was from five borrow areas located offshore which consisted primarily of calcarious material, whereas approximately 35% of the native material was quartz sand. The Miami Beach area is interesting because the narrowed beach width, which resulting in the need for nourishment, was substantially due to encroachment of the upland hotels on the beach. These hotels were initially constructed with a small but reasonable set-back from the waterline. As tourism developed, the hotels responded to a need for a deck

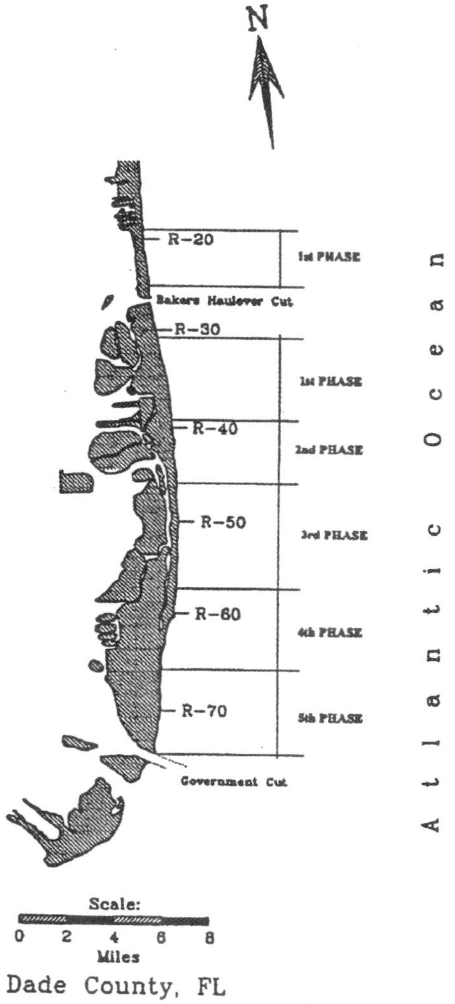

Fig. 9.4 The five different phases of the miami beach nourishment project.

and swimming pool by expanding their facilities seaward and incorporating much of the remaining dry beach width. This seaward encroachment occurred, of course, prior to the time that the State of Florida had instituted adequate restrictions regarding such development. The hurricane of September 1926 caused considerable erosion of the Miami Beach area and resulted in construction of many long groins in an attempt to stabilize the beach. Under the action of the relatively small net longshore sediment transport (estimated at approximately 10,000 cubic meters per year to the south) in conjunction with the excavation of and construction of jetties at Bakers Haulover Inlet at the north end of Miami Beach in 1925, the beach gradually narrowed such that there was little beach left in many locations at normal tide and no beach at high-tide. The Miami Beach tourist industry commenced to decline consistently with the condition of the beach. This decline and the recognition of the importance of the beach to the economy of the area resulted in support for and implementation of the beach nourishment project.

The borrow areas for this project were located between shore parallel coral reefs located approximately 5 km offshore. There was a considerable amount of fine material in at least one of these borrow areas resulting in the placement of considerable silts and clays (greater than 10%). Borrow area quality requirements in Florida now preclude the use of nourishment materials with greater than approximately 4% to 5% silts and clays.

Surveys have shown that as of 1996, some 15 years after nourishment, the total volume remaining within the Miami Beach area was within 9.5% of the total amount placed, see Fig. 9.5, a consequence of the substantial terminal structures at the north and south ends of the project making it difficult for sand to be transported from the project area. However, two erosional hot-spots have occurred which are areas that erode more rapidly than anticipated and/or more rapidly than the adjacent beaches. The occurrence of these erosional hot-spots has led to the requirement

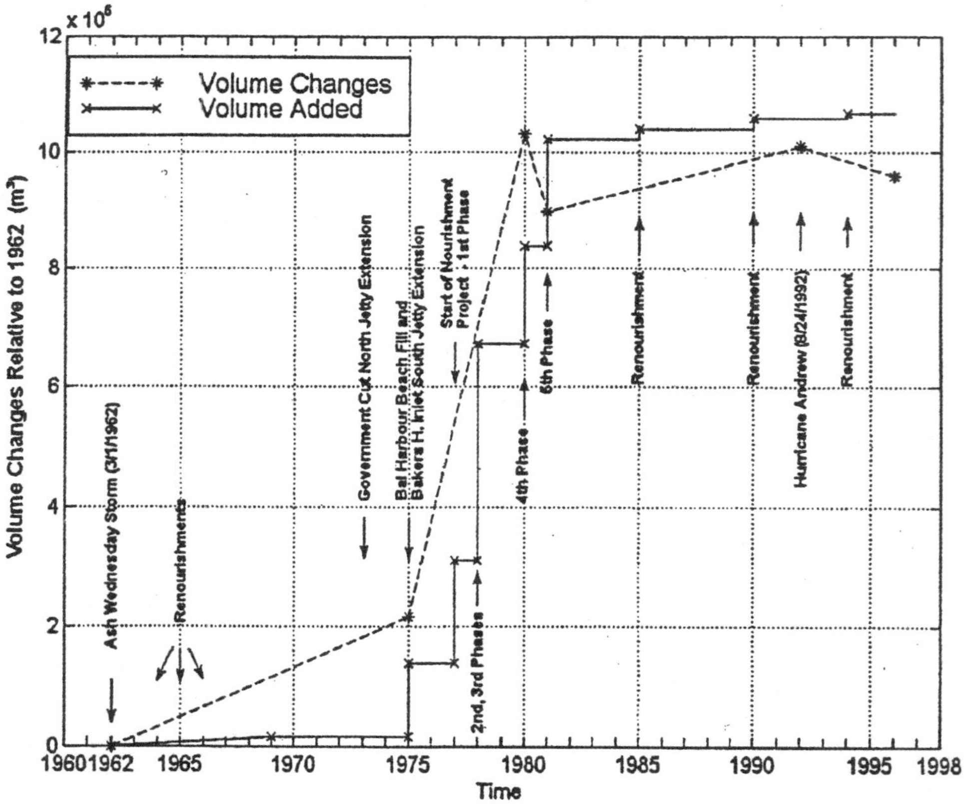

Fig. 9.5 Total volume changes and volumes added relative to 1962 between bakers haulover cut and government cut (based on shoreline changes).

for additional limited renourishments totaling approximately 400,000 cubic meters, about 6% of the original amount placed.

The Miami Beach project is credited with revitalizing the economy of Miami Beach. Today when airplanes arrive at the Miami International Airport from South America they are greeted with the recorded message: "Welcome to Miami and its beaches." An excellent summary of the Miami Beach nourishment project is provided by Wiegel (1992).

(2) Ocean City, Maryland

The Ocean City, Maryland project was constructed in two phases with the first and second phase constructions occurring in 1988 and 1990–1991, respectively, and comprising a total of 5 million cubic meters. Soon after the completion of the Ocean City, Maryland beach nourishment project, the so-called "Halloween Storm of 1991" occurred which caused significant modifications to the project especially in terms of reduction in beach width. The project was immediately hailed as a great success by its proponents and a failure by its detractors. The proponents noted that the damages prevented by the beach nourishment project due to this one storm alone were substantially greater than the cost of the project itself. Thus the benefits of the project had exceeded the project cost within the first year of the project life. The critics of the project cited the reduction in beach width considered as a "loss of sand". However, surveys documented that the reduction in beach width was due dominantly to profile adjustment; most of the placed sand remained in the project limits. The Halloween Storm of 1991 was clearly a rare storm with a number of vessels lost at sea. A valuable series of articles documenting the construction and performance of the Ocean City, Maryland project appears in the January 1993 issue of Shore and Beach (Kraus; Grosskopf and Stauble, 1992; Grosskopf and Behnke, 1992; Jensen and Garcia; Stauble and Grosskopf; and Kraus and Wise).

(3) Santa Monica Bay, California

Santa Monica Bay is located in the vicinity of Los Angeles, California and is some 64 km in length. Near the southern end of the Bay the head of a submarine canyon approaches close to shore. The natural beach was quite narrow, with widths ranging from 15 m to 45 m (Leidersdorf *et al.*, 1993). As a result of harbor excavation projects with the sand placed on the beach and through compartmenting the beach by a number of structures the beach widths today are of the order of 60 m to 150 m. (An excellent paper describing the history of this beach is Leidersdorf *et al.*, 1993.) With the subsequent construction of King Harbor in 1939 an additional volume of material was placed on the beach; however, the construction of King Harbor also caused erosion to the beaches south of the harbor. It was concluded that the King Harbor breakwater blocked the southerly directed longshore sediment transport and that the submarine canyon was draining the sand from the now shorter compartment to the south of King Harbor. A revetment and south breakwater to King Harbor were constructed updrift of the submarine canyon thereby preventing additional losses to the canyon which functioned as a sand sink. As a result of the placement of a total of 22.2 million cubic miles of sand, the present beaches of Santa Monica Bay are much wider than they were naturally. Similar comments apply qualitatively to Huntingdon Beach, California and other California beaches.

4. Simplified Methodology for Representing Project Evolution

4.1 *General*

Referring to Fig. 9.3, it is seen that with the placement of a beach nourishment project two components of sediment transport are induced; cross-shore and longshore sediment transport. The longshore sediment transport is due to the planform anomaly associated with the project and this transport will be termed here "spreading out losses". The time-scales associated with these two components of transport are quite important to the decision to finance a particular project and methodologies to develop estimates of the project longevity will be presented in this chapter. Although the cross-shore and longshore processes occur simultaneously, for a project of reasonable length the cross-shore equilibration time-scales are considerably shorter than the longshore equilibration time-scales, thus providing a reasonable basis for separate consideration of these two components of evolution.

4.2 *Cross-shore considerations*

(1) Profile Closure Depth

Construction of a beach nourishment project usually occurs at a slope that is steeper than equilibrium which is a function of the grain size and water depth. A consequence of the steeper slope is a beach width that is initially greater than that associated with an equilibrated profile. A necessary parameter for the calculation of the equilibrated dry beach width is the depth to which the waves will mobilize the placed sand. Greater depths of equilibration will result in smaller additional dry beach widths for a given placed sand volume per unit beach length and vice versa. This depth of equilibration is referred to as the "closure depth" or "limiting depth of motion" and for purposes here will be denoted as h_*, which is referenced to mean low water datum. The closure depth concept represents an idealization of the actual processes; however, it is very useful in design and prediction of the performance of beach nourishment projects.

Hallermeier (1977, 1978, 1980, 1981) describes the results of an examination of profiles from laboratory and field environments and characterizes the relevant depths in terms of an inner limit and an outer limit of motion. The inner limit is considered to be the depth corresponding to seasonal changes of the profile whereas the changes at the outer limit occur less frequently. Hallermeier recommended that the inner limit be used for engineering purposes and this limit is usually considered to represent the closure depth for engineering applications. Hallermeier recommended the following for the annual closure depth

$$h_* = 2.28\ H_e - 68.5\left(\frac{H_e^2}{gT_e^2}\right) \tag{1}$$

in which H_e is the significant wave height which is exceeded 12 h per year, T_e is the associated wave period and g is gravity. It is noteworthy that the sediment size does not appear in this or other expressions for h_*.

The Hallermeier equation was modified by Birkemeier (1985) using more accurate measurements of field profiles, resulting in

$$h_* = 1.75\ H_e - 57.9\ \frac{H_e^2}{gT_e^2} \tag{2}$$

with a reasonable estimate provided by

$$h_* = 1.75\ H_e \tag{3}$$

Finally Nichols *et al.* (1996) using accurate profile data from Duck, NC generalized these results by considering a multi-year approach. The original Hallermeier recommended relationship (Eq. (1)) for the closure depth, h_*, was found to provide a conservative estimate for multi-year considerations; however, it overpredicted h_* increasingly with longer periods of time considered. For multi-year considerations H_e and T_e are now to be determined on the basis of a 12 h exceedance for the multi-years considered. In summary, the closure depth should be regarded as a concept providing a useful approximation for engineering applications but is not an absolute limit of sediment motion. Also it is a simplification of processes which are poorly understood.

(2) EBPs and the Role of Sediment Size

Two principal measures of the success of a beach nourishment project are the resulting equilibrated dry beach width and total plan-form areas both within the project area and including the adjacent beaches. For our immediate purposes we will limit our discussions to the equilibrium dry beach width. The most simple available EBP methodology provides an excellent method to evaluate the additional dry beach width associated with the placement of a particular volume of material of a specific grain size. Although beach nourishment material may consist of a fairly wide range of sizes, in most of our discussions here we will characterize the material in terms of the median grain size which is the size separating the weight of the sample equally, that is, one-half of the sample by weight is coarser and one-half of the sample by weight is finer than the median grain size. EBP methodology has established (Bruun, 1954; Dean, 1977, 1991) that natural beach profiles can be represented approximately by the following equation

$$h(y) = Ay^{2/3}, \tag{4}$$

in which $h(y)$ is the depth at a distance, y, from the mean water line and, A, is the so-called "sediment scale parameter" with dimensions of length to the one-third

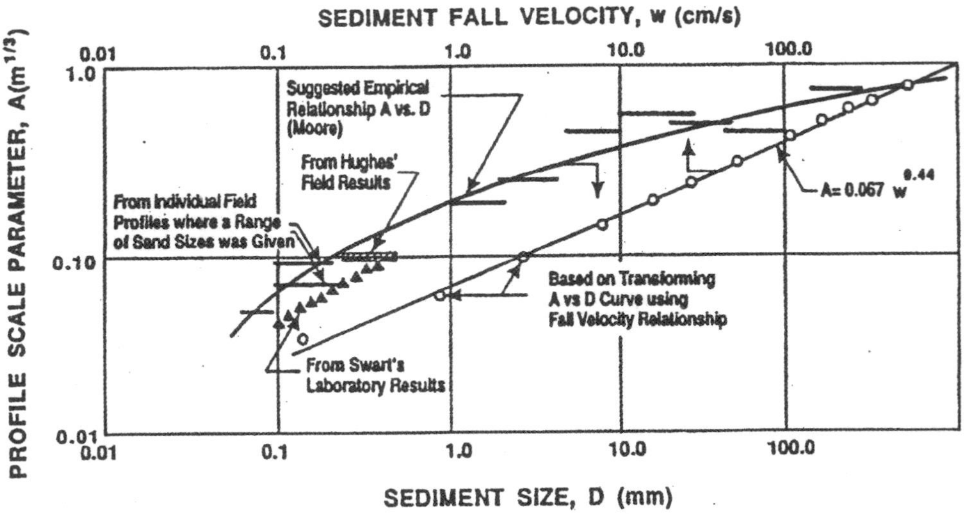

Fig. 9.6 Variation of sediment scale parameter, A, with sediment size and fall velocity (Dean, 1987; modified from Moore, 1982).

power. This form of the equation has two disadvantages and one major advantage. The two disadvantages are the infinite slope at the shoreline and the monotonic nature of the profile; that is, this form cannot represent the presence of bars. The major advantage of this form of EBP is its simplicity and its ready applicability to many significant problems of coastal engineering interest.

The sediment scale parameter, A, has been investigated for more than a 1000 beach profiles and the recommended relationship developed by Moore (1982) and modified by Dean (1987) is shown in Fig. 9.6, where it is seen that the sediment scale parameter increases with median grain size. Table 9.1 presents a tabulation of the sediment scale parameter, A, versus sand sizes for the normal beach material size range for sand size increments of 0.01 mm. For example, the A value for a sand size of 0.27 mm is 0.119 m$^{1/3}$. Since the sediment scale parameter, A, increases with sediment size, D, equilibrated profiles composed of sediments which are coarser and finer than the native will be steeper and more mildly sloped than the native, respectively.

Applying EBP methodology as represented by Eq. (4), it can be shown that the three generic types of equilibrated nourished profiles shown in Fig. 9.7 are possible: intersecting profiles, non-intersecting profiles and submerged profiles (Dean, 1991). Intersecting profiles require a nourishment sediment that is coarser than the native. Non-intersecting profiles occur for nourishment sediments that are of the same size as the native or can also occur for sediments that are coarser or finer than the native. The third type, submerged profiles, requires that the nourishment sediment be finer than the native. Figure 9.8 presents an example in which the same volume of different sized sediments is placed as nourishment. In all four examples presented in Fig. 9.8, the native sand size and the nourishment volume are the same.

Table 9.1 Summary of recommended A values ($m^{1/3}$).

D(mm)	0.00	0.01	0.02	0.03	0.04	0.05	0.06	0.07	0.08	0.09
0.1	0.063	0.0672	0.0714	0.0756	0.0798	0.084	0.0872	0.0904	0.0936	0.0968
0.2	0.100	0.103	0.106	0.109	0.112	0.115	0.117	0.119	0.121	0.123
0.3	0.125	0.127	0.129	0.131	0.133	0.135	0.137	0.139	0.141	0.143
0.4	0.145	0.1466	0.1482	0.1498	0.1514	0.153	0.1546	0.1562	0.1578	0.1594
0.5	0.161	0.1622	0.1634	0.1646	0.1658	0.167	0.1682	0.1694	0.1706	0.1718
0.6	0.173	0.1742	0.1754	0.1766	0.1778	0.179	0.1802	0.1814	0.1826	0.1838
0.7	0.185	0.1859	0.1868	0.1877	0.1886	0.1895	0.1904	0.1913	0.1922	0.1931
0.8	0.194	0.1948	0.1956	0.1964	0.1972	0.198	0.1988	0.1996	0.2004	0.2012
0.9	0.202	0.2028	0.2036	0.2044	0.2052	0.206	0.2068	0.2076	0.2084	0.2092
1.0	0.210	0.2108	0.2116	0.2124	0.2132	0.2140	0.2148	0.2156	0.2164	0.2172

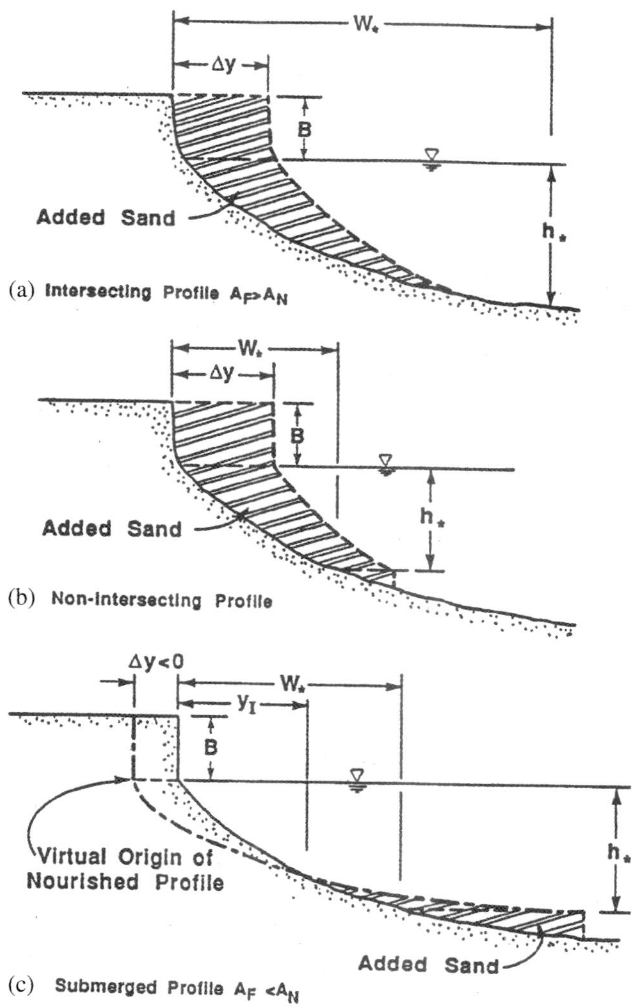

Fig. 9.7 Three generic types of nourished profiles. (Dean, 1991).

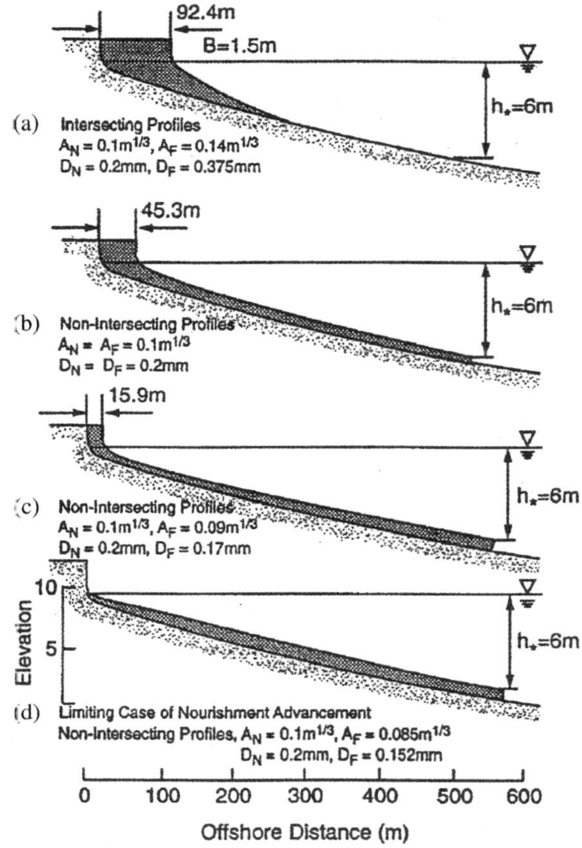

Fig. 9.8 Effect of nourishment scale parameter, A_F, on width of resulting dry beach. Four examples of decreasing A_F with same added volume per unit beach length. Berm height, B, and closure depth, h_* are 1.5 m and 6 m, respectively (Dean, 1991).

In the upper panel (Fig. 9.8(a)), the native sand size is 0.2 mm and the nourishment sand size is 0.375 mm. It is seen that through the placement of 340 cubic meters per meter of beach length, intersecting profiles occur with an equilibrated dry beach width of 92 m. In Fig. 9.8(b), a nourishment sand size of 0.2 mm (the same as the native) is considered, non-intersecting profiles occur, and the equilibrated dry beach width is 45 m, approximately one-half of that, as shown in Fig. 9.8(a). In Figures 8(c) and 8(d), nourishment sand sizes of 0.17 m and 0.152 m, respectively, are considered; and it is seen that the equilibrated dry beach widths decrease to 16 m and 0 m, respectively. The example in Fig. 9.8(d) represents an incipient dry beach in which all of the placed sand is configured as an underwater "submerged" profile, the seaward end of which is located at the so-called "depth of closure", h_*. Figure 9.9 shows a similar example of placement of different volumes of the same size sand which is finer than the native. Progressing from the upper to the lower panel, greater and greater volumes of nourishment sand are placed. It is seen that with a volume

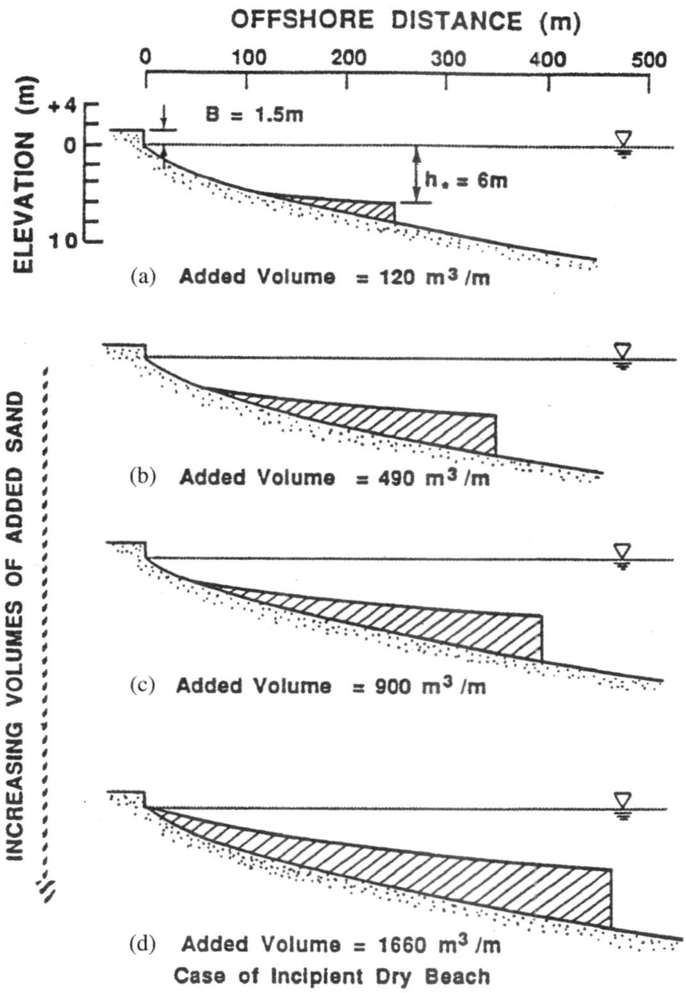

Fig. 9.9 Effect of increasing volume of sand added on resulting beach profile, $A_F = 0.1\,\mathrm{m}^{1/3}$, $A_N = 0.2\,\mathrm{m}^{1/3}$, $h_* = 6.0\,\mathrm{m}$, $B = 1.5\,\mathrm{m}$ (Dean, 1991).

of 1,660 cubic meters per meter an incipient dry beach occurs. In all four cases shown in Fig. 9.9 the seaward depth of the nourishment profile is 6 m. The relative importance of sand size can be illustrated further as shown in Fig. 9.10, where the additional dry beach width is plotted as the ordinate and the volume of nourishment sand added is plotted as the abscissa. In this plot, nourishment with sands of three different sizes is represented with a native sand size of 0.2 mm for all three cases. For a nourishment sand that is 0.275 mm, some 38% larger than the native sand, it is seen that placement of 400 cubic meters per meter increases the dry beach width over that which would be obtained with sand of the same size as the native by approximately 69%. It is also noted that for the coarser material, the additional dry beach width increases rapidly for small volumes added and then the relationship

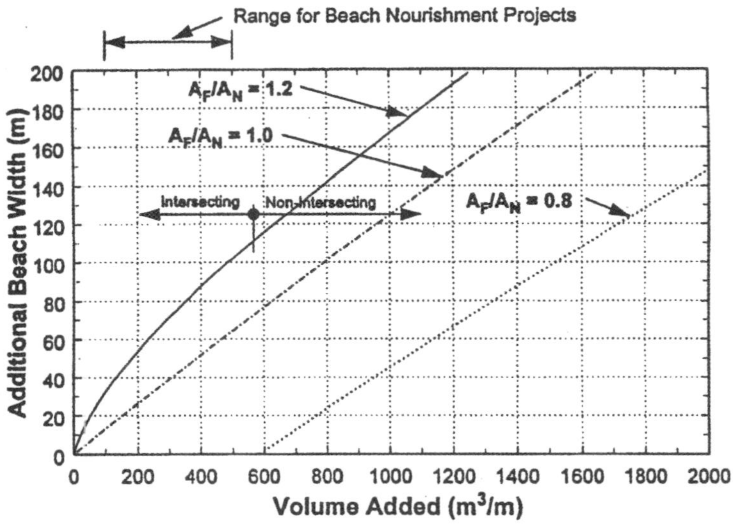

Fig. 9.10 Additional dry beach width versus volume of sand added per unit length of beach, $h_* = 6\,\mathrm{m}$, $B = 1.5\,\mathrm{m}$, $D_N = 0.2\,\mathrm{mm}$.

becomes approximately parallel to the case in which the nourishment and native sand sizes are the same. The volume at which the two relationships become parallel corresponds to the transition from intersecting profiles to non-intersecting profiles. It is at this condition that the depth of intersection of the nourished profile on the native profile becomes h_*. For a nourishment sand that is significantly finer than the native ($D_F = 0.14\,\mathrm{mm}$), approximately $600\,\mathrm{m}^3$ per meter, is required before an additional dry beach width occurs. This threshold volume is required due to the small equilibrium slopes associated with the fine nourishment sand and the need for the profile to extend from the closure depth landward to the mean water line. As indicated in Fig. 9.10, the volume density of most nourishment projects range from $100\,\mathrm{m}^3/\mathrm{m}$ to $500\,\mathrm{m}^3/\mathrm{m}$.

It can be shown from EBP methodology that the non-dimensional beach width Δy' can be related to the following three non-dimensional quantities: non-dimensional volume V', non-dimensional sediment scale parameter A_F/A_N , and non-dimensional berm height, $B' = B/h_*$, i.e.,

$$\Delta \mathrm{y}' = \mathrm{f}(\mathrm{V}', \mathrm{A_F}/\mathrm{A_N}, \mathrm{B}/\mathrm{h}_*) \tag{5}$$

in which $\Delta y' = \Delta y/W_*$ and $V' = V/(W_* B)$ and W_* is the distance from the mean water line to the depth of closure, h_* , on the native profile. For nourishment sands which are the same as the native, the shoreline advancement, Δy, can be expressed approximately for any profile form as

$$\Delta \mathrm{y} = \frac{\mathrm{V}}{\mathrm{h}_* + \mathrm{B}} \tag{6}$$

or

$$\Delta \mathrm{y}' = \frac{\mathrm{V}'}{1 + \frac{1}{\mathrm{B}'}}. \tag{7}$$

Returning to idealized profiles described by Eq. (4), there are three specific equations of the form of Eq. (5), one for each of the nourished profile types shown in Fig. 9.7. These are:

a. Intersecting Profiles:

$$\forall' = \Delta y' + \frac{3}{5B'}(\Delta y')^{5/3}\frac{1}{\left[1-\left(\frac{A_N}{A_F}\right)^{3/2}\right]^{2/3}} \tag{8}$$

b. Non-Intersecting Profiles:

$$\forall' = \Delta y' + \frac{3}{5B'}\left\{\left[\Delta y' + \left(\frac{A_N}{A_F}\right)^{3/2}\right]^{5/3} - \left(\frac{A_N}{A_F}\right)^{3/2}\right\} \tag{9}$$

c. Submerged Profiles:

$$\forall' = \frac{3}{5B'}\left\{\left[\left(\frac{A_N}{A_F}\right)^{3/2} + \Delta y'\right]^{5/3} + \frac{(-\Delta y')^{5/3}}{\left[\left(\frac{A_N}{A_F}\right)^{3/2} - 1\right]^{2/3}} - \left(\frac{A_N}{A_F}\right)^{3/2}\right\} \tag{10}$$

where in the last equation $\Delta y'$ is negative and represents the "virtual shoreline" position relative to the original shoreline (Fig. 9.7(c)). The critical volume, $(\forall')_{c1}$, associated with intersecting/non-intersecting profiles is

$$(\forall')_{c1} = \left(1 + \frac{3}{5B'}\right)\left[1 - \left(\frac{A_N}{A_F}\right)^{3/2}\right] \tag{11}$$

and applies only for $A_F/A_N > 1$. In Eq. (11), if the volume $(\forall')_{c1}$ is exceeded, the profiles are non-intersecting. The critical volume $(\forall')_{c2}$ which will just yield a finite shoreline for non-intersecting profiles is

$$(\forall')_{c2} = \frac{3}{5B'}\left(\frac{A_N}{A_F}\right)^{3/2}\left(\frac{A_N}{A_F} - 1\right) \tag{12}$$

and, of course, only applies for $A_F/A_N < 1$.

Because the non-dimensional equilibrated dry beach width, Δy', is a function of three variables (Eq. (5)) it is not possible to present a general solution on a single graph; however, solutions for values of $B' = B/h_* = 0.25$ and 0.5 are presented in Figs. 11 and 12, respectively. The general character of these solutions can be understood by discussing Fig. 9.11 which is applicable for a value of $B' = 0.25$. The horizontal axis is the ratio of fill to native sediment scale parameters, A_F/A_N, and the vertical axis is the non-dimensional additional dry beach width, Δy'. The isolines on this figure represent$\forall'$, non-dimensional volume of sand added per unit beach length. The form of the non-dimensional volume can be interpreted as the number of layers of sand of the berm height, B, which are spread out to W_* which is the beach width out to h_* on the native profile. For example, $\forall' = 0.1$ is equivalent

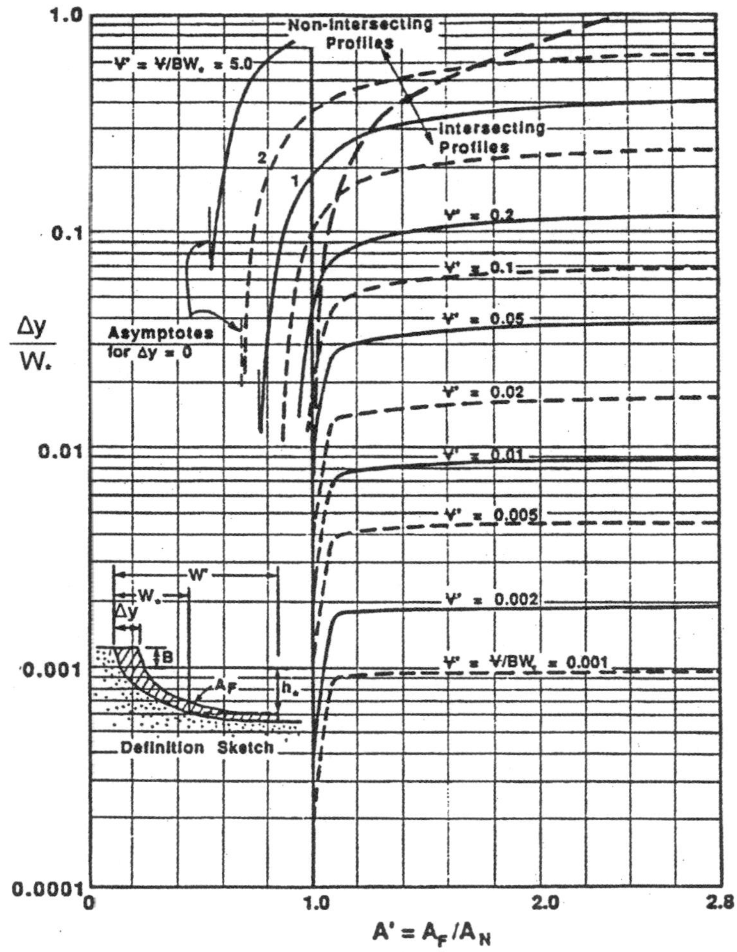

Fig. 9.11 Variation of non-dimensional shoreline advancement $\Delta y/W_*$, with A' and V'. Results shown for $B'(= B/h_*) = 0.25$ (Dean, 1991).

to a volume of 1/10th of a berm height spread out to a depth of closure on the native profile over a distance W_*.

It can be shown that the consideration of a single grain size in the application of EBPs tends to overpredict the effect of sand size. For example, the effect of smaller sediment sizes is qualitatively as predicted; however, submerged profiles will not occur if the nourishment sediment contains some sediment with a size equal to that of the native. See Dean (2002) for more detail.

(3) Profile Evolution

The evolution of a beach profile placed steeper than its natural slope is of interest because there will be a wider beach for some length of time as the profile approaches equilibrium. This evolution is also important as it is necessary to inform the general

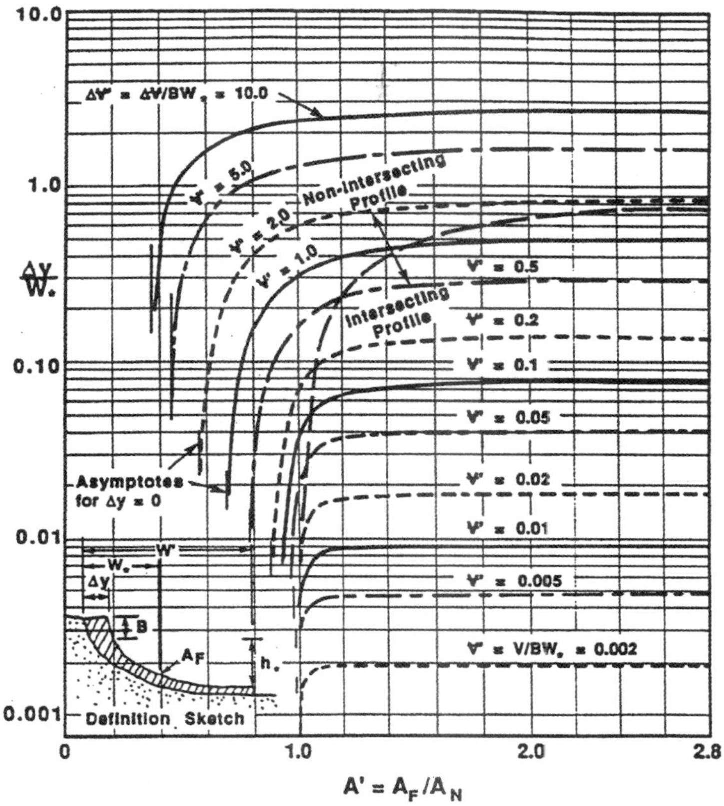

Fig. 9.12 Variation of non-dimensional shoreline advancement $\Delta y/W_*$, with A' and V'. Results shown for $B'(= B/h_*) = 0.5$ (Dean, 1991).

public of the magnitude of this anticipated narrowing of the beach and the probable range of time-scales. The available monitoring results indicate that the time-scales required for an equilibration of 50% to the native profile is on the order of 2 to 5 years and depends on the wave climate.

There are no known numerical or analytical models that can reasonably predict the time-scales of equilibration. There are many numerical models that have been developed to represent cross-shore sediment transport; however, most of these are more applicable for the case of erosion associated with a condition that is substantially out of equilibrium such as an elevated water level during a storm and the accompanying large wave heights. To appropriately represent the evolution of an over-steepened profile it is necessary that the slope play a role in the direction of cross-shore transport. This characteristic is present in some of the available profile evolution models, for example in the EDUNE model by Kriebel (1982) and Kriebel and Dean (1985) but not present in the SBEACH model by Larson and Kraus (1989) in which the direction of sediment transport is controlled entirely by the wave and sediment fall velocity characteristics. In the application of a model to represent a nourished (over-steepened) profile evolution toward equilibration, it would be necessary to represent the variability in wave height and period since the larger waves of shorter period are dominantly responsible for the seaward transport.

(4) Underwater Mounds

The placement of material as an underwater mound is of interest due to the lower costs of placement in this manner and also the possibility of using material that would not otherwise be acceptable for direct placement on the beach for nourishment. Figure 9.2 has presented one example of nourishment on the beach and in an underwater mound. One of the major questions that arises in association with placement of an underwater mound is whether the mound will be stable or unstable and if it is unstable whether the mound will move landward or seaward. There are two effects by which the beach system can benefit from the presence of an underwater mound. The first of these is if the mound is unstable and will gradually move shoreward and thus act as a feeder of sand to the beach. The second possibility is if the mound is stable and causes sufficient wave energy dissipation such that the waves are smaller when they reach the shore and the mound thus functions as an offshore breakwater thereby contributing to the stability of the beach. On an intuitive basis, if a profile is initially in approximate equilibrium, it would seem that the placement of an underwater mound on that profile would result in either a mound that is stable or one that moves only landward. The rationale for this is that after placement of the underwater mound, the profile would have an excess of sediment relative to equilibrium and this excess can only be reduced through landward sediment transport.

Andrassy (1991) reported on the monitoring of an underwater mound placed off Imperial Beach, California. The mound was placed in approximately 6 m of water depth and was documented to migrate landward over a period of approximately one year. Figure 9.13 presents a series of profiles from approximately the center of this project. It is seen that the placed sediment appears to have been transported shoreward as a dampened wave of translation. A second well-documented underwater mound project is that off the west coast of Denmark near Torsminde, (Laustrup *et al.* 1996). This project has been monitored exceptionally well including some 17 surveys over a period of 2 years. There are no know documented incidences of mounds being transported offshore after placement (Ahrens and Hands, 1998).

Hands and colleagues (1980, 1988, 1991, 1994) and Ahrens and Hands (1998) have conducted analyses to develop methodology addressing the question of stability or instability of underwater mounds. As noted previously, the methodology of Hallermeier (1978) includes two limits (the inshore and offshore limit) of depths for sediment motion. Hands found that if the mound crest was inside the depth of inner motion as determined by Hallermeier's predictions, then the mound always moved landward and if the crest of the mound was at a depth greater than the outer limit, the mound was always stable. For mound crest depths that were between the inner and outer depths as predicted by Hallermeier's methodology the mound could

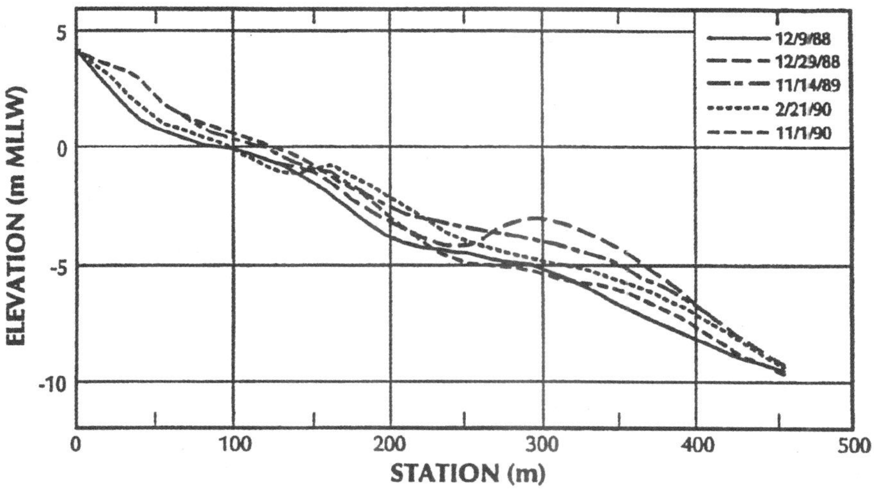

Fig. 9.13 Berm evolution at imperial beach, California (Andrassy, 1991).

be either stable or unstable (Hands and Allison, 1991). More recently, Ahrens and Hands (1998) have investigated the appropriate wave-induced water particle velocities for correlation with movement of the mound. These velocities were based on the non-linear numerical stream function wave theory (Dean, 1965; Dalrymple, 1974). Quite surprisingly, Ahrens and Hands found that the best correlation with movement of an underwater mound was with the trough velocities as predicted by the stream function wave theory.

Otay (1994) has reported on the monitoring of a substantial underwater mound placed in the Gulf of Mexico (Fig. 9.2) and has applied the earlier methodology of Hands and Allison (1991).

This methodology predicted that the mound should move; however, detailed monitoring results conducted 7 years after placement found that although the irregular surface of the mound tended to be smoothed out, there was no discernable movement of the centroid of the underwater mound.

McClellan and Kraus (1991) have applied a criterion for offshore bar formation to the movement of an underwater mound. Their criterion would admit both onshore and offshore mound movement. It is not clear that this application is appropriate since usually the underwater mound lies well outside the limits of the normal surf zone and the motion of an offshore bar, for which the methodology was developed, is associated with the mechanics of breaking waves.

Douglass (1995) has developed a method for predicting the migrational characteristics of a mound placed underwater. The method employs Stokes second-order wave theory and Bagnold's method for sand transport (Bailard and Inman, 1981) and results in a convective-diffusion equation, thereby predicting both the speed of movement and dispersion of the mound, both of which were found to depend strongly on water depth.

Bodge (1994) has documented the performance of an underwater mound off the east coast of Florida and concluded that the method of Hands and Allison (1991) predicted the onshore movement.

(5) Plan-form Evolution

The following sections present the available methodologies for conducting simple predictions of evolution for a number of different initial plan-forms and conditions. Later, the use of numerical models will be reviewed. Basically, the use of numerical models allows a greater variety of initial plan-forms, a more detailed wave climate, more flexibility in the internal and boundary conditions and the capability to include complex details of the wave/sediment interaction.

(a) Methodology/Tools for Plan-form Evolution

In the simplest models of plan-form evolution, it is necessary to assume that the profile response to an increase or decrease of volume is a seaward or landward displacement, respectively, of the profile without change of form, Eq. (6). Thus the profile is displaced landward or seaward in accordance with gradients in the longshore sediment transport. These models are termed "one line models" as they track only one contour, usually the mean sea level contour. The simplest analytic basis for dealing with plan-form evolution combines the linearized longshore sediment transport equation with the continuity equation which results in the Pelnard Considere (1956) diffusion equation.

$$\frac{\partial y}{\partial t} = G\frac{\partial^2 y}{\partial x^2} \tag{13}$$

This equation is recognized as the familiar heat conduction equation and the quantity G in the above equation is referred to as the "longshore diffusivity" and can be defined in terms of the breaking wave conditions (denoted by subscript "b") as

$$G = \frac{KH_b^{5/2}\sqrt{\frac{g}{\kappa}}}{8(s-1)(1-p)(h_* + B)} \tag{14}$$

or in terms of deep water conditions (denoted by subscript "0") as

$$G = \frac{KH_0^{2.4}C_{G_0}^{1.2}g^{0.4}}{8(s-1)(1-p)C_*\kappa^{0.4}(h_* + B)} \tag{15}$$

in which s is the ratio of sediment density to that of the water in which it is immersed, p is the in-place porosity, C_G is the group velocity, C is the wave celerity, κ is the ratio of breaking wave height to breaking water depth ($\approx$0.78) and the subscript "$_*$" denotes conditions at the depth of closure. The coefficient K in the equations above is dimensionless and is the so-called "sediment transport coefficient" and is often taken as 0.77 as determined by Komar and Inman (1970).

Based on limited field data, Dean (1987) developed an inverse relationship between the sediment transport coefficient, K, and sediment size, D. Komar had

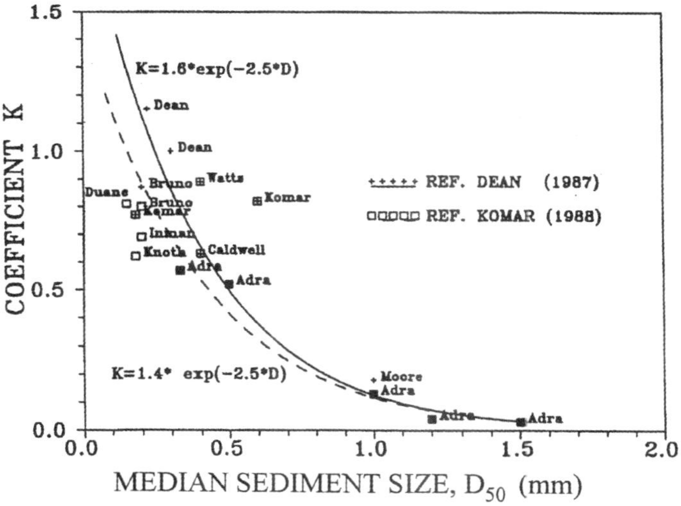

Fig. 9.14 Relationship of sediment transport coefficient with median sediment size, D_{50} (delValle *et al.*, 1993).

earlier (1977) considered such a relationship and concluded that over the limited sand size ranges present on most coastlines, a relationship between K and D could not be determined. Komar (1988) reviewed the data employed by Dean, pointed out uncertainties and again concluded that no relationship existed over the range of sand sizes for which data were available, delValle *et al.* (1993) reported on a field investigation of the erosion of the Adra River Delta (Spain) in which transport coefficients were determined for several grain sizes along the delta. Their results are presented in Fig. 9.14 using adjustments in the data recommended by Komar (1988). The solid and dashed curves are the relationships originally recommended by Dean (1987) and that by delValle *et al.* (1993), including the adjusted original and additional data. In summary, although there is uncertainty in the K versus D relationship, it appears that K is inversely related to D. In view of the significance of K to the performance of beach nourishment projects it is recommended that the dashed line in Fig. 9.14 be employed.

Returning to the Pelnard Considère equation (Eq. (13)); its two main advantages are: (1) linearity (allowing superposition of various solutions), and (2) frequent occurrence in classical physics, resulting in a number of solutions that can be immediately adopted to represent various cases of general relevance to beach nourishment. We will review some of those cases in the following sections.

(b) Examples of the Pelnard Considère Solutions

The simplest and perhaps most useful solution is that of an initially rectangular beach nourishment project on a long straight beach with no background erosion considered. The resulting evolution is described by the following well-known

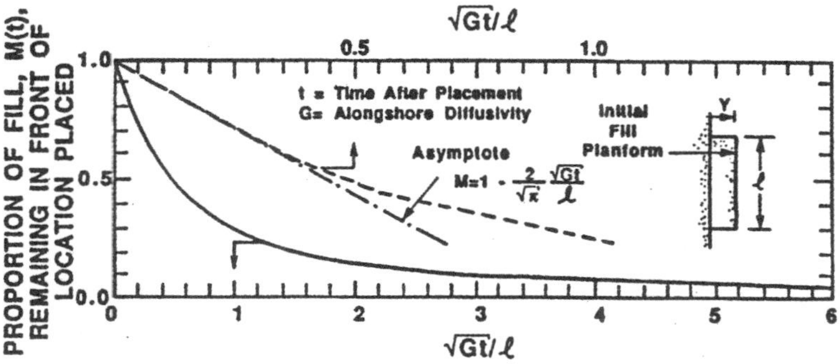

Fig. 9.15 Proportion of material remaining in region placed versus the parameter $\sqrt{\mathrm{Gt}}/\ell$.

solution.

$$y\,(x,t) = \frac{Y}{2}\left\{\mathrm{erf}\left[\frac{\ell}{4\sqrt{Gt}}\left(\frac{2x}{\ell}+1\right)\right] - \mathrm{erf}\left[\frac{\ell}{4\sqrt{Gt}}\left(\frac{2x}{\ell}+1\right)\right]\right\}, \tag{16}$$

in which "erf" is the so-called error function, Y is the initial beach width, x is the base line in the longshore direction, and t is time. It can be shown by integrating Eq. (16) over the beach segment in which the rectangular nourishment was placed, that the proportion of material remaining, $M(t)$, at any time, t, is given by

$$M(t) = \frac{\sqrt{4Gt}}{\ell\sqrt{\pi}}\left(e^{-(\ell/\sqrt{4Gt})^2} - 1\right) + \mathrm{erf}(\ell/\sqrt{4Gt}) \tag{17}$$

Figure 9.15 is a plot of Eq. (17) and it is seen that initially sediment is lost fairly rapidly from the project limits and then later the losses occur much more slowly. Referring to either Eqs. (16) or (17), if the group of terms $\sqrt{Gt}/\ell$ is the same for two projects, the projects will evolve in a geometrically similar form. Thus the longevity of a project depends only on this parameter. Referring to Fig. 9.15, we can develop a simple and useful equation for the time at which X% of the material placed will remain within the placement area. Thus, this time is a reasonable measure of the longevity of the project. As an example, selecting a value of 50%, that is, the half-life of the project, we see by referring to Fig. 9.15 that the associated parameter, $\sqrt{Gt}/\ell$, is equal to 0.46 which results in the following equation

$$t_{50}\% = \frac{(0.46)^2\ell^2}{G}. \tag{18}$$

Selecting reasonable values for most of the parameters, it can be shown that the half-life is given by

$$t_{50}\% = K''\frac{\ell^2}{H_b^{5/2}}, \tag{19}$$

where K" is equal to 0.179 years — $m^{5/2}/km^2$ for $t_{50}\%$ in years, H_b in meters, and ℓ in kilometers. Thus, if an initially rectangular beach nourishment project is placed in an area where the effective wave height is 1 m and the project length, ℓ, is 1 km, the half-life will be 0.18 years, slightly more than 2 months. If the project length is doubled, the half-life is quadrupled due to the square of the project length appearing in Eq. (19). For a project length of 10 km and a wave height of 1 m, the projected half-life is nearly 18 years! With this as background we can now return to Fig. 9.15 and see why the losses from the project area occur rapidly at first and then later much more slowly. As losses occur from the project area, the sand from these losses is deposited adjacent to the project area and, in effect, the project is effectively longer than it was initially.

The recognition that long projects lose sand more rapidly than short projects can be applied immediately to the case of repeated nourishments, a problem first considered by Dette *et al.* (1994). It can be shown through reasonably simple numerical modeling that for a project which is nourished and then allowed to evolve until a threshold proportion of the initial nourishment remains within the project area and then the project area is renourished back to its original nourished volume, the required time intervals between renourishments increases with renourishment number. The explanation for this follows from the above discussion. After the first renourishment, there are two nourishments in place that are losing sand (recall the linearity of the Pelnard Considère equation and resulting superimposability of solutions). The first nourishment, now functioning as a longer project, is losing sand more slowly than after the initial nourishment whereas the second nourishment (the first renourishment) loses sand at the same proportional rate as the initial nourishment. Thus, the composite losses are slower than they were for the initial nourishment. With continued renourishments at any particular time, the rate at which sand is lost from the nourishment area is a composite of each of the nourishments, including the initial nourishment and all of the preceding renourishments. The longer a project has been in place, the more slowly the losses from the project area. However, in the presence of background erosion, the renourishment intervals initially increase with renourishment number, and then decrease for larger renourishment numbers. The interpretation for this, though rational, is somewhat complicated. Figure 9.16 presents an example of multiple nourishments for a zero background erosion rate.

A second specific example of a Pelnard Considère solution for beach nourishment demonstrates the significance of boundary conditions such as might occur on a barrier island. Consider an initially rectangular plan-form on a barrier island in which the nourishment can extend to the limits of the barrier island at an inlet or the nourishment can be set back from the inlet. In this case, approximating the shoreline displacement as zero at the inlet is the most logical boundary condition. The results for an initial nourishment extending along an entire barrier island are presented in Fig. 9.17 in which the results are compared with a shoreline of infinite length. The advantages of nourishment on a long beach are clearly evident since the effect of an inlet is to cause the sand to be drained off much more rapidly. Figure 9.18 shows the proportion remaining for different lengths of the nourishment relative to the length of the barrier island.

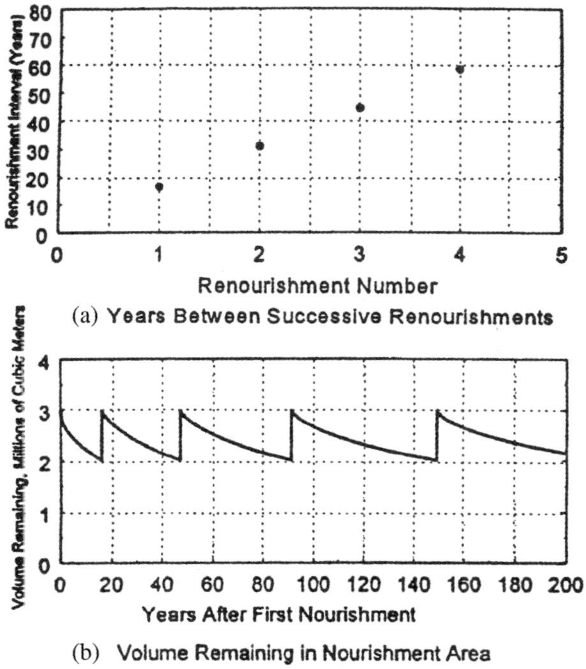

Fig. 9.16 Renourishment characteristics. No background erosion. Breaking wave height = 0.5 m, project length = 6 km, $h_* + B = 9\,\mathrm{m}$, $V_{\max} = 3,000,000\,\mathrm{m}^3$, $V_{\min} = 2,000,000\,\mathrm{m}^3$ (Dean, 1995.)

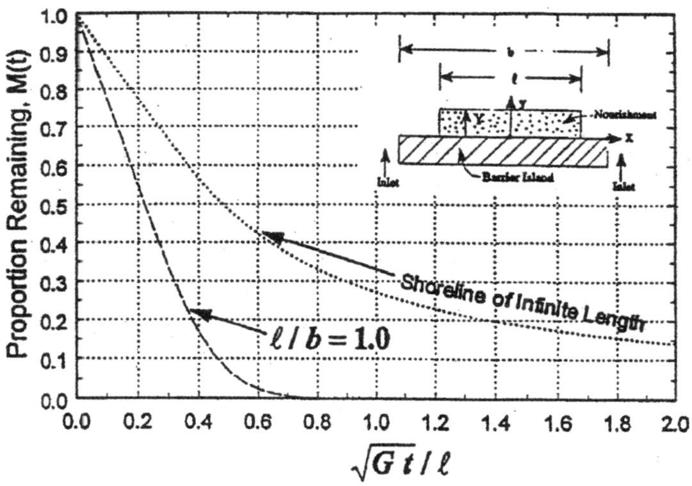

Fig. 9.17 Proportion remaining, $M(t)$, for nourishment along full length of barrier island compared to nourishment on an infinitely long beach, "ℓ" = nourishment project length (Dean, 1997).

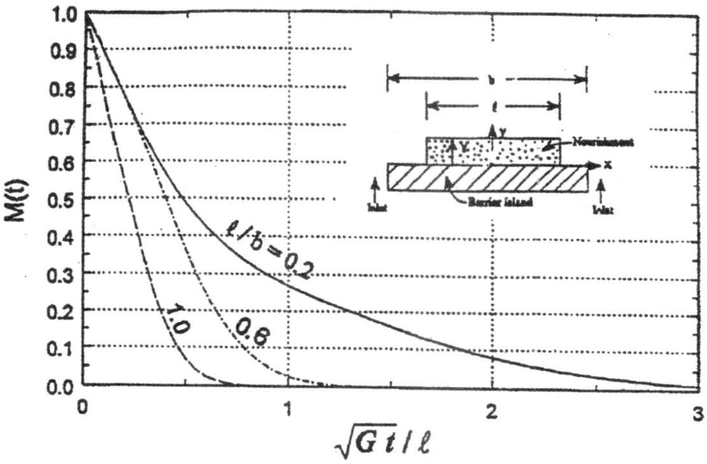

Fig. 9.18 Proportion remaining, $M(t)$, for nourishment ends set back from ends of a barrier island (Dean, 1997).

Description	Illustration	Solution
Initially Rectangular Planform on a Long Straight Beach	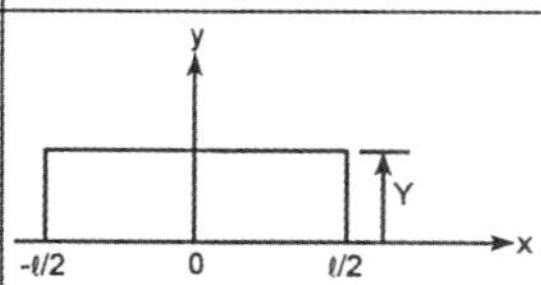	$y(x,t) = \frac{Y}{2}\left\{ \operatorname{erf}\left[\frac{\ell}{\sqrt{4Gt}}\left(\frac{2x}{\ell}+1\right)\right] - \operatorname{erf}\left[\frac{\ell}{\sqrt{4Gt}}\left(\frac{2x}{\ell}-1\right)\right]\right\}$
Initially Trapezoidal Planform on a Long Straight Beach	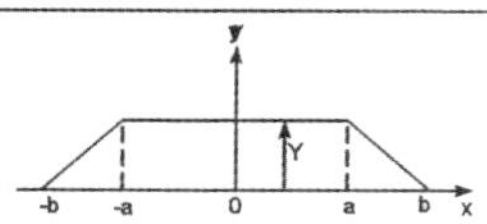 $A = \frac{a}{2\sqrt{Gt}}, B = \frac{b}{2\sqrt{Gt}}, X = \frac{x}{a}$	* $y(x,t) = \frac{Y}{2(B-A)}\left\{ \begin{array}{l} (A-AX)\operatorname{erf}(AX-A) - (A+AX)\operatorname{erf}(AX+A) \\ +(B+AX)\operatorname{erf}(AX+B) - (B-AX)\operatorname{erf}(AX-B) \\ +\frac{2}{\sqrt{\pi}}\left[e^{-(A^2X^2+B^2)}\cosh(2AXB) - e^{-(A^2X^2+A^2)}\cosh(2A^2X)\right] \end{array} \right\}$
Initially Rectangular Planform Centered on a Barrier Island	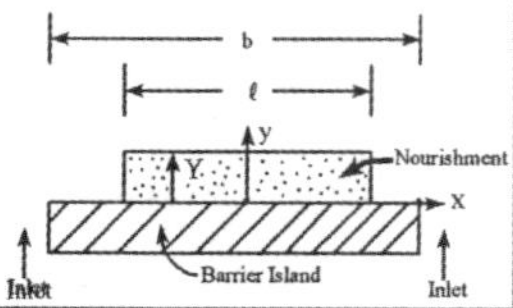	$y(x,t) = 4Y\left(\frac{\ell}{b}\right)\sum_{n=1}^{\infty}\frac{1}{\mu_n}\sin\left(\frac{\mu_n}{2}\right)\cos\left(\mu_n\frac{x}{\ell}\right)e^{-\mu_n^2 Gt/\ell^2}$ $\mu_n = (2n-1)\frac{\ell}{b}\pi$
Initially Rectangular Planform Near Inlet on a Long Barrier Island	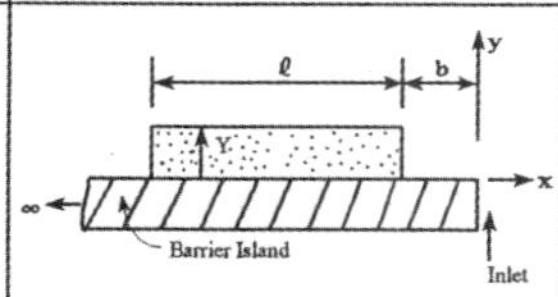	$y(x,t) = \frac{Y}{2}\left\{ \operatorname{erf}\left(\frac{x+b+\ell}{\sqrt{4Gt}}\right) - \operatorname{erf}\left(\frac{x+b}{\sqrt{4Gt}}\right) + \operatorname{erf}\left(\frac{x-b-\ell}{\sqrt{4Gt}}\right) - \operatorname{erf}\left(\frac{x-b}{\sqrt{4Gt}}\right)\right\}$

*Walton, 1994

Fig. 9.19 Solutions for four initial planform geometries and boundary conditions.

Larson and Kraus (1997) have presented a number of solutions to the Pelnard Considère equation which are of interest to coastal engineering practice. Walton (1994) has presented an analytic solution to the case of an initially trapezoidal beach plan-form. Figure 9.19 presents the solutions to four initial plan-forms of coastal engineering interest, including Walton's trapezoidal plan-form. Recognizing that trapezoidal plan-forms lose sediment more slowly than rectangular plan-forms, Hanson and Kraus, (1993) following an earlier study by the US Army Corps of Engineers (1982) have investigated the possible benefits of commencing with an initially trapezoidal plan-form. It can be shown that the apparent benefits of employing an initially trapezoidal plan-form rather than a rectangular plan-form are an artifact of the manner in which the benefits are calculated. For example, if we consider a particular length of shoreline, ℓ, and place an amount of fill in a rectangular plan-form from $-\ell/2$ to $+\ell/2$ and compare the evolution with that of an initially trapezoidal plan-form with the same volume of material with the mid-point of the sloping sides at $+\ell/2$ and $-\ell/2$, it can be shown that at any future time the initially rectangular plan-form will maintain a greater proportion of material within the longshore limits of the initial trapezoidal plan-form than the initially trapezoidal plan-form. This is consistent with intuition since the initially trapezoidal plan-form is very similar to the evolved rectangular plan-form at some later time.

There are a number of interesting and useful generic results that can be developed from the Pelnard Considère linearization and associated solutions. One result is that a beach nourishment project on a long straight beach is relatively insensitive to wave direction. The reason is that if a project is located on a long straight beach, the effect of an oblique wave direction is to superimpose a uniform transport with the spreading out effect due to the plan-form anomaly. This follows from the superimposability of the Pelnard Considère equation. In this case, one solution is simply the background transport and the second is the result of the perturbation planform causing a symmetric spreading out of the initial plan-form. Note that Eq. (16) represents a symmetric evolution and wave direction does not occur in the longshore diffusivity, G, which is the only parameter in which wave characteristics occur. (Actually, there is a very weak dependency on wave direction in G; however, this has been neglected.)

Using the Pelnard Considère formulation, the evolution of a beach nourishment project can be shown to be independent of storm sequencing. As an example, considering a large storm and a small storm, the resulting evolution due to both storms does not depend on which storm occurs first. This can be considered in greater detail by noting that in Eq. (16) or Eq. (17) the relevant parameter is equal to the product of the longshore diffusivity, G and time t. If the longshore diffusivity varies with time, usually dominantly due to varying wave heights, then the relevant quantity is $\int_0^t \mathrm{G(t)dt}$.

The effect of wave refraction around the nourished plan-form is to reduce the losses from a beach nourishment project. Intuitively it is evident that if a beach nourishment project is placed on a long, straight beach and the waves break landward of the perturbed contours then the contours will cause some refraction around the perturbation such that the breaking wave obliquity is reduced. The result is a smaller transport out of the placed plan-form. To a first approximation, this result can be represented simply by multiplying the longshore diffusivity by the ratio of C_b/C_* where C_b represents the wave celerity at breaking and C_* represents the wave celerity at the toe of the equilibrated nourished profile. In some cases this effect can be substantial, increasing the predicted longevity of a project by 20% to 40%.

(c) Representation of Background Erosion

In almost all cases where implementation of beach nourishment is being considered, a background erosion rate exists. This background erosion rate can be relatively uniform within the longshore segment of interest or can vary substantially in the longshore direction. In the interest of realistic representation of the natural conditions, background erosion must be represented in beach nourishment design and prediction methodologies. Dean and Yoo (1992) have recommended that the background erosion be included in the prediction process simply by considering the background erosion to be associated with a gradient in longshore sediment transport and adding this background transport to the transport associated with the nourishment project in the following form

$$Q_T(x) = Q_B(x) + Q_P(x), \tag{20}$$

where Q_T, Q_B and Q_P represent the total, background and "project" transport, respectively.

If there are no structures within the shoreline segment considered, it can be shown that it is only the transport gradients $\partial Q_B/\partial x$ that are important and not the background transport itself.

However, if internal or terminal structures are present within the project area, it is necessary to quantify the background transport, Q_B, which requires a more complete knowledge of the system. The background transport $Q_B(x)$ can be obtained from the pre-project (background) shoreline change rates, $\partial y_B/\partial t$ as

$$Q_B(x) = Q_B(x_0) - \frac{1}{(h_* + B)} \int_{x_0}^{x} \frac{\partial y_B}{\partial t} dx, \tag{21}$$

where $Q_B\ (x_0)$ is the background transport rate at some reference location, x_0. As noted, for a long straight unstructured beach, the value of $Q_B\ (x_0)$ is arbitrary.

A second approach to including the effects of the background transport is to attempt to calibrate a numerical model such that the predicted shoreline changes over some previous period are consistent with the historical erosion rates and distributions. Unless the background erosion is due to a recent large-scale perturbation introduced into the system such as a jetty or groin, in most cases, the causes of erosion are so subtle and possibly variable that it is not possible to calibrate a numerical model such that the background erosion is correctly related to the cause.

Figure 9.20 presents isolines of the proportion of sand remaining in the nourishment area, $M(t)$ versus $\sqrt{Gt}/\ell$ and non-dimensional uniform background erosion

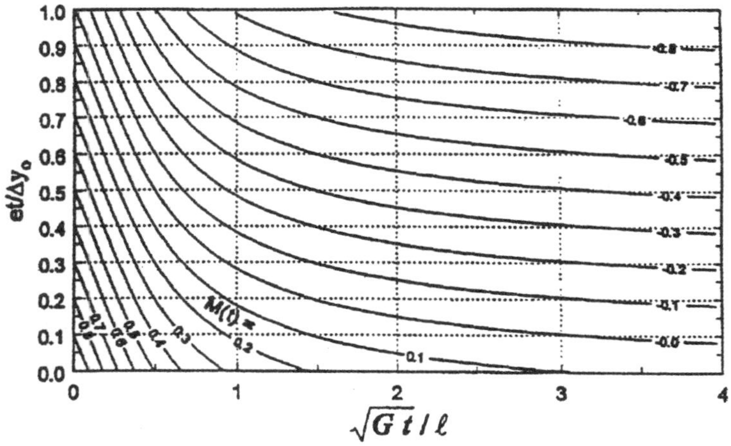

Fig. 9.20 Isolines of $M(t)$ versus spreading and background erosion parameters, e = erosion rate, Δy_0 = initial equilibrium shoreline displacement, G = longshore diffusivity and t = time. Long straight beach (dean, 1997).

rates, $et/\Delta y_0$, where e is the uniform background erosion rate. For $et/\Delta y_0 = 0$ (along the abscissa), the values of $M(t)$ are the same as plotted in Fig. 9.15. It is noted that for non-dimensional erosion rates greater than 0.2 the portion of material remaining in the project area, $M(t)$, becomes negative after a non-dimensional time (horizontal axis) of approximately 1.5.

(d) Nourishment Plan-form Evolution in Seawalled Areas

This generic problem represents conditions in which a nourishment project is placed in an area where there is no other sediment available to be transported. This might be approximated in a case where a seawall has been placed and there are rocky reefs seaward of the seawall. This problem has been evaluated analytically, numerically and in the laboratory by Dean and Yoo (1994) and by Yoo (1993). It was found that the wave direction for this problem is quite important. Since for oblique waves, there is no "feed" of sediment from the updrift direction, the beach nourishment is cannibalized on the updrift end of the nourishment and transported to the downdrift direction where it is deposited, thus resulting in a downdrift migration of the planform centroid. For normal wave incidence it can be shown from idealized considerations that the project simply spreads out at the same rate as if it were located on a long, straight beach of compatible sand. For an oblique wave approach the centroid of the project migrates in the downdrift direction as has been mentioned earlier and the project becomes longer and the rate of centroid migration increases with time. See Dean and Yoo (1994) for additional details.

(e) Erosional Hot and Cold Spots

Erosional hot spots (EHS) are defined as areas which erode more rapidly than anticipated and/or more rapidly than the adjacent shoreline segments. EHS can occur on natural or nourished beaches and they can be stationary or migrate. EHS on nourished beaches are of concern as they are usually interpreted as indicative of a generally poor project performance and may require equipment mobilization and renourishment earlier than would have been required otherwise. It is not unusual for an area of accretion to occur adjacent to an EHS. Such locations may be referred to as "erosional cold spots" (ECS).

Bridges (1995) has identified eight possible causes of EHSs. Some of these causes are related to the manner in which the nourishment was carried out and thus can be remedied whereas the effects of other causes are unavoidable. However, the anticipation of the EHS may allow minimization of their consequences, possibly by the placement of additional sand during nourishment.

The causes of EHS in beach nourishment that can be avoided include those due to wave transformation over bathymetric anomalies associated with the project. Combe and Soileau (1987) have documented two salients (ECS) that occurred landward of two borrow areas off Grand Isle, LA, see Fig. 9.21. Adjacent to these ECS were narrow beaches, considered to be EHS. One possible interpretation of the effects in this area, advanced by Gravens and Rosati (1994), is that wave refraction over the borrow areas and adjacent bathymetry resulted in low waves landward of the borrow areas with a concentration of energy adjacent to the areas of low energy. The resulting currents due to gradients in wave set-up, from the regions of high to low waves resulted in sediment transport to and deposition in the salient locations. A second explanation is that the offshore borrow areas reflect and dissipate wave energy thus acting much like detached breakwaters causing the waves on either side to diffract with the curved crests causing transport into the lee of the dredged holes. Horikawa *et al.* (1977) have conducted computer simulations and laboratory studies of the influence of dredged holes on the shoreline. The shoreline was found to

Fig. 9.21 Erosional "hot" and "cold" spots on grand isle, Louisiana (Combe and Soileau, 1987).

advance in the lee of a dredged hole. The interpretation was that "quiet water" was created in the lee of the dredged hole resulting in sand accumulation which is consistent with both of the mechanisms described above. The effects of offshore holes on the wave field have been investigated by McDougal *et al.* (1996) and Bender and Dean (2002, 2003a, 2003b, 2004a, 2004b, 2005, 2006) and were shown to reflect and diffract wave energy.

The previous example has considered the effects of the borrow area on the equilibrium plan-form. The configuration of the placed material can also affect the equilibrium shoreline. Consider the case in which sand placement occurs in an irregular manner to depths greater than those rapidly configured by the waves. Dean and Yoo (1993) have shown that the orientation of the shoreline, $\Delta\beta_2$, is related to that of the seaward nourished contour $\Delta\beta_1$ by

$$\Delta\beta_2 = \Delta\beta_1 \left[1 - \left(\frac{C_2}{C_1}\right)\right] \tag{22}$$

where C_1 and C_2 are the wave celerities at the outer depths of the nourishment and at the closure depth, respectively. $\Delta\beta_1$ and $\Delta\beta_2$ are the angular deviations of the offshore contours and shoreline, respectively, from uniform alignment. Dean and Yoo refer to this as "residual bathymetry" effect.

A second and probably more frequent cause of EHS is due to uneven alignment of the prenourished shoreline due to hardened shoreline segments. As an example, consider an eroding shoreline along which a valuable building is located. A shore parallel seawall of 60 m length is constructed to protect the building; however, the adjacent shoreline continues to erode to a position which is 20 m landward of the seawall. At this time, there is additional sufficient economic development of the upland that nourishment is justified and it is decided to add a uniform beach width of 40 m along the entire coast. The 40 m advancement adjacent to the seawalled coast represents a 20 m projection relative to the adjacent nourished shoreline. This projection of 20 m in the cross-shore direction by 60 m in the longshore direction is equivalent to a beach nourishment project of the same dimensions. Previous discussion has shown the very substantial effect of nourishment length on project longevity. The beach fronting the seawall would erode rapidly from a 40 m width to a 20 m width, to establish uniform alignment with the adjacent beaches. The sand eroded adjacent to the seawall would spread out to the adjacent beaches with the interpreted result of an EHS fronting the building and an ECS, on each side of the EHS. In such cases, a groin field or detached breakwater could be considered for stabilization of the protruding shoreline.

Other causes of EHSs may include: (1) not providing sufficient fill adjacent to seawalls at which there is initially no beach present. A "threshold" amount of sand must be added to these profiles in order to fill the profile up to the mean water line and application of commonly used rules of thumb will underestimate the volume required to achieve a desired equilibrated beach width (Fig. 9.10), (2) profiles oversteepened by a groin field which, through trapping sand from the longshore sediment transport, has maintained the shoreline more or less fixed; however, the seaward portion of the profile has continued to erode. As in (1) above, if the deepened offshore portion of the profile is not accounted for in the nourishment volumes, the equilibrated profile will be narrower than suggested by the usual rules of thumb, and (3) the use of sand that is finer than the native. As described earlier in the chapter, sand size is extremely significant to the width of the equilibrated profile. If a borrow area includes areas of sand which are finer than the average and if the contractor is not constrained to those portions where the coarser sand is located, since the costs of pumping fine sand are less than coarse sand, it is to the contractor's advantage to excavate the finer sand.

(f) Effects of Sediment Size on Dry Beach Area Evolution

The additional plan-form area resulting from a beach nourishment project is the dominant measure applied by the layperson in evaluating and judging its success and thus is an important performance measure. The preceding section has considered that the beach is nourished with compatible sands; however, in the earlier discussions of EBPs, we have seen that the equilibrium width of the nourished profile depends substantially on the size of the nourishment sediment relative to that of the native profile. Thus it can be appreciated that if the same volume of nourishment material is used, the initial equilibrated plan-form area will be substantially larger for a coarser sand than for a smaller nourishment sand size. However, a surprising result is that if the nourishment sand is coarser than the native sand, the equilibrated dry beach total plan area increases with time whereas if the nourishment sand is smaller than the native sand, the equilibrated dry beach plan-form decreases with time. This can be shown readily by referring to Fig. 9.10 which presents the relationship between the additional dry beach width versus volume per unit length of beach for three different sand sizes. It is seen that for the coarser sand, the ratio of dry beach width to volume added is greatest for the smaller volume densities. Thus, as the beach nourishment project spreads out on a long beach there will be smaller and smaller values of volume density resulting in the ratio of additional dry beach width to volume density increasing with time. In the limit for infinite time it can be shown that for the case of nourishment sand characterized by a single size which is coarser than the native, the dry beach width, Δy is related to the volume density, V, by $\Delta y(x) = V(x)/B$. Thus, the total plan area, at infinite time, $\mathrm{PA}(t = \infty)$ is

$$\mathrm{PA}(\mathrm{t} = \infty) = \int_{-\infty}^{\infty} \frac{\mathrm{V(x)}}{\mathrm{B}} \mathrm{dx} = \frac{\mathrm{V_{TOT}}}{\mathrm{B}} \tag{23}$$

where V_{TOT} is the total volume added. Following the same argument, the total plan area for a nourished beach composed of sand which is compatible with (of the same size) as the native will result in a plan area that does not change with time and is given by (see Eq. (6))

$$\mathrm{PA(t)} = \int_{-\infty}^{\infty} \frac{\mathrm{V(x)}}{(\mathrm{h}_* + \mathrm{B})} \mathrm{dx} = \frac{\mathrm{V_{TOT}}}{(\mathrm{h}_* + \mathrm{B})}. \tag{24}$$

Considering now the situation where a beach is nourished with sand which is finer than the native and referring again to Fig. 9.10 it can be seen that for the smaller

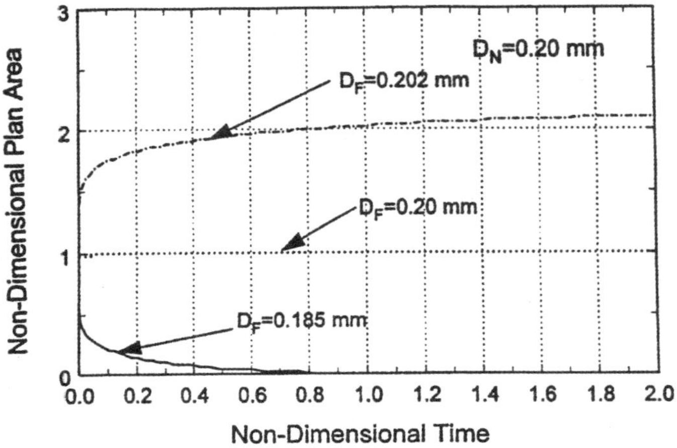

Fig. 9.22 Non-dimensional plan area versus non-dimensional time for various sizes of fill material. Single nourishment size sand (Dean, 1998).

volume densities the additional dry beach width approaches zero at the threshold volume. Thus, for this case the total plan area would evolve to a limit of zero (no additional dry beach area) at a time when the greatest alongshore volume density equals the threshold volume density. These two limits for nourishment sands that are finer and coarser than the native are somewhat unrealistic since they predict limiting plan areas which are independent of the amount by which the nourishment and native sand sizes differ. The only condition required for these limits to occur is that the sizes should be coarser or finer than the native. Figure 9.22 presents an example illustrating these results in non-dimensional form for three different sand sizes.

A more realistic representation is when the nourishment sands are considered to be composed of two sizes as shown in Fig. 9.23. In this case, a recognition of the sorting of the nourishment sediments is considered by representing the nourishment sediments as composed of two fractions of different proportions. One of the fractions must be of the same grain size as the native and the second can be either coarser or finer than the native. It is assumed, consistent with findings in nature, that the coarser of the fractions is located in the nearshore region. Characterization of the sand size distribution into these two sizes requires a procedure which will not be discussed here but is presented in Dean (2002). The profiles for each of the fractions follow Eq. (4). The results of this representation for the nourishment sediments are more realistic since now the limiting plan area ($t \to \infty$) is dependent on the proportion and size of the fraction of sand which differs from the native. In this case, the evolution of the beach nourishment plan-form can be determined using a somewhat more complicated numerical model. Several examples will be presented for illustration. In Fig. 9.23, since the nourishment material is generally coarser than the native, the sands of the same size as the native are located in the seaward

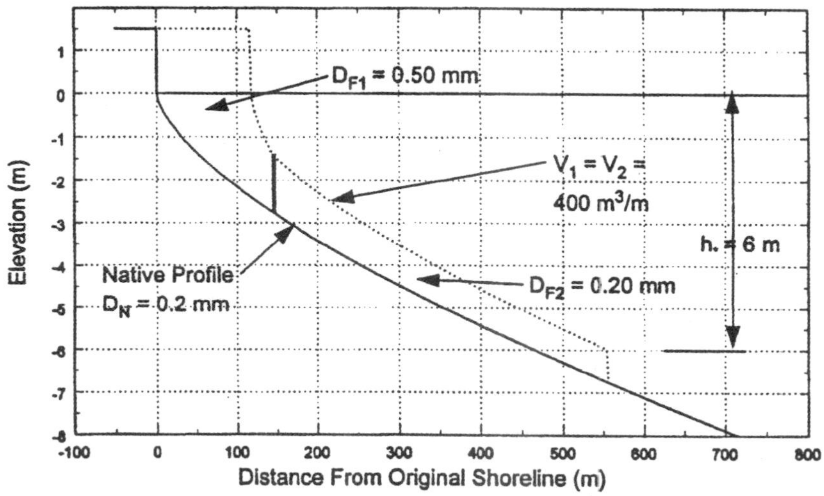

Fig. 9.23 Example of $V_1 = V_2$ with $D_N = 0.20\,\text{mm}$, $D_{F1} = 0.50\,\text{mm}$, $D_{F2} = 0.20\,\text{mm}$, $h_* = 6\,\text{m}$, $B = 1.5\,\text{m}$ (Dean, 1998).

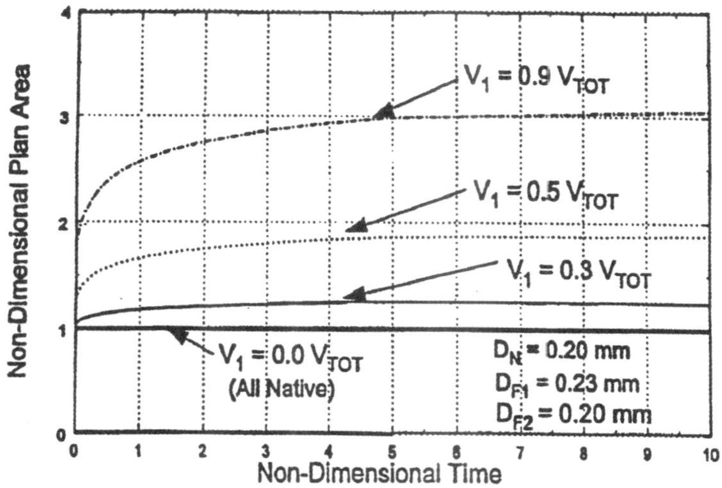

Fig. 9.24 Example calculation for nourishment sand coarser than the native. Effect of proportions of coarser sand. Two nourishment sizes considered (Dean, 1998).

portion of the profile. Figure 9.24 shows the results for a case in which the native sand is 0.20 mm and nourishment sands are generally coarser than the native such that the finer size component is the same size as the native (0.20 mm) and the second component is coarser, in this case 0.23 mm. The proportion of the coarser (0.23 mm) sand ranges from 0.0 to 0.9 of the total volume placed. Figure 9.25 illustrates the plan-form evolution for cases in which the nourishment sands are generally coarser than the native with equal proportions of nourishment sand equal to and coarser than the native. In this example the size of the coarser fraction is varied. Again, the sensitivity of both initial and asymptotic plan areas on sand size are substantial. Figures 26 and 27 are the counterparts of Figs. 24 and 25, respectively, for the case in which the nourishment sand is generally finer than the native.

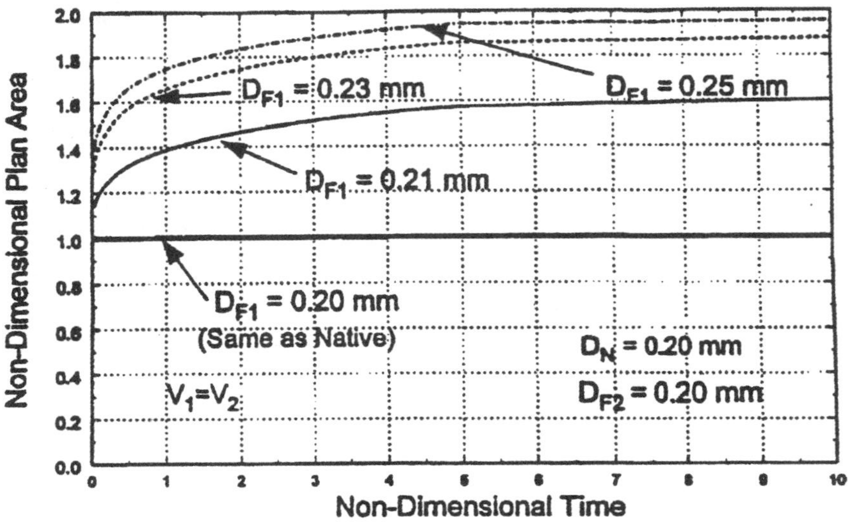

Fig. 9.25 Example calculation for nourishment sand coarser than the native. Effect of sand size. Two nourishment sizes considered (Dean, 1998).

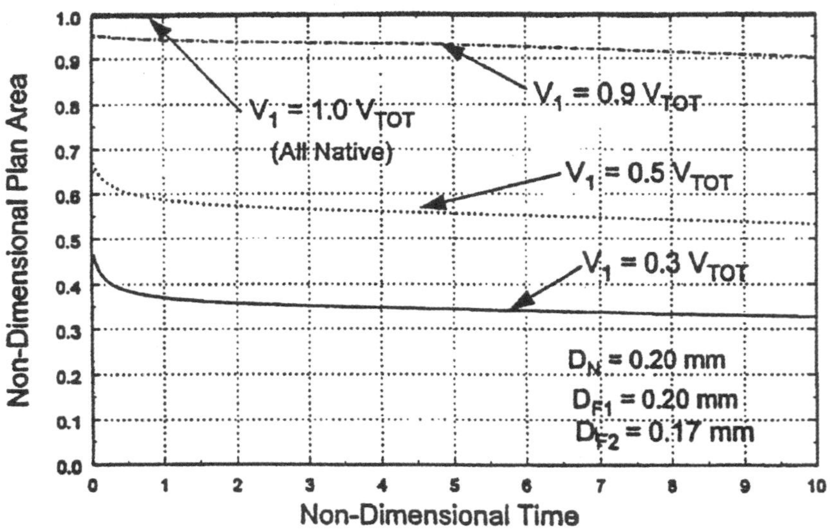

Fig. 9.26 Example calculation for nourishment sand finer than the native. Effect of proportions of finer sand. Two nourishment sizes considered (Dean, 1998).

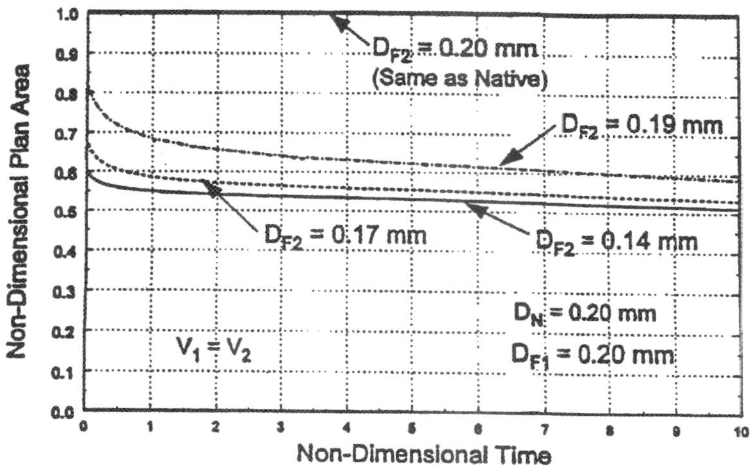

Fig. 9.27 Example calculation for nourishment sand finer than the native. Effect of sand size. Two nourishment sizes considered (Dean, 1998).

5. Detailed Methodology for Representing Project Evolution

The preceding sections have reviewed simplified methods for predicting the cross-shore and plan-form characteristics of beach nourishment projects. These methods provide a good basis for understanding the general characteristics of project evolution, an approximate estimate of the time-scales and an appropriate framework for preliminary design. However, these simple methods are limited in the detail of forcing that can be represented and in the boundary and internal boundary conditions that can be simulated. The present section describes numerical models which allow greater flexibility and realism to be incorporated into the predictions. The best basis for developing an overall understanding of a particular design is through a combination of simple and more detailed models.

5.1 *Models of profile evolution*

Cross-shore sediment transport processes are so complex that practically all models of profile evolution require numerical solution. Two types of profile evolution exist which can be termed as "open loop" and "closed loop." All such models employ a transport equation and the conservation of sand equation.

(1) Open Loop Profile Evolution Models

Open loop profile evolution models require detailed or idealized representation of the hydrodynamics and sediment transport processes. As an example, the suspended sediment distribution and mean velocity may be a component of, or the dominant transport mechanism. Some models include bed load transport as a principal component. Most, if no all, of these open loop models if allowed to run indefinitely under the action of steady forcing will become unstable unless some artificial "smoothing" of the results is introduced. Examples of open loop models include: Bailard and Inman (1981), Dally and Dean (1984), Watanabe (1988), Southgate and Nairn (1993) and Nairn and Southgate (1993). Roelvink and Hedegaard (1993) have presented a review of several open-loop models.

(2) Closed Loop Profile Evolution Models

Closed loop profile models are characterized by the specification of a particular "target profile," to which the computed profile will converge if the forcing mechanisms are held constant indefinitely, thus, ensuring computational stability. Equation (4) has been used as the target profile in at least three of these models. Since Eq. (4) is consistent with uniform wave energy dissipation per unit water volume, D_*, within the surf zone, (where the subscript "$_*$" denotes the value for equilibrium conditions), the seaward transport, q, is related to the actual (non-equilibrium) wave energy dissipation per unit volume by

$$\mathrm{q} = \mathrm{K}'(\mathrm{D} - \mathrm{D}_*)^{\mathrm{m}}, \tag{25}$$

where K' and m are constants and in two of the closed loop models, $m = 1$. Examples of closed loop profiles include: EDUNE (Kriebel and Dean, 1985), and SBEACH (Larson and Kraus, 1989). In EDUNE, the cross-shore positions of the contours are specified and it is thus not efficient to represent non-monotonic profiles. SBEACH represents the offshore distance as a series of grids which readily allows representation of offshore bars. The closed loop models are much more effective and reliable for the beach and dune erosion phase than for the recovery phase which occurs on a longer time-scale and can be characterized by the sediment moving onshore in "pulses". Closed loop models are considered much more suitable for engineering applications as compared to the open loop models which are more in the research and development arena.

More complex open-loop profile models include those of Kobayashi *et al.* (2010) and Delft 3-D http://oss.deltares.nl/web/opendelft3d, which is now an open source.

5.2 *Models of plan-form evolution*

The simplest model for plan-form evolution is the "one-line" model, so-called because the cross-shore position $y(x, t)$ of one contour line is represented as in the Pelnard Considère methodology discussed previously. A basic assumption in the one-line model is that the profile is displaced landward or seaward in response to gradients in the longshore transport, Q as represented by Eq. (26). All one-line models incorporate a transport equation and an equation representing the conservation of sand. The solution to these equations may be explicit or implicit. One equation for sediment transport is the so-called CERC equation

$$\mathrm{Q} = \mathrm{K}\frac{(\mathrm{EC_G})_\mathrm{b}}{2\rho\,\mathrm{g}(\mathrm{s}-1)(1-\mathrm{p})}\sin 2(\beta - \alpha_\mathrm{b}) \tag{26}$$

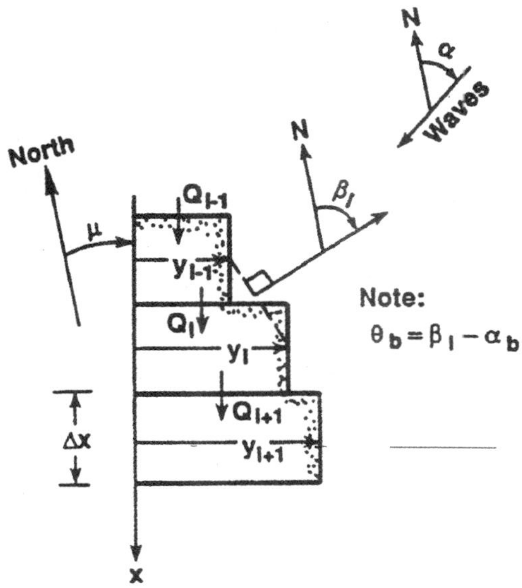

Fig. 9.28 Definition sketch for one-line numerical model.

in which E is the wave energy density, C_G is the group velocity, β and are defined in Fig. 9.28 and the subscript "b" denotes that the subscripted variables are to be evaluated at wave breaking. The continuity equation is

$$(\mathrm{h}_* + \mathrm{B})\frac{\partial \mathrm{y}}{\partial \mathrm{t}} + \frac{\partial \mathrm{Q}}{\partial \mathrm{x}} = 0 \tag{27}$$

written in finite difference form as

$$\mathrm{y}_\mathrm{i}^{\mathrm{n}+1} = \mathrm{y}_\mathrm{i}^\mathrm{n} - \frac{1}{(\mathrm{h}_* + \beta)}\frac{\Delta t}{\Delta x}\left(\mathrm{Q}_{\mathrm{i}+1}^\mathrm{n} - \mathrm{Q}_\mathrm{i}^\mathrm{n}\right) \tag{28}$$

in which the superscript represents the time level and the subscript the cell for which the computations are being carried out. Note that the "i" index on the transport represents the grid cell onto which positive transport occurs. The grid system on which these equations are solved is shown in Fig. 9.28, where it is seen that the Q values are defined on the grid lines and the y values at the center of the cells. The explicit method solves the transport and conservation equations sequentially whereas the implicit method solves these two equations simultaneously. A second difference is that there is a limiting time step, Δt, associated with the explicit method of solution

$$(\Delta t)_{\max} < \frac{\Delta x^2}{2\mathrm{G}}, \tag{29}$$

where G is defined by Eq. (14) or Eq. (15). The solution of the explicit method is described first. In the explicit method, the shoreline displacements are held fixed and the transport values, $\mathrm{Q}_\mathrm{i}^{(\mathrm{n})}$, computed. For this purpose the β_i values are

$$\beta_\mathrm{i}^\mathrm{n} = \tan^{-1}\left(\frac{\mathrm{y}_\mathrm{i}^\mathrm{n} - \mathrm{y}_{\mathrm{i}-1}^\mathrm{n}}{\Delta x}\right) + \mu \tag{30}$$

With the new values of $Q_i^{(n)}$, the y_i values are then updated from time level n to $(n+1)$ by Eq. (28). The process is then repeated until the desired time is reached. One useful characteristic of the explicit method is that it is relatively straightforward to incorporate boundary and internal conditions on shoreline displacements or discharges. An example would be a terminal structure at one or both ends of a computational region which blocks the sediment transport completely and this would simply require that the transport be assigned the value of zero. A second boundary condition could be at a groin at the downdrift end of a computational domain at which the shoreline is at the groin terminus. In this case, the shoreline displacement at this point would be specified. Structures such as groins can alternately require representation as a boundary condition on y or Q depending on wave direction and shoreline position relative to the structure. Also, since a limitation of the one-line model is the previously noted displacement of the profile without change of form, it may be appropriate to prescribe a gradual transition between full by-passing and no by-passing depending on the shoreline position relative to the groin terminus. Such a scheme would better mimic nature. GENESIS is an example of a one-line model that incorporates substantial flexibility in terms of boundary and internal conditions (Hanson, 1989; Hanson and Kraus, 1989). Implicit models of shoreline change are usually represented as a tri-diagonal matrix which can be solved by relatively simple numerical procedures.

5.3 *Combined (three-dimensional) models*

Combined models include consideration of both the cross-shore and longshore sediment transport components simultaneously and incorporate any interaction of these two components. This interaction can be especially important in the vicinity of structures whereby the profile steepens and flattens on the updrift and downdrift sides of the structure, respectively, inducing seaward and landward cross-shore transport. As discussed for the case of cross-shore sediment transport, the cross-shore and longshore transport components of the model can be at the engineering or research level. The engineering level would incorporate a model of the CERC type (Eq. (26)) for longshore sediment transport along with a distribution function that spreads out the total transport across the active zone. Examples of this cross-shore distribution function have been presented by Komar (1977) and a comprehensive review has been published by Bodge (1989). For an engineering level model, the cross-shore element would usually be of the "target" type such that the profile would converge to a prescribed form. Solution of the engineering type models can be by an explicit procedure in which the cross-shore and longshore transport components are solved for sequentially by an implicit procedure in which the two components are determined simultaneously, or a combination implicit/explicit procedure, which may be an improvement over the other two methods. An example of the combined engineering models is that of Bakker (1968) which is a "two-line" model, that is, two depth contours are monitored in the process. Perlin and Dean (1985) presented an "n-line" model in which the solution was fully implicit and an arbitrary number (n) of contour lines was represented. Dean (1996) presented the results of applying an n-line model that was solved by a combined implicit–explicit method. A disadvantage of monitoring contours in which the shore-normal coordinates to several contours are represented is that it is very difficult to simulate the presence of bars which cause the profile to be non-monotonic such that there are several distances from the shoreline to the same depth contour. For situations in which bars are present, a grid system is much more effective in which the depth is the dependent variable and x and y are the independent variables.

The research level models are generally of the type that calculate the suspended sediment transport based on calculated fields of hydrodynamics and suspended sediment. The contribution of bedload is calculated separately. The cross-shore and longshore sediment transports are thus components of a transport vector. It should be noted that several of these models are used for engineering problems and some are only available commercially and require purchase and/or a licensing agreement.

6. Summary

A review has been presented, of the available methodologies, for the design and prediction performance of beach nourishment projects. Both "simple" and "detailed" methodologies have been examined. The simple methodologies consider the cross-shore and longshore processes separately, represent the forcing (waves and tides) as constant or by representative values and are useful for preliminary design and for demonstrating the interrelationships between the various parameters. The detailed methodologies employ numerical models and are available at various levels of complexity and detail and offer advantages of time varying input of the forcing (waves and tides) and greater flexibility with boundary conditions and internal conditions such as a groinfield. For the simple treatment of the cross-shore processes, the focus is on the equilibrium beach width resulting from placement of a specified amount of sand per unit beach length. Using EBP methodology, it is shown that three generic types of nourished profiles can result: intersecting, non-intersecting and submerged. The equilibrium beach width varies substantially with the size of the nourishment sediment relative to that of the native. The plan-form evolution of nourishment projects is investigated through the Pelnard Considère equation and the associated solutions. For the case of nourishment on a long straight beach, the longevity of a project varies with the square of the project length and inversely as the 2.5 power of the wave height. Other results and/or methodologies that are demonstrated at the simple level include: background erosion, the increase of renourishment intervals for a project on a long straight beach with renourishment number where the background erosion is small and where a minimum threshold of sediment volume is specified, a simple method to account for refraction around a nourished plan-form, effects of nourishing with sands coarser and finer than the native, plan-form migration of nourishment in an area which is otherwise devoid of sediment, and the invariance of total evolution to the sequencing of storms. Moreover, the total plan area of a nourishment project on a long straight beach is shown to be dependent on the nourishment sand size relative to the native sand size and to evolve to larger and smaller total plan areas for nourishment sands coarser and finer than the native, respectively.

A brief review of the general characteristics of detailed calculation methodology is presented. This methodology is carried out with numerical models which are developing rapidly. All models of this type can be solved by explicit, implicit or a mixed explicit–implicit approach. Results are best simulated for systems which are substantially out of equilibrium due to a large beach nourishment project, the imposition of a structure which interrupts a significant longshore sediment transport or a profile that is subjected to a storm surge. All models require one or more transport equations and a conservation equation. Two types of methodologies are available for profile evolution models involving cross-shore sediment transport. In this chapter, these models have been referred to as engineering models which specify a target profile to which the solution will converge if the conditions are maintained to be constant and a "research" methodology which is based on calculated wave and suspended sediment transport fields and bedload transport. This latter class is not yet considered to be appropriate for engineering applications. Profile response models represent the erosion mode of profile evolution much better than the recovery mode. The simplest plan-form evolution models are based on a gross transport model and a continuity equation that assumes that the profile responds to gradients in sediment transport by being displaced landward or seaward without change of form. More complex models are based on calculations of the hydrodynamics and suspended sediment fields and the bedload transport. Combined (three-dimensional) models compute the profile and plan-form evolution simultaneously and generally parallel to the characteristics, limitations and features of the individual models for profile and plan-form evolution.

Acknowledgments

The author is indebted to the Florida Sea Grant Program and the Bureau of Beaches and Coastal Systems of the Florida Department of Environmental Protection which have provided funding over the years for the work which he and his students have conducted on nearshore processes and related engineering applications.

References

Ahrens, J. P. and E. B. Hands (1998): Velocity parameters for predicting cross-shore sediment movement, *J. Waterway, Port, Coastal and Ocean Engrg. ASCE*, **124**(1), 16–20.

Andrassy, C. J. (1991): Monitoring of a disposal mound at silver strandstate park, *Proc. Coastal Sediments* '91, ASCE, 1970–1984.

Bailard, J. A. and D. L. Inman (1981): An energetics bed load model for a plane sloping beach: Local transport, *J. Geophys. Res.* **86**(C3), 2035–2043.

Bakker, W. T. (1968): The dynamics of a coast with a groyne system, *Proc. 11th Int. Conf. on Coastal Engrg.* 492–517.

Bender, C. J. and R. G. Dean (2002): Erosional hot spot prediction through wave analysis, *Proc. ASCE Conf. on Ocean Wave Measurement and Analysis*, 1306–1315.

Bender, C. J. and R. G. Dean (2003a): Wave field modification by bathymetric anomalies and resulting shoreline changes: A review with recent results, *Coastal Engrg.* **49**, 125–153.

Bender, C. J. and R. G. Dean (2003b): Wave transformation by two-dimensional bathymetric anomalies with sloped transitions, *Coastal Engrg.* **50**, 61–84.

Bender, C. J. and R. G. Dean (2004a): Potential shoreline changes induced by 3-dimensional bathymetric anomalies with gradual transitions in depth, *Coastal Engrg.* **51**, 1143–1161.

Bender, C. J. and R. G. Dean (2004b): Wave field modifications by elliptic bathymetry with gradual transitions in depth, *Proc.* 29th *Int. Conf. on Coastal Engrg.*, Copenhagen, Lisbon, Portugal, Summer, **2004**, 243–255.

Bender, C. J. and R. G. Dean (2005): Wave transformation by axisymmetric three-dimensional bathymetric anomalies with gradual transitions in depth, *Coastal Engrg.* **52**, 331–351.

Bender, C. J. and R. G. Dean (2006): Wave field modifications by bathymetric anomalies with gradual transitions in depth: A review, *Proc.* 30th *Int. Conf. on Coastal Engrg.*, Lisbon, Portugal Summer, 2006, 412–424.

Birkemeier, W. A. (1985): Field data on seaward limit of profile change, *J. Waterway, Port, Coastal and Ocean Engrg. ASCE*, **111**(3), 598–602.

Bodge, K. R. (1989): A literature review of the distribution of longshore sediment transport across the surf zone, *J. Coastal Res.* **5**(2), 307–328.

Bodge, K. R. (1994): Performance of nearshore berm disposal at Port Canaveral, Florida, Dredging '94, 1183–1191.

Bridges, M. H. (1995): Analysis of the Processes Creating Erosional Hot Spots in Beach Nourishment Projects, MSc. Thesis, Dept. Coastal and Oceanogr.Engrg., Univ. Florida, Gainesville, FL.

Bruun, P. (1954): Coast erosion and the development of beach profiles, US Army Corps of Engrs., Beach Erosion Board, Tech Memo No. 44.

Combe, A. J. and C. W. Soileau (1987): Behavior of man-made beach and dune, Grand Isle, Louisiana, *Proc. Coastal Sediments* '87, ASCE, 1232–1242.

Dally, W. R. and R. G. Dean (1984): Suspended sediment transport and beach profile evolution, *J. Waterway, Port, Coastal and Ocean Engrg.*, ASCE, **110**(1), 15–33.

Dalrymple, R. A. (1974): A finite amplitude wave or a linear shear current, *J. Geophys. Res.* **79**(30), 4498–4504.

Dean, R. G. (1965): Stream function representation of nonlinear ocean waves, *J. Geophys. Res.* **70**(18), 4561–4572.

Dean, R. G. (1977): Equilibrium beach profiles: U.S. Atlantic and Gulf Coasts, Dept. Civil Engrg., Ocean Engrg. Report No. 12, Univ. Delaware, January.

Dean, R. G. (1987): Coastal sediment processes: Toward engineering solutions, Coastal Sediments '87, Specialty Conf. on Advances in Understanding of Coastal Sediment Processes, New Orleans, LA, ASCE, 1–24.

Dean, R. G. (1991): Equilibrium beach profiles: Principles and applications, *J. Coastal Res.* **7**(1), p. 53–84.

Dean, R. G. (1995): Beach nourishment: Planform considerations, *Proc. Coastal Dynamics* '95, ASCE, 533–546.

Dean, R. G. (1996): Interaction of littoral barriers and adjacent beaches–Effects on profile shape and shoreline change, *J. Coastal Res.*, Special Issue, No. 23, 103–112.

Dean, R. G. (1997): Models for barrier island restoration, *J. Coastal Res.* **13**(3), 694–703.

Dean, R. G. (1998): Beach nourishment: A limited review and some recent results, *Proc. 26th Int. Conf. on Coastal Engrg.*

Dean, R. G. and C.-H. Yoo (1992): Beach nourishment performance predictions, *J. Waterway, Port, Coastal, and Ocean Engrg. ASCE*, **118**(6), 567–586.

Dean, R. G. and C.-H. Yoo (1993): Predictability of beach nourishment performance, Proceedings of the Coastal Zone Management, Special Volume on Beach Nourishment Engineering and Management Considerations, 86–102.

Dean, R. G. and C.-H. Yoo (1994): Beach nourishment in presence of seawall, *J. Waterway, Port, Coastal, and Ocean Engrg. ASCE*, **120**(3), 302–316.

Dean, R. G, R. Chen and A. E. Browder (1997): Full scale monitoring study of a submerged breakwater, Palm Beach, Florida, USA, Coastal Engrg., **29**, 291–315.

Dean, R. G. (2002): Beach Nourishment: Theory and Practice: World Scientific Press, 396 pp.

Delft3D http://oss.deltares.nl/web/opendelft3d . Last accessed on 5 October 2012.

delValle, R, R. Medina and M. A. Losada (1993): Dependence of coefficient K on grain size, Technical Note No. 3062, *J. Waterway, Port, Coastal, and Ocean Engrg.*, ASCE, **119**(5), 568–574.

Dette, H. H., A. Führböter, A. J. Raudkivi (1994): Interdependence of beach fill volumes and repetition intervals, *J. Waterway, Port, Coastal, and Ocean Engrg. ASCE*, **120**(6), 580–593.

Douglass, S. L. (1995): Estimating landward migration of nearshore, constructed sand mounds, *J. Waterway, Port, Coastal and Ocean Engrgs. ASCE*, **121**(5), 247–250.

Gravens, M. B. and J. D. Rosati (1994): Numerical model study of breakwaters at Grand Isle, Louisiana, Miscellaneous Paper CERC-94-16, U. S. Army Corps of Engrs., 75 pp.

Grosskopf, W. G. and D. K. Stauble (1992): Atlantic Coast of Maryland (Ocean City) Shoreline-protection plan, *Shore and Beach* **61**(l), 3–7.

Grosskopf, W. G. and D. H. Behnke (1992): An emergency remedial beach fill design for Ocean City, Maryland, *Shore and Beach* **61**(l), 8–12.

Hallermeier, R. J. (1977): Calculating a yearly limit depth to the active beach profile, Coastal Engineering Research Center, U. S. Army Corps of Engrs., Technical Paper No. 77-9.

Hallermeier, R. J. (1978): Uses for a calculated limit depth to beach erosion, *Proc. 16th Int. Conf. on Coastal Engrg.,* ASCE, Hamburg, 1493–1512.

Hallermeier, R. J. (1980): Sand motion initiation by water waves: Two asymptotes, *J. Waterway, Port, Coastal, and Ocean Engrg. ASCE*, **106**(3), 299–318.

Hallermeier, R. J. (1981): A profile zonation for seasonal sand beaches from wave climate, *Coastal Engrg.* **4**(3), 253–277.

Hands, E. B.(1991): Unprecedented migration of a submerged mound off Alabama Coast, Proceedings of the 12th Annual Conference of the Western Association and the 24th Annual Texas, A&M Dredging Seminar, Las Vegas, NV.

Hands, E. B. and M. C. Allison (1991): Mound migration in deeper waters and methods of categorizing active and stable depths, *Proc. Coastal Sediments* '91, ASCE, 1985–1999.

Hands, E. B. and S. R. DeLoach (1980): An offshore mound constructed of dredged material, *Proc. Dredging and Dredged Material Disposal*, Dredging '84, New York, NY, ASCE, **2**, 1030–1038.

Hands, E. B., S. R. DeLoach and R. Vann (1988): Post-construction adjustment of an offshore mound. *Abstracts, 21st Coastal Engrg. Conf.*, New York, NY, ASCE, 262–263.

Hands, E. G. and D. T. Resio (1994): Empirical guidance for siting berms to promote stability or nourishment benefits, *Proc. 2nd Int. Conf. Dredging and Dredged Material Placement*, Dredging '94, New York, NY, ASCE, 220–228.

Hanson, H. (1989): GENESIS–A generalized shoreline change numerical model. *J. Coastal Res.* **5**(2), 1–28.

Hanson, H. and N. C. Kraus (1989): GENESIS: Generalized model for simulating shoreline change, Report 1, Technical reference, Technical Report No. 89–19, Coastal Engineering Research Center, Waterways Experiment Station, Vicksburg, MS.

Hanson, H. and N. C. Kraus (1993): Optimization of beach fill transitions, In: Beach Nourishment Engineering and Management Considerations (D. K. Stauble and N. C. Kraus, eds.), Coastal Zone '93, ASCE, pp. 103–117.

Horikawa, K., T. Sasaki and H. Sakuramoto (1977): Mathematical and laboratory models of shoreline changes due to dredged holes, *J. Faculty of Engrg.*, Univ. Tokyo, **34**(5), 49–57.

Jensen, R. E. and A. Garcia (1992): Wind, wave and water level assessment for the 4 January 1992 Storm at Ocean City, Maryland, *Shore and Beach* **61**(l), 13–22.

Kobayashi, N., A. Farhadzadeh, J. A. Melby, B. Johnson and M. Gravens (2010): Waveovertopping of levees and overwash of dunes. *J. Coastal Res.* **26**(5), 888–900.

Komar, P. D. (1977): Beach sand transport: Distribution and total drift, *J. Waterway, Port, Coastal, and Ocean Division*, ASCE, **103**(WW 2), 225–239.

Komar, P. D. (1988): Environmental controls on littoral sand transport, *Proc. 21st Int. Conf. on Coastal Engrg.*, ASCE, 1238–1252.

Komar, P. D. and D. L. Inman (1970): Longshore sand transport on beaches, *J. Geophys. Res.* 5914–5927.

Kraus, N. C.(1993): The 4 January 1992 Storm at Ocean City, Maryland, *Shore and Beach* **61**(l), 2.

Kraus, N. C. and R. A. Wise (1992): Simulation of January 4, 1992 Storm Erosion City, Maryland, *Shore and Beach* **61**(l), 34–40.

Kriebel, D. L.(1982): Beach and Dune Response to Hurricanes, MSc. Thesis, Univ. Delaware, 349 pp.

Kriebel, D. L. and R. G. Dean (1985): Numerical simulation of time-dependent beach and dune erosion, *Coastal Engrg.* **9**(3), 221–245.

Larson, M., H. Hanson and N. C. Kraus (1997): Analytical solutions of one-line model for shoreline change near coastal structures, *J. Waterway, Port, Coastal and Ocean Engrg. ASCE*, **123**(4), 180–191.

Larson, M. and N. C. Kraus (1989): SBEACH: Numerical model for simulating storm-induced beach change, Report 1 : Empirical foundation and model development. U. S. Army Coastal Engrg. Research Center, Waterways Experiment Station, Technical Report CERC-89-9.

Laustrup, C., H. T. Madsen, P. Sorensen and I. Broker (1996): Comparison of beach and shoreface nourishment, Torsminde, Tange, Denmark, *Proc. 25th Int. Conf. on Coastal Engrg.*, ASCE, Orlando, FL, 2927–2940.

Leidersdorf, C. B., R. C. Hollar and G. Woodell (1993): Beach enhancement through nourishment and compartmentalization: The recent history of Santa Monica Bay, in: Beach Nourishment Engineering and Management Considerations (D. K. Stauble and N. C. Kraus, eds.), ASCE, pp. 71–85.

McDougal, W. G., A. N. Williams and K. Furukawa (1996): Multiple pit breakwaters, *J. Waterway, Port, Coastal and Ocean Engrg. ASCE*, **122**, 27–33.

McLellan, T. N. and N. C. Kraus (1991): Design guidance for nearshore berm construction, *Proc. Coastal Sediment* '91, ASCE, 2000–2011.

Moore, B. D.(1982): Beach Profile Evolution in Response to Changes in Water Level and Wave Height, MCE Thesis, Dept. Civil Engrg., Univ. Delaware, 164 pp.

Nairn, R. B. and H. N. Southgate (1993): Deterministic profile modelling of nearshore processes. Part 2. Sediment transport and beach profile development, *Coastal Engrg.* **19**(1/2), 57–96.

National Research Council (1995): Beach Nourishment and Protection, National Academy Press, Washington, DC.

Nichols, R. J., W. A. Birkemeier and R. J. Hallermeier (1996): Application of the depth of closure concept, *Proc. 25th Int. Conf. on Coastal Engrg.*, ASCE, 3874–3878.

Otay, E. N.(1994): Long-term Evolution of Nearshore Disposal Berms, PhD. Dissertation, Dept. of Coastal and Oceanogr. Engrg., Univ. Florida, Gainesville, FL.

Pelnard-Considère, R. (1956): Essai de theorie de involution des formes de rivate en plages de sable et de galets, 4th Journees de *l*'Hydraulique, Les Energies de la Mar, Question III, Rapport No. 1 (in French).

Perlin, M. and R. G. Dean (1985): 3-D model of bathymetric response to structures, *J. Waterway, Port, Coastal, and Ocean Engrg. ASCE*, **111**(2), March, 153–170.

Rijkswaterstaat (1987): Manual on Artificial Beach Nourishment, Centre for Civil Engineering Research, Codes and Specifications.

Roelvink, J. A. and I. B. Hedegaard (1993): Cross-shore profile models, *Coastal Engrg.*, Special Issue No. 21, Part 1/3, 163–192.

Southgate, H. N. and R. B. Nairn (1993): Deterministic modelling of nearshore processes, Part 1. Waves and currents, *Coastal Engrg.* **19**(1), 2, 27–56.

Stauble, D. K. and W. G. Grosskopf (1992): Monitoring project response to storms: Ocean City, Maryland Beachfill, *Shore and Beach* **61**(l), 23–33.

U. S. Army Corps of Engrs. (1982): Beach Fill Transitions, Coastal Engineering Technical Note CETN-II-6, U. S. Army Waterways Experiment Station, Coastal Engineering Research Center.

Walton, T. L.(1994): Shoreline solution for tapered beach fill, *J. Waterway, Port, Coastal, and Ocean Engrg.*, Technical Note, ASCE, **120**(6), 651–655.

Watanabe, A. (1988): Modeling of sediment transport and beach evolution, in: Nearshore Dynamics and Coastal Processes; Theory, Measurement and Predictive Models (K. Horikawa, ed.), Univ. Tokyo Press, pp. 292–302.

Wiegel, R. L. (1992): Dade County, Florida, Beach nourishment and hurricane surge protection, *Shore and Beach* **60**(4), 2–28.

Yoo, C. H. (1993): Realistic Performance of Beach Nourishment, PhD dissertation, Dept. Coastal and Oceanogr. Engrg., Univ. Florida, Gainesville, FL, 150 pp.

Chapter 10

Dredging of Navigational Channels

John B. Herbich, Ph.D., P.E.

W. H. Baver, Professor Emeritus
Ocean Engineering Program, Department of Civil Engineering
Texas A & M University, College Station, TX USA

1. Need for Navigational Channels

1.1 *Introduction*

Over 4000 years ago, the Phoenicians established a port on the open sea coast at Tyre. The Romans built a naval port on the Tiber River in Ostia. Waterborne traffic developed in both ancient Greece and the Roman Empire. In recent decades, many shipwrecks have been discovered in the Mediterranean Sea which confirms the existence of ship traffic carrying goods between ports. Many ports were built in relatively sheltered areas on rivers, in estuaries and bays, examplified by the harbors of New York, Boston, New Orleans, Houston, London, Rotterdam and Hamburg. Navigation channels were dredged to connect the actual port with the open sea. Other ports have been constructed along the coast and are protected by breakwaters extending into the ocean from the coast.

The growing demand for deep water ports in more exposed sea areas has prompted the installation of larger breakwaters with a breakwater head located in depths up to about 30 m (98 feet) and with the main part of the breakwater located in depths exceeding 20 m (66 feet). The total height of such breakwaters may be about 40 m (131 feet) with a crown of 10 m (33 feet) and side slopes of 1 on 1.5. The cost may be \$50,000 to \$150,000 per meter length, depending on location (Bruun, 1992).

PIANC (1985) cautions that a number of deep-water breakwater accidents occurred in the past — "a breakwater constructed in water with no depth limitations to the design wave load poses greater risks".

In the case of breakwaters constructed along the coast, navigational channels need not be dredged unless the underwater slope is very flat such as in the Gulf of Mexico. An example of such a harbor is located in Freeport, Texas.

1.2 *Economics of depth*

The development of very large crude carriers (VLCC) and ultra large crude carriers (ULCC) forced many government planners and port managers to evaluate existing navigational channels and port facilities with a purpose of determining what improvements, deepening and/or widening must be done to accommodate larger vessels. The shortfall in port and harbor facilities capable of handling these vessels, coupled with growing volume of imports and exports, has resulted in a critical need for improved port facilities in the United States (Berriman and Herbich, 1977). Several studies conducted by the Marine Board, National Research Council (Marine Board 1981, 1983) pointed out that the US should act expeditiously to increase its capacity to handle large ships. Improvement of port facilities has been vigorously pursued by many countries, notably the Netherlands (Europort-Rotterdam), France (Fox-Marseille), Belgium (Zeebrugge), Taiwan (Keelung, Taichung), South Africa (Richards Bay, Saldanha) as surveyed by Herbich (1983).

Large ships offer lower cost transportation with the resulting economic advantage to the consumer. Transportation costs per ton of cargo for a tanker and a bulk carrier are shown in Fig. 10.1 and the costs per container space in Fig. 10.2. The cost savings are significant up to vessels of 50,000 deadweight tons (DWT). Increases in the size of dry bulk carriers have been less dramatic as compared to oil tankers. In 1982, 75% of the worlds' oil was carried in tankers 100,000 DWT and larger (Cargo Systems, 1982). Bulk carriers, 100,000 DWT or more, increased their share of ocean borne commerce 29% between 1971 and 1980 (Drewry, 1982). Figures 10.3–10.5 show vessel characteristics of bulk carriers, tankers and general cargo vessels (US Army Corps of Engineers, 1983).

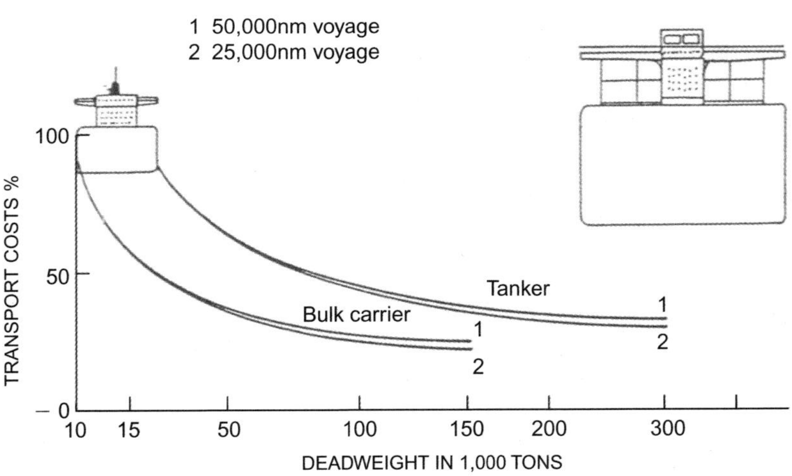

Fig. 10.1 Relationship of tanker and bulk carrier vessel size to transport costs per ton of cargo (Schonknecht *et al.*, 1983).

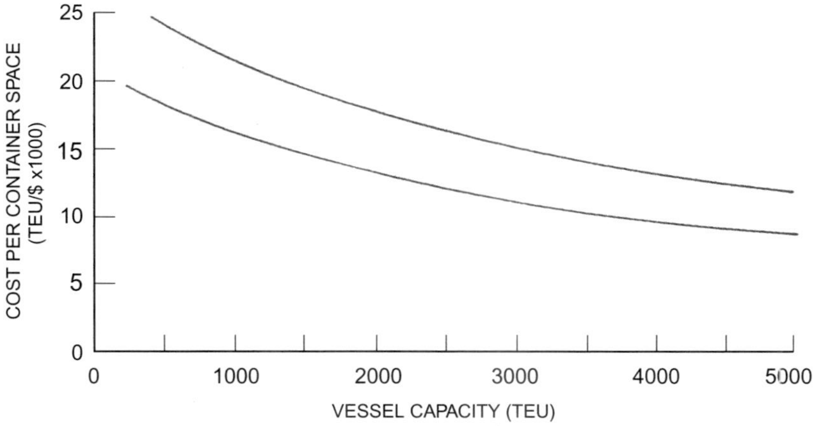

Fig. 10.2 Relationship of container-carrying capacity to cost/container (Cushing 1984).

Many countries are paying higher prices for the transportation of goods because of their limited capacity to accommodate large ships. Due to the limited water depths, many vessels call at ports less than fully loaded. For example, the landed price of coal in foreign markets can be heavily influenced by transportation costs. Graves *et al.* (1984) indicated the potential advantages of using large bulk carriers for transporting coal between Hampton Roads and Rotterdam where the transportation differential costs per ton could vary from $10.33 to $6.95 (Marine Board, 1985). The US' consumers also pay an economic penalty because the water depth in most ports is insufficient to allow the large bulk carriers and tankers into its ports. Freight rates for bulk carriers between different ports are shown in Table 10.1.

Because the capital costs of onshore facilities are substantial, research and development efforts regarding offshore ports (such as LOOP, off Louisiana) have been underway for many years. In the case of liquid cargo, no port facilities are required as the liquid is pumped through a pipeline from the ship to onshore.

Additional capacity is required for three reasons: (1) economics, (2) national security and defense, and (3) the need to be able to respond rapidly and flexibly to future changes in the character of the ocean transportation system.

Two categories of deficiency in port capacity have been identified in studies conducted by the US Army Corps of Engineers and by the ports of the US: (1) the limited ability of the US to handle large bulk carriers and (2) the need by some ports to handle medium size vessels; in particular, the latest-generation containerships, as well as, other specialized or general vessel types of Panamax dimensions (Table 10.2). The limits of the Panama Canal are 900 feet length, 106 feet beam and 42 feet draft (draft limits vary with water supplied to the canal and may in some seasons be less). These two identified needs have been taken as defining "large ships."

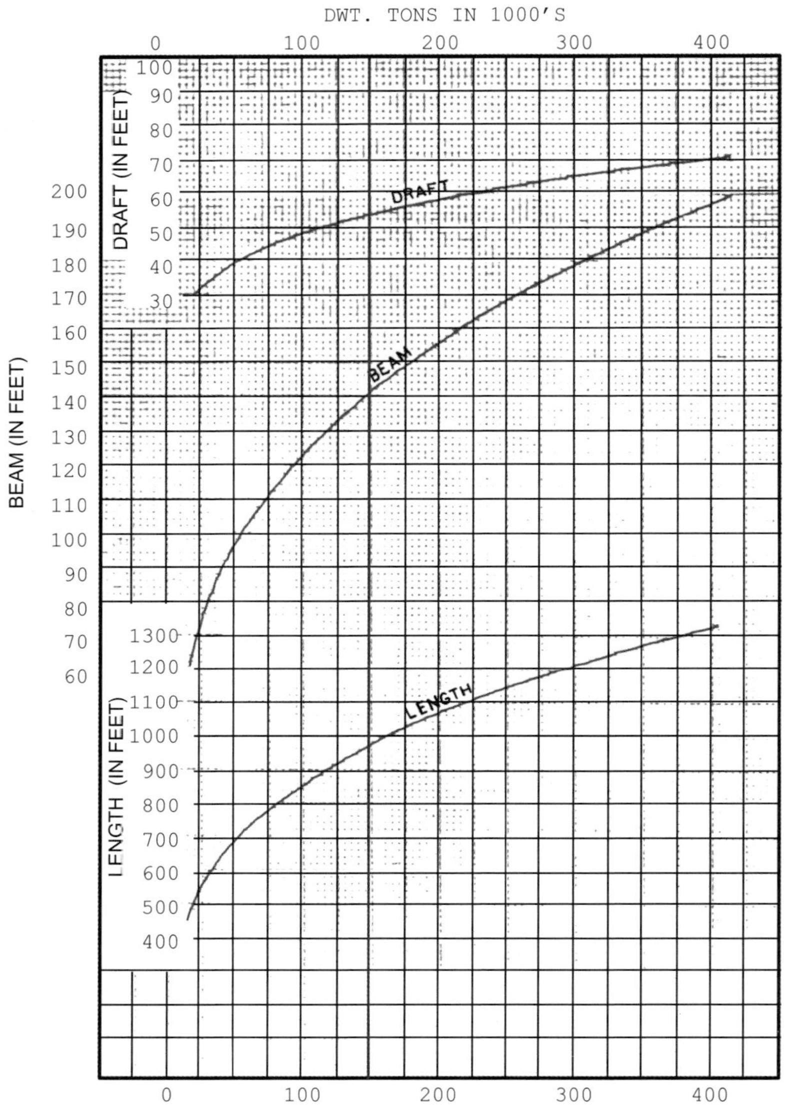

Fig. 10.3 Bulk carrier characteristics (US Army Corps of Engineers, 1983: Data taken from memo of Chief of Ship Const., Maritime Admin., Washington DC, 17 October 1967. Subject: Ship Sizes of the Future and the Distribution Thereof, July 1981).

1.3 *Design of navigational channels*

In 1972, the Permanent International Association of Navigation Congresses (PIANC) organized a commission to develop criteria for the reception of large vessels (IOTC, 1973; Marine Board, 1985). Six years later, a PIANC working group again studied the requirements of large vessels (200,000 DWT and greater), and

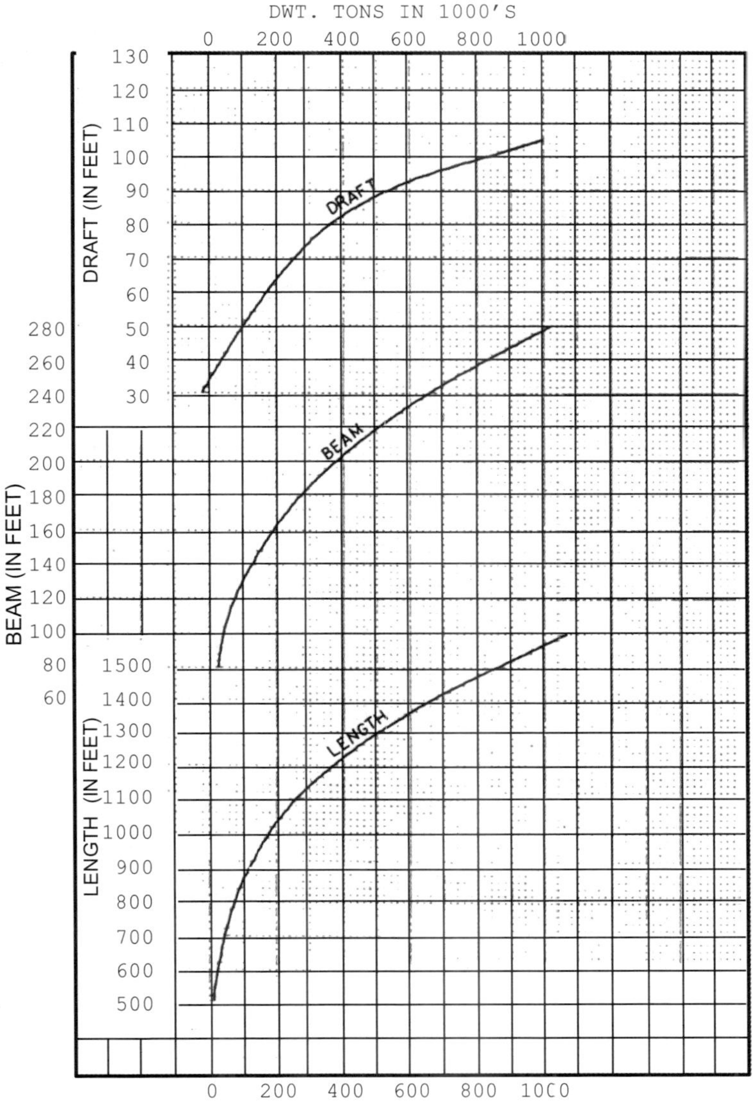

Fig. 10.4 Tanker characteristics (US Army Corps of Engineers, 1983: Data taken from memo of Chief of Ship Const., Maritime Admin., Washington DC, 17 October 1967. Subject: Ship Sizes of the Future and the Distribution Thereof, July 1981).

published recommendations for port design (ICORELS, 1980), as did a working group of the International Association of Ports and Harbors (COLS, 1981).

The recommendations of these international organizations include guidelines for the dimensions of channels and maneuvering areas, and address forces of the physical environment, equipment and training.

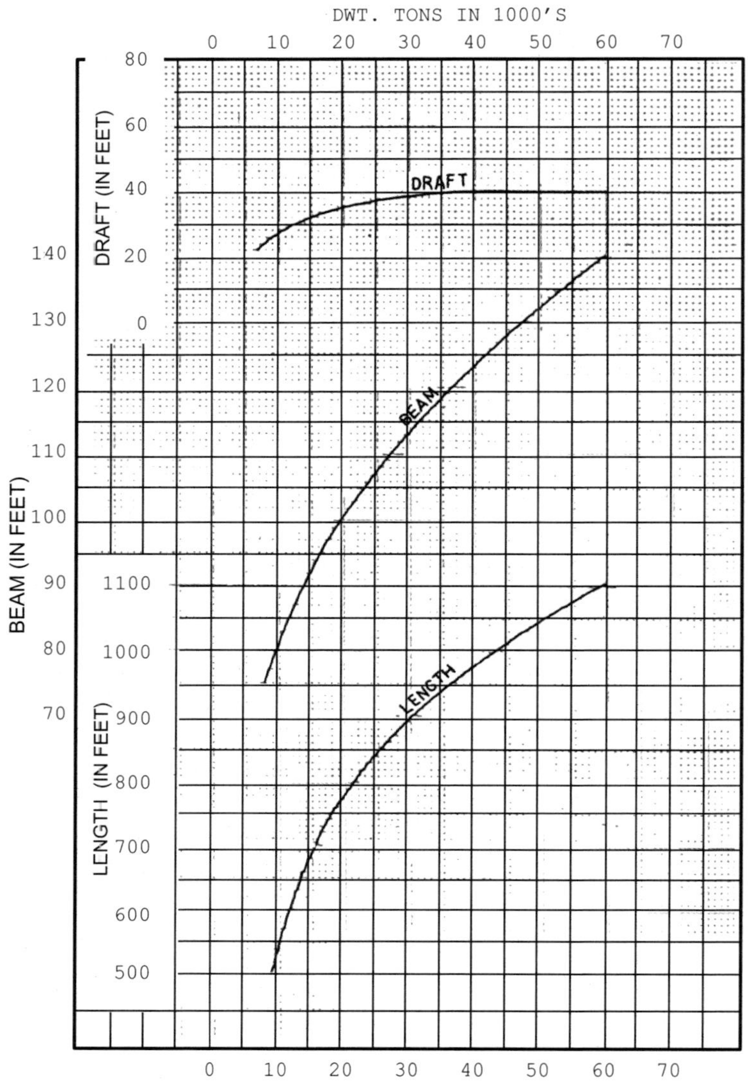

Fig. 10.5 General cargo ship characteristics (US Army Corps of Engineers, 1983: Data taken from memo of Chief of Ship Const., Maritime Admin., Washington DC, 17 October 1967. Subject: Ship Sizes of the Future and the Distribution Thereof, July 1981).

Maritime nations have developed general design criteria: those of Canada — TERMPOL — are based on prevention of oil pollution from marine casualties (Canadian Coast Guard, 1977). The general guidelines developed for ports and harbors in Japan (Bureau of Ports and Harbours, 1980) are detailed, reflecting the economic significance of ports to the country, its challenging natural environment coupled with the need to balance economical design and construction and safety margins.

Table 10.1 Freight rates for bulk carriers (Maritime Transport Committee, 1984).

Highest and Lowest Rates		In $/ton of cargo 1981	1982	1983
Grain	High	22.00	12.00	9.00
US Gulf Ports to Rotterdam or Antwerp	Low	8.75	5.74	7.00
Coal	High	28.50	19.60	17.50
Hampton Roads to Japanese Ports	Low	17.50	10.80	12.35
Iron Ore	High	15.00	7.00	6.50
Brazil to Northwest European Ports	Low	7.00	4.45	5.95

Table 10.2 Transportation savings: cost vs. price (Poten and Partners, 1983; Marine Board, 1985).

		$/ton			
		Panamax		125,000 DWT	
Coal		1980	1983	1980	1983
US East Coast to Rotterdam/Antwerp	Cost	12.50	10.97	9.64	8.57
	Savings for larger vessel			2.94	2.40
	Market rate	10.29	5.51	7.45	4.12
	Savings for larger vessel			2.84	1.39
US East Coast or Gulf to Japan	Cost	40.23	35.993	28.63	25.35
	Savings for larger vessel			11.60	10.58
	Market rate	33.08	18.08	21.91	11.57
	Savings for larger vessel			11.17	6.51

The general design criteria used in the US are developed by the US Army Corps of Engineers. These were updated in 1983 (US Army Corps of Engineers, 1983).

Succeeding tables and figures describe and compare these general criteria. All these sets of criteria acknowledge the importance of "(1) the several site-specific factors of great importance to design, (2) the need for consultations with ship-owners, pilots, and others, and (3) the need to employ analysis and design tools" (Crane, 1983; Marine Board, 1985).

Figure 10.6 and Tables 10.3–10.5 summarize the design criteria developed by different agencies and associations in different countries. Sample estimated costs for various US Ports are shown in Table 10.6.

1.4 *Probabilistic vessel admission policy in the Netherlands*

The vessel admission policy for deep-draft vessels at Rotterdam and Western Scheldt Harbors has been determined by the employment of the probabilistic method rather than the deterministic method. All possible wave and water level conditions as well as ship characteristics are considered and used for determining the optimal accessibility. But, instead of using discrete allowances, each factor is translated into

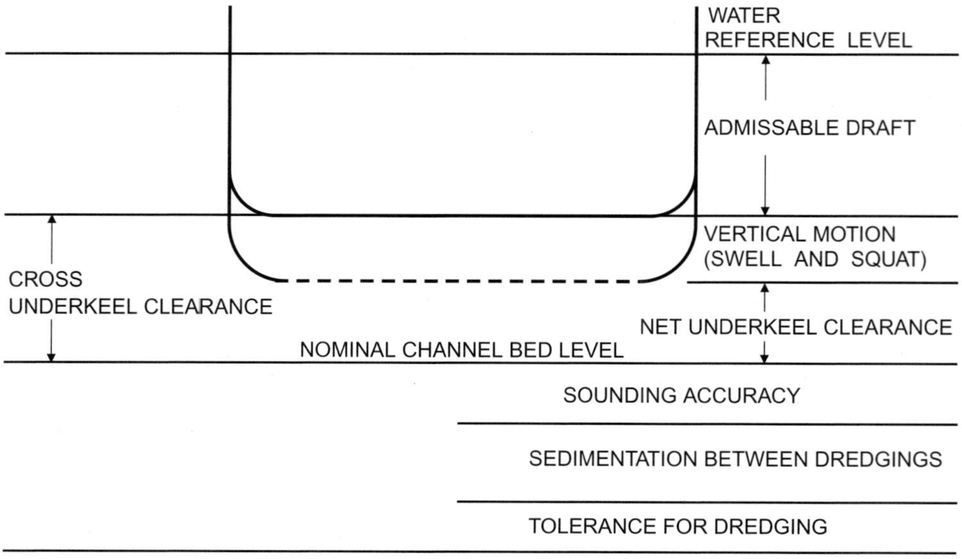

Fig. 10.6a Conventional net under-keel clearance calculation, definitions from PIANC. (*Source*: Permanent International Association of Navigation Congresses).

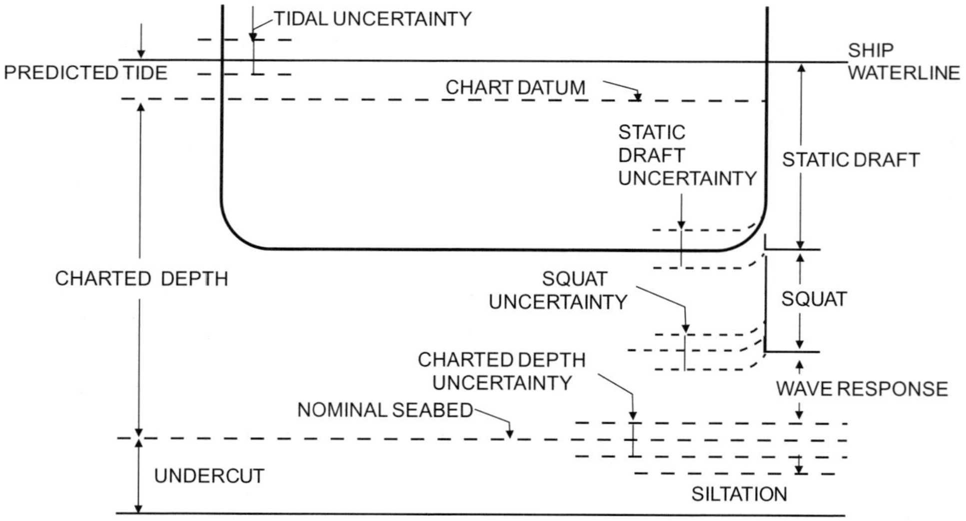

Fig. 10.6b Statistical under-keel clearance calculation. (*Source*: C. Lincoln Crane, Jr.; Marine Board, 1985).

Table 10.3 Comparison of Port Design Guidelines for Channel Width (Referred to Dimensions of Design Vessel(s)).

	One-Way Traffic			Two-Way Traffic			
	COE*[a]	TERMPOL[b]	PIANC[c]	COE*[a]	TERMPOL[b]	PIANC[c]	Definitions and Comments
Maneuvering Lane							Maneuvering Lane (ML):
Straight channel							— lane in which a single vessel maneuvers
Bends of	2.0 beam	2.0 beam	Additional width $\Delta W = L^2/BR+ =$ length$^{2/}$ (beam × radius)+	2.0 beam	2.0 beam		— accounts for uncertainty in vessel position and time lag to correct position
26°							— applies to straight and turning, not additive
40°	4.2 beam 4.9 beam			4.2 beam	4.0 beam		— required for each ship (i.e., MLX2 for 2-way traffic) — can reduce if operational limits applied (e.g., two design ships do not pass each other and no passing in turns)
Bank Clearance							Bank Clearance (BC):
Each side	0.6 beam +	1.0 beam	Safety margin .5	0.6 beam +	1.0 beam		— clearance to avoid bank suction — required on both sides, BC × 2 for all channels
Passing Ship Clearance							Passing Ship Clearance (PSC):
Between lanes				0.8 beam	1.0 beam		— between MLs to avoid interference

(*Continued*)

Table 10.3 (*Continued*)

	One-Way Traffic			Two-Way Traffic			
	COE*[a]	TERMPOL[b]	PIANC[c]	COE*[a]	TERMPOL[b]	PIANC[c]	Definitions and Comments
Total Width w/o Weather, Current							Total straight channel width without current clearance (W w/o WCC)
Straight Channels	3.2 beam	4.0 beam		6.0 beam +	7.0 beam		— artificial since cannot usually be used without
Bends of							— WCC, even for mild conditions
26°	5.4 beam	6.0 beam		10.4 beam	11.0 beam		
40°	6.1 beam +			11.8 beam			
Weather and Current Clearance	0.9 beam Each side	1.0 beam total both sides		$(0.2 + 0.9 \times 2)$ beam total both sides	1.0 beam total per ship lane		Weather and Current Clearance (WCC): — for beam wind, current, and waves which cause away angle (10–15 max.) — additional clearance should be made for varying conditions (gusts) and other factors — for design, must correspond to limiting environmental operation condition
Total Width							Total Width of Channel (W) at Design Depth:
Straight Channels	5.0 beam	5.0 beam	5 beam	8.0 beam + (2)	9.0 beam	8 beam	— in general, minimums are shown for design, however, local conditions must be considered
Bends of							
26°	7.2 beam	7.0 beam	5 beam + length^2/(beam × radius)	12.4 beam +	13.08 beam		
40°	7.9 beam +			13.8 beam			

(*Continued*)

Table 10.3 (*Continued*)

	One-Way Traffic			Two-Way Traffic			
	COE*[a]	TERMPOL[b]	PIANC[c]	COE*[a]	TERMPOL[b]	PIANC[c]	Definitions and Comments
Parallel Bends	> 5 length	w/o tugs		SAME AS ONE WAY			Radius of Turn (R):
Radius for	1:10						Radius of the channel centerline for bends
bends, of							
< 25°		3 length					TERMPOL gives max rudder angle (δ) of 15° (related to the design ships' turning diameters at $\delta = 15°$)
26°–35°		5 length					
> 36°		8 length					
Width transition (total)		1:10	5–10 length				Transition: The ratio of widening (sum of both sides) to length along channel
Approximate Ship Size**	Small/ medium size tanker	Not specified	Very ultra large crude carrier	Small/ medium size tanker	Not specified	Very/ ultra large crude carrier	Ship Size: — the approximate ship size for which the specific guide is intended (in terms of tankers)
General Equation	Width = maneuvering lane + 2 bank clearance + weather & current clearance			Width = 2 maneuvering lane + 2 bank clearance + passing ship clearance + weather and current clearance			General equation relating the individual components to the total width

*Using values for "poor vessel controllability".
[a]US Army Corps of Engineers.
[b]Canadian Coast Guard (1977).
[c]Permanent International Association of Navigation Congresses (1980).
Medium-size tanker: 60,000–120,000 DWT.
Very large crude carrier: 120,000–320,000 DWT.
Ultra large crude carrier: more than 320,000 DWT.
**Small-size tanker: 16,000–60,000 DWT.

Table 10.4 General criteria for turning basins and anchorages.

Turning Basins

US Army Corps of Engineers

Equal to area of circle with radius = 1.5 × ship's length overall + allowances for congestion, sedimentation, current, etc.

Side parallel to channel longer, ends angled 45° to channel boundary.

PIANC

Equal to circular area with diameter = ship's length overall.

Elliptical shape recommended.

Japan

Equal to circular area with radius = 1.5 × ship's length overall.

Anchorages

US Army Corps of Engineers

Free-swinging.

Area = area of circle with radius = ship's length overall + anchor chain (5 × to 6 × water depth).

Fixed dolphins, berths.

Width = 1.5 × ship's beam.

Parallel to channel.

PIANC

None

Japan

Design Objective	**Mooring**	**Seabed/Wind**	**(LOA = ship's length overall) Radius**
Offshore; waiting	Swinging	Good anchoring	LOA + (6 × water depth)
		Bad anchoring	LOA + (6 × water depth + 30 m (99 ft)
Mooring in storm	Mooring with 2 anchors	Good anchoring	LOA + (4.5 × water depth)
		Bad anchoring	LOA + (4.5 × water depth) + 25 m (82.5 ft)
		Wind vel. = 20 m/sec (40 km)	LOA + (3 × water depth) + 90 m (297 ft)
		Wind vel. = 30 m/sec (60 km)	LOA + (4 × water depth) + 145 m (478.5 ft)

a probability distribution with a mean value and a variance. All the probabilistic distributions are combined and together, determine the probability of touching the channel bottom.

The results of the probabilistic calculations are tidal windows, inaccessibility percentages and downtimes. These tidal windows and the according probability of touching the channel bottom conform to redefined safety criteria.

The water level is first determined. Information about astronomical and meteorological water levels, current, vessel speed and wave climate is used. This results in the calculation of the keel clearance across the channel. Next, the vertical movements of the different types of vessels are added which depend on the wave climate. The final result at this stage is one optimal tidal gate.

Table 10.5 General criteria for depths of dredged navigational facilities.

US Army Corps of Engineers.

Ship's draft and sinkage + allowances for wind, waves, currents, type of bottom, etc.

PIANC

Gross under-keel clearance 1.20 × ship's draft, exposed.
1.15 × ship's draft, waiting area, exposed.
1.07 × ship's draft, calmest area, least ship speed (berthing).

Net under-keel clearance at least 0.5 m (1.7 ft).

SHIP OWNERS

STATISTICAL

TERMPOL (CANADA)

1.15 × ship's draft, exceptions require special under-keel clearance survey,

Japan

Depth of maneuvering basins (1.10 × ship's draft) + allowances for wind, waves, currents, type of bottom, etc.

The Netherlands

Keel clearance has been established for different channels as shown in Table (Savenije, 1996).

Table 10.6 Estimated costs and trade by selected ports, 1990.

	(million 1990 dollars)			(million short tons)		
Port	Existing operations and maintenance costs[a]	Total costs, new construction dredging[b]	Incremental operations and maintenance costs[c]	Estimated total trade in 1990[d]	Estimated coal exports in 1990	Deepening plans
Hampton Roads[e]	3.2	438.5	6.1	80.6	58.9	55 ft
Great Lakes[f]	4.5	0.0	0.0	31.5	19.7	None
Baltimore	2.1	383.7	1.6	74.9	29.2	50 ft
New Orleans/ Baton Rouge	14.9	479.6	125.1	173.5	8.6	55 ft
Mobile	4.6	371.8	2.8	25.8	4.7	55 ft
Los Angeles/ Long Beach	0.1	420.2	0.0	81.4	2.8	80 ft
Philadelphia	5.8	0.0	0.0	67.7	12.0	None

Sources: Office of Policy, U.S. Army Corps of Engineers, for cost estimates.
1990 total trade by port estimated from U.S. Maritime Administration data and forecasts from the Federation of American Controlled Shipping.
1990 coal trade by port estimated using regional coal export forecasts form the International Coal Trade Model, existing port capacity and capacity under construction from U.S. Maritime Administration.

[a]Converted from 1982 dollars using GNP deflator 1.641.
[b]Converted from 1981 dollars using GNP deflator 1.0946.
[c]Converted from 1981 dollars using GNP deflator 1.0946.
[d]Exports, imports and coastwise movements.
[e]Norfolk and Newport News, Virginia.
[f]Includes Ohio Ports of Ashtabula, Conneaut, Sandusky and Toledo.

Table 10.7 Minimal gross keel clearance.

	Keel Clearance	
Channel	Outer area	Inner area
Euro–Maas Channel	20%	10%
Western Scheldt	15%	15%
IJ Channel	17.5%	15%

This optimal tidal window consists of one time: the optimal time to start the channel transit with minimal probability to touch the bottom of the channel, given all the circumstances at that moment.

Using the keel clearance percentages, shown in Table 10.7, the accessibility of the channel can be determined. For the Western Scheldt, the maximum draft was determined for the predicted water level at the most critical location along the channel. For the Euro-Mass Channel and the IJ-channel, tidal windows were calculated at different locations along the channel, taking into account the changing water level during the channel transit. The process is repeated for different combinations of tidal data. Using the probabilistic method, the accessibility and safety of the harbors has been increased without incurring additional dredging costs (Savenije, 1996).

1.5 *Estimated costs*

Estimated costs and trade by selected ports (1990) are summarized in Table 10.6. Dredging and maintenance costs are very site specific and vary considerably for US ports (Marine Board, 1985).

1.6 *Navigable (nautical) depth*

Navigable depth is defined as the maximum depth of water from the chart datum that is safe to accept as the bed of the channel, such that damage does not occur to a ship's hull and maneuverability is not significantly affected (Figs. 10.7 and 10.8). These criteria are directly dependent on the rheological properties of the sediment

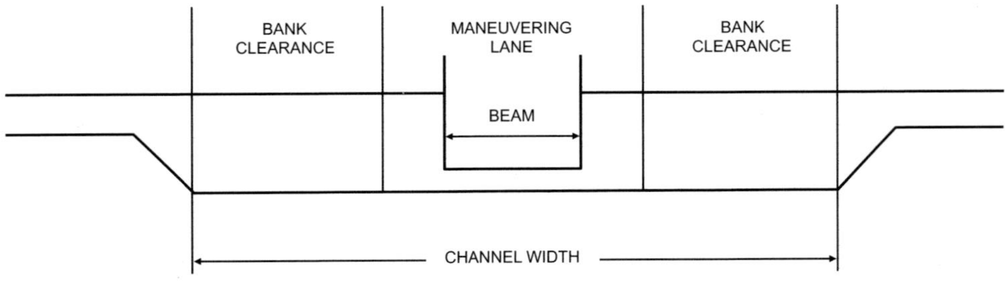

Fig. 10.7 Definition sketch: channel width for one-way traffic.

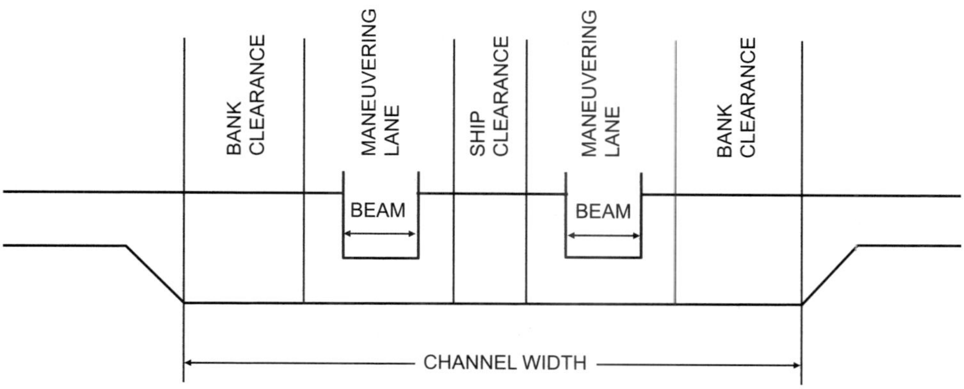

Fig. 10.8 Definition sketch: channel width for two-way traffic.

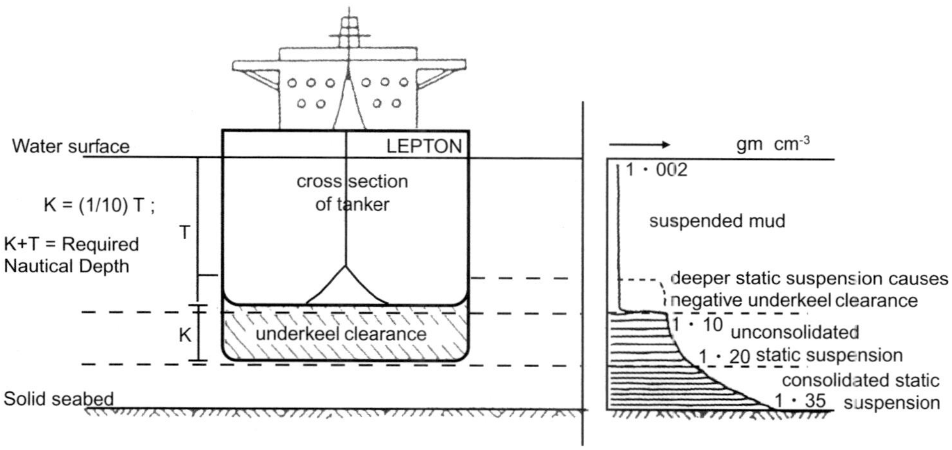

Fig. 10.9 Comparison of density profile and tanker cross-section to illustrate concept of nautical depth (Nederlof, 1980).

within the navigation channel. It may be possible to use specific gravity of the material; however, no single value is acceptable for use at different geographical locations. The US has not formally accepted the navigable depth concept, and no definite criteria has been established to determine how to measure it.

The "nautical depth" concept evolved from studies of the behavior of vessels in Europort and Rotterdam Harbors as part of an effort to minimize the maintenance dredging costs. The normal design depth of ship channels in sheltered waters includes a 10% under-keel clearance to provide safe navigation and maneuverability. The concept of nautical depth is depicted in Fig. 10.9.

The behavior of vessels sailing close to the surface and in the upper layers of dense suspensions was investigated. Extensive model tests were conducted in the

Netherlands Ship Model Basin with a two-layer system to study the sailing and maneuvering characteristics of supertankers in a channel with a soft bed. Prototype field tests were made with the 240,000 DWT supertanker Lepton sailing in the Europort channels. Extensive investigations were also made in the Chao Phraya River, Bangkok, and along the coast of Surinam where ships sail in mud with a negative under-keel clearance. On the basis of these investigations and a search of the literature, it was found that fluids of specific gravity to 1.2 had only a slight influence on maneuverability (Marine Board, 1983).

Comparison of the navigable behavior of ships in and without the fluid silt layer present, showed that where a silt layer is present, the following has been observed:

1. The resistance is much greater, requiring higher propeller revolutions to maintain a constant speed.
2. The stopping distance is substantially reduced.
3. In general, dynamic movements (course change) are faster; static movements (forward velocity, drift and turning velocity) are slower.
4. The effectiveness of the rudder for a given propeller rpm is reduced so that the turning circle increases.
5. When navigating a turn at the usual speed of about 5 knots, an additional angle of about 10° of rudder is required, as compared with clear water, whereas at lower speeds there is need for a considerably greater rudder angle.
6. When navigating on a straight course, the ship will show quicker response to the rudder thus reducing the required track width.
7. Application of a power surge for increase in rudder effectiveness results in a lesser increase in speed of the ship, so that more surges can be applied.
8. The squat and trim values are reduced.

Thus, channel depth can be increased if the upper layers of static suspensions are included in the under-keel clearance.

Based on the results of the investigations and studies, the concept of nautical depth was developed (Nederlof, 1980): "a density within the suspension above whose altitude vessels can safely sail" (Fig. 10.9), and the density was defined as $1.2\,g/cm^3$, or having specific gravity of 1.2.

Use of the nautical depth concept to define channel depths depends critically on frequent (bi-weekly) and accurate density measurements. The navigable depth (or nautical depth) concept has been adopted at Rotterdam Harbor and other harbors, where the navigable depth has been defined "safe" when the specific gravity of the fluid mud/water mixture is 1.20. Germany has also adopted a "nautical" depth concept in several ship channels (Rechlin, 1995). Depending on the location, the recommended values of specific gravity vary between 1.17 and 1.24.

A field study was conducted at the Calcasieu River Ship Channel in Louisiana to determine echo sounder records and rheological properties of the fluid mud. The design depth for the channel is 42 feet, as defined by the US Army Corps of Engineers (USACE). This depth is typically measured with echo-sounding equipment. The ability to interpret echo sounder output to determine the navigable depth is directly dependent on the local sediment rheological properties and the frequency at which the echo-sounding equipment is operated. Inaccuracies in interpreting the echo-sounding records may lead to underestimating the safe navigable depth and over-estimating the required amount of material to be dredged (Herbich and Wilkinson, 1993).

Understanding the relationships between echo sounder records and the rheological properties of the fluid mud is necessary in defining the safe navigable depth. Each harbor or navigational channel may have different relationships between echo sounder records and rheological characteristics because the sediment sizing and gradation may be different. The rheological properties may differ among US ports but the specific gravity at which maneuverability of ships is affected is most probably similar.

Investigations into the relationships between echo sounder records and rheological properties of fluid muds found in the Calcasieu River Ship Channel were made. The echo sounder records obtained indicated that significant quantities of fluid muds exist in this location. Echo sounder records from both high and low-frequency systems correlated well as to the depth of the water/fluid mud interface. The comparisons of low-frequency echo sounder records (i.e. 28 kHz) and density measurements of the fluid mud layer showed that the reflections related to changes in the density gradient of the layer and not to a specific value of density. This could be a problem if the "nautical depth" concept were adopted on the basis of specific gravity. To avoid such a problem, an instrument (i.e. density gauge) capable of detecting a datum level inside the fluid mud layer (i.e. specific gravity value) would be required if accurate measurements were desired.

The combination of echo sounder records and density measurement devices would assist in correlating the acoustic charts to the necessary safe level of density inside the fluid mud stratification. However, a value of specific gravity would need to be determined based on the viscosity of the fluid mud. In arriving at such a value, rheological tests could be performed to represent the viscous behavior of the fluid. Then, model tests could be conducted to find the maximum shear rates sustainable by a vessel's hull without causing damage or maneuverability restrictions. Hence, a definition of the safe navigable depth of the Calcasieu River Ship Channel could be implemented, and maintenance dredging costs could be reduced.

2. Dredging Equipment

The dredging of sedimentary deposits within ports and navigational waterways is accomplished by one of two primary techniques, hydraulic or mechanical. Within each class, a number of functionally different systems are available (see Fig. 10.10). The ultimate selection of the operating system is based primarily on the sediment

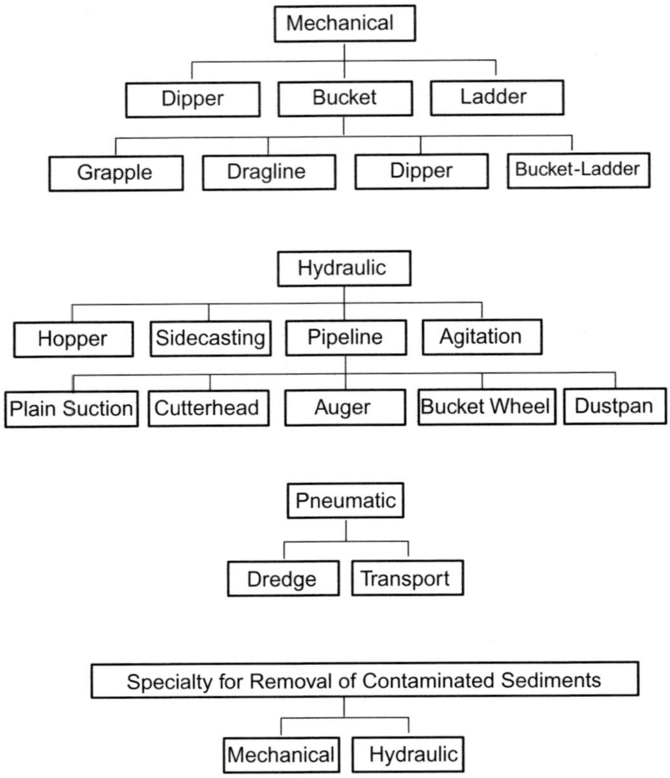

Fig. 10.10 Dredging systems.

type, water depth, sea conditions, location and proximity of the disposal area and, to some extent, the availability of equipment. In addition, the contamination levels of the sediment and the need to minimize near-field resuspension and far-field dispersion may be considered.

All dredges can be classified as either mechanically operating or hydraulically operating.

2.1 *Mechanical dredges*

Mechanical dredges are used because of their simplicity and analogy with land-based equipment. The dredges can be classified as dipper, bucket, ladder, grapple, dragline, and bucket ladder.

The grapple dredge consists of a derrick-mounted on a barge and equipped with a clamshell bucket. The dragline bucket dredge consists of a steel bucket suspended from a movable crane. The bucket is dragged through the soil toward the crane by a steel cable.

The dipper dredge mounted on a barge is similar to a dipper dredge operated on land.

The bucket-ladder dredge consists of an endless chain of buckets; each bucket digs its own load and carries it to the surface. Since the work cycle is continuous, bucket-ladder dredges are more efficient than either the grapple or dipper dredge.

2.2 *Hydraulic dredges*

Hydraulic dredges are self-contained units that handle both phases of the dredging systems. They not only dig the material but dispose of it either by pumping the material through a floating pipeline to a placement area or by storing it in hoppers which can be subsequently emptied over the disposal area. Hydraulic dredges are more efficient, versatile and economical to operate because of the continuous, self-contained digging and disposal principle of operation.

In a hydraulic dredge, the material to be removed is first loosened and mixed with water by a cutter head or by agitation with waterjets. Then it is pumped as a slurry, which is a mixture of water and sediment. The three basic units in a hydraulic dredge are the dredge pumps, the agitating machinery, and the hoisting and hauling equipment. The latter is used primarily to raise and lower the cutter in a cutterhead dredge.

The self-propelled trailing-suction (hopper) dredge was originally developed in the US. European manufacturers have made refinements in recent decades. The development has revolutionized the dredging industry by drastically reducing dredging costs. Maintenance of European channels and ports is virtually dominated by trailing-suction hopper dredges that are suitable for all but hard materials and are by far the best-suited dredges for offshore work. The dredge consists of a ship-type hull with hoppers to hold material dredged from the bottom. The material is brought to the surface through a suction pipe and draghead. The configuration of the draghead varies with the type of material. Many improvements have been effected in the design of dragheads to make them suitable for even compact material. Hopper dredges have been built with hopper capacities ranging from several hundred cubic meters to 18,000 m^3 (23,543 yd^3) (Dredge AMSTERDAM). The notable design improvements achieved in recent years in the Netherlands are: (1) distribution system, (2) integral suction system, (3) submerged dredge pump, (4) active draghead with rotating cylinder, (5) modular draghead, (6) draghead winch control, (7) split-trail, (8) slick-trail hopper dredges and (9) multi-purpose dredges. The hoppers are usually unloaded through the bottom doors. Pump-out facilities are also provided in modern dredges: they are extremely mobile and do not require sheltered areas. The maximum dredging depth is 59.1 to 69 feet (18 m to 21 m). By fitting a submerged pump at the draghead, it has been possible to increase the dredging depth up to 131 feet (40 m). The effect of trailing-suction hopper dredges on the environment depends on the type of bed material. Dragheads can resuspend the sediments at the bottom in fine materials. The overflow water may also carry fine material, creating a turbidity plume; although some improvements have been made to reduce the turbidity (Herbich, 1992).

Sidecasting dredges have the advantage of continuous operation since the material is dredged and disposed of some distance from the navigation channel, usually in an open water disposal site. Because of environmental concerns, only one sidecasting dredge is operating in the US.

Pipeline cutterhead dredge is a very efficient and versatile dredge. It is equipped with a rotating cutter surrounding the intake end of the suction pipe. Cutterhead dredges can efficiently dig and pump all types of alluvial materials and compacted deposits such as clay and hardpan. The larger, more powerful machines are used to dredge rock-like formations such as coral and the softer type of basalt and limestone without blasting. The cutterhead dredge is generally equipped with two stern spuds which are used to advance the dredge into the cut. Modern dredges have a spud carriage to increase the overall dredging efficiency. One of the important developments is the installation of a submerged dredge pump on the ladder; it can increase the concentration of the slurry and allows dredging at greater depths.

Preliminary selection guide for dredges is shown in Fig. 10.11; for example a 14-inch dredge can pump 320 cubic meters of solids per hour (350 cubic yards per hour) a distance of 457 meters (1500 feet).

An auger dredge (Mud Cat) has a horizontal cutterhead equipped with knives and a spiral auger that cuts the sediment and moves it laterally towards the center of the auger where it is picked up by suction.

Agitation dredging was practiced in ancient times and in the Middle Ages. A newly developed method in the Netherlands calls for the use of water jets to resuspend the sediment which is then carried away by prevailing water and turbidity currents. Two water pumps discharge water through the nozzles as close as possible to the channel bed. The waterjet injection system with a capacity of up to 12,000 m^3/hour (15,695 cubic yards/hour) can resuspend up to 5000 m^3 (6,539 cubic yards) of solids per hour according to the owner of the JETSED. The vessel is most efficient in silt and fine sand sediments, which typically accumulate in navigation channels. A density and turbidity current is set up which can transport the fine sediments out of the dredge area. The production is said to be up to 4,000 m^3/hour (5,232 cubic yards/hour) (Herbich, 1992).

The dustpan dredge is so named because its suction head resembles a large vacuum cleaner or dustpan. The dustpan dredge is a hydraulic, plain suction vessel. It consists essentially of a dredge pump which draws in a mixture of water and dredged materials through the suction head, which is lowered by winches to the face of the deposit to be removed. The suction head, which is about as wide as the hull of the dredge, is outfitted with high-velocity water jets for agitating and mixing the material. After sucking the mixture to the surface, the dredge pumps it to a disposal area, either at sea or shore, through a floating pipeline. Because it does not have a cutterhead, which loosens hard compact materials, the dustpan is suited mostly for high-volume, soft material dredging. The dustpan dredges are employed for maintenance dredging in large rivers, such as the Mississippi.

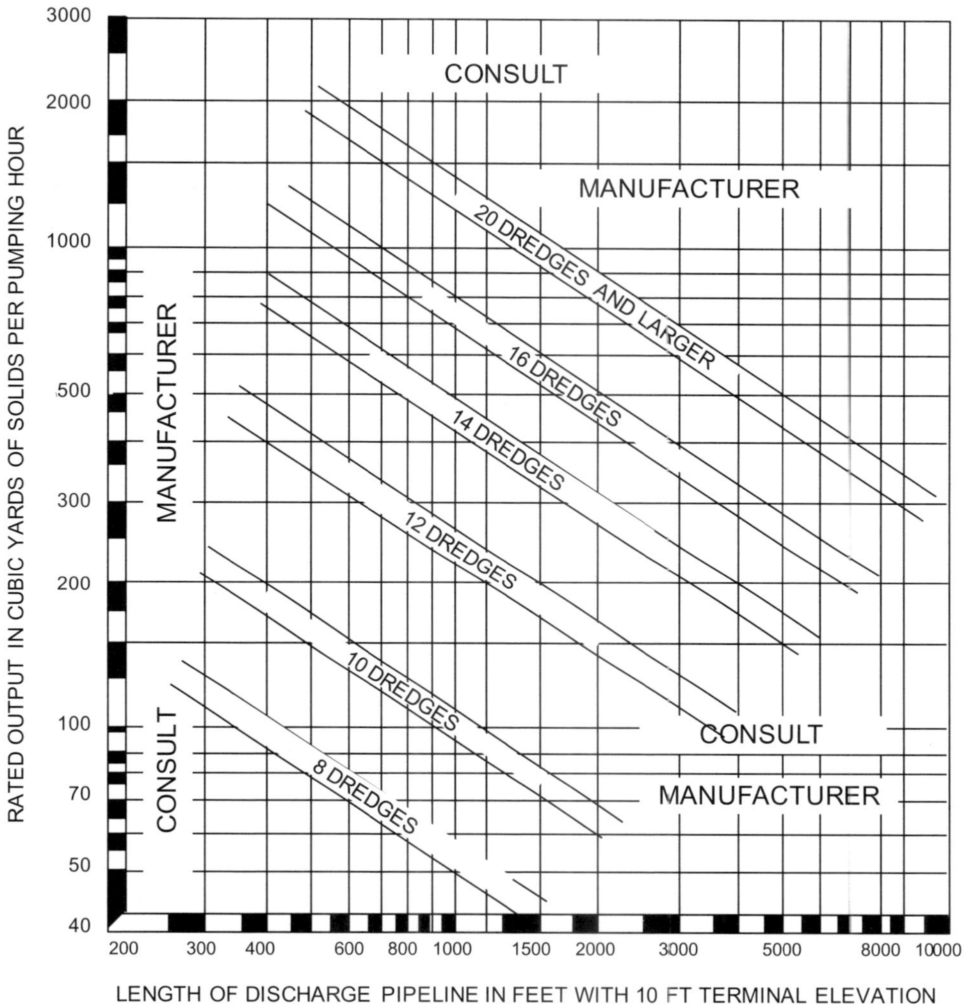

Fig. 10.11 Preliminary selection guide for dredges.

3. Instrumentation and Automation

The required information in a dredging operation is the accurate determination of the amount of solid material passing through the dredge pump. If this information is known, many other parameters for the dredge system can be accurately determined and controlled. To know the amount of material pumped requires knowledge of the density of the water mixture passing through the pump and the mixture flow velocity, or rate of flow. Numerous methods have been developed and employed to determine these values.

Without this information there can be no efficient dredging operation. Almost 20 or 25 years ago, very few dredges had this instrumentation and if they did, much

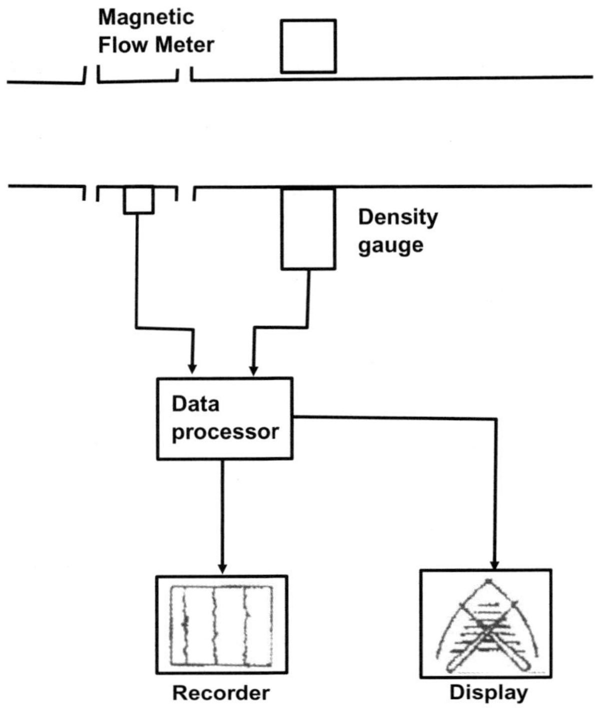

Fig. 10.12 A production meter system with a nucleonic density gauge and a cross-pointer display (Erb, 1995).

of the equipment was unreliable and subject to frequent breakdowns. The reliability of instrumentation equipment has considerably improved since then.

To determine the amount of solid materials being dredged, one requires a magnetic flow meter, a nuclear density meter and an instrument showing the total solids flow as well as velocity and slurry density. A production meter system is shown in Fig. 10.12.

3.1 *Other instrumentation*

(1) Echo sounders

The echo sounders may be used for surveying inland waterways, harbors and navigation channels, dredging and sediment studies.

(2) Heave, roll and pitch compensators

Typical applications are echo sounder and multi-beam sounder compensation and vessel motion analysis and navigation correction.

(3) Dual axis profiling

Dual axis profiling is a new method that allows mapping of large areas underwater from a static platform. A dual axis sonar head rotates both horizontally and vertically scanning the dredged area.

(4) Profile module

A profile module plots the survey results on the same screen as the required contract dredging template. Volume calculations can also be obtained from the data and plotted on the screen.

(5) Suction winch controller

A suction winch controller provides instantaneous position of the draghead.

(6) Solids content in the hopper

The monitor determines the amount of dry solids, controls the overflow, and controls the discharge of light mixture (say 1.2 specific gravity) overboard.

The number of dredges in the US with instrumentation to monitor production has increased from 7% in 1980 to 46% in 1989; however, the percentage of dredges equipped with production meters is only 17%.

The magnetic flow meter is reliable and relatively maintenance free as compared to the Doppler flow meter. The nuclear density meter is accurate and reliable when installed at the proper location.

There are several manufacturers of dredge production instrumentation worldwide who are constantly upgrading their products (Herbich *et al.*, 1992).

4. Project Planning

Major ports and navigational channels have been constructed and maintained by the US Army Corps of Engineers. Since federal funds are involved, the planning process is very elaborate and long. Table 10.8 describes the planning, approval, authorization and funding process for major navigation by the Corps of Engineers. This process takes, on the average, about 21.6 years (Table 10.9).

5. Dredging Methods

The US Army Corps of Engineers is responsible for maintaining navigation channels in the US. Since the Corps only maintains a small fleet, most of the dredging is done under a contract with private dredging industry. Approximately 300 million cubic yards of sediments are dredged every year sponsored by the Corps of Engineers and about 75 million cubic yards of sediments are dredged by private industry (oil and gas), counties and small harbors. These volumes are removed from navigable

Table 10.8 Planning, Approval, Authorization and Funding Process for Major Navigation Projects (US Army Corps of Engineers).

1. Congress authorizes study.
2. Congress appropriates funds.
3. Following appropriation of funds, District Engineer conducts initial public meeting to review draft plan of study. This provides opportunity to identify and discuss local problems and alternatives emphasizing national economic efficiency and environmental quality.
4. District Engineer.
   - Investigates all alternatives.
   - Performs limited
     - Technical feasibility studies.
     - Environmental assessments.
   - Proposes most feasible solutions in preliminary feasibility report.
5. Formulation stage — stage public meeting to discuss most feasible alternatives.
6. District Engineer.
   - Investigates formulation stage alternatives.
   - Performs detailed
     - Technical feasibility.
     - Environmental assessments.
   - Selects plan for proposal in detailed Feasibility Report (FR).
   - Distributes draft Environmental Impact Statement (EIS) and FR (15 days prior to state public meeting).
   - Files Draft EIS with EPA.
7. Public meeting — tentative plan proposed and discussed.
8. States, agencies, interest groups, public respond to draft EIS and draft FR.
9. District Engineer.
   - Reviews comments to draft EIS & FR.
   - Prepares recommended
     - Final EIS.
     - Final FR.
10. Division Engineer.
    - Reviews.
    - Modifies as appropriate.
      - Final FT as appropriate.
      - Final EIS.
    - Issues public notice requesting public views be sent to Board of Engineers for Rivers and Harbors (BERH).
    - Forwards recommendations to BERH.
11. Board of Engineers for Rivers and Harbors.
    - Considers views of:
      - Public.
      - States.
      - Agencies.
    - Reviews and provides recommendations:
      - Final EIS.
      - Final FT.
    - Transmits to chief of engineers.

(Continued)

Table 10.8 (*Continued*)

12. Chief
    - Reviews board report.
    - Prepares his draft recommendations.
    - Distributes for outside review.
    - Files final EIS with EPA.
    - Circulates to public for 30 day review period and to governors, federal departments (90 day review period).
13. Chief.
    - Reviews comments received.
    - Modifies report as appropriate.
    - Prepares record of decision (ROD).
14. Chief.

    Forwards recommendations to Secretary of the Army for consideration.
    - Final Report.
    - Final EIS.
    - ROD.
15. Secretary of the Army
    - Reviews
    - Coordinates with OMB.
    - Prepares his recommendations.
    - Forwards final FT, final EIS.
    - ROD to Congress (6 months).
16. Project Authorization.
    - Congress holds hearings.
    - Congress includes in Water Resources Development Act or other legislation
    - President signs.
17. OMB
    - Reviews Corps budget.
    - Submitsto Congress.
18. Project Funding.
    - Congress includes in Appropriations Act.
    - President Signs.
19. Local interests guarantee to fulfill obligations required by law (e.g., real estate, cost sharing, maintenance, operation, flood zoning).
20. District Engineer.
    - Formulates pre-construction planning general design memoranda
        - Updates EIS as required for Sec. 404 compliance, obtains necessary Water Quality certificates.
        - Issues public notice and conducts at least one public meeting (36 months).
    - Obtains additional congressional authorization as appropriate (24 months).
    - Initiates and completes construction (60 months).
    - Operates and maintains.

Table 10.9 Average Schedule for Navigation Projects (U.S. Army Corps of Engineers).

Activity	Years
Survey/study authorized	
Funds for study appropriated	4.9
Study/survey sent to division	4.1
Report sent to Congress	1.5
Project authorized	0.6
Initial funds appropriated for preconstruction planning and engineering	1.9
Initial construction funds appropriated	5.8
First contract award	2.8
TOTAL TIME	21.6

Source: General Accounting Office, 1984.

rivers, bays, estuaries and coastal areas. Dredging permits are issued by the Corps after review by the Environmental Protection Agency (EPA), other federal and state agencies, environmental groups and the general public.

Dredging in sheltered areas is generally performed by cutterhead dredges, and by hopper dredges in areas exposed to wave action (wave heights over 0.60 m (2 feet)).

6. Disposal and Placement of Dredged Material

6.1 *Disposal alternatives*

The disposal of dredged material has taken a variety of methods. In the past (prior to 1970), most dredged material had been discharged in deeper water. Because of growing environmental concerns, a great volume of sediments have been placed in confined disposal areas. The advantages and disadvantages of the placement of sediments in water or on land may be evaluated from an economical and environmental point of view.

What may be advantageous from an economic point of view (e.g. placement of material in open water close the channel being dredged), may be disadvantageous from an environmental point of view (e.g. changes in water circulation patterns, dispersion of fine sediment over a wide area under high energy wave and current conditions).

Figure 10.13 presents disposal alternatives.

(1) Disposal in lowland areas without the use of containment dikes

This method is seldom used today because of the uncontrolled flow of dredged material damaging vegetation and adjoining waters. The area affected would be larger than if dikes were used.

(2) Disposal in lowland areas using containment dikes

This method might be used if pumping distances to upland sites are so far away as to make them impractical. In many cases, the actual laying of pipelines across the

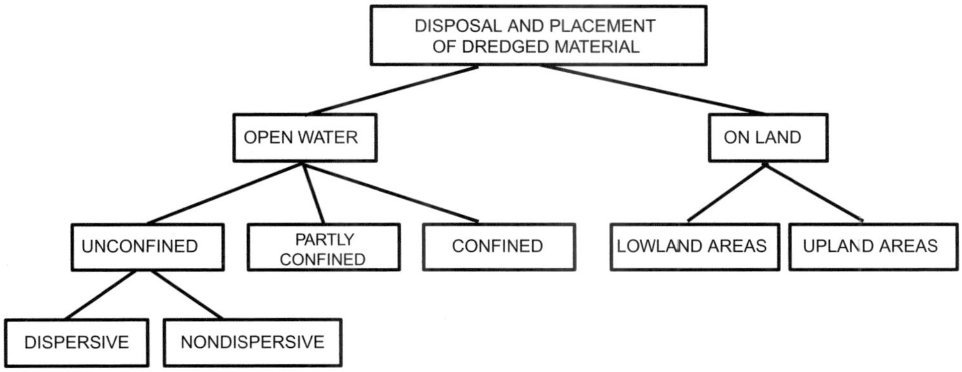

Fig. 10.13 Disposal alternatives.

lowland areas to upland disposal sites might be as detrimental to the lowland areas as the dredged material would be.

(3) Disposal in open water without the use of containment dikes

This method is often used for virgin material and maintenance dredging as well. In some cases the disposal of dredged material in this manner may actually be beneficial to the environment. The release of nutrients may more than compensate for the production lost by plants and animals covered by the dredged material. The creation of shallower water after completion of dredging may also be beneficial in some areas. The material may or may not become emergent. Emergent islands are often constructed from virgin dredged material but seldom from maintenance dredged material. These islands may prove to be beneficial to the environment by providing additional nesting areas for birds. A disadvantage of this method is that the underwater slope formed with fine-grained dredged material may be very flat (thus requiring large areas for disposal) and storms in the area may completely destroy the islands before the disposal islands have had time to stabilize.

(4) Disposal in open water using containment dikes

The use of containment levees in open water is justified when it is desired to prevent the dredged material from spreading under the action of waves and currents. For this to be effective, the islands should be emergent, and the exterior surface of the levees should be protected to prevent erosion. To prevent the levees from forming flat slopes and covering large areas, they should be constructed of suitable material such as clay balls, coarse to medium sands, etc. (Herbich, 1992).

Diked containment areas are used to retain dredged material solids while allowing the carrier water to be released from the containment area. The release of effluent is governed by the standard adopted by the Corps of Engineers Districts (from 8–13 grams per liter; or from 5 to 50 Jackson Turbidity Units (JTU)). Two objectives in the design and operation of a containment area are to provide storage capacity to meet dredging requirements and to attain the highest possible efficiency in retaining solids during the dredging operation in order to meet effluent standards (US Army Corps of Engineers, 1987).

Containment dikes are usually constructed from the available soil at the site by the contractor — major projects such as the Hart–Miller disposal area in Chesapeake Bay or disposal areas in Mobile Bay. The dikes are engineered and combined with rip-rap or other types of revetments to keep the dikes from breaching during major storms.

An example of a large disposal area is Pleasure Island, constructed in Sabine Pass, Texas. Part of the disposal area is now used by the Corps of Engineers Area Office and as a large recreational facility.

A very large disposal area called Shifter was constructed in 1986 and 1987 in the North Sea to accommodate some of the 30 million cubic yards (23 million cubic meters) of sediments that have to be dredged from the Rotterdam and Europort area (Terra et Aqua). The design criteria were as follows:

1. The disposal area will be constructed partly by excavation and partly by raising the existing land level.
2. The dikes will be constructed of sand.
3. The disposal area should be located in shallow water.
4. The shape of the disposal area should conform to the existing coastline.
5. The sand for the dike construction should be excavated from inside the disposal area.

The western side of the dikes will reach an elevation of 79 feet (24 m) above MSL (the Netherlands Reference Datum — NAP). The total area of the disposal site is about 642 acres (260 hectares).

The success of any disposal operation on land depends on proper planning, designing, operating, and managing containment areas. The goal is to provide maximum storage volume and to meet required effluent solids standards. Design procedures include consideration of dredged material sedimentation and consolidation behavior, as well as consolidation of foundation soils.

The purpose of containment area management is to promote natural dewatering of fine-grained dredged material and thus reduce the volume of containment required. Factors to consider are:

a. Careful planning: Careful planning is needed because different types of soil are encountered in dredging operations. Plans should include not only a logical sequence of operations for (1) dredging, (2) disposing and (3) dewatering of dredged material, but also an estimate of the dredging requirements and an estimate of adequate storage capacity for the material dredged during that period.
b. Shape of disposal area: This can be any shape from square to oval to long and narrow, generally dictated by the shape of land available. It should be sloping to provide natural gravity drainage of water. There may be some advantage to long

and narrow areas to increase the sedimentation (settling rates) and to facilitate the use of draglines and/or clamshells for constructing dewatering trenches.

c. Sizing of containment areas: The containment areas are generally designed on the basis of a bulking factor (swell factor) or on experience from previous dredging of similar materials. The factor relates the volume of material *in situ* (to be dredged) to the volume it is expected to occupy after being pumping into a containment area.

   The sizing of the containment area is also a function of (1) lift thickness of the placed dredged material, (2) flow rate of dredged material and (3) minimum effluent standards.

   Dewatering, and thus reducing the volume of containment areas can be accomplished by (1) placement of thin layers (lifts), (2) division of large areas into several compartments, and (3) cross and spur dikes.

d. Compartments: The use of cross dikes to divide a containment area is common practice on many projects. The general purpose of cross dikes is to reduce the velocity of dredged material being discharged into a containment area. Lower velocities, or calm conditions, promote settlement of particles, thus improving effluent water quality.

The compartments may be used in series (Fig. 10.14) or in parallel (Fig. 10.15). When a series operation is employed, the first compartment acts as a primary sedimentation basin. Compartments, as needed, can be used to meet the effluent quality standards (Fig. 10.15).

PIANC (1990) published a report recommending a strategy for selection of environmentally-sound and cost-effective disposal options. A survey conducted by an *ad hoc* Dredging Commission in 1981 is shown in Table 10.10. The distribution of dredged material disposal in different categories varies among different countries, but it is clear that very large volumes of dredged material must be handled annually.

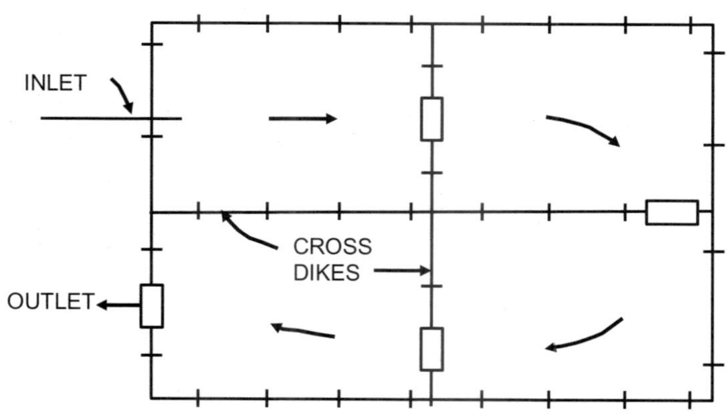

Fig. 10.14 Series compartments, Chicago District (Bartos, 1977).

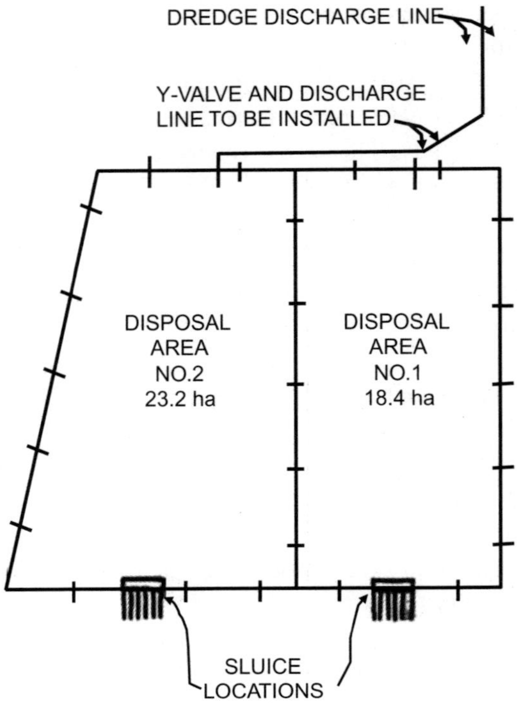

Fig. 10.15 Parallel compartments, Mobile District (Bartos, 1977).

Table 10.10 Worldwide method of disposal (volume in thousands of cubic yards) (Ad Hoc Dredging Commission, 1981 (PIANC, 1990)).

Region	Number of responses	Upland	Near wetlands*	Shore	Ocean	Other	Total
Northern Europe (21%)	26	39,196	59,502	42,936	62,044	29,412	233,091
Mediterranean (3%)	3	0	13,774	15,001	664	0	29,421
Africa (24%)	2	0	152,942	76,471	25,549	0	254,963
Southern Asia (27%)	12	62,484	11,197	121,831	89,149	0	284,661
Southeast Asia (2%)	8	0	3,078	3,698	15,190	0	21,966
East Asia (13%)	16	5,783	32,220	102,451	4,323	0	144,777
South Pacific (5%)	18	3,972	2,687	26,335	32,588	0	65,582
North America (4%)	18	6,012	9,696	8,459	16,549	159	40,875
Caribbean (1%)	5	820	646	0	2,484	0	3,950
TOTAL (100%)	108	118,267	285,742	287,202	248,522	29,571	3,079,286

*Includes estuaries, marsh, shallow water, etc.

The majority of dredged material is uncontaminated and should be treated as such (e.g., it is estimated that over 95% of dredged material in the US is suitable for placement in open water). The contaminated material must be handled differently. The selection of an appropriate disposal method depends on a number of variables:

1. Size of the project.
2. Physical characteristics of dredged material.
3. Level of contaminants present in the material to be dredged.
4. Available dredging equipment.
5. Site-specific conditions.
6. Potential environmental impacts.
7. Economical considerations.
8. Social, political and regulatory considerations.

Herbich *et al.* (1990) indicated that open water disposal areas in the US numbered 193; a number of these designated areas may actually diminish in the future.

6.2 *Beneficial uses of dredged material*

US Army Corps of Engineers (undated) provides guidelines for planning, designing and managing dredged material for beneficial uses, incorporating environmental concepts and engineering designs with biological, economical and social feasibility.

Dredged material disposal provides opportunities for a number of environmental, economic and aesthetic beneficial uses. Innovative beneficial uses appear to be unlimited; over 1,300 cases of beneficial uses of disposal sites have been documented in North America alone.

Ten broad categories of beneficial uses have been identified, based on their functional use of dredged material at disposal sites (US Army Corps of Engineers). They are:

1. Habitat development (wetland, upland, island, aquatic, including migratory and nesting use by water birds, shorebirds, waterfowl, and other groups).
2. Beach nourishment.
3. Aquaculture.
4. Parks and recreation (commercial and noncommercial).
5. Agriculture, forestry and horticulture.
6. Strip mine reclamation and solid waste management.
7. Shoreline stabilization and erosion control.
8. Construction and industrial use (including port development, airports, urban and residential).
9. Material transfer (fill, dikes, levees, parking lots, roads).
10. Multi-purpose.

Recognition of the ecological value of many areas being historically used as dredged material disposal sites resulted in severe environmental constraints on location and placement of disposal sites, especially those in open water and wetlands. These constraints have increased the values placed on coastal and riparian wetlands and aquatic areas and, have increasingly accented the need for alternate methods of dredged material disposal. As land uses have changed and areas once available for dredged material disposal have become scarce, the concept and beneficial use of dredged material disposal sites, such as land improvement and habitat development, has become more attractive economically and more environmentally acceptable.

Dredged material is a manageable, valuable soil resource, with beneficial uses of such importance that plans for ultimate use of disposal sites should be incorporated into project plans and goals at the project's inception.

The known and potential effects of dredging and dredged material disposal on the environment in and around US waterways have led to considerable research efforts and inter-agency and intra-agency coordination. Many waterway projects involving dredging have purposes which require consideration of ecological effects. While maintenance of navigable channels is the prime objective, the development and application of beneficial alternatives for dredged material disposal must receive appropriate consideration (US Army Corps of Engineers).

7. Removal of Contaminated Sediments

7.1 *Introduction*

The problem of contaminated marine sediments has emerged as an environmental issue of national importance. Harbor areas in particular have been found to contain high levels of contaminants in bottom sediments due to wastes from municipal, industrial and riverine sources.

Contamination of marine sediment in all areas of the world, particularly in shallow water areas, poses a potential threat to marine resources and human health. Improving the capability to assess, manage and remediate these contaminated sediments is critical not only to the well-being of the marine environment but as well as to its use for navigation, commerce, fishing and recreation.

There are many definitions for contamination of sediments. The Marine Board of the National Research Council (1985) defined contaminated sediments as follows:

> "Contaminated sediments are those that contain chemical substances at concentrations which pose a known or suspended environmental or human health threat."

Many contaminated marine sediments are found along all coasts of the contiguous US and Canada, both in local "hot spots" and distributed over large areas. There is a wide variety of contaminants including: heavy metals, polychlorinated biphenyls (PCBs), DDT, and polynuclear aromatic hydrocarbons (PAHs).

Legislative authority for the management of contaminated marine sediments falls largely under three statutes: (1) the Comprehensive Environmental Response, Compensation, and Liability Act of 1980 (CERCLA), (2) the Marine Protection Research and Sanctuaries Act (MPRSA), and (3) the clean Water Act (CWA). MPRSA was amended by the Superfund Amendments and Reauthorization Act

(SARA) of 1986 and is principally aimed at the cleanup and remediation of inactive or abandoned hazardous waste sites, regardless of location.

Superfund sites have been ranked by the US Environmental Protection agency (EPA) based on the hazard that the contaminants may pose to human health and the environment through release of contaminants or toxic materials to groundwater, surface water and air.

As a result of legislative responsibility and programmatic interests, a wide variety of federal agencies have shown active interest in this subject. EPA's responsibilities under Superfund and the CWA are the source of its interests in water quality concerns and remediation of uncontrolled hazardous waste sites. The US Army Corps of Engineers (COE) is involved because of its responsibility to dredge and maintain navigable rivers and harbors. The COE also assists in the design and implementation of remedial cleanup actions under Superfund. NOAA has responsibility for assessing the potential threat of Superfund sites to coastal marine resources as a natural resources trustee. The US Fish and Wildlife Service have legal authority for various endangered coastal species, food chain relationships, and habitat considerations, all of which are potentially impacted by contaminated sediments.

Although the dredged material management strategy developed by the Corps of Engineers may be relevant to severely contaminated sediments, it is important from a management standpoint to differentiate them from less contaminated sediments. In particular, the most highly sophisticated remedial technologies (i.e. those involving treatment or destruction of associated contaminants) are likely to be cost-effective only in small areas and for sediments with relatively high contamination levels. Sediment contamination problems often involve large volumes of sediments with relatively low contamination levels. As a result, some highly sophisticated technologies may be inapplicable or inefficient for remediating contaminated sediments.

"No action" may be the preferred alternative in cases in which the remedy may be worse than the disease — *e.g.*, where dredging or stabilizing contaminated sediments results in more biological damage than leaving the material in place. Contaminants generally accumulate in depositional zones and, if the source is controlled, new clean sediments will deposit and cap the contaminated material over time. In effect, no action alternatives in such cases may result in natural capping.

The procedures may be summarized as follows:

1. "No action" may be an acceptable option if the contamination degrades or is buried by natural deposition of clean sediment in a relatively short period of time (6–12 months). In using the "no action" strategy as a form of natural capping of contaminated material, consideration should be given to the length of time it takes for contaminants to be isolated from the food chain.
2. In-place capping may be a useful option if the sediments are not in a navigation channel or if groundwater is not flowing through the site.

3. Removal and subaqueous burial off-site may be a viable option, although experience with this technique is limited to relatively shallow water (less than 30.5 m (100 fit)).
4. Incineration seems to be viable only for sites with relatively small amounts of sediments containing high concentrations of combustible contaminants.
5. Other techniques to assist in remediation of contaminated sediment may be appropriate in special cases. Examples include a variety of sediment stabilization or solidification techniques, and biological and/or chemical treatment.
6. Additional evaluation should be conducted to determine the applicability of the Corps of Engineers' dredged material management strategy to more severely contaminated sediments.

7.2 *Remedial technologies*

From a remediation standpoint, the most important factors are likely to be a definition of the clean-up target, technical and cost feasibility, natural recovery estimates, and ability to distinguish and/or control continuing sources of contaminants.

Dredging technology exists that is capable of greatly reducing turbidity and resuspension during the dredging of bottom sediments; however, special equipment has to be deployed and modified operational methods must be used (Herbich and Brahme, 1991; Herbich, 1993).

A number of specialty dredges have been developed, principally in Italy, the Netherlands, Japan and the US for the removal of contaminated sediments. These include the Pneuma pump (available in Italy), the Oozer dredge, the Cleanup dredge and Refresher system (available in Japan), the Dutch-designed "Matchbox-head" dredge and the horizontal-auger dredges (MUD CAT) and blackhole dredge developed in the US and Cable Arm dredge developed in Canada. A new combination dredge employing mechanical, hydraulic and pneumatic concepts has been developed in Japan. This high-density pneumatic conveying system has been used in Kumamoto Port, Japan.

8. Summary and Recommendations

(1) Dredging deeper channels and maintaining navigational channels should be a high priority for governments, as it affects the economic well-being of nations.
(2) Dredging equipment should be fully instrumented, well-maintained, and modernized. Two goals to keep in mind about depth are:
    (i) the percentage of solids in the dredging mixture should be as high as possible, and
    (ii) the amount of sediment resuspension should be minimized.
(3) It is recommended that countries not having sufficient depth for modern vessel traffic, plan to deepen their navigational channels and ports to achieve economic benefits.

(4) Countries, worldwide, should pay more attention to the environmental effects of dredging, consider dredged material as a resource for beneficial uses (e.g. building of artificial islands for transitory bird refuges, beach nourishment, recreational areas, foundations for expanded port facilities, etc).

References

Bruun, P. (1992): Marine terminal technology, in: *Handbook of Coastal and Ocean Engineering* (J. B. Herbich, ed.): Gulf Publishing Company, Houston, TX.

Canadian Coast Guard (1977): Code of Recommended Standards for the Prevention of Pollution in Marine Terminal Systems. Canada.

COLS (Committee on Large Ships) (1981): Guidelines for Safety and Environmental Protection of Ports,: International Association of Ports and Harbors, Tokyo.

Erb, T. L. (1995): Production Meter Systems for Suction Dredges (unpublished dredging short course notes), Center for Dredging Studies, Texas A&M University, College Station, TX.

Estourgie, A. L. P. (1988): A New Method of Maintenance Dredging. IRO Journal.

Harlow, E. H. (1992): Harbor planning and design, in: *Handbook of Coastal and Ocean Engineering* (J. B. Herbich, ed.): Gulf Publishing Company, Houston, TX.

Herbich, J. B, J. Y. Lee, D. R. Trivedi, G. Wilkinson and D. O. DeHert (1992): Survey and Evaluation of Production Meter Instrumentation and Uses (2 Volumes), Texas A&M Research Foundation, Center for Dredging Studies, CDS Report No. 313.

Herbich, J. B., ed. (1992): *Handbook of Dredging Engineering*: McGraw-Hill, New York.

Herbich, J. B. (1992): Dredged navigation channels, in: *Handbook of Coastal and Ocean Engineering* (J. B. Herbich, ed.): Gulf Publishing Company, Houston, TX.

Herbich, J. B. (1993): Dredging Equipment for the Removal of Contaminated Sediment, state-of-the-art. *Sixth Int. Symp. on the Interaction between Sediments and Water*, Santa Barbara, CA, December 5–8.

Herbich, J. B. and S. B. Brahme (1991): A Literature Review and Technical Evaluation of Sediment Resuspension During Dredging, Contract Report HL-91-1, U.S. Army Engineer Waterways Experiment Station, Vicksburg, MS.

Herbich, J. B. and G. L. Wilkinson, (1993): Defining the Safe Navigable Depth in the Calcasieu River Ship Channel, Louisiana, *Proc. 8th Symp. on Coastal and Ocean Management*, ASCE.

Herbich, J. B., D. O. DeHert, K. M. Krafft, G. L. Wilkinson and D. R. Trivedi (1990): Inventory of Open Water Disposal Areas. Texas A&M Research Foundation, Center for Dredging Studies, CDS Report No. 315-B.

ICORELS (International Commission for the Reception of Large Ships) (1980): Optimum Layout and Dimensions for the Adjustment to Large Ships of Maritime Fairways. Report of Working Group IV, Permanent International Association of Navigation Congresses.

IHC Holland. (1998): Brochures, Sliedrecht, the Netherlands.

Marine Board, National Research Council. (1983): *Criteria for the Depths of Dredged Navigational Channels*: National Academy Press, Washington, DC.

Marine Board, National Research Council. (1985): *Dredging Coastal Ports*: National Academy Press, Washington, DC.

Nederlof, L. (1980): Sailing Through Water Rich in Floating Silt: A Vessel Behaves Differently but Remains Manageable. Rotterdam Europort Delta.

PIANC. (1985): The Stability of Rubble Mound Breakwaters in Deeper Water, Supplement to Bulletin No. 48.

PIANC (1990): Ad Hoc Dredging Commission.

Rechlin, D. (1995): Definition of the Nautical Depth in the Main Muddy Areas of the Federal Waterways Board, PIANC Bulletin, No. 86.

Savenije, P. P. (1996): Probabilistic Admittance Policy Deep Draught Vessels, PIANC Bulletin No. 91.

U.S. Army Corps of Engineers. (1983): Engineer Manual: Hydraulic Design of Deep Draft Navigation Projects. EM1110-2-1613, Washington, DC.

U.S. Army Corps of Engineers. (1987): Engineer Manual. Confined Disposal of Dredged Material. EM1110-2-5027, Washington, DC.

U.S. Army Corps of Engineers (1987): Beneficial Uses of Dredged Material. EM1110-2-5026, Washington, DC.

Chapter 11

Enhancement of Marine Habitats

Makoto Nakamura
Professor Emeritus,
Fukui Prefectural Univ. & Tokyo University of Fisheries

Shinya Otake
Professor, Fukui Prefectural University, Japan

1. The Needs of Marine Habitat Enhancement

1.1 *Increasing roles of creating habitats for marine ecosystems*

Man has created an enormous negative stress on the environment. Unless the present destructive trend is reversed, a catastrophe would arise for the planet's entire ecosystems, including our human societies. There is a global awareness today that the earth's capacity to absorb man's loading is limited, so that we must immediately reduce the loading that we have created and we must do it now.

We envision that, in the 21st century, our marine environments will inevitably be transformed into a wholly human-dependent and human-managed ecosystem. Under this scenario, the technology that could help create marine ecosystems artificially will become a cornerstone for our strategy to balance utilization and preservation of the earth's natural environment.

Humans have nurtured the agro-ecosystem in trying to feed its growing population, which now stands at 5.7 billion (1997). It seems inevitable that we are destined to rely continuously on habitat-control technologies on a progressively expanding scale. Thus, we eventually will not be able to help but embrace the aquaculture ecosystem as a source of sustainable food supply.

Two pressing issues that we are facing today are to supply food for rapidly growing populations and to maintain a balance between development and the preservation of natural environments.

1. To supply food for rapidly growing populations, we must endeavor to "create" harvestable ecosystems such as artificial fishing grounds and release of nurtured larvae.

2. To maintain a balance between development and natural environments, we must endeavor to "preserve" environments in the form of "artificially managed systems".

Thus, shallow marine geography may be altered by inevitable modification in terms of water quality and substrates to accommodate harvestable resources.

1.2 *Assessment of ecosystems in marine environments*

For the purpose of either utilizing or preserving marine ecosystems as resources, their value to humans and their contents must be quantified. Thus, the functions of the environment are divided into the three categories in Fig. 11.1 as a guide to the ecosystem-creating technology in perspective:

Fisheries landing statistics can be used to quantify reproductive functions of biological resources.

To understand biological purification functions, we must quantify the removal of suspended organic matter by filter feeders in aerobic situations, digestive degradation of organic mud by bottom feeders, uptake and decomposition of organic matter by bacterial growth, and transfer of nutrients to larger animals.

Physical purification mainly consists of disposal of nutrient-rich water into the outer ocean, creation of aerobic environments through flushing exchange of seawater, and quantification of the functions of various man-made marine habitats such as artificial mud flats, artificial kelp forests, artificial rock beds, and artificial reefs.

Amenity functions may vary widely by the location, as they are sensitive to specific types of environments such as urban, rural, and so-forth.

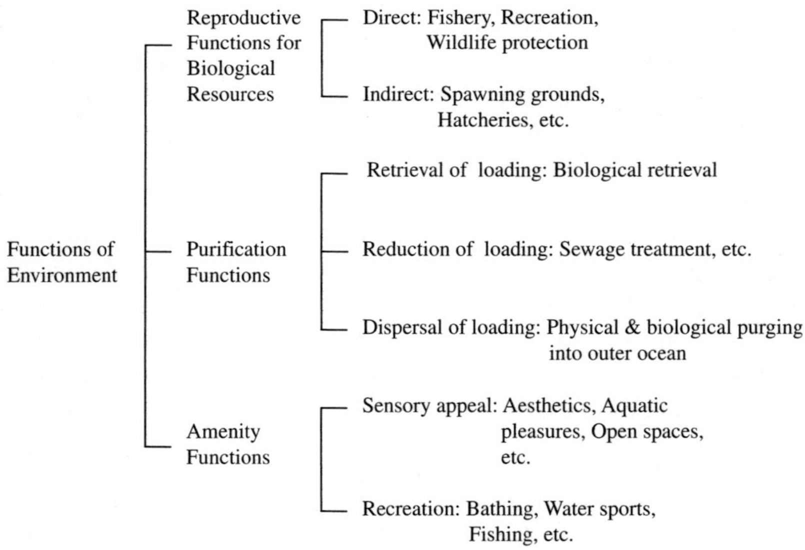

Fig. 11.1 Functions of marine environments.

All the individual components in this classification are amenable to quantification in economic terms.

1.3 *Pros and Cons for biodiversity, and wildlife protection*

Marine life suitable for human utilization belongs to the high echelon of the food chain. Their reproduction relies on ecosystems. On the other hand, the ecosystems useful to man do not necessarily ensure a high-level of biodiversity. What we need is a technology that will preserve species useful to man as preferred components in ecosystems. Also needed is an ability to quantify the roles of marine micro-organisms in the context of marine environments.

It is imperative that we protect endangered species. All our efforts essentially consist of conservation of genetic resources and are motivated by our love for all things living.

1.4 *Biological purification functions*

1. Removal of suspended organic matter in the water by filter feeders

The filtration mechanism by clams (see Figs. 11.2 and 11.3) is defined by

$$q = 0.012t^{1.25}w^{0.25}$$
$$t = 10 \sim 35^\circ, \quad w = 0.6 \sim 15\,\text{gr}.$$

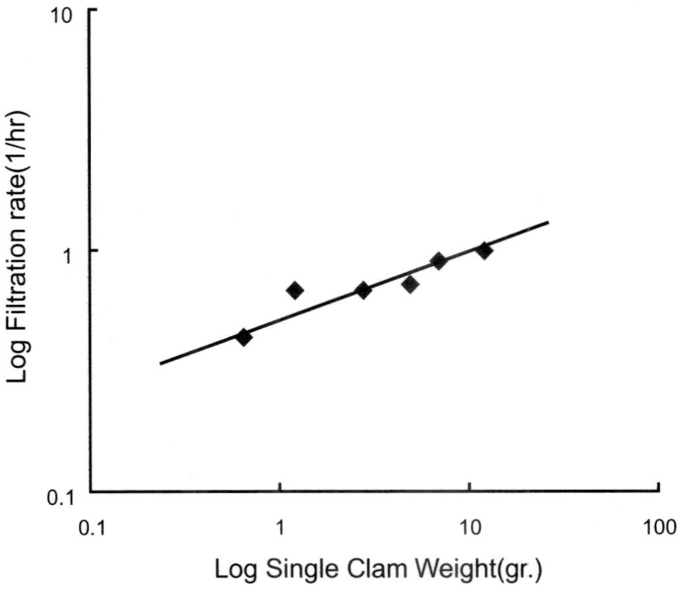

Fig. 11.2 Relationships between single clam weight and filtration rate. The filtration rate is proportional to the single clam weight in the logarithmic axes.

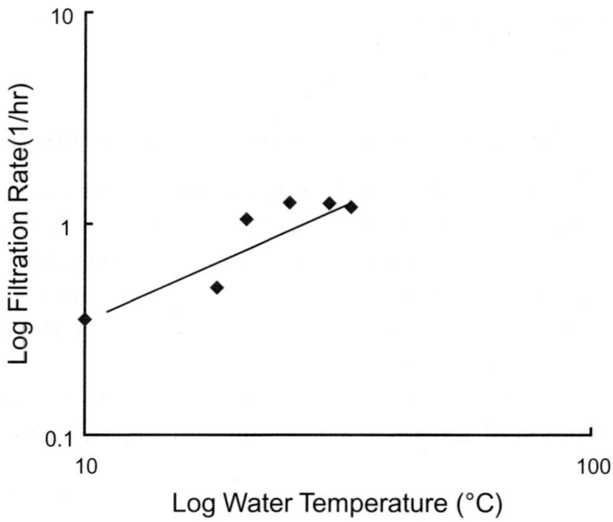

Fig. 11.3 Relationships between filtration rate of clam and temperature.

where q is the filtration rate (liter/hr) per clam, w (gr.) is the weight, including the shell, and t is water temperature (°C).

The creation of artificial tidal flats for clam fishing grounds leads to acceleration of the filtration and removal of suspended organic matter in near-shore waters. Thus, it results in on-land removal of nitrogen and phosphorous loading.

2. Decomposition of organic detritus by aerobic bacteria

Decomposition of organic detritus is used to measure oxygen consumption by aerobic bacteria (i.e., BOD: Biochemical Oxygen Demand).

Oxygen consumption velocity in nutrient-rich seabed (see Fig. 11.4) is given by

$$v = K_0 t^{1.33} \mathrm{C}$$

$$K_0 = 0.0031\,(\text{summer}) \sim 0.0021\,(\text{winter})$$

provided $C \geqq 3\,\text{ppm}$ or $2.01\ \mathrm{O_2/m^3}$ and $K_0 t^n$ is in the unit m/day, v is BOD (in $1\ \mathrm{O_2/m^2/day}$ or $\mathrm{grO_2/m^2/day}$) and C is the dissolved oxygen concentration in the sea-bottom water.

In order to understand the economic implications of v, we may compare it to the values in a modern sewage treatment plant.

2. The Plan and Design Criteria of Artificial Habitats

2.1 *Planning process for ecosystem creation*

As an analytical tool to rationalize the choice of desirable environmental parameters for a selected species in created ecosystems, we have developed a procedure

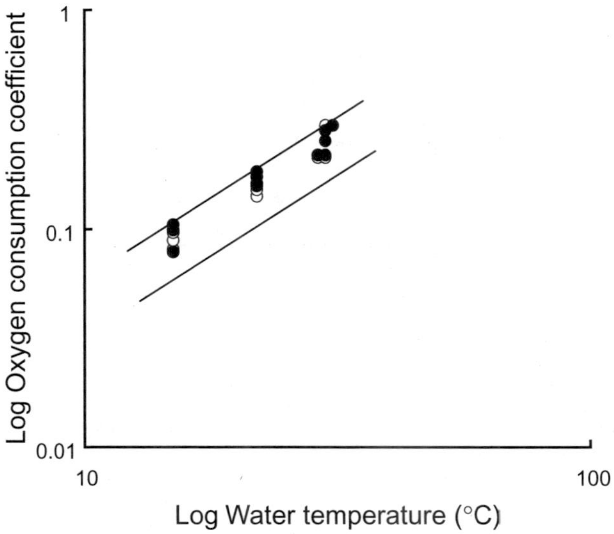

Fig. 11.4 Relationships between oxygen consumption coefficient and water temperature in the nutrient-rich seabed.

named Environmental Resistance Analysis (ERA). This analysis identifies a set of parameters that we must target to fulfill the necessary and sufficient conditions for enhancing and/or creating an environment of highly desirable state for a given species "y" at a given location.

First of all, let us consider a species denoted by "y". We assume that species "y" is monitored at all "m" stations, and that at each of these stations a set of environmental parameters, denoted by "X_N" ($N = 1, 2, \ldots$), were investigated simultaneously. Then we group the stations depending on the degree of ecological well-being of species "y". Group "A" represents the stations where this species is found in a desirable state A, group "B" represents the stations where species "y" is in an ordinary state B, and group C represents the stations where the species is in an undesirable state C.

We now compare the environmental parameters between groups "A" and "C", in order to identify a set of environmental parameters " X_{Ia}" which constitute a necessary and sufficient condition to elevate the species to the desirable state A. Figures 11.5 to 11.7 are referred to for the statements mentioned above.

In Fig. 11.5, a parameter "X_I" is unrelated to species "y" as far as this monitoring is concerned. In Fig. 11.6, "X_{Ia}" refers to necessary conditions for species "y" to be in state A. In Fig. 11.7(a), "X_{Ia}" may appear to be the necessary conditions for the species "y" to be in state A. However, this impression is false since there are also points where species "y" is found in state C. Under this condition, we examine the relationship between species "y" and other environmental parameters "X_J". As shown in Fig. 11.7(b), if we find that species "y" belonging to state C within the range of "X_{Ia}" in Fig. 11.7(a) is not in state A due to the condition of "X_J" in

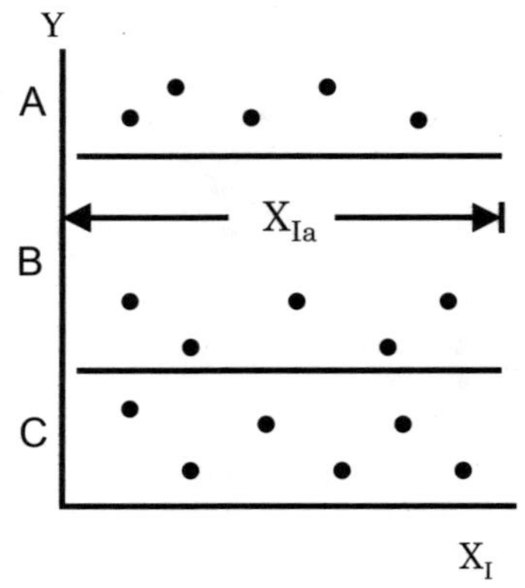

Fig. 11.5 Environmental Resistance Analysis: Case (1). The environmental parameter "X_I" is unrelated to the species "y".

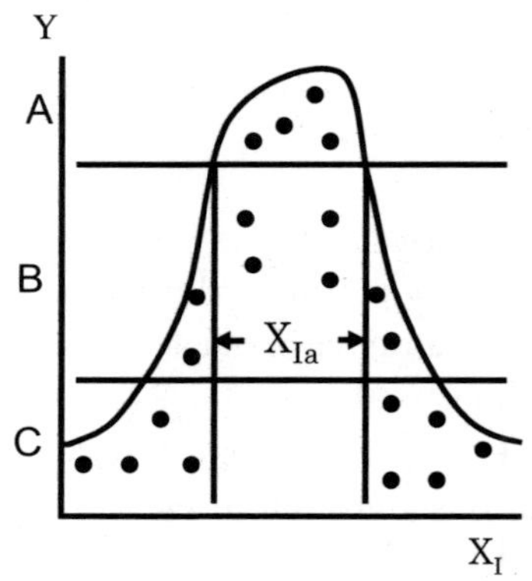

Fig. 11.6 Environmental Resistance Analysis: Case (2). The environmental parameter "X_I" is a necessary condition for the species "y" to be in the state "A" within the range "X_{Ia}".

Fig. 11.7(b), this "X_{JC}" would be a restrictive condition that prevents species y from entering state A. In this case, "X_{Ia}" is a necessary condition for species "y" to be in state A. If a restrictive condition "K" in "X_{JC}" can be improved to be a sufficient condition "K"' in "X_{Ja}", species "y" can be moved into state A.

We may carry on this analysis to involve all the individual parameters in the set $X_N(N = 1, 2, \ldots)$. Thus, we can eventually select the optimum set of the parameters that offers the necessary and sufficient conditions to elevate species "y" into state A.

2.2 *Artificial fish reef habitat*

(1) Function

Artificial fish reefs protect and propagate fish populations, thereby contributing to increasing the catch and the efficiency in fisheries operations. In the construction

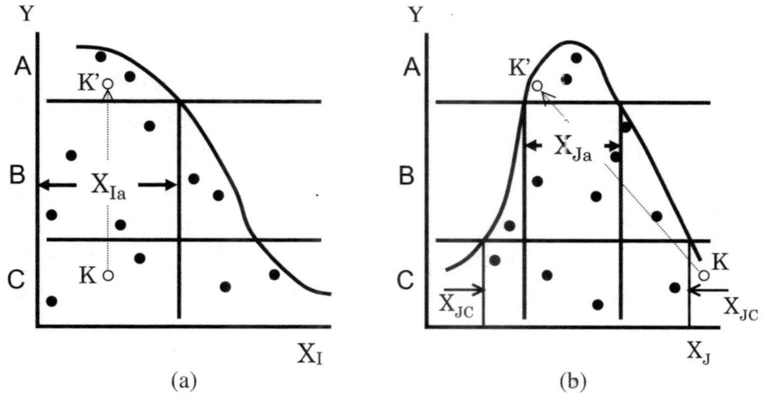

Fig. 11.7 Environmental Resistance Analysis: Case (3). Fig. 11.7(a) shows that "X_{Ia}" may appear to be a necessary condition for the species "y" to be in the state "A", but this impression is false since there are also points where the species "y" is found in the state "C". Figure 11.7(b) shows that the presence of other environmental parameters "X_J" is included in the "C" in spite of the necessary condition "X_{Ia}". This parameter "X_J" is called a restrictive condition.

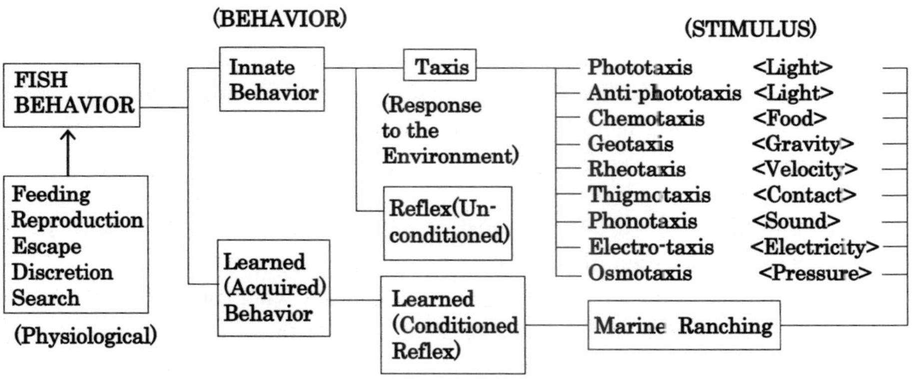

Fig. 11.8 Environmental stimuli and fish behavior.

of artificial fish reefs, we take advantage of behavioral characteristics of fish such as their tendency to aggregate in and around natural fish reefs and sunken ships.

Reefs in a widely varying topography can be attractive habitats for various species of fish. When we look at behaviors of fish, they show certain taxes, or naturally guided movements in response to stimulus. Examples of various forms of taxes are listed in Fig. 11.8.

Animal behavior is triggered by natural intuitive impulses and displayed when the animal's sense organs are stimulated. There are various forms of taxes, e.g., phototaxis caused by fish's sensitivity to light, chemotaxis triggered by the smell of chemical components in the feed, geotaxis (a fish intuitively keeps its abdomen down, but this may be interpreted as its anti-phototaxis) and rheotaxis, caused by the animal's sense of movement. Animals determine their body positions in relation to the direction of surrounding objects or their movements (thigmotaxis — position is determined by sense organs on the skin and the lateral line in reference to the position or movements of other objects), or certain sounds (phonotaxis). This behavioral characteristic can be used in the acoustic learning of fishes.

Intuitive behavior is a set of inborn behavioral patterns triggered by inner and outer conditions, including activities such as feeding (searching for food), reproduction, escape, imitation, and research. Reflexes are quick, regardless of innate or learned reactions. An innate reaction to any stimulus is known to be an unconditional reflex. Learned reactions (through a learning process) are called conditional reflexes.

Schools of fish congregate in and around fish reefs because of their taxis and intuition. When designing the structure of a fish reef, it is important to know the functions of the sense organs of fish. In an effort to create environments that are attractive to fish, fisheries engineers take advantage of these behavioral patterns of fish.

The reaction to fish reefs differs from fish to fish. Although there is no quantitative definition, the tendency of fish to gather around fish reefs is well-known and is called "reef affinity."

By means of the relative positions that fishes maintain within the reef, we can classify them into three to five categories. But it should be noted that such relative positions, and therefore, fish classification based on them, change according to the fishes' stages of growth (see Fig. 11.9).

Type I: The fish keeps a part or most of its body in close contact with the reef (examples: greenling, rockfish, black rockfish).

Type II: The fish does not directly come into contact with the reef, but the school swims around the reef or positions itself around the reef at the seabed (examples: red sea-bream, crimson sea-bream, chicken grunt, parrot bass).

Type III: The fish swims close to the sea surface or in the intermediary waters far from fish reefs (example: yellowtails, amberjack, skipjack, dorado, horse mackerel, and mackerels).

Type IV: The fish aims around reefs at the seabed (examples: right-eyed flounder, sole, bastard halibut).

(2) Structure

The most important issue in artificial reef design is how to make the best use of the target fish's taxis. For fish that gathers around reefs, spacing is an important concern because they select habitats according to their body size. For example, Type I fishes such as the black rockfish, which shows thigmotaxis, will need space and clearance for all sizes.

As for the Type II fish, e.g., sea-breams, chicken grunts and common nibblers, their phototaxis behaviors get stimulated by vision and their phonotaxis is activated

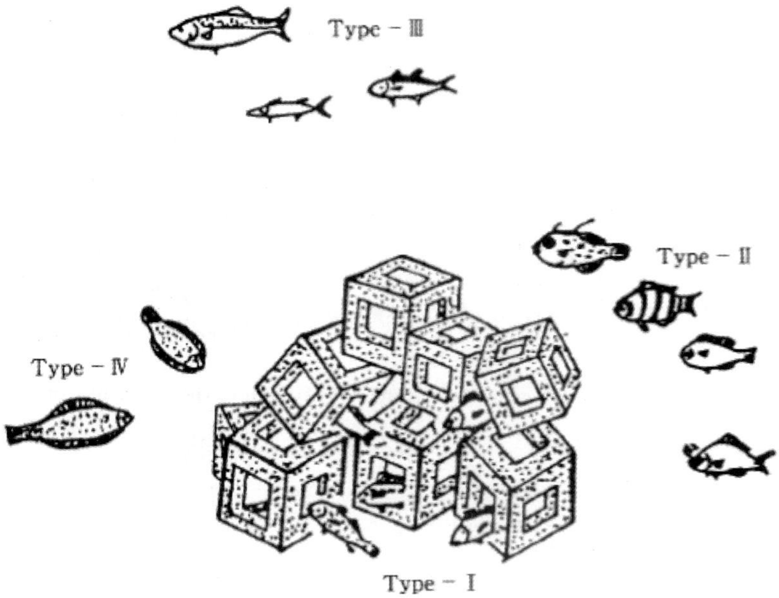

Fig. 11.9 Fishes classified according to their typical position relative to the reef.

by the lateral line. For them, the best clearance size is from 1.5 m to 2.0 m so that and they can continuously keep objects in view. Fish eyes are true spheres and can see quite a wide range, as far as within 150° around the eye. However, their length of vision is only about 1 m, due to very poor resolution. Therefore, ideal fish reefs for Type I and II fishes, in other words, bottom fishes, should have space and clearance of various sizes, but most preferably a distance of less than 2 m.

Many fishes leave reefs at night when their vision deteriorates. Those remaining in reef areas position themselves by detecting the differences in water pressure caused by turbulence in the flow that arises from the physical structures of the reefs. The minimum requirements for driving fish away from turbulence caused by the components of the reefs are shown in the following equation.

$$Bu > 100 \text{ (Unit: cm, s.)} \tag{1}$$

where B is the width of the component and "u" is the flow velocity.

When the velocity increases, for example during tidal flow, fishes with a low swimming capacity move to the downstream of the reefs. If the downstream flow velocity is shown by u', and the upstream flow velocity by u, then the following equation holds:

$$u' = u\left(1 - \frac{C_D A}{2S}\right) \tag{2}$$

where C_D is the drag coefficient of the fish reef, A is the actual projected area which receives fluid pressure, and S is the area of the upstream flow tube that corresponds

Fig. 11.10 Wakes generated in the downstream of the artificial reef.

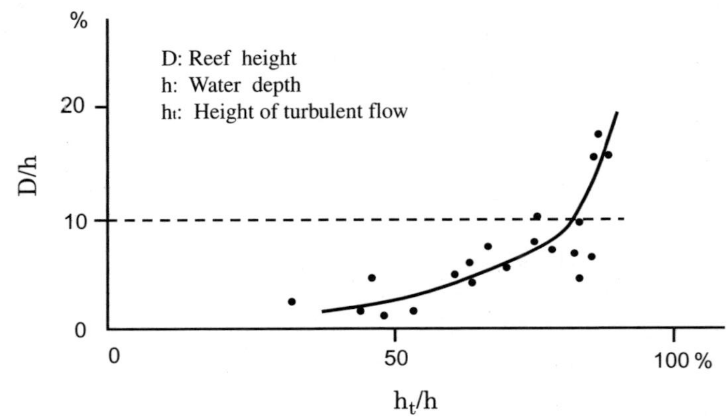

Fig. 11.11 Correlation between the height of the reef and the height of turbulent flow.

to the flow line of the hind-stream formed by the entire reef. The variable S is almost the same as the nominal projected area of the entire fish reef, and this area includes all the internal areas, including empty space (see Fig. 11.10).

In summary, artificial fish reefs targeted at bottom fishes (Type I and Type II) should have varying spaces of no larger than 2 m apart so that they have the width of the component obtained from Eq. (1). Also, they should be capable of creating a flow rate u' behind the reefs, as obtained from Eq. (2).

In the case of pelagic fishes (Type III), it is important to place artificial reefs at the surface and in intermediate sea areas. The reefs should have sufficient height to be able to interrupt the water flow and create fluid noise. Interruption of the flow and fluid noise can be calculated from Eqs. (1) and (2).

Figure 11.11 shows the height of turbulent flows in reference to the reef height/water depth ratio. The figure was prepared by estimating the heights of turbulent flows from records of fish sounders. In this figure, we can see that turbulent flows develop rapidly up to about 80% of the water depth. if the height of the reefs is about one-tenth (10%) of the water depth. Turbulent flows do not develop further when the reef height exceeds 10% of the depth. (Turbulence recorded by fish sounders are considered internal waves.)

From Fig. 11.11, we can conclude that the optimal height of artificial reefs for Type III fishes is about one-tenth of the depth of the water.

Sometimes the flow conditions at the seabed influence the upper layers. One example is shown by topographic waves (Lee waves). When the density gradually increases from the surface to the bottom, and if there is only a mild flow, artificial reefs located at the seabed create topographic waves, a type of internal waves, as shown in Figs. 11.11 and 11.12. Topographic waves frequently develop when the reef height/water depth ratio is 1/10, and when the Froude number F_r equals 0.09. The Froude number is defined by the following equation:

$$\left.\begin{aligned} F_r &= \frac{u}{\sqrt{\varepsilon g h}} \\ \varepsilon &= \frac{\rho_2 - \rho_1}{\rho_2} \end{aligned}\right\} \tag{3}$$

where u is the velocity, g is the acceleration of gravity, h is the water depth, and ρ_1 and ρ_2 are the densities of seawater at the upper and lower layers.

At the maximum flow rate, $F_r > 0.32\ (=1/\pi)$, a topographic wave will be pushed back and the area behind the reef will flow as shown in Fig. 11.13. This is provided by the movement of pelagic fishes to deeper waters in bad weather.

The correspondence between topographic waves and the behavior of Type III fishes has been assumed by the present authors but it has not yet been proven. Gathering the data for research in the real sea is quite difficult. Nevertheless, applying our assumptions here, we can use the following criteria to determine reef location:

If $F_r \leqq 0.32$, a submerged reef is better. (4)
If $F_r > 0.32$, a floating reef is better.

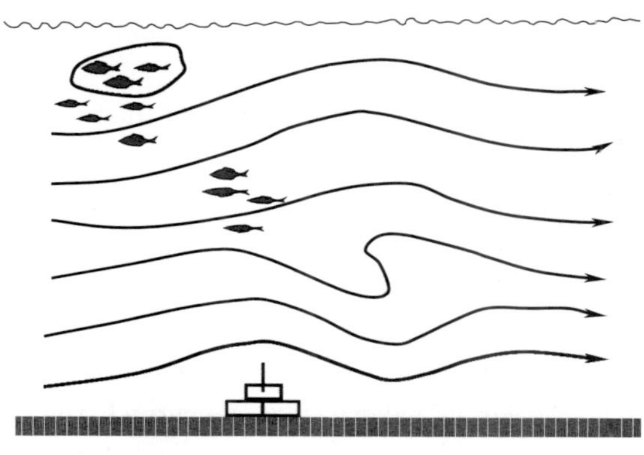

Fig. 11.12 Formation of topographic waves generated by artificial reefs.

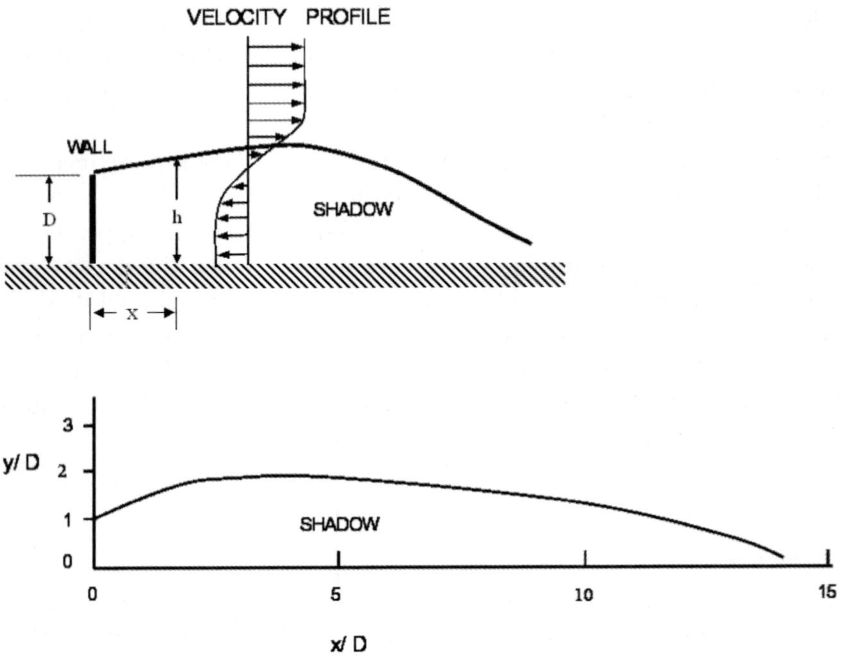

Fig. 11.13 Disturbance range of current caused by a two-dimensional obstacle.

(3) Construction

Artificial reefs should be laid out as shown in Fig. 11.14. The terms used in artificial reef construction are defined as follows:

Fish reef unit: The single, smallest unit structure used for constructing artificial fish reefs.

Unit fish reef: The smallest fish reef habitat created by placing one or more fish reef units together.

Fish reef group: An artificial reef habitat created by combining more than two inter-related fish reef units.

Fish reef zone: An expansive area containing more than two fish reef groups.

2.3 *Rocky reef habitat*

(1) Function and Formation

A sea area with a heavy growth of seaweed is called a seaweed bed, and beds filled with larger seaweed, e.g., edible seaweed and kelp, (*Ecklonia cava*) , are sometimes called marine forests. As far as fisheries are concerned, there are three types of seaweed beds that grow in and around rocky reefs: (1) Seaweed collection areas where fishermen can harvest seaweed for human consumption or for use as feed.

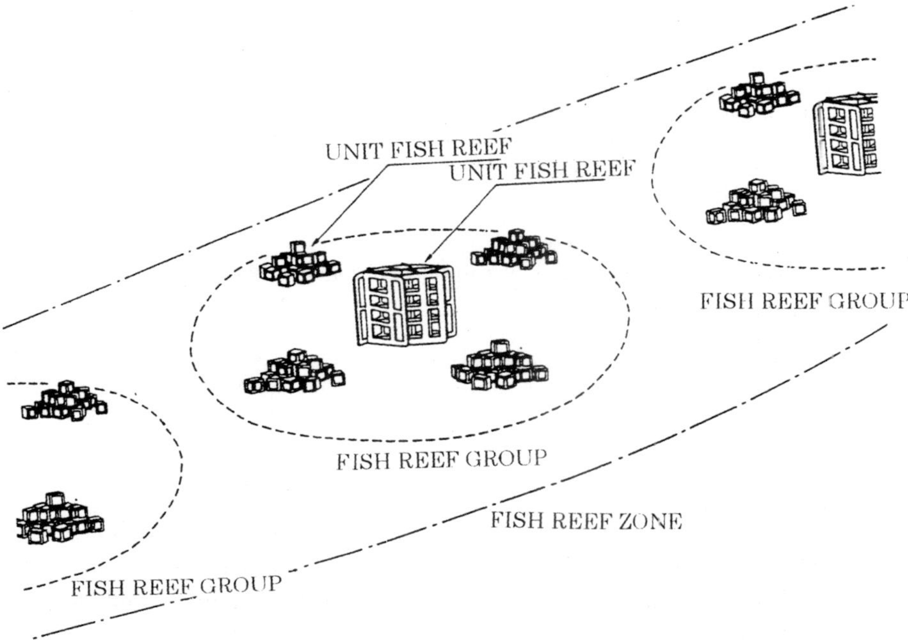

Fig. 11.14 Conceptual sketch of artificial fish reefs.

Kelp and undaria (Wakame seaweed) species are major plants in this category. (2) Seaweed feed areas which can support abalone and sea urchins by providing abundant food, and which often make good fishing grounds. (3) Nursery seaweed beds in mild environmental conditions, such as mild current flows and large amounts of suspended nutrients, e.g., detritus, microscopic plankton and benthos.

Nursery seaweed beds can support small life forms by serving as hideaways and resting grounds. They are ideal for nursing larvae and young fish since they offer a rich source of organic nutrients, thus supporting primary producers in the food chain. Seaweed beds produce large amounts of dissolved oxygen by means of photosynthesis so that they are quite effective at keeping the marine environment clean.

The purpose of seaweed bed construction is two-fold: restoration of lost or damaged beds and the creation of new ones.

Loss of or damage to seaweed beds can occur because of changes in the axial flow of ocean currents, geographical conditions by natural forces such as floods and typhoons, and man-induced changes such as the development of rivers, ports and waterways, and water quality changes. Once these beds are lost, they usually do not return to the original form even if the causes are removed and natural physical and chemical states are restored. We notice that the existence of a seaweed bed is itself a propagation factor.

Table 11.1 Restoration and creation of seaweed bed in a rocky reef.

Restoration of lost seaweed beds.
Case of going loss or damage to seaweed beds.
Creation of a new seaweed bed after removal of the causes of loss or damage.
(Installation of anchor ropes to induce spore attachment,
Transplantation of mother seaweed, Scattering of spores, etc.)
Case of delay in restoration after removal of the causes of loss or damage.
Improvement of seaweed bed environments.
Interference with ecosystem.
Creation of environments similar to those of original beds.
Creation of new beds.
Improvement of seaweed bed environments.
Settlement of base materials for attachment.
Creation of environments similar to those of original beds.

Seaweed beds greatly contribute to creating a calm sea area, which is necessary to prevent the dispersion of spores and larvae. Small water-mass movement in the seaweed ground helps maintain stable water temperature, salinity and tidal flows. Thanks to this feature, seaweed beds often retain abundant nutrients, while minimizing the effect of littoral sediment transport. Seaweed beds form a stable biosphere where the population of herbivorous animals and the growth of sea vegetation are in equilibrium. Thus, we can only help restore the life-supporting functions of seaweed beds by improving their environments and other essential factors.

If seaweed beds diminish owing to adverse natural conditions, and herbivorous fish begin to eat sprouting weeds, it will eventually lead to a total destruction of seaweed grounds. In similar cases, Volterra's equation applies to terrestrial animals but it is not meant to apply to marine forests.

Volterra's equation explains the natural food chain in the following manner. In North America, for example, as the hare population increases, the number of their predators, wolves, also increases. As a result, the hare population drops significantly. However, this then leads to a reduction in the number of wolves, which in turn helps restore the hare population.

It is known that abalone and sea urchin populations will not decrease even after the seaweed which supports them has perished. This is one of marine life's major differences from animal life on land.

As described above, seaweed beds can hardly be restored naturally once they are lost, and this is why it is so important to exert effort to preserve or restore them by artificial means. If seaweed restoration does not succeed even after dedicated efforts, it is then necessary to implement careful management and control measures, such as installation of anchor ropes to induce spore attachment.

For growth, seaweed beds require an adequate supply of seeds (e.g., spores and larvae), substrates or base materials to which spores can attach themselves, mild life-supporting environments, and protection against predators.

In consideration of these requirements, seaweed bed restoration efforts usually focus on the following:

- Propagation of existing seaweed beds if any part of the original beds has remained intact.
- If seaweed beds have been lost, creation of environments similar to those for original beds.
- Elimination of herbivorous predators.

(2) Submerged Dike for Creating Circular Flows: Formation of Abalone and Sea Urchin Propagation Grounds.

The use of a submerged dike to obtain circular flows is one of the common techniques used to improve seaweed bed environments. The basic requirement, as stated earlier, is a mild current flow to check the loss of spores and seeds. Figure 11.15 shows a submerged dike which reduces the effect of waves and creates circular flows by tapping the wave force. The stationary water area thus created minimizes the loss of spores and zoospores.

Generally speaking, stationary waters are significant for fisheries. A technique which encloses a part of the sea to protect spores without deteriorating the water quality is very important. However, as we well know, it is quite difficult to contain the flow of water and a technique for achieving this is one of the major challenges for fisheries engineering.

If stationary water areas can be developed, it will stop both spores and nutrients from dispersing. The fact that only a little change occurs in the water temperature and salinity in stationary waters makes the technology to obtain calm waters one of the most useful, not only for fishing ground propagation but also for the fisheries industry in general.

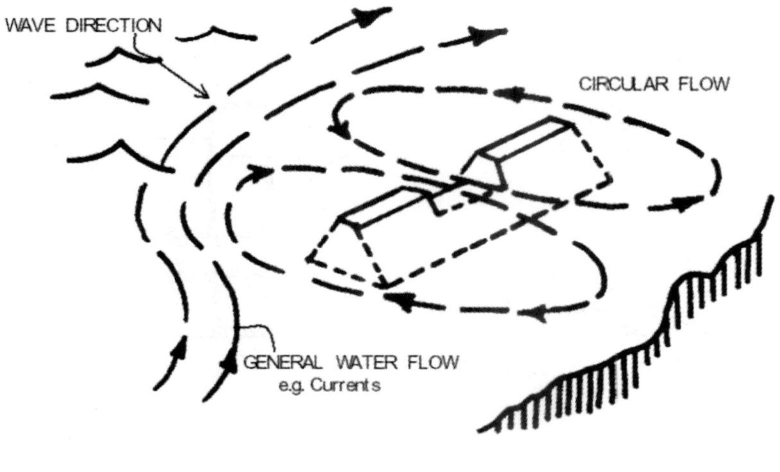

Fig. 11.15 Circulation created by a submerged dike.

When waves move into shallow areas, water particles display reciprocal motions so that they move in the direction of the waves, since waves carry on them. When waves break, they create violent streams and rush toward the shore. It is possible to induce circular flows if we use this principle. First, submerged dikes will be built to break waves at the top of the bank. Then water is led through a waterway set up between adjacent banks. This engineering technique is illustrated below.

(3) Current generation technique by wave energy concentration

In yet another technique, wave energy can be turned into the flow of seawater. Figure 11.16 shows several methods for achieving this. In (a), waves are focused onto a narrow area to tap their energy. Thus, if wave movements are carefully directed, the flow energy increases the height of the waves. Water flows over the intake weir and turns into gushing streams. In (b), water intake weirs are installed below the average water surface. In this design there is a wave setup or surging of the water level behind the weir. This surge is led into waterways to form streams of water.

Design (c) takes advantage of the reciprocating flow of waves. Blocking the return flow can increase the water level.

It is possible to introduce seawater into aquaculture beds, fish shelters or propagation grounds by raising the water level through any of the aforementioned methods.

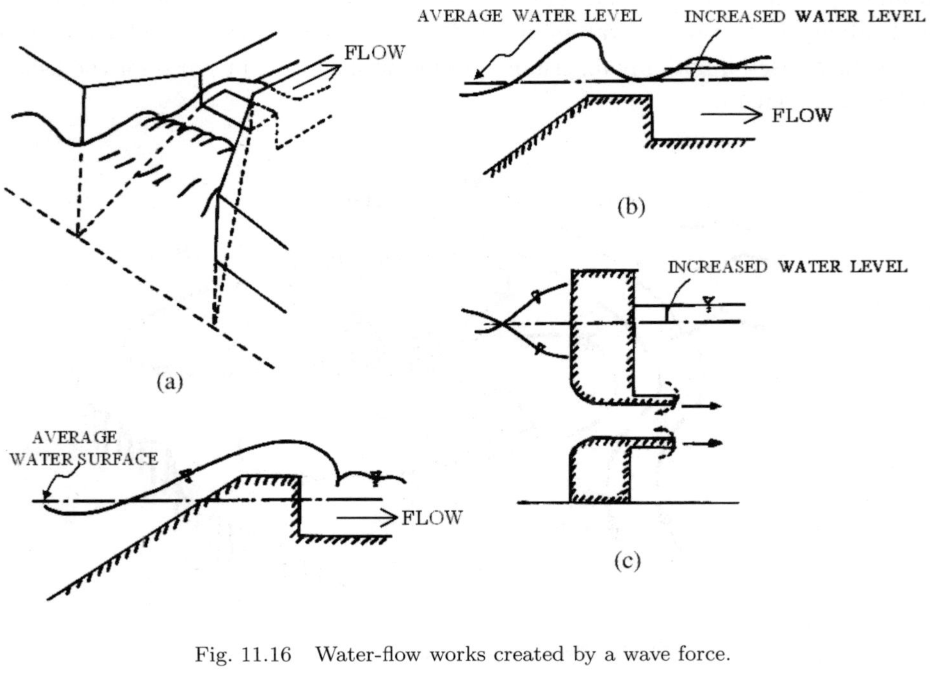

Fig. 11.16 Water-flow works created by a wave force.

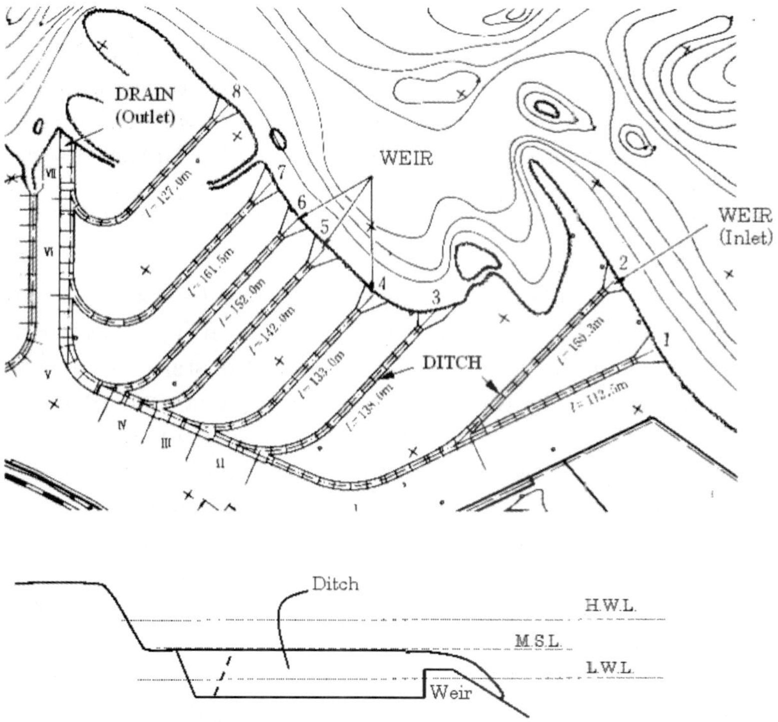

Fig. 11.17 Nursery grounds of abalone and sea urchin created in Ohama Coast (Taneichi), Iwate Prefecture, Japan.

Figure 11.17 is a cross-section of propagation ditches dredged in a flat seabed in Taneichi, Iwate Prefecture. The seabed is made of soft rocks and lies below the average tide water level. During ebb tides, however, the reef bed is exposed, so this area did not support any seaweed. Dredging several rows of ditches with submerged dikes (see Fig. 11.17) can develop formerly useless flat seabed areas.

Seawater intake weirs were built in the area facing the open sea. After the dikes were installed, it became possible to obtain streams of water with a flow velocity of 0.15 m/s during ebb tides. Kelp was planted in the area to culture abalone and sea urchins. This project turned out to be of great success.

In a smaller project, 98 ditches were dredged in other areas. Altogether, 42 hectares of culture grounds were created in Iwate Prefecture.

2.4 *Sandy coast habitat*

(1) Bivalve habitat

Among bivalves living in sandy coasts, short-necked clams and hard clams favor semi-enclosed water areas such as inlets, while hen clams, *Gomphina melanaegis*, and shield clams prefer sandy shores which are open to the sea. The blood ark shell and ltaya scallops require substrates to attach their spawn and larvae. Because of the great variation in the habitats and growth patterns of bivalves, fisheries engineering targeted at bivalve farming naturally varies with the specific purpose of each project. Some of the common factors in bivalve culture are discussed below.

(a) Determining the Volume of Mother Shells Suited for Larval and Juvenile Diffusion Characteristics

Larval loss is expected to be less intense in inlets and tidal basins, so our major concern is in maximizing the survival rate of larvae and their growth into adulthood. However, in areas where the larval population is small to start with, or where larvae can be easily carried away by the water, proper arrangements must be provided to reduce loss. If any sea area has experienced several years of stable year classes, it is reasonable to expect a large number of larvae growing in the area and attached to substrates or nursing beds. If a dominant year class has been observed, it is important to enhance larval growth (e.g. protection of mother shells, creating of spawning grounds) and to take measures to prevent dispersion of the spawn, for example by creating stationary water areas.

(b) Implantation — Environmental Conditions

For some bivalve species whose larvae directly implant themselves in sand at the bottom of the sea, clean environments with little mud are indispensable. On the other hand, larvae, which attach to pieces of wood, seashells, etc., need materials that will function as substrates.

A reciprocating flow pattern is another prerequisite for suspended larvae to congregate in certain areas or to move reciprocally.

(2) Water Route-making Engineering

As shown in Fig. 11.18, water route-making engineering involves excavation of tidal basins or shallow water areas to produce waterways or water routes. Waterways enhance the flow of seawater and so promote the exchange of water and improve

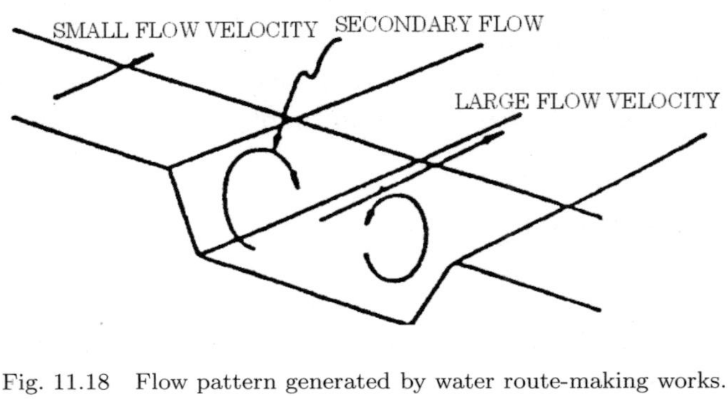

Fig. 11.18 Flow pattern generated by water route-making works.

the water quality at the seabed. A large flow velocity prevents sedimentation of suspended mud. This keeps the silty and sandy bottoms of the water route clean and enhances the attachment of young shellfish.

A greater flow velocity and flow volume can carry large amounts of larvae so that they increase the chances for attachment to substrates. Figure 11.18 shows the formation of secondary streams which promote this action further. The environment created by water route-making engineering is good for feeding shellfish if they use siphons to absorb nutrition. This engineering technique is mainly used to form nursing grounds for larvae and young shellfish.

(3) Waterway Engineering for Creating Circular Flows

As illustrated in Fig. 11.19, waterway engineering for creating circular flows aims at creating nursery grounds for young shellfish by dredging waterways in tidal basins. The figure shows the technique actually used in hard bed formation in Buzen Sea, Oita Prefecture.

When the tide level is low during the early stages of a rising tide or the final stages of ebb tide, short training walls narrow the flow of water and turn it into a greater movement of seawater. This naturally increases the tractive force that contributes to creating a waterway. At the same time, circular flows appear behind the training walls.

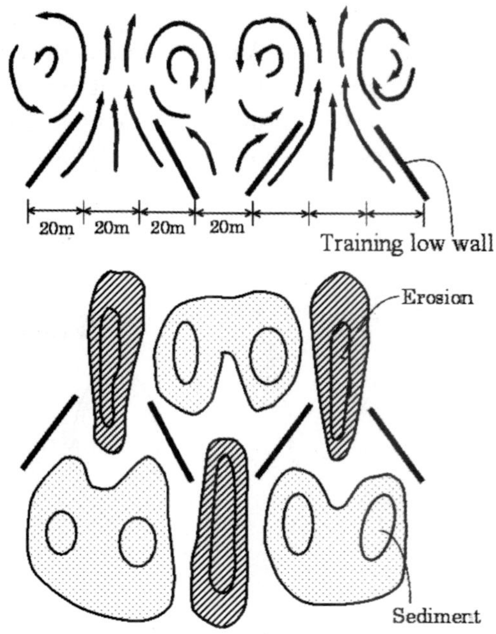

Fig. 11.19 Flow pattern and sedimentation in a tidal basin generated by short training walls works.

(4) Tidal Basin Engineering — Excavation and Filling

This technique is used to adjust the height of a tidal basin when it becomes inappropriate for the growth of shellfish, due to land settlement, natural development of the basin or other causes.

3. Artificial Upwelling

The technology to generate artificial upwelling is efficient for the case of the creation of offshore fishing grounds. Some practical examples of artificial upwelling are introduced here. Figure 11.20 shows the conical artificial reef set up in the Sea of Enshu, off Shizuoka Prefecture. Figure 11.20(a) shows the paired vortices which are created by crashing currents against the conical fish reef. These vortices have their own drivingen force, as shown in Fig. 11.20(b), and produce a horseshoe-shaped upwelling behind the reefs. Post-larval fishing grounds have been made around the upwelling. Figure 11.21(b) shows the artificial upwelling structure, called an artificial mountain on the bottom of the sea. It is often set up on the continental shelf at a depth of about 100 m. The purpose is to build an area that increase fishes which live in juvenile. We had the experience of carrying out one experimental case off Nagasaki Prefecture. The same project is also being carried out at several places for fishing grounds on several Japanese coasts. The upwelling generation mechanism can be explained as follows: The flow beyond the reef is fast, whereas the flow detouring around the reef is slow, as shown in Fig. 11.21(a), which represents the phenomenon occurring in natural reefs. Also, the flow at the central area behind the reef receives the detouring current. Thus, the flow gathering from three directions makes an upward flow. Furthermore, this flow promotes the upwelling caused by the occurrence of upward vortices as shown in Fig. 11.20(b).

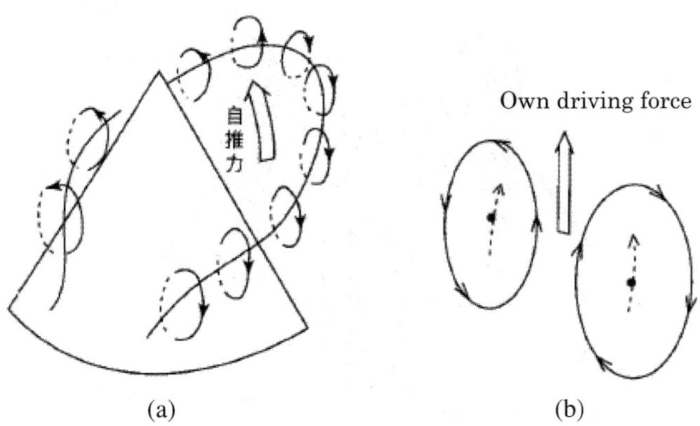

Fig. 11.20 Generation mechanism of upwelling behind the conical reef. (a) own driving force by horseshoe vortex; (b) own driving force by paired vortices.

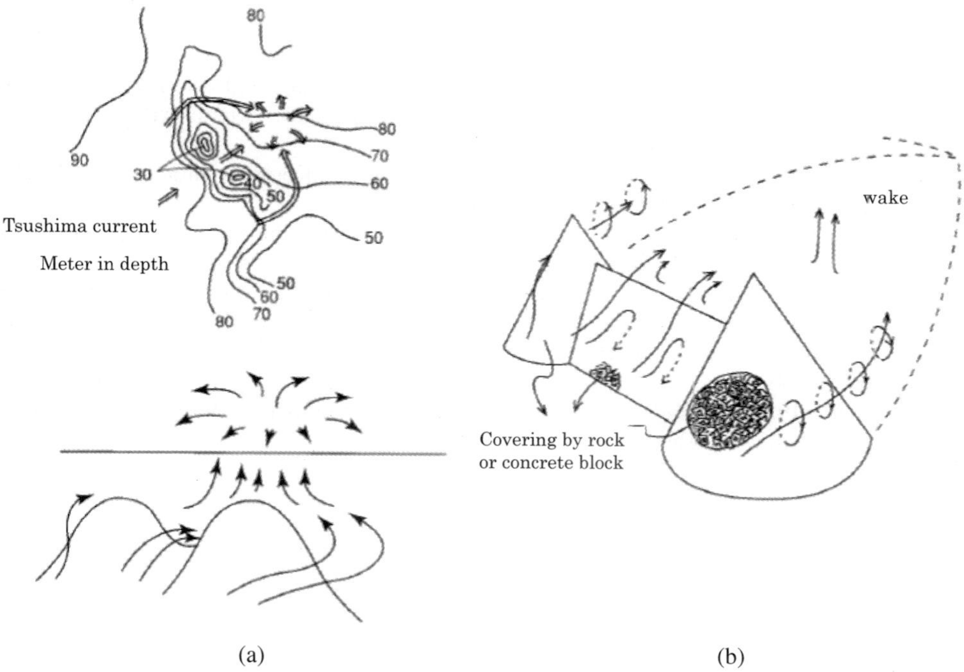

Fig. 11.21 Natural reef and artificial mountain on the seabed. (a) Shiomaki-sho, off Yamaguchi Prefecture (natural reef), and the upwelling generation mechanism, (b) Sea mountain set up off Ikitsuki Island, Nagasaki Prefecture (82 m in depth) and the upwelling generation mechanism.

4. Conclusions

The ecosystems in marine environments are destined to become a system which is impacted and manipulated by mankind. However, mankind is still far from understanding the functions of natural environments. Our ignorance of the functions of natural environments may lead us to manipulation of natural ecosystems by trial and error, but any serious mistake should be avoided by all means.

The future prospect and research needs are as follows:

- The study of behaviors of marine organisms, particularly their response to physical environments.
- The identification, formation and control of environments in which marine life proliferates.
- Hydraulic designs to enhance accident prevention and required functions.
- Marine environment control technology which makes the best use of natural marine energy such as tides, waves, currents, wind, internal waves and so forth.

Let us hope that it is not too late to start a serious effort to understand the ecosystems.

References

Nakamura, M., ed. (1980): Fisheries Engineering Handbook, Fisheries Engineering Research Subcommittee, Japan Society of Agricultural Engineering, Tokyo (in Japanese).

Nakamura, M. (1985): Evolution of artificial fishing reef concepts in Japan, *Bull. Marine Science* **37**, 271–278.

Nakamura, M. (1991): Fisheries Engineering, marine ecosystem marine environmental engineering, (Suisan Dobokugaku), Tokyo, 561 (in Japanese).

Otake, S., H. Imamura, H. Yamamoto and K. Kondou (1991): Physical and biological conditions around an artificial upwelling structure, in proceedings, Japan-US Symposium on Artificial Habitats for Fisheries, Southern California Edison, Rosemead, CA, 299–310.

Volterra, V. (1928): Variation and fluctuations of the number of individuals in animal species living together, *Animal Ecol.* 409–448.

Chapter 12

Hydro-Environmental Assessment of Coastal and Estuarine Systems

H. Lee Butler
Principal Officer
Veri-Tech, Inc., formerly USAE
Waterways Experiment Station, USA

Carl F. Cerco
Research Hydrologist
USACE Engineer Research and
Development Center, USA

1. Introduction

As a consequence of rapid pace of development and population growth in coastal watersheds, many of the world's coastal seas are subject to "cultural eutrophication" — excessive discharge of nutrients into the shallow sea stimulates algal production which leads to decreased water transparency and loss of submerged aquatic vegetation (SAV). Dead algae settle to bottom waters, where their decay consumes dissolved oxygen. Loss of food and shelter provided by SAV and the absence of vital dissolved oxygen leads to decline in fish population and other valuable living resources.

The Chesapeake Bay is the prime example of an ailing coastal ecosystem. Although nature provides significant oxygen input to the Bay (atmosphere and photosynthesis from aquatic plants), significant portions of the Bay contain little or no dissolved oxygen during summer, which in turn threatens marine life. The eggs and larvae of fish may die; growth and reproduction of oysters and clams are impaired; and adult fish find their habitat reduced and their feeding affected. An overwhelming reason to restore the Bay is that its living resources are under direct assault. This chapter discusses hydro-environmental modeling by example of a recent application to Chesapeake Bay, on the mid-Atlantic coast of the United States.

1.1 *Chesapeake bay a trend setter*

A few years ago, the literature was expressing the view that in striving toward better and more sophisticated models, one day three-dimensional model technology would be a standard tool. That day has come and, in a sense, passed. Once success has been achieved, such as in the case of Chesapeake Bay described here, it becomes impossible to turn the clock back and use a less capable technology. It is likely the judicial or a public intervenor court system will demand the best when sensitive environmental issues are involved.

The Chesapeake Bay study (Robey and Lower, 1991) represented the state of the art in eutrophication modeling. It achieved a major first in the planning, conduct, and overwhelmingly satisfactory completion of a large practical study coordinated and evaluated by a selected group of representatives formed from Federal, State, and local agencies and academia. The comprehensive model met or exceeded the goals specified at the outset, exhibiting a high degree of correspondence to observations of the system and to current consensus regarding eutrophication processes in the Bay. The comprehensive model package was fully accepted and the first to be used as a primary guide for cleanup strategies.

Significant advances over the previous state of the art were achieved. The first is coupling of fully three-dimensional hydrodynamic and eutrophication models together with a comprehensive watershed model. Then, there is the coupling to a fully-predictive sediment oxygen demand and nutrient flux model. The use of Lagrangian residual hydrodynamics (Dortch *et al.*, 1992) for inter-tidal transport modeling was crucial for allowing cost-effective simulation of long-term environmental conditions in an accurate manner.

Long-term hydrodynamic and water quality simulations demand substantial computer resources, both computational cycles and memory for online data storage. Completion of multi-year three-dimensional hydrodynamic simulations had not been achieved when this study was initiated in 1989, and the thought of multi-decade eutrophication simulations captured the imagination. It was questioned whether or not one could actually hindcast 30 years of Bay decline. Realistically, comprehensive data sets for such a long period did not exist; however, various field data collection efforts were periodically conducted, and the data did support qualitative acceptance of the results of such simulations.

1.2 *Political and social environment*

The restoration of most degraded water bodies, especially true for the Chesapeake Bay, does not lie with any one agency or just within government. What is required is that all relevant governmental agencies and the general public do their part to communicate with each other to ensure that programs and activities are in harmony and lead to the mutual goal — a healthy and productive resource and its wise use.

Fortunately, for America's largest estuary, the Chesapeake Bay Program (CBP) was formed in 1983 at a time when a greater awareness and growing concern was felt by the States and local communities surrounding the Bay. A call to arms was sounded when a summary report concluded the volume of anoxic bay water increased by an order of magnitude from 1950 to 1980 (Flemer *et al.*, 1983). Other indicators, including diminished fisheries harvest and shrinking distribution of SAV, supported the judgement that natural resources of the Bay were deteriorating. The potential loss of Bay resources spawned increased activity in Bay monitoring, ecological research, and modeling for management purposes.

The CBP had the wisdom to develop a computerized data management system to compile and evaluate data collected by individual CBP projects as well as those conducted by other research efforts. The information assembled in the CBP database is considered to be the most extensive body of scientific knowledge on any single estuary in the world. More important, the database provides a common set of knowledge about the Bay's ecological problems, a pre-requisite necessary to carry out individually and collectively the most urgent task of establishing common goals for action to improve the Bay. It goes without saying that this rich resource of data is also necessary for verifying the suite of models to be used for management purposes.

Historically for the US, the CBP was a unique cooperative effort between State and Federal agencies to restore the health and productivity of America's largest estuary, forming a strong lobby for a favorable Presidential decision to make the Bay "clean up" a national issue. Success was achieved when President Reagan announced his commitment to begin restoration of the Bay in his 1984 State of the Union message. This action precipitated a Forum meeting (Chesapeake Bay Forum, 1985) to bring together leading government, academic, and private-sector decision makers concerned with solving the problems in and around the Bay. Its stated purpose was to acquaint the community of decision makers with the full range of programs and activities that affect the Bay and establish relationship among programs. The goals of the Forum were to foster awareness that the Bay is endangered; identify problems that affect the Bay; identify near- to long-term solutions to the problems; and identify those who could play a role in achieving these solutions.

The CBP took the lead in the President's initiative and approved a modeling strategy that called for phased development of a series of numerical models to evaluate alternative control strategies and to guide the establishment of pollutant reduction goals. The management issues at stake were extremely costly, either financially or environmentally, and therefore required the highest level of confidence in these technical tools. The CBP established a network of committees to aid in the oversight of the Bay studies. Also established was a Scientific and Technical Advisory Group comprised of representatives from Federal, State, academic, and public interest agencies and groups. This advisory panel provided peer review, evaluation, and direction setting for the studies. They were supported by a select group of nationally recognized experts (called the Modeling Evaluation Group, MEG). The MEG consisted of knowledgeable engineers and scientists covering the disciplines of fluid dynamics and estuarine water quality, contaminant, and biological and chemical

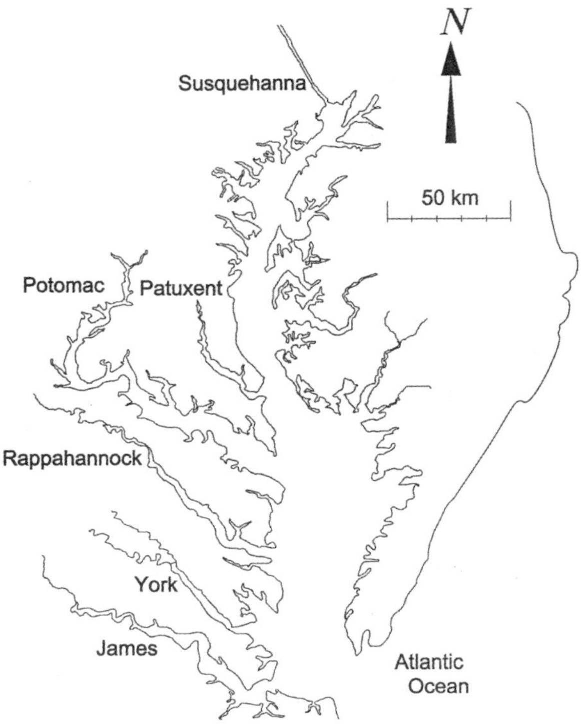

Fig. 12.1 The Chesapeake Bay.

processes. The MEG played a strong role in building public confidence by reaching consensus agreement on the progress and eventual culmination of the study effort.

1.3 *Chesapeake bay characteristics*

Physical. The Chesapeake Bay system (Fig. 12.1) consists of the mainstem bay, five major western-shore tributaries, and a host of lesser tributaries and embayments. Urban centers along the Bay and tributaries include Norfolk and Richmond Va., Washington, D.C., and Baltimore, Md. The mainstem is roughly 300 km long, 8 km to 48 km wide, and has 8 m average depth. A deep trench with depth of 50 m runs up the center of the mainstem.

Total drainage area of the Bay is 166,000 km^2. The primary source of freshwater to the system is the Susquehanna River ($\approx$64% of total gauged freshwater flow) which empties into the northernmost extent of the Bay. Other major freshwater sources are the Potomac ($\approx$19%) and James Rivers ($\approx$12%). The remaining western-shore tributaries, the York ($\approx$3%), the Rappahannock ($\approx$3%), and the Patuxent ($<$1%) contribute only small fractions of the total freshwater flow to the Bay.

An annual runoff cycle exists in the major bay tributaries. Peak flows typically occur in March or April, whereas minimum flows occur in August and September. Both floods and droughts cause daily and monthly flows to deviate from the long-term pattern. The occurrence of maximum annual flow is particularly susceptible to flood events. As a result of tropical storms, maximum flow may occur in late summer or in autumn rather than in spring.

Eutrophication. Severity of eutrophication in the Bay is typically manifest in (1) bottom water hypoxia, (2) spring phytoplankton bloom, and (3) sediment-water interaction. Longitudinal and lateral extents of hypoxia are determined by the geometry of the trench that runs up the center of the Bay. On occasion, hypoxic water extends from the bottom to within a few meters of the surface but summer-average oxygen concentration within the surface mixed layer is usually $6\,\mathrm{g\,m^{-3}}$ or greater.

Bottom-water hypoxia typically begins in late May when spring warming enhances decay of material deposited in spring and in previous years, removing oxygen from bottom water. Density stratification prevents mixing of oxygenated surface water downward. Low-oxygen conditions at the bottom continue through the summer, until mid-September, when autumn winds end the hypoxic period by mixing surface water down to the bottom.

The spring phytoplankton bloom primarily consists of three genera of marine diatoms: *Cerataulina*, *Rhizosolina*, and *Thalassiosira*. The bloom usually commences in February, reaches a maximum in April, and ends precipitously in May (Fig. 12.2). The bloom is characterized by high chlorophyll concentrations throughout the water column. At times, a subsurface chlorophyll maximum occurs. Elevated chlorophyll levels throughout the water column mark the spring bloom as the period of maximum algal biomass. During summer, chlorophyll concentrations are generally less than in spring and restricted to surface waters.

Over time-scales of years, benthic sediments are sinks of oxygen, nitrogen, phosphorus, and silica removed from the water column. Oxygen is consumed, directly or indirectly, in the oxygenation of organic carbon and in the nitrification of ammonium. Substantial fractions of particulate nitrogen, phosphorus, and silica that settle into surficial sediments are buried to deep sediments from which recycling to the water column is impossible.

Over seasonal time-scales, sediments can be significant sources of dissolved nutrients to the overlying water. The role of sediments in the system-wide nutrient budget is especially important in summer when seasonal low flows diminish riverine nutrient input. During summer, warm temperature enhances biological processes in the sediments. Diagenesis (decay) of organic matter produces ammonium, phosphate, and silica that are released to the water column. Sediment ammonium release, supplies an estimated 13% to 40% of algal nutrient requirements during August in Chesapeake Bay (Boynton and Kemp, 1985).

2. Model Development

As in many estuaries, long-term net circulation in Chesapeake Bay is influenced by longitudinal density gradients. Net circulation tends to be upstream near the

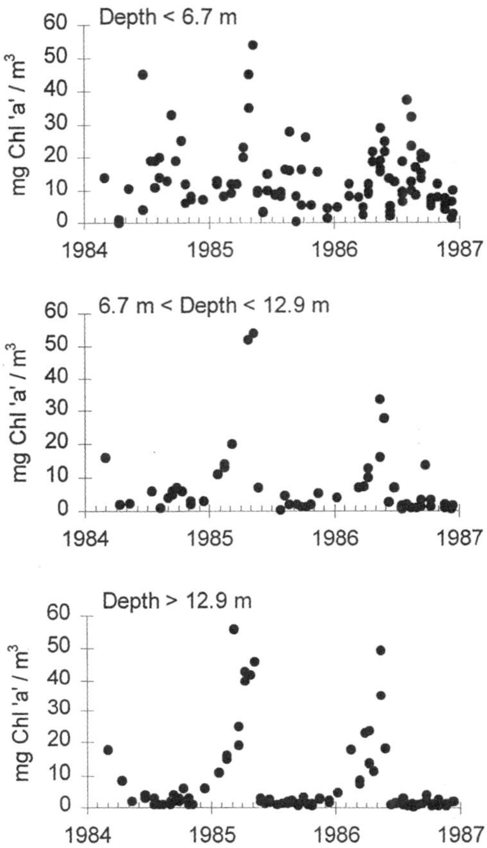

Fig. 12.2 Chlorophyll measured in the upper Bay.

bottom of the estuary and downstream at the surface. Rotational effects cause a tilting of the level of no net motion, such that it occasionally breaks the surface. The complex vertical and lateral circulation field and density structure mandate the use of a fully three-dimensional model.

The hydrology, hydrodynamics, and water quality are necessarily linked and a high level of compatibility is required among them, especially between the latter two. The water quality model must consider nutrient kinetics, oxygen dynamics and anoxia, point and nonpoint loadings, and sediment/water quality interactions at time and space scales compatible with transport processes determined by the hydrodynamic model.

2.1 *Requirements*

The EPA Chesapeake Bay Program Office formulated the following requirements for the Chesapeake Bay model: the ability to analyze the current origins and extent of anoxia, the ability to predict the future of the Bay as a result of management action or inaction, and the ability to reconstruct the processes that led the Bay to its present condition. Eight capabilities were envisioned in order for a model to meet the requirements. These were:

- Simulation of responses of water quality and sediment processes to point and nonpoint-source control actions.
- Performance of short-term (annual) and long-term (decades) simulations.
- Determination of the effect of spring runoff events on summer anoxia.
- Simulation of lateral water quality variations.
- Determination of Bay response to area-specific management activities.
- Determination of response time of the Bay to management actions.
- Evaluation of frequency of critical water-quality events.
- Evaluation of historical changes in anoxia.

2.2 *Chesapeake bay model package*

A package of interactive models (Fig. 12.3) was assembled to comply with requirements and recommendations for the eutrophication model study. The CH3D-WES hydrodynamic model (Johnson *et al.*, 1991) produced three-dimensional predictions of velocity, diffusion, surface elevation, salinity, and temperature on an intra-tidal (five-minute) time-scale. A processor appended to CH3D-WES filtered intra-tidal details from hydrodynamic output but maintained inter-tidal (twelve-hour) transport (Dortch *et al.*, 1992). Inter-tidal flows, vertical diffusion, and surface elevation were written to disk for subsequent use by the water quality model, named CE-QUAL-ICM. Computation of eutrophication processes within the water quality model was conducted on a two-hour time-scale. The water quality model interacted dynamically with a predictive sediment diagenesis model (DiToro and

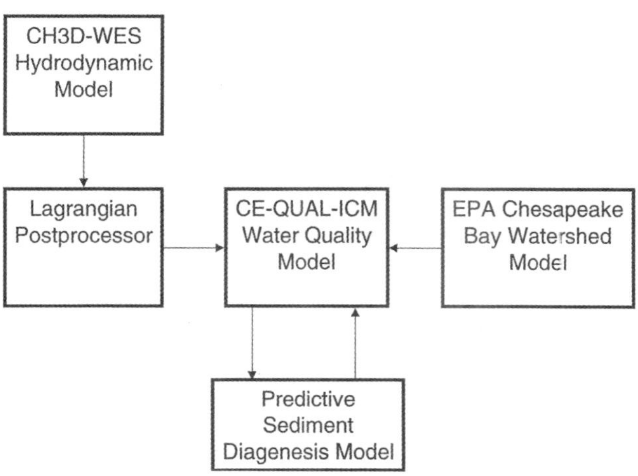

Fig. 12.3 Chesapeake Bay model package.

Fitzpatrick, 1993). Fall-line loads and below-fall-line nonpoint source loads were supplied to the eutrophication model by the EPA's Chesapeake Bay Watershed model (Donigian *et al.*, 1991).

2.3 *Watershed model*

Nutrient loads which enter the Bay from numerous sources consist of loads carried by the rivers which drain the upland watershed (fall-line loads), loads carried directly through runoff to tidal waters (nonpoint-source loads), loads from industries and municipal waste treatment plants (point-source loads), loads deposited by rainfall and dryfall from the atmosphere to the water surface, and loads introduced through exchange with the ocean.

The Watershed Model (Donigian *et al.*, 1991) provided the sole means by which the actions of management practices on fall-line and nonpoint-source loads could be evaluated through identification of the origins and nature of the upland loads which reach the Bay. The nonpoint-source loads were difficult to supply by almost any means except a model, as these loads enter the Bay at no distinct location and are impossible to measure in their entirety.

The Watershed Model is a much-revised version of HSPF (Bicknell *et al.*, 1993). The model operates by dividing the Bay watershed (Fig. 12.4) into more than 60 sub-basins. Within each sub-basin, edge-of-stream loads are computed by a variety

Fig. 12.4 Chesapeake Bay watershed.

Table 12.1 Long-term-average nutrient loads.

Source	Nitrogen, kg day^{-1}	Phosphorus, kg day^{-1}
Fall–line	226,895	15,012
Below fall-line	64,836	4,846
Point source	86,357	7,359
Atmospheric	43,025	1,823
Oceanic	164,000	23,500
Total	585,113	52,540

of modules that simulate land uses including cropland, pasture, forest, and urban development. Loads are routed from their originating sub-basins through a network of continuously-stirred-tank reactors which simulate the network of upland rivers which drain the watershed.

State variables simulated by the watershed model are sediment, temperature, dissolved oxygen, nitrate, ammonium, organic nitrogen, phosphate, organic phosphorus, chlorophyll 'a', and total organic carbon. From these, loads of nitrogen, phosphorus, and organic carbon were input to the Bay model.

Other loading sources to the Bay are point-source, atmospheric, and oceanic. The EPA CBP Office provided monthly loads from nearly 150 municipal and industrial point sources. Atmospheric loads to the water surface were computed using data from the National Atmospheric Deposition Program (NADP, 1989) and other published sources (USEPA, 1982). Gross oceanic loads were derived from a mass-balance algorithm applied at the mouth of the Bay (Cerco and Cole, 1994).

Table 2.1 summarizes long-term-average loads of nitrogen and phosphorus to the Bay. The major fall-lines provide the largest quantity of nitrogen to the Bay, followed by loads from continental shelf waters (oceanic loading) and point sources. Continental shelf waters provide the largest quantity of phosphorus followed by fall-line and point-source loading.

Although, direct atmospheric loading to the water surface is the least of the nutrient sources, atmospheric loading of nitrogen, especially, to the entire watershed is tremendous. Note that nitrogen loading to the watershed shows up as fall-line loadings in the summary presented here. As to the nutrient loads to the Bay from the ocean, the Bay is a net exporter of nitrogen and a net importer of phosphorus.

2.4 *Hydrodynamics of the bay*

A time-varying, three-dimensional (3D) hydrodynamic model called CH3D-WES (Curvilinear Hydrodynamics in 3D-Waterways Experiment Station) computes on a curvilinear or boundary-fitted planform grid, physical processes impacting water body circulation and vertical mixing driven by tides, wind, density effects (salinity and temperature), freshwater inflows, turbulence, and the effect of the earth's rotation. The basic model (CH3D) was originally developed by Sheng (1986) and was extensively modified at WES (Johnson *et al.*, 1994).

Adequately representing the vertical turbulence is crucial to a successful simulation of stratification/destratification. A second-order turbulence model based upon the assumptions of local equilibrium of turbulence is employed. The model's boundary-fitted coordinates provide the ability to adjust the solution grid to fit deep navigation channels and irregular shoreline geometries, and permit adoption of an accurate and economical grid schematization. The solution algorithm employs a mode-splitting procedure whereby the external mode consists of vertically-averaged equations to provide the solution for the free surface to the internal mode consisting of the full 3D equations.

Governing Equations. To better resolve complex geometries in the horizontal directions, the Bay model makes computations on the boundary-fitted planform grid shown in Fig. 12.5. The use of a curvilinear grid necessitates the transformation of the governing equations into boundary-fitted coordinates (ξ, η). However, not only are the x, y-coordinates transformed into the (ξ, η)-curvilinear coordinates, but also the velocity is transformed such that its components are perpendicular to the (ξ, η)-coordinate lines, i.e., contravariant components. With the governing equations written in terms of the contravariant components of the velocity, boundary

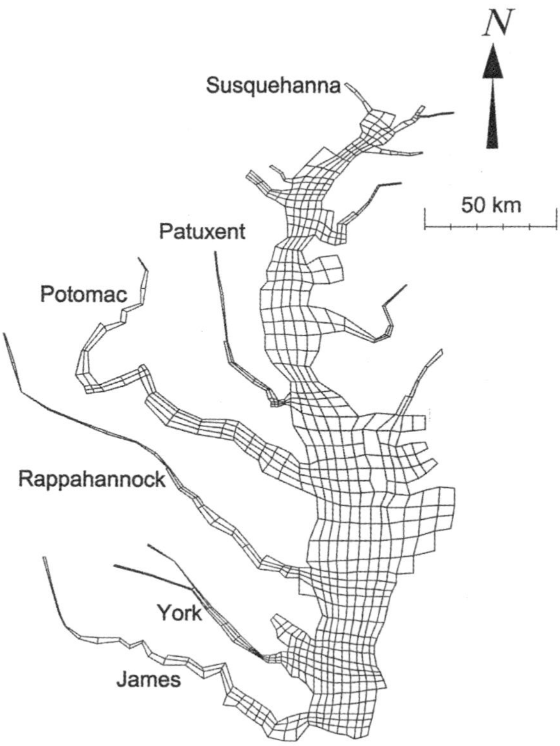

Fig. 12.5 Boundary-fitted planform grid.

conditions can be prescribed on a boundary-fitted grid in the same manner as on a Cartesian grid.

As noted above, in the Sheng (1986) model, the vertical dimension was represented through the use of what is commonly called a sigma-stretched grid whereby layers within the water column conform to the contour of the local bathymetry. However, with the grid resolution shown in Fig. 12.5, stratification in the deep channels could not be maintained during long-term simulations on Chesapeake Bay. With a sigma-stretched grid, the bottom layer in one column communicates with the bottom layer in an adjacent column. Because depth changes were coarsely resolved, channel stratification could not be maintained. As a result, the governing equations were written for solution on the Cartesian or z-plane in the vertical direction. The reader should refer to Johnson *et al.* (1994) for details on the equations and numerics for CH3D-WES.

The numerical grid used for the Chesapeake Bay application contained 729 active horizontal cells and a maximum of 15 vertical layers, resulting in 4,073 computational cells. The grid resolution is 1.52 m vertical and approximately 10 km longitudinal and 3 km lateral. Major tributaries, i.e., the James, York, Rappahannock, Potomac, Patuxent, and Susquehanna rivers, are modeled fully as 3D in the lower reach and two-dimensional in the upper reach with a single cell spanning the width of the tributary.

Realizing the Bay study began in 1988, a choice was made at the time to balance the requirement for computer resources (ever increasing with the number of grid cells, time step, and length of simulations) with the degree to which the model could refine the simulation of the physical processes. Although supercomputers (CRAY Y-MP class) were available for use in this study, the requirement to make three-year-long hydrodynamic simulations and water quality simulations spanning three decades resulted in careful choice of grid spacing. The grid selected proved sufficient for investigating Bay response to control actions, but as will be discussed later, resolution in the tributaries was insufficient to resolve details in mid to upper reaches where questions on industrial use of the river were being posed.

2.5 *Interfacing hydrodynamics and eutrophication modeling*

The hydrodynamic and water-quality models are separate codes that are independently executed. The hydrodynamic model operates on a structured grid where each node on the hydrodynamic grid is located in Cartesian space. The water-quality model operates on an unstructured grid, in which model cells are numbered sequentially. The unstructured grid was selected to allow maximum flexibility in linking the water quality model to a variety of hydrodynamic model formats including finite-difference, finite-element, and link-node.

Mapping hydrodynamic information from the structured to the unstructured grid and the averaging of hydrodynamics create the need for a linkage module, a set of subroutines appended to the CH3D-WES model code (Dortch *et al.*, 1992). Development of a linkage module requires continuous testing of the linkage between the two models. Comparison of transport of a conservative substance provides an ultimate test that all quantities are correctly passed between the two models. Two types of transport comparisons were conducted. Simulated dye dumps were the most demanding. Secondary comparisons were conducted of salinity computed by the two models.

2.6 *Water quality*

Water quality is modeled by an integrated-compartment model, CE-QUAL-ICM, developed for this study. CE-QUAL-ICM solves, for each grid cell and for each state variable, the conservation of mass equation:

$$\frac{\partial V_i C_i}{\partial t} = \sum_{j=1}^{n} Q_j C_j^* + \sum_{j=1}^{n} A_j D_j \frac{\partial C}{\partial x_j} + \sum S_i \tag{1}$$

in which: V_i = volume of ith compartment; C_i = concentration in ith compartment; Q_j = volumetric flow across flow face j of ith compartment; C_j^* = concentration in flow across flow face j; A_j = area of flow face j; D_j = diffusion coefficient at flow face j; n = number of flow faces attached to ith compartment; S_i = external loads and kinetic sources and sinks in ith compartment; and t, x are the temporal and spatial coordinates, respectively. Finite-difference solution to Eq. (1) employs the QUICKEST algorithm (Leonard, 1979) in the horizontal directions and a Crank–Nicolson scheme in the vertical direction.

CE-QUAL-ICM incorporates 22 state variables: temperature, salinity, inorganic solids, cyanobacteria, diatoms, green algae, dissolved organic carbon, labile particulate organic carbon, refractory particulate organic carbon, ammonium, nitrate, dissolved organic nitrogen, labile particulate organic nitrogen, refractory particulate organic nitrogen, total phosphate, dissolved organic phosphorus, labile particulate organic phosphorus, refractory particulate organic phosphorus, chemical oxygen demand, dissolved oxygen, particulate biogenic silica, and dissolved silica. Kinetic interactions among the state variables are described in 80 partial differential equations (Cerco and Cole, 1994) that employ over 140 parameters. The state variables can be categorized into six groups or cycles. The physical group consists of salinity, temperature, and inorganic suspended solids. Salinity is included primarily to verify the linkage to CH3D-WES. When the models are properly linked, salinity predicted by CE-QUAL-ICM replicates predictions by CH3D-WES. Temperature, predicted by an equilibrium-temperature approach (Edinger *et al.*, 1974), is included for its fundamental impact on the rate of biochemical processes. In the present model configuration, inorganic solids are a surrogate for iron and manganese that adsorb phosphate (PO_4) and dissolved silica (DSil) in the water column.

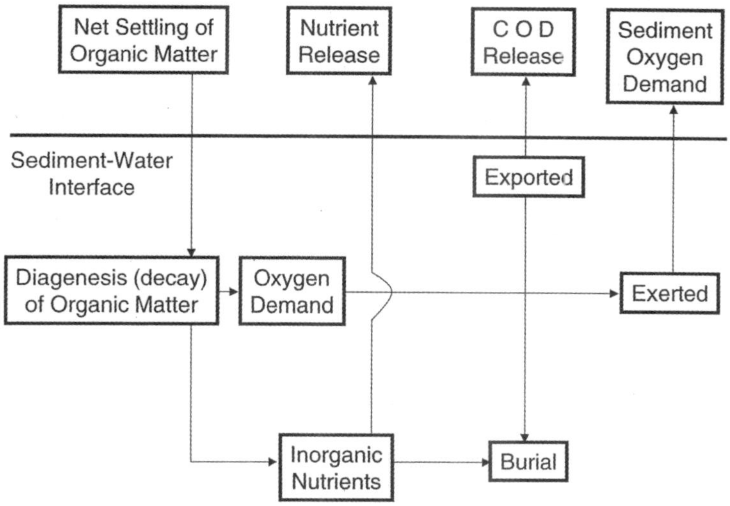

Fig. 12.6 Diagenetic sediment model.

Sediment Processes Model. The water quality model is directly coupled to a predictive benthic sediment model (Fig. 12.6). The sediment model (DiToro and Fitzpatrick, 1993) is an expansion of diagenetic principles established for fresh-water sediments (DiToro *et al.*, 1990) into the estuarine environment. CE-QUAL-ICM simulates the settling of particulate matter to the sediments. Diagenesis of accumulated organic matter is computed within the sediment model. The oxygen demand created by carbon diagenesis is fractionated into sediment oxygen demand, chemical oxygen demand released to the water column, and loss through burial. The NH_4 produced by nitrogen diagenesis is routed into three pathways: recycle to the water column, conversion to gaseous nitrogen through coupled nitrification–denitrification reactions, and burial. The actions of nitrification–denitrification also determine sediment–water fluxes of NO_3. The PO_4 produced by phosphorus diagenesis is recycled to the water column, maintained in the sediment as dissolved and particulate (sorbed) PO_4, or lost through burial. The particulate biogenic silica in the sediments undergoes dissolution to DSil or else is buried. Burial also removes a small fraction of reactive organic matter.

The water quality and sediment models interact on a time-scale equal to the integration time step of the water quality model. After each integration, the predicted particle deposition, temperature, nutrient and dissolved oxygen concentrations are passed from the water quality model to the sediment model. The sediment model computes sediment–water fluxes of dissolved nutrients and oxygen based on predicted diagenesis and concentrations in the sediments and water. The computed sediment–water fluxes are passed to CE-QUAL-ICM and incorporated into appropriate mass balances and kinetic reactions.

3. Model Performance

All models used in the study were validated by an extensive calibration and verification process. Both short-term events and long-term behavior were simulated and the results were analyzed statistically to judge the model performance. Once the entire suite of models was calibrated and verified, the next step was to develop Bay responses to varying point and nonpoint source control actions and other area-specific management activities. To accomplish this, a host of management scenarios was developed and simulated.

3.1 *Calibration and verification*

The first step in validating any model is the acquisition of sufficient measurements from periods where important processes and key events occurred. With the Bay being very large, collection of synoptic data is very expensive, and no Bay-wide comprehensive field study for collecting both hydrodynamic and water quality parameters has ever been attempted. Three one-month duration datasets for periods characterized by low freshwater inflow, strong spring runoff, and a fall wind mixing event were available for calibrating the hydrodynamic model.

An intense measurement program was undertaken during calendar years 1984–1986, providing data for water quality model validation. Observations were made in Bay-wide surveys conducted twice monthly in March through October and once monthly in the remaining months of 1984–1986. Water-quality samples were collected by multiple agencies and institutions and analyzed in several laboratories. Careful, continual quality control measures were adopted to ensure comparability among samples collected and analyzed by differing agencies. Calibration of the water quality model and presentation of results required both temporal and spatial aggregations of observations and predictions that allowed for meaningful direct comparison. Aggregation by season was defined through examination of characteristic, repetitive Bay behavior. Spatial aggregation was accomplished by subdividing the Bay into 10 horizontal zones (approximately 30 km in length) and into three vertical layer zones depicting the surface and bottom layers as well as a pyncnocline region where vertical transition was greatest.

The Watershed Model was verified in two fashions. The first was a conventional comparison of observed and computed concentration time series. These were conducted at key locations in the upland watershed and at the major bay fall lines. A second verification was comparison of loads computed by the watershed model with loads computed by a minimum-variance-unbiased-estimator regression model (Cohn *et al.*, 1989). This second verification was based on the premise that loads computed by the regression model are the most accurate attainable. At the final stage of watershed model calibration, differences between watershed model loads and regression loads were insignificant.

The strategy adopted for verifying the hydrodynamic model was to use the three datasets described above to calibrate and skill assess the model. The final version of the model and values of model coefficients used with these datasets were employed to simulate the year-long periods of 1984 through 1986 to complete the hydrodynamic model verification.

Since the goals for the water quality model were to skill assess the model by hindcasting previous decades of perceived Bay decline and to use the model to investigate long-term scenarios for developing guidance on Bay management strategies, the challenge was to devise hydrodynamic simulations which could be concatenated in such a way as to represent typical conditions over any period of years. Fortunately, calendar years of 1984 through 1986 represented high, low, and average freshwater inflows, respectively. Hydrodynamics for these years were simulated and used for driving the water quality model for calibration and verification. Sequences of these hydrodynamics were also constructed to drive long-term (decades) water quality simulations.

To calibrate the water quality model, predictions and observations for the mainstem Bay and the major western tributaries were compared in a variety of formats and on several spatial and temporal scales. Computations of water quality behavior through the years 1984 to 1986 were able to replicate the seasonal cycling of chlorophyll, nutrients, and dissolved oxygen.

Summer is a critical season for Bay water quality due to the occurrence of bottom-water anoxia. Example results are given for the summer of 1986; salt water (Figs. 12.7a and 12.7b) intrudes from its source at the mouth of the Bay (0 km) nearly to the Susquehanna fall line (300 km). Chlorophyll (Figs. 12.7c and 12.7d) is higher near the surface, in the photic zone, than near the bottom, and achieves a maximum around 250 km. This location is immediately downstream of the turbidity maximum and situated above a sediment source of nutrients. The Susquehanna is the major source of nutrients to the upper Bay. For example, the distribution of nitrogen (Figs. 12.7e and 12.7f) is characterized by a gradient from the source at the Susquehanna River to the sink represented by the ocean. Nitrogen concentrations in bottom waters indicate a sediment source of this nutrient as well, especially in the portion of the estuary characterized by anoxic bottom water. Surface DO (Fig. 12.7g) is near saturation everywhere. However, at the bottom (Fig. 12.7h) observed DO exhibits a minimum with average concentration near zero, from 250 km, at the head of the trench that runs up the mainstem, down to 200 km. Intermittent anoxia extends another 100 km down the Bay. Details can be found in Cerco and Cole (1994).

3.2 *Assessment scenarios*

Models are used to develop and test load-management strategies aimed at limiting eutrophication processes. Scenario analysis in estuaries presents multiple challenges. One challenge is that steady-state analysis is not appropriate. Estuaries are dynamic environments in which water quality varies on time scales from tidal cycles to decades. A second challenge is the activity of benthic sediments. The sediments are a reservoir for nutrients that are recycled to the water column seasons, years or

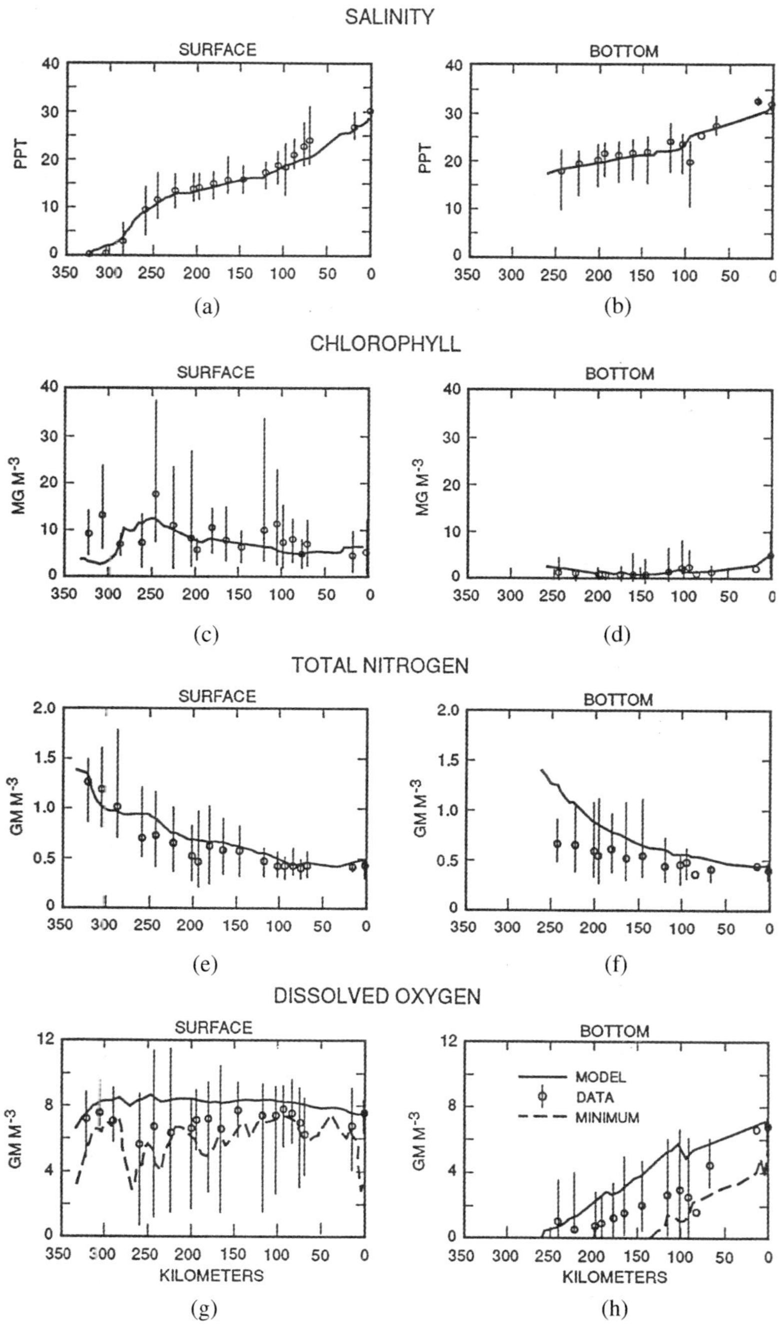

Fig. 12.7 Longitudinal comparisons of simulated and observed: (a, b) salinity; (c, d) chlorophyll; (e, f) total nitrogen; (g, h) dissolved oxygen (predictions and observations are averaged temporally, June–September 1986, and into two layers: surface 0–6.7 m and bottom >13 m).

decades following their introduction to the system. Consequently, estuarine scenario analysis must account for sediment activity and simulate a period sufficient for sediments to react to load changes. A third challenge is representing the transitional estuarine environment that is nearly riverine at the fall-line and approaches oceanic conditions at the mouth.

Two categories of scenarios were performed: "Management" scenarios which determined the response of the Bay to feasible nutrient control strategies and "Sensitivity" scenarios which examined the behavior of the bay under a wide range of load reductions. Management scenarios were analyzed by Thomann *et al.* (1994) and sensitivity scenarios were presented by Cerco (1995). The present document examines two scenarios of particular interest: limit-of-technology (LOT) nutrient load reduction and all-forest conditions. The LOT scenario represented the optimal conditions obtained through management controls of point-source and nonpoint-source nutrient loads above and below the system fall-lines. The all-forest run comprised the best estimate of Bay water quality under natural conditions.

Most scenarios were run for 10 years. The 10-year duration allowed water-column conditions to attain roughly 95% of complete response to load reductions. This 10-year period was employed for all but a few scenarios for which the highest degree of completion was desired. The LOT scenario was executed for 30 years. The all-forest scenario was executed for 20 years beyond conditions attained by LOT load reduction. For reference, the CRAY C-90 computational time for a 10-year scenario was 20 h.

Chlorophyll content during the spring bloom in the all-forested condition was roughly half the bloom biomass during the existing condition (Fig. 12.8a). Chlorophyll content under the LOT scenario was midway between the extremes of the existing and all forested conditions. Effects of the LOT reductions were less noticeable in the lower 150 km of the Bay, however.

Under existing conditions, during the spring bloom, the model indicates phosphorus is the most limiting nutrient from the fall-line down to 125 km (Fig. 12.8b). Below 125 km, silica is the most limiting nutrient. Under LOT controls (Fig. 12.8c), phosphorus remains the most limiting nutrient in the upper bay but nitrogen replaces silica as most limiting in the lower bay. The replacement occurs as silica, no longer utilized by algae in the upper bay, travels downstream to a previously silica-limited area. Under all-forest conditions (Fig. 12.8d), the regions of phosphorus and nitrogen limitations remain the same as LOT but phosphorus limitation is more severe in the upper regions of the bay.

Summer chlorophyll diminished under both LOT and all-forest conditions (Fig. 12.9a). The reduction of chlorophyll was pronounced near 150 km, above the region of greatest sediment release under base conditions.

During summer, under existing conditions, the model indicates that phosphorus is most limiting from the fall-line down to 100 km (Fig. 12.9b). Below 100 km, nitrogen is most limiting. Careful examination of time-series plots for multiple years indicates that the transition from phosphorus to nitrogen limitation is not clearly

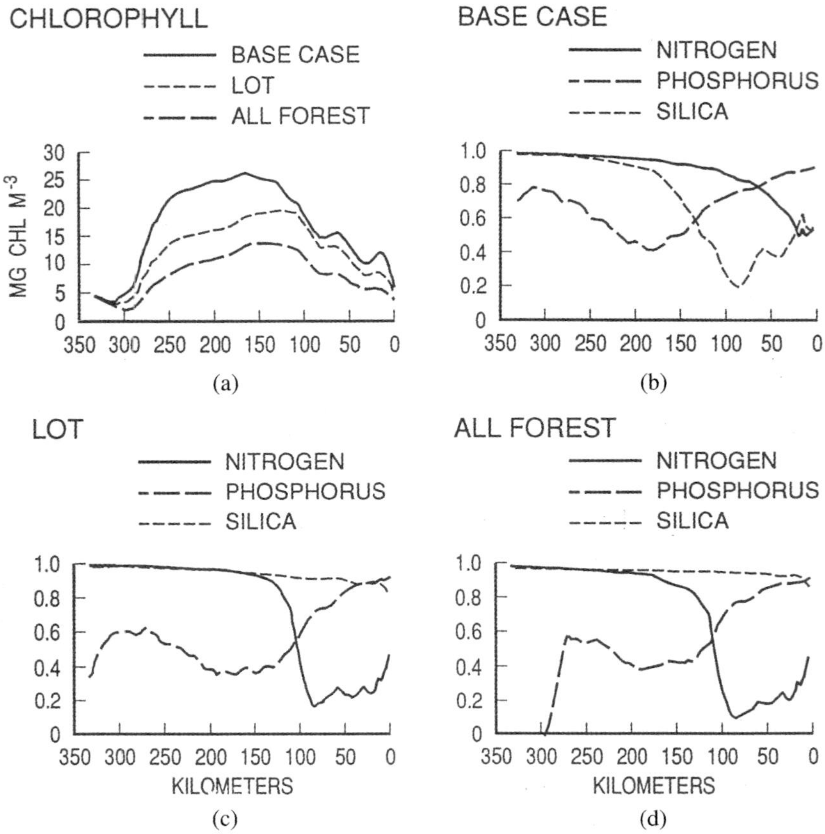

Fig. 12.8 Spring surface chlorophyll concentration (a) and nutrient limitations under (b) existing or base conditions; (c) LOT load reductions; (d) all-forest conditions.

defined. The location depends on hydrology and time of the season. A more accurate description of summer is that phosphorus clearly limits above 250 km, nitrogen clearly limits below 100 km and nitrogen and phosphorus limit from 250 km to 100 km (Figs. 12.9c and 12.9d). Silica limit in summer is negligible due to sediment release and presence of algal species which require no silica.

Total nutrient concentrations were much more sensitive to load reductions near the fall-line than near the mouth (Fig. 12.10). Although, concentration at the mouth was free to fluctuate in response to load reductions, the concentration change was minimal due to the overwhelming influence of adjacent continental shelf waters.

Under LOT nutrient controls, average bottom dissolved oxygen increased by $\approx 0.5\,\mathrm{g\,m^{-3}}$ throughout the mainstem from 100 km to 260 km (Fig. 12.10c). The improvement occurred primarily at depths greater than 12 m, in the trench that runs from the mouth to 260 km. A substantial volume of water with dissolved oxygen concentration below $1\,\mathrm{g\,m^{-3}}$ persisted.

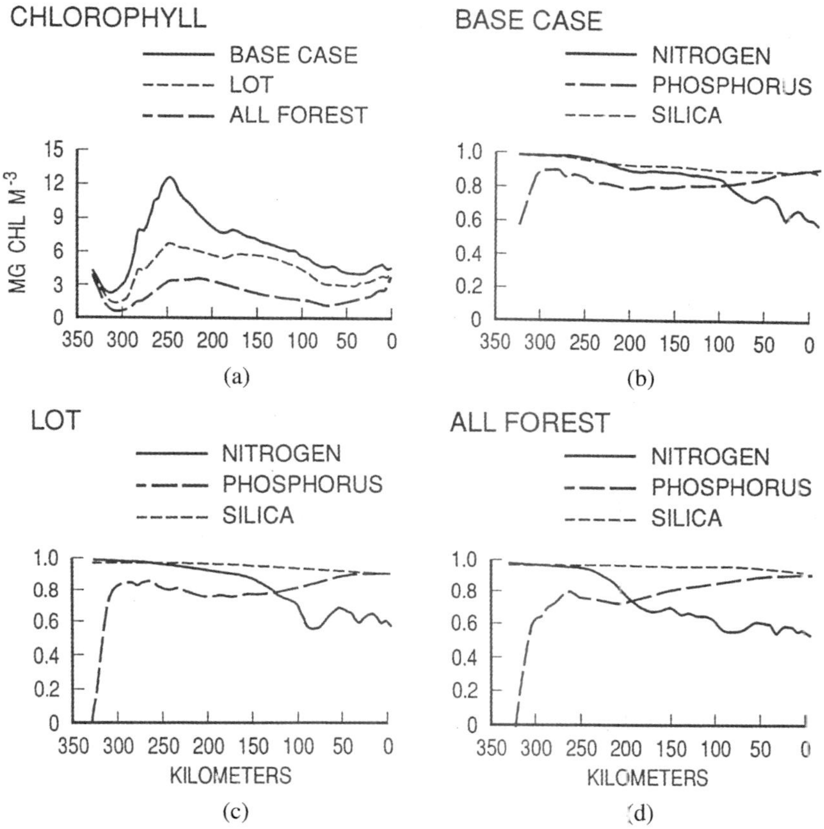

Fig. 12.9 Summer surface chlorophyll concentration (a) and nutrient limitations under (b) existing or base conditions; (c) LOT load reductions; (d) all-forest conditions.

Under all-forest conditions, minimum seasonal-mean dissolved oxygen increased to $\approx 2.7\,\mathrm{g\,m^{-3}}$ (Fig. 12.10c). Dissolved oxygen improved in the pyncnocline as well as near the bottom. Occurrences of dissolved oxygen less than $1\,\mathrm{g\,m^{-3}}$ were nearly banished under average hydrologic conditions. Dissolved oxygen less than $1\,\mathrm{g\,m^{-3}}$ still occurred under wet hydrology but the volume was greatly diminished from base conditions. The sensitivity and all-forest scenarios indicated that the model conforms to current paradigms regarding nutrient limitations (Fisher *et al.*, 1992; Conley and Malone, 1992) and prehistoric conditions in the Bay (Cooper and Brush, 1991). Agreement between the model and these and other paradigms provided confidence that conclusions reached from management scenarios were correct.

4. Model Enhancements

The Bay study represented a tremendous leap toward adopting model technology as a tool for management of eutrophication in the region. As with any study,

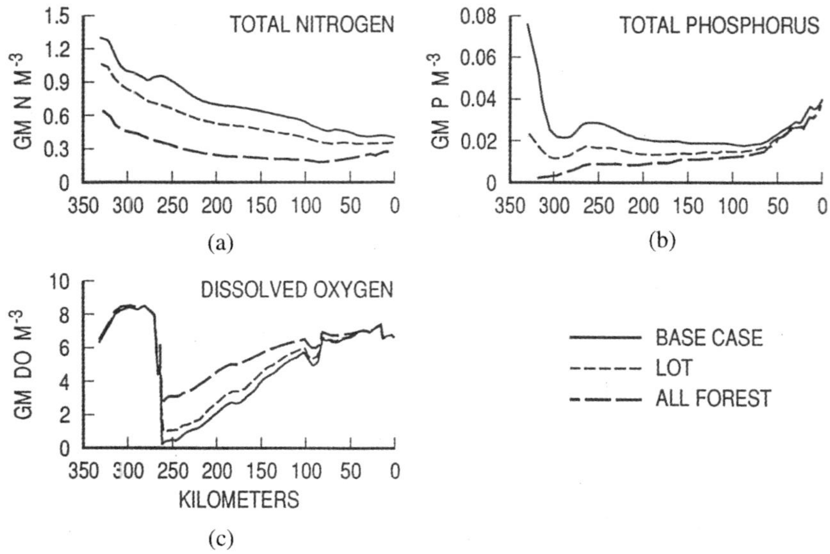

Fig. 12.10 Summer concentrations of: (a) surface total nitrogen; (b) surface total phosphorus; (c) bottom dissolved oxygen.

weaknesses were identified as the study progressed. In the initial Bay study, tributaries were rather coarsely resolved by the numerical grid, resulting in questionable results on the tributaries. Thus, it was decided to initiate a follow-on study to more accurately model processes on the major tributaries. In addition to the need for increased grid resolution, the incorporation of a SAV model was considered crucial in the modeling of water quality processes in these rivers. To aid in removing uncertainty on boundary conditions at the mouth (especially boundary conditions on the water quality variables), the numerical grid was extended onto the continental shelf.

The vertical turbulence closure model within CH3D-WES was improved to give an even better representation of stratification processes. Very shallow littoral zones, where SAV flourishes, were accounted for in the enhanced hydrodynamic model. Also, a direct linkage to the watershed model was accomplished, accounting for daily averaged runoff from land-bordered grid cells along the Bay and its tributaries.

Are model improvements still needed? Of course they are. For the physical side, we can do much better in modeling mixing processes. Adaptive gridding can be used in the vertical to resolve pycnoclines and vertical mixing together with the use of improved turbulence schemes. Perhaps the inclusion of a full 3D sediment transport model is the most challenging task for investigators. Such a capability will be required to fully model contaminant transport and associated impacts. On the environmental side, there is a need to include relationships of higher trophic levels. How do we evaluate how society's changes to a system impact not only the marine life but eventually the human species?

Computer resources will most likely not be a concern in the future. We initiated the Chesapeake Bay study knowing that we had access to a CRAY Y-MP supercomputer. Eight years later, the engineer has that power in a desktop UNIX workstation. The future will only bring greater processing speed and memory. The move afoot is to convert codes to massive parallel processors to achieve even greater throughput. Perhaps, a more important issue is to improve graphics and animation of results for more rapid analysis as well as a means of expressing the message of model results to the decision makers.

Finally, success can only be achieved with a parallel commitment to continued improvement in instruments for obtaining real data for better model verification. Techniques of data assimilation to improve model predictions will require field data for proper implementation and will most likely be required to further enhance management's view of model confidence and acceptability. Such techniques in turn should permit optimization of monitoring programs as confidence in the model results increases.

References

Bicknell, R., J. Imhoff, A. Kittle, A. Donigian and R. Johanson (1993): Hydrological simulation program — FORTRAN: User's manual for release 10, EPA/600-R-93/174, US Environmental Protection Agency, Athens, GA.

Boynton, W. and W. Kemp (1985): Nutrient regeneration and oxygen consumption along an estuarine salinity gradient, *Marine Ecology Progress Series*, 23, 45–55.

Cerco, C. and T. Cole (1994): Three-dimensional eutrophication model of Chesapeake Bay, Technical Report EL-94-4, US Army Engineer Waterways Experiment Station, Vicksburg, MS.

Cerco, C. (1995): Response of Chesapeake Bay to nutrient load reductions, *J. Environmental Engineering*, 121(8), 549–557.

Chesapeake Bay Forum (1985): Proceedings of the National Defense University, Fort McNair, Washington, DC, October 30–31.

Cohn, T., L. DeLong, E. Gilroy, R. Hirsch and D. Wells (1989): Estimating constituent loads, *Water Resources Research*, 25, 937–942.

Conley, D. and T. Malone (1992): Annual cycle of dissolved silicate in Chesapeake Bay: Implications for the production and fate of phytoplankton biomass, *Marine Ecology Progress Series*, 81, 121–128.

Cooper, S. and G. Brush (1991): Long-term history of Chesapeake Bay anoxia, *Science*, 254, 992–996.

DiToro, D., P. Paquin, K. Subburamu and D. Gruber (1990): Sediment oxygen demand model: Methane and ammonia oxidation, *J. Environmental Engineering*, 116(5), 945–986.

DiToro, D. and J. Fitzpatrick (1993): Chesapeake Bay sediment flux model, Contract Report EL-93-2, US Army Engineer Waterways Experiment Station, Vicksburg, MS.

Donigian, A., B. Bicknell, A. Patwardhan, L. Linker, D. Alegre, C. Chang and R. Reynolds (1991): Watershed model application to calculate Bay nutrient loadings, Chesapeake Bay Program Office, US Environmental Protection Agency, Annapolis, MD.

Dortch, M., R. Chapman and S. Abt (1992): Application of three-dimensional Lagrangian residual transport, *J. Hydraulic Engineering*, ASCE, 118(6), 831–848.

Edinger, J., D. Brady and J. Geyer (1974): Heat exchange and transport in the environment, Report 14, Department of Geography and Environmental Engineering, John Hopkins University, Baltimore, MD.

Fisher, T., E. Peele, J. Ammerman and L. Harding (1992): Nutrient limitation of phytoplankton in Chesapeake Bay, *Marine Ecology Progress Series*, 82, 51–63.

Flemer, D., G. Mackiernan, W. Nehlsen and V. Tippie (1983): Chesapeake Bay: A profile of environmental change, US Environmental Protection Agency, Region III, Philadelphia, PA.

Johnson, B., K. Kim, R. Heath and L. Butler (1991): Development and verification of a three-dimensional numerical hydrodynamic, salinity and temperature model of Chesapeake Bay, Technical Report HL-91-7, US Army Engineer Waterways Experiment Station, Vicksburg, MS.

Johnson, B. H., K. W. Kim, R. E. Heath and H. L. Butler (1994): Development and application of a three-dimensional hydrodynamic model, In: *Computer Modeling of Free-Surface and Pressurized Flows* (eds. Chaudhry and Mays): Kluwer Academic Publishers, The Netherlands, pp. 241–280.

Leonard, B. (1979): A stable and accurate convection modelling procedure based on quadratic upstream interpolation, *Computer Methods in Applied Mechanical and Engineering*, 19, 59–98.

National Atmospheric Deposition Program (IR-7)/National Trends Network (1989): NADP/NTN Coordination Office, Natural Resources Ecology Laboratory, Colorado State University, Fort Collins, CO.

Robey, D. L. and L. J. Lower (1991): A time-varying, three-dimensional hydrodynamic and water quality model of Chesapeake Bay, The Military Engineer, August 1983, pp. 544.

Sheng, Y. (1986): A three-dimensional mathematical model of coastal, estuarine and lake currents using boundary fitted grid, Report 585, ARAP Group of Titan Systems, Princeton, NJ.

Thomann, R., J. Collier, A. Butt, E. Casman and L. Linker (1994): Response of the Chesapeake Bay water quality model to loading scenarios, CBP/TRS 101/94, Chesapeake Bay Program Office, US Environmental Protection Agency, Annapolis, MD.

US Environmental Protection Agency (1982): Chesapeake Bay Program Technical Studies, A Synthesis, USEPA, Washington, DC.

Chapter 13

Coastal Zone Management (1) — U.S. Perspective

Peter Grenell
General Manager
San Mateo County Harbor District, USA

1. Introduction

The US (US) has developed into a world leader in coastal engineering over the past 70 yrs. The US coastline is long — 19,800 km or 12,300 miles (Cicin-Sain and Knecht, 1998), and its coastal engineering problems are varied and extensive. This nation's experience with coastal engineering and coastal management provides valuable insights into the role, limits and opportunities of engineering in addressing coastal problems. Coastal engineering has many accomplishments to its credit, but some coastal problems will be solved only through multi-disciplinary approaches. The US has pioneered this multi-disciplinary outlook through the collaborative efforts of coastal engineers, environmental scientists and activists, and government interests.

Notwithstanding the considerable growth in our understanding of coastal processes and ecosystems, their dynamic and complex nature and human actions driven by economic, political, and social motivations, will continue to limit our knowledge of these interactions. Engineers' ability to routinely predict results with high degrees of certainty will remain less than perfect. Technical solutions alone cannot resolve social issues. Yet coastal engineering's positive contributions are even now being enhanced by more experiential learning, continued receptivity to alternative methods, and further integration into more coordinated coastal management policy.

This chapter first presents a brief historical perspective on the US experience with coastal engineering (for a more comprehensive treatment, see Wiegel and Saville, 1996). Recent responses to implications of climate change and particularly potential sea level rise impacts are considered. It highlights the importance of the State of California's experience and implications for coastal engineering. We also discuss different engineering and policy approaches, and "hard" and "soft" engineering alternatives; the role of public participation in, and its implications for, coastal engineering; some lessons of the past 40 yrs of US coastal zone management; and a suggested multi-disciplinary framework within which to enhance the current level of integration of coastal engineering and multi-disciplinary coastal zone management.

1.1 *Historical perspective*

Engineers have played a major role in coastal management in the US for many years. The founding of the American Shore and Beach Preservation Association and the Beach Erosion Board during the late 1920's and early 1930's had great influence on the evolution of coastal engineering (Wiegel and Saville, 1996). Many observers mark the rise of coastal engineering in the US in the 1930's stemming from beach erosion studies occasioned by the construction of Santa Barbara Harbor (Walker, 1991; Corps of Engineers, 1986). Even before that, the US Army Corps of Engineers were delegated by Congress, the authority to administer the "Commerce Clause" of the US Constitution (Article I, Section 8, Cl. 3). This authority has been held by the Supreme Court to include regulation of navigation and control of navigable waters (Beatley *et al.*, 1994). To accomplish this aim, the Corps have executed many civil work projects over the years including navigation channel dredging and construction of breakwaters and harbor facilities. (see, for e.g. US Army Corps of Engineers, 1986) The Corps' regulatory power is also derived from Section 10 of the Rivers and Harbors Act of 1899. The Act requires the approval of the Secretary of the Army for excavation or construction activities in navigable waters (*ibid.*); and Section 404 of the Clean Water Act, which governs alterations of and discharges into "waters of the US." Regulation of alterations to coastal wetlands and riparian areas has been a very influential, if limited, element of the Corps' coastal development permit authority.

Amphibious military operations during World War II stimulated development of modern analytical techniques, and the First Conference on Coastal Engineering in 1950 brought forth an extensive review of the state of coastal engineering's "art and practice" (Wiegel and Saville, 1996).

The 1950s and 1960s saw the growth of research programs, data collection, analytical techniques, and coastal engineering projects; and the late 1960s and 1970s led to greater awareness of environmental impacts and the need for coastal management (*ibid.*). This development, and the widespread recognition of the importance of conserving the nation's coastlines that followed, was spurred by intensive public concern that surfaced in the 1960s in California.

More than one-third of San Francisco Bay had been diked and filled for development and agriculture since European settlement began in the late 18th century. Unrestricted filling of the Bay became a major issue after the Corps of Engineers published a map in 1959 which graphically portrayed large areas of the Bay "susceptible of reclamation" (Bodovitz, 1985; Travis, 1987). A vigorous citizen effort led to passage by the State Legislature of the McAteer–Petris Act of 1965, which created the San Francisco Bay Conservation and Development Commission (BCDC). The

Commission, BCDC, had a mandate to prepare a Bay Plan, regulate Bay filling and shoreline development and provide for public shoreline access. BCDC became permanent in 1969.

The success in San Francisco Bay emboldened citizen efforts to pass legislation to regulate development along California's 1,770 km-long (1,100 miles) coast. After several fruitless attempts by environmental interests to get coastal legislation through the State Legislature, an intensive public signature campaign placed a coastal initiative, Proposition 20, on the 1972 state ballot. This passed by a comfortable margin.

Proposition 20 created the California Coastal Commission with a unique set of regulatory and planning powers (Public Resources Code Section 27,000 et seq). The State Commission and its six District Commissions temporarily took total control of the development permit process in the 914 m wide (1,000 yards) coastal zone established by Proposition 20 from the 68 (later 72) coastal city and county governments. Meanwhile, the Commission prepared a comprehensive Coastal Plan for the state's coastline (Commission, 1975). The Plan identified scores of resource issues and made recommendations for remedial and anticipatory action. The California Coastal Act of 1976 continued the Commission as a permanent agency and brought local government back into the process by requiring each coastal jurisdiction to prepare a Local Coastal Program (LCP) for its area within the coastal zone. LCPs were to consist of issue identification, a land use map and accompanying zoning ordinances. LCPs would be approved locally and certified by the Commission, after which each local government with a certified LCP would resume its permit authority.

As public anxiety over protection of San Francisco Bay turned to action, environmental concern nationwide prompted formation by Congress of the federal Commission on Marine Science, Engineering, and Resources, known after its chairman as the Stratton Commission. The Commission's 1969 report "Our Nation and the Sea: A Plan For National Action", attracted public, official and scientific attention to the importance and value of coastal resources and the current management vacuum, and led to the establishment of the National Oceanic and Atmospheric Administration (NOAA). The federal Coastal Zone Management Act (CZMA) of 1972, created a voluntary grant program administered by NOAA within the Department of Commerce. The program acts to support activity by coastal states to "promote a balance between development and the environment through coastal land use planning and management" (Beatley *et al.*, 1994).

An important provision of the CZMA requires consistency of federally sponsored coastal projects with state Coastal Zone Management (CZM) programs. Thus, for example, a Federal Corps of Engineers breakwater project to protect a coastal harbor must be found consistent with the CZM Program of the state within which the project is to be built. To cite another instance, maintenance dredging of a federal channel into a California harbor by the Corps of Engineers must not cause adverse impacts to aquatic habitats in violation of California Coastal Act policies.

There are now 32 federally approved state coastal zone management programs of a total of 35 coastal and Great Lakes states and island territories, including California. These programs protect almost 99% of this country's 95,000 miles of oceanic and Great Lakes coastline. Two remaining eligible states are now formulating their programs; one state is inactive.

Until recently, however, responsibilities for CZM in the US at the federal level have been scattered among many agencies besides the CZM Program, each derived from different legislative mandates and with no coherent mechanisms for coordination and conflict resolution. Recognition of the importance of addressing ocean and coastal issues on a more coherent and organized basis led to the formation of the US. Commission on Ocean Policy and the Pew Oceans Commission. The Pew Commission's 2003 report, America's Living Oceans: Charting a Course for Sea Change, and the US Commission's 2004 report, an Ocean Blueprint for the 21st Century, with their numerous recommendations for more coordinated action, paved the way for development of federal ocean policy.

An Interagency Ocean Policy Task Force was created under presidential direction, and in July 2010, the President by executive order accepted the Task Force's recommendations and established a National Ocean Policy (White House, 2010). Policy implementation of nine priority objectives occurs under the guidance of a National Ocean Council. Among the objectives are ecosystem-based management, coastal and marine spatial planning, and adaptation to climate change.

Meanwhile, coastal states have been developing their own ocean policy initiatives under the national policy umbrella. Especially noteworthy among these efforts is the regional action of California, Oregon, and Washington, where the West Coast Governors' Agreement on Ocean Health of 2006 led to the preparation of a 2008 Action Plan. The West Coast Governors' Alliance is now taking action on seven major priority areas identified in the action plan.

As part of this overall effort, and consistency with a US Ocean Policy Commission recommendation, the three west coast states are moving toward managing sediment regionally. This is a major departure from the traditional project-by-project approach to managing coastal sediments. The Alliance produced a Regional Sediment Management Action Plan in May 2010 to guide their respective efforts. One example of this activity is discussed in Sec. 1.2.

1.2 *The importance of the California model*

The spirit behind the popular move to "save the coast" is articulated in Proposition 20 of 1972. Section 27001 declares that (Note: The authors view are marked in italics):

> "... the California coastal zone is a distinct and valuable natural resource *belonging to all the people and existing as a delicately balanced ecosystem*"

> "...permanent protection of the remaining natural and scenic resources... is a paramount concern to present *and future* residents;"
>
> "...it is necessary to *preserve the ecological balance of the coastal zone* and prevent its further deterioration and destruction;"
>
> "...that it is the policy of the state to preserve, protect, and, where possible, to restore the resources of the coastal zone for the enjoyment of the current *and succeeding* generations;"
>
> "...and that to protect the coastal zone it is necessary:... (b) To prepare... *in full consultation with all affected governmental agencies, private interests, and the general public* , a comprehensive, coordinated, enforceable plan for the orderly, long-range conservation and management of the natural resources of the coastal zone."

The framers of Proposition 20 clearly viewed the coast and its resources as a public trust, consisting of a sensitive and unique ecosystem that requires long-term stewardship for present and future generations. This was considered to involve full and active public participation in planning and implementation of the permit process. These elements: the coast as a public trust with the underlying assumption of social equity, an ecosystem-based outlook, long-term stewardship, and full public participation comprised a new basis for public policy that had not been set forth so directly before. This new outlook embraced coastal engineering as an integral part of overall coastal resource protection, and placed the US at the forefront of coastal management.

The federal CZMA, passed by Congress the same year that Proposition 20 was passed by California voters, was less forthright in its policy declarations. Although it, too, declared national policy "...to preserve, protect, develop, and where possible, to restore or enhance, the resources of the Nation's coastal zone for this and succeeding generations," the CZMA did not explicitly acknowledge the critical importance and primacy of preserving coastal ecosystem balance (as distinguished from protection of individual resources). Nor did the CZMA mandate full public involvement; it only sought "...to *encourage* the participation and cooperation of the public, state and local governments" and other agencies in implementing the CZMA.

The California Coastal Commission's first four years of control over permitting created much controversy. Private property and development interests were disturbed by what they thought were onerous impacts of regulation. Many problems remained unsolved even with the Commission's total control over development. The Commission could only respond to permit applications. It had neither the mandate nor funding to undertake coastal engineering, restoration, acquisition, or beneficial development projects. For example, it could prevent filling of wetlands for development, but it could not restore a degraded marsh to biological health. It could require dedication of land for public access as a permit condition, but it could not construct the access way. It could regulate construction in coastal hazard areas, but it could not compensate a land owner, who it felt should not build by buying his property to prevent development outright.

The Commission's 1975 Coastal Plan thus recommended that a separate non-regulatory mechanism be created to address these and other issues not amenable to the Commission's development permit process. A companion act to the Coastal Act passed in 1976 (which made the Commission permanent) created a new state government agency, the State Coastal Conservancy (Public Resources Code Section 31,000 et seq).

The Coastal Conservancy was independent of the Commission, and its purpose was to protect, restore and enhance coastal resources by implementing programs for provision of public access, preservation and enhancement of wetlands and other natural habitats, protection of agriculture lands, consolidation of inappropriately sited and designed subdivisions, restoration of urban waterfronts, and acquisition of open space for public recreation, scenic and habitat values. By designing and funding coastal projects cooperatively with local governments, other public agencies, developers, ranchers, environmental organizations and community groups, the Conservancy was a non-confrontational facilitator, conflict resolution agent and problem-solver whose discretionary intervention could forestall and often eliminate the often adversarial character of the permit process (Grenell, 1988b). In 1980, the Conservancy's jurisdiction was extended to the San Francisco Bay Area, where it complemented BCDC as it did the Coastal Commission.

As attention to ocean policy grew at the federal level, California likewise moved to develop its own approach. In 2004, the State Legislature established the California Ocean Protection Council to "maintain healthy, resilient, and productive ocean and coastal ecosystems." A chief Council purpose is to coordinate the actions of ocean-related State agencies. In particular, through its five-year Strategic Plans, the Council supports research, projects, and planning to prepare for and reduce harmful impacts of climate change. Its 2011 resolution on sea-level rise, calls for state agencies and other entities implementing state-funded projects or those on state lands to consider sea level rise risks in their decisions, and to follow the recommendations on sea level rise of the 16-state agency Coastal and Ocean Working Group of the California Climate Action Team (Council, 2011).

California's ongoing contribution to the West Coast Governors' Alliance goal regarding regional sediment management includes the formation of the California Coastal Sediment Management Workgroup, a multi-agency effort including the Army Corps of Engineers. The Working Group is focused on preparing regional sediment management plans, based on littoral cell areas, as parts of a federally funded Coastal Sediment Master Plan. The underlying view of these efforts is of sediment as a resource and not a waste, or "spoils", when considering areas with sediment deficits or surpluses. Beneficial use of coastal sediments is becoming recognized as an acceptable approach to managing this valuable asset.

California's CZM program is perhaps the nation's best example of (a) comprehensive coastal land use regulation and development policy, and (b) an effective combination of regulatory and non-regulatory management methods. The California experience is particularly relevant to coastal engineering concerns with beach erosion and recreation, shoreline armoring and development policy, harbor dredging and expansion, and habitat protection.

Other state CZM programs provide important and instructive contrasts to the California situation. For those interested, state coastal programs may be accessed via the internet through the Office of Ocean and Coastal Resource Management. These are worthy of study because of the extremely varied physical conditions to be found along the US' extensive coastlines.

Coastal engineers in the US must contend with a broad array of physical issues and institutional structures, which highlight the question of how coastal engineering can be even more effectively integrated with government and non-government coastal management efforts than it is now.

1.3 *Summary of Section 1*

Coastal engineering's traditional technical approach to specific situations has evolved during the past 40 years to address more directly the natural resource, socio-economic and political issues that increasingly infuse coastal decision-making. Along with heightened concern for climate change and especially sea level rise impacts, incorporating coastal engineering into a broader and more coordinated management approach that addresses the cumulative impacts of development, management, and engineering decisions over wider geographic areas is the direction in which coastal engineering is now moving at federal and state levels.

2. Policy and Practice: Implications for Coastal Engineering's Future

2.1 *Importance of the coast and development impacts*

The contributions of coastal areas to national employment and the economy are immense. Studies by the nonprofit Coast Alliance (Weber, 1995; Holing *et al.*, 1995) showed that:

- US coasts provide over 28 million jobs, 34% of the nation's employment.
- In 1985, almost one-third of all jobs and nearly one-third of the nation's Gross Domestic Product were generated in approximately 10% of land that comprises coastal counties except Alaska.
- Over 180 million Americans visit ocean and bay beaches annually.
- US coastal areas yield 70% of the nation's commercial and recreational fisheries harvest annually.
- An estimated 94 million Americans engage in boating and fishing along the coast each year.
- The recreational fishing industry generates almost $15 billion annually.

These figures reflect the coast's demographic magnetism as well. Over half of the nation's population resides in coastal watershed counties, which comprise less than 20% of total land area of the US, excluding Alaska and Hawaii (US Census, 2011). To cite one graphic example, roughly 84% of California's population (in 2010) of over 38 million resides within coastal watershed counties (US Census, NOAA 2011).

These population and economic pressures are increasingly affecting the coast's natural habitats, fish and wildlife species, water quality, scenic areas, and recreational amenities. Increasing pollution of coastal waters and wetlands, especially from non-point agricultural and urban run-off, is causing reductions in shell-fish harvests and numerous beach closures and advisory warnings. In addition, conflicts between private landowners and public agencies over public shoreline access, protection of private property, and development rights continue, as more people seek to locate on and use the coast.

2.2 *Onshore impacts on coastal zone resources*

Onshore impacts on coastal resources from grading and construction for urban development, dams and flood-control measures, agriculture, non-point source run-off, and timber harvesting are primary reasons for declining water quality, habitat and species damage, beach erosion, and negative impacts on coastal business, employment and income. The adverse impacts cover commercial and sport fishing, recreation and tourism-related activities, and harbor operations and development. Many sources of these impacts are located far from the shoreline, yet they have significant influences on human activities and coastal processes. To note one example, the destruction of salmon spawning gravel by stream-bank erosion caused by timber harvesting in northern California, along with fresh water diversions for agriculture and urban use, have severely affected commercial and sport fishing industries and tourism along this part of the US coastline.

2.3 *Policy implications*

Coastal management experience under the CZMA has shown that major adverse impacts are caused mainly by human actions rather than natural ones (Pernetta and Milliman,1995); and that the most critical problems with integrated approaches stem from economic, social, and institutional factors rather than scientific and technical ones (IPCC, 1994). These concerns generate several major issues:

- the nature of human responses to impacts of climate change on sea-level rise, lack of uniformity in sea-level change, changes in intensity and frequency of storms, and affects of shifts in weather patterns;
- controversy over use of shoreline protection strategies of managed retreat versus hard edged, engineered solutions or soft-edged approaches for total protection;
- cumulative impacts of development, resource exploitation, and engineered coastal defense measures on regional shorelines, ecosystems, and natural resources;
- cost-benefit factors including evaluation of public costs for shoreline protection;

- ownership conflicts over private property rights versus public rights of access to, and use of, coastal shorelines and resources; and
- sustainability of alternative coastal-development strategies.

(1) Armoring, Managed Retreat and Cumulative Impacts. In coastal areas of the US such as southern California, rapid growth occurred over several decades causing a virtual "walling off" of the coast by exclusive, high-value developments (Boyd, 1985). This growth blocked public access to the shore, accelerated erosion of coastal bluffs and beaches, damaged fish and wildlife habitats, and degraded scenic areas. The situation resulted in intense resistance to managed-retreat policies in favor of uncoordinated, privately initiated coastal armoring methods where coastal defense has proven necessary to protect private property. Cumulative impacts of these individual, uncoordinated efforts have aggravated erosion, beach loss, and habitat damage. These conditions are particularly evident along the California coast, much of which is unstable but whose attractiveness has drawn development to the water's edge.

In its 1990 amendments to the CZMA, Congress identified cumulative impacts as a key concern requiring improved management by the states. In one instance, California's Coastal Commission undertook a pilot project to develop a process for integrated regional cumulative impacts management for the whole California coast (California Coastal Commission, 1995). Present California policy allows development in high-hazard areas. California's emergency permit process has often been used to approve armoring projects to protect individual private properties. Approvals have frequently occurred with little technical analysis, examination of alternatives, or mitigation for negative impacts on resources (California Coastal Commission, 1995). The Commission's pilot project in the Monterey Bay area has made clear the need for regional, long-term planning to address erosion and related development impacts. The engineering of private shoreline-protection devices is becoming unacceptable without considering the larger regional context of the littoral cell.

Requiring setbacks is a common coastal policy approach that has been used in North Carolina, California and elsewhere to enable development without armoring. In North Carolina, for example, new development must be

> "the farthest landward of the following alternatives:
>
> - A distance equal to 30 times the long-term erosion rate,
> - The crest of the "primary" dune,
> - The landward toe of the "frontal" dune,
> - Sixty feet landward of the "vegetation line" (Cicin-Sain and Knecht, 1998, from North Carolina 1985)."

Setbacks cannot address all risks in all cases, however, such as building on infill lots or very small lots, or with existing development. Setbacks have been found in California to be most effective if incorporated into a planned development at the subdivision stage (California Coastal Commission, 1995). In built-up areas containing wetlands, though, there is little if any scope for allowing wetland "creep" and buffer areas to absorb the affects of sea-level rise.

In Florida, erosion has tended to be deemed "critical" only where development is threatened. Florida's beaches suffer from significant erosion and with their great importance as recreational and tourist resources, beach nourishment and other measures have been implemented to protect and maintain them. Recent evidence suggests that beach fill is economic, given the consequent positive return in tourism revenue, even though beach nourishment must be sustained. In contrast, armoring structures in developed areas have resulted in losses of public lands and beaches. In already built-up and rapidly developing areas, political pressures have largely precluded use of reduced development density and managed retreat measures such as abandonment and replacement (Schmahl and Conklin, 1991). In less developed areas, the state has acquired coastal land to preclude future development. This approach is limited by the high cost of shoreline property and limited public funds.

One alternative for protecting public rights to beach access and preventing loss of public lands now being examined by several states is use of the public trust doctrine. Briefly put, the principle of the public trust, which arose from our English common law heritage, holds that all tidelands and submerged lands under navigable waters are owned by the states; states hold these lands in trust for the benefit of the public subject to specific use rights related to maritime commerce, navigation and fisheries; and that grants of such lands to private owners remain subject to that public trust. This mechanism exists in addition to the traditional police power of local and state government. The public-trust doctrine is viewed as promising by many because of its basis in present public ownership rights. Public trust rights have been expanded recently in Texas, New Jersey, North Carolina and Massachusetts (Schmahl and Conklin, 1991). A related proposal would be the concept of sand rights, which would apply the public trust to ensure the supply of sand to beaches by requiring mitigation of development impacts that adversely affect the flow of beach sand within a littoral cell (Stone and Kaufman, 1989; Gustaitis, 1996).

Another potentially powerful mechanism to encourage voluntary development setbacks is elimination of federal flood and hazard insurance. This approach could have considerable impact especially along the low-lying southern Atlantic seaboard and the hurricane-prone Gulf of Mexico coast. Regulation alone remains inadequate to address the problem.

(2) Sea-Level Rise. In the US, until recently, sea-level rise has been less of a factor in coastal change than more immediate coastline subsidence and recession resulting from human-caused and natural impacts (Pernetta and Milliman, 1995). Current research and projections now indicate that sea-level rise impacts over the next several decades could be considerable.

For example, a 2009 report from the nonprofit Pacific Institute, Impacts of Sea-Level Rise on the California Coast, estimated that 480,000 people, critical infrastructure, wetlands and other habitats, and nearly $100 billion in property on the California coast would face increased flooding risk from a 1.4 m sea-level rise (Pacific Institute, 2009).

In 2009, the Secretary of the Interior signed Secretarial Order No. 3289, addressing the Impacts of Climate Change on America's Water, Land, and other Natural and Cultural Resources, to establish a coordinating framework regarding climate change activities within the department.

More directly related to coastal engineering, the US Army Corps of Engineers Circular No. 1165-2-211 of July 2009 requires that potential sea-level rise changes must be considered in every Corps coastal activity as far inland as the extent of estimated tidal influence. Further, future sea-level rise projections must be incorporated into Corps project management, planning, engineering, design, construction, operation and maintenance.

To cite the California example again, in 2001, the California Coastal Commission compiled a report on sea-level rise and its possible effects on the California coast. The report discusses response actions including "hard" engineering, e.g. seawalls, revetments, breakwaters, levees, etc., "soft" engineering, e.g. beach nourishment, buffer areas, accommodation and retreat. In addition to its regulatory role, the Commission is coordinating with the Coastal Conservancy on encouraging property acquisition in high risk areas.

Recognizing these concerns, the California State Lands Commission, which administers public trust lands in the State, produced a Report on Sea-Level Rise Preparedness in December 2009. The Commission's concern is that lands under its jurisdiction are developed and managed with consideration of potential sea-level rise impacts.

Combinations of regulatory restrictions on development and armoring, land acquisition, and reduction of public investments that promote unwise development will be much more successful, economical, and environmentally sound. These should be joined with new financial and other incentive methods to provide less harmful directions for economic development to move. Integrated regional coastal planning, together with these measures, can provide a more potent response to these issues. Public pressure for budgetary restraint may generate sufficient political will to reorient coastal policies. We have enough workable methods and a wealth of practical experience, generated by coastal engineering and other disciplines over many decades.

(3) Cost/Benefit Considerations. These have particular relevance to preserving public resources such as erosion-prone beaches. Breakwaters and groins are expensive, and beach nourishment less so, although it is ongoing. Local revenues from beach-related tourism and water-oriented recreational business are considerable and pressures to maintain heavily used beaches subject to erosion, however costly, are intense, as in Florida and southern California. Cost-benefit methodology does have limitations, including difficulties sometimes encountered in quantifying or calculating environmental benefits; no internalization of some costs, such as external negative impacts, which may occur in some analyses; and use of frequently controversial discount rates in present-value calculations (Cicin-Sain and Knecht, 1998). The basis for more informed decision-making on these and other situations is being strengthened by development of methods to value coastal resources and coastal defense impacts on the shoreline environment, particularly on recreational and natural habitat resources. (See, for example, Spurgeon and Brooke, 1995; Green *et al.*, 1990; and other authors cited in Cicin-Sain and Knecht, 1998.)

A current California challenge to the Corps of Engineers concerns a demonstration project to test solutions to beach erosion caused by scouring from a breakwater built to protect a harbor of refuge. Analysis of comparative costs and benefits of "soft" alternatives like beach replenishment and artificial reefs versus "hard" alternatives like installing groins must now also consider sea-level rise impacts, and whether saving a public beach is, in the final analysis, financially feasible.

(4) Inland Impacts. Upstream actions are a significant source of adverse affects on coastal resources. They are often neglected in coastal management and frequently are beyond the purview of coastal engineering. For example, sand budgets in many coastal areas have decreased because of damming and diversion of coastal streams and rivers that traditionally replenished the beaches. Concerns about flooding and demands for more fresh water for urban and agricultural use have been the chief influences on this cause of beach loss. Other examples include the degradation of fish spawning areas and coastal marshes caused by upstream timber logging and cattle grazing impacts on stream banks and water quality. Non-point-source agricultural pesticide and urban storm-water run-off generate substantial coastal water quality impacts. These have replaced industrial point sources as the chief sources of pollution of coastal waters.

(5) Alternatives to "Hard" Engineering. In recent years, the "soft" engineering approach has increasingly been used in designing coastal defenses to reduce both costs associated with traditional "hard" solutions and adverse impacts on coastal ecosystems and other resources. "Soft" engineering seeks to "work with nature by manipulating natural systems which can adjust to the energy of the waves and tides to good effect" (Richardson, 1995), rather than confront this energy with rigid engineering structures. "Soft" engineering thus attempts to respond to concerns for sustainability of coastal defenses through designs that conserve natural resources and the coastal environment.

Other innovative coastal engineering methods have emerged to address the resolution of competing demands on coastal wetlands. For example, in California the Coastal Conservancy has employed new hydrological engineering approaches to design wetland projects with multiple goals including fish and wildlife habitat enhancement, use of treated urban waste water, flood management, and public access (Josselyn *et al.*, 1993; Grenell, 1995). At Buena Vista Lagoon on the southern California coast, the Conservancy resolved a conflict over flood-control impacts on coastal wetland habitat by finding an alternative to the traditional engineering solution of a concrete-lined channel. This had led to costly dredging of the downstream lagoon by the state and loss of aquatic habitat. The Conservancy organized the three cities through which flood-prone Buena Vista Creek flowed into a joint powers committee to find a cooperative solution, studied the lagoon and its watershed for the first time as a single integrated system, then designed and financed a "soft-edged" engineering treatment of the creek. The cooperative solution involved drop structures to reduce flood flow acceleration and willow plantings and sediment detention basins to diminish sediment deposition in the lagoon (Grenell, 1988a).

(6) Public participation, consultation, and consensus. The coastal shore is a unique place; as Rachel Carson so eloquently put it: "... The primeval meeting place of the elements of earth and water, a place of compromise and conflict and eternal change" (Carson, 1955). The public's right of access to the shore is an ancient principle traceable at least to Roman times. Public trust doctrine holds that tidelands belong to the state, which manages them in trust for all the people. Living on the coast is a privilege, ideas of private property rights notwithstanding. Coastal resource users are likewise privileged and have a stewardship responsibility to the wider community; and they should pay for their privileged use of these resources.

Cicin-Sain and Knecht (1998) emphasize the importance of public participation and consensus-building in ensuring the effectiveness of integrated coastal management (ICM) programs. They stress the need to bring "coastal users and the public" into the ICM process as early as possible because:

> "1. The input of those who use and rely on the coastal zone is of great importance in the design of the ICM process. Users have valuable insights regarding both the management needs and the economic opportunities an ICM program should address.
> 2. Support of the users for development and implementation of an ICM program is crucial to its success.
> 3. Increasingly, governments are finding it necessary to develop private-public partnerships to fully accomplish the goals of their resource management programs. This is especially true with economic development initiatives related to coastal and ocean resources such as those that might be included in an ICM program."

Coastal engineers ask themselves who benefits from their designs for shoreline protection, flood management, and harbor development, and what are the impacts on other coastal users and resources of their schemes? These concerns reinforce the need for coastal management policy and coastal engineering to maintain a broad basis for public input to the decision-making process, and for more integration of coastal engineering with overall coastal management.

The validity of this viewpoint is substantiated by experience. Coastal management policy development and administration in the US has largely come about as a result of public concern and action. This is nowhere more readily apparent than in California (Galanter, 1983). Coastal management has been most effective when public agencies have encouraged public participation in policy formulation, project development, and the regulatory process (Petrillo and Grenell, 1985; Galanter, 1983).

Bringing citizens and technical experts together, in more productive ways, to get the most from coastal engineering solutions must be an essential element in the evolution of future coastal management and engineering policy and practice (Child, 1995; Grenell, 1995). Coastal engineers need to work more closely, and earlier in the design process, with the public and other experts including planners, lawyers, scientists, and economists. An example of such outreach is the approach of the Corps of Engineers' San Francisco District office, which actively encourages coastal development permit applicants to confer with agency staff as soon as possible in the project development process.

2.4 *Lessons of 40 years of US coastal management*

Coastal zone management in the US has achieved much during the past 40 yrs. More public access has been provided, the rate of coastal wetland loss has in most instances been reduced, development of sprawling rural subdivisions has been greatly checked, and offshore energy development has been controlled in some highly sensitive areas. Many of these accomplishments are readily visible; one becomes aware of others by what has not happened.

Of particular significance is the observation borne out by coastal management experience that regulation is necessary but not sufficient for guiding coastal development and resource stewardship. In particular, the experience of California with its regulatory Coastal and San Francisco Bay Commissions and its non-regulatory Coastal Conservancy illustrates the necessity, and the effectiveness, of a combined approach. Regulation may be able to halt certain undesirable development or resource exploitation, but it cannot stimulate, facilitate, or initiate restorative or anticipatory action. Examples abound of new roles for government as catalyst, facilitator, and demonstrator of more sensitive coastal management methods that link environmental and economic needs in a unified approach.

In the US, coastal zone management has evolved as a partnership of the federal government and the states. The US approach has allowed the states considerable flexibility to develop their own management strategies. Performance has varied significantly, however, among the states.

Major coastal issues are now seen to require broader cooperative action between federal, state, and local governments with built-in public participation, under the aegis of a national ocean policy. The basis for this action needs to be geographical: trans-boundary; regional; and based on a spatial continuum from inland watersheds to the outer continental shelf limits. It must also be multi-sectoral. We have become increasingly aware of the need to consider coastal issues, including engineering concerns such as coastal defense, within the scope of watershed and regional ecosystem.

Coastal issues must also be considered within littoral-cell thinking; and in terms of partnerships among public and private interests, business, non-profit, and community elements working cooperatively with government. The states themselves need to have minimum management requirements that take these concerns into account.

Local government has great potential ability to beneficially influence coastal development patterns. But local governments often lack time and resources; especially financial. They need help in capacity building and support to participate more effectively in trans-boundary management initiatives. They also need to move beyond traditional zoning and land use controls. Last, they must develop a stronger will to act more vigorously to fulfill their broader stewardship responsibilities. These concerns are part of the context within which US coastal engineers must function.

2.5 *Summary of Section 2*

The coast's vital economic importance is reflected in continuing demographic and economic pressures on coastal resources, usually with adverse impacts. These impacts are mostly caused by human actions, for example, fragmented approval of coastal armoring, strong opposition to managed-retreat policies, and cumulative impacts on public and private coastal lands and resources.

Better scientific and technical information is enabling more accurate cost-benefit and environmental assessments of coastal engineering and other management alternatives. "Soft" engineering solutions to coastal-defense and resource-management issues show promise, although they, too, require extensive testing and monitoring to adequately determine their effectiveness and unintended environmental consequences. The above experience suggests the need for more coordinated coastal management strategies and fuller integration of coastal engineering into overall coastal zone management practice, as is now being recognized in national and regional policy initiatives.

3. A New Framework to More Effectively Integrate Coastal Engineering with Coastal Zone Management

Public concern for environmental protection is rising with continuing population and development pressures on coastlines. Awareness of long-term implications of climate change on coastal resources is becoming more widespread. The drive to reduce federal expenditures in the interest of budgetary control and debt reduction has directly affected the availability of funds for shoreline protection, harbor and channel dredging, and beach preservation. Reduced federal funding is placing much larger fiscal burdens on state and local governments for these functions. Concurrently, greater public concern for conserving remaining wetlands and other coastal habitats has generated opposition to seawalls, breakwaters, groins, dams, and other "hard-edged" methods typically used in the past; this concern also has stimulated exploration of alternative approaches.

Following is an outline for a coastal-resource management framework that maintains coastal engineering as an essential element within a broader basis for decision-making. This framework must have the characteristics explained in the following subsections.

3.1 *Strategic*

Incorporates coastal engineering for shoreline protection, erosion control and development into broad-based, long-range coastal policy that covers all coastal resources. It recognizes a range of concerns including population and development pressures, ecosystem influences on coastal change, sea-level rise and other climate change factors, and threats to unique and irreplaceable natural and scenic resources from development, inadequately planned coastal defense measures, and other human interventions. This strategic approach also takes into greater account the interdependence of decisions and actions of competing interests over wider areas (Huggett, 1995).

3.2 *Pragmatic*

Recognizes human motivations while promoting alternative choices for decision-making that acknowledge uncertainty and limitations of scientific knowledge. The policy basis for coastal decision-making must focus more on ways to channel human actions into less environmentally and socially destructive paths.

3.3 *Comprehensive*

Provides a basis for decisions concerning *the past* — restoration to correct and repair past mistakes; *the present* — pre-empting unwise actions and favoring cautious preventive action that enables monitoring and adjustments; and *the future* — anticipating impacts of physical and economic change and providing incentives for behavioral change that confront human motivations (Grenell, 1991).

3.4 *Flexible*

Adapts to a variety of circumstances in which priorities for action may differ. Coastal geomorphology, onshore impacts on shoreline and near shore marine resources, sea-level rise, development pressures and public use demands vary greatly between different coastlines. The framework should have regulatory and non-regulatory components, but regulation should be seen as a set of parameters for action rather than as the principal solution for all issues. Coastal engineering designs should be adaptive and capable of being tested and modified on the basis of monitoring actual experience.

3.5 *Regional*

Based on manageable units derived from coastal processes and taking into account cumulative impacts of individual decisions. For example, littoral cells are now being considered as appropriate divisions to guide coastal engineering and management planning, such as regional sediment management, and including cumulative impact issues. Institutionally, decentralization of coastal management activities, as for example in California and Massachusetts, can facilitate links between local decision-makers and citizens, state, and federal agencies, and promote regional cooperation and information distribution (Courtney, 1991).

3.6 *More realistic in its expectations of science*

Capable of yielding readily usable data on problem recognition to decision-makers who must act rapidly on the basis of uncertainty and lack of scientific consensus. Our understanding of coastal processes and the ecosystems within which they occur remains partial, in spite of recent advances in our knowledge and research methods.

Coastal engineers and scientists must communicate their *current* understanding of coastal problems clearly to decision makers, and place it within the social context of overall coastal decision-making. Recommended solutions should be seen as experiments or demonstrations whose performance is monitored and evaluated to provide continuing input to the learning and decision-making processes. Solutions claiming sustainability should be carefully accessed in view of the lack of sustainability of much past shoreline protection and resource exploitation.

3.7 *Summary of Section 3*

This integrated coastal-management approach should be *strategic* in considering multiple issues in a coherent framework, *pragmatic* in addressing human motivations as the key to effective decision-making, *comprehensive* in its past, present and future orientation, *flexible* in its adaptability to different conditions and constellations of competing interests, *regional* in analyzing and designing technical solutions, and more *modest and pragmatic* in its quest for scientific certainty.

Coastal engineering has been, and continues to be, an essential component of integrated coastal management. Coastal engineers can offer technical alternatives to address specific physical conditions; and engineering solutions are necessary in certain circumstances. Efforts now being made to expand knowledge of more environmentally sensitive and cost-effective engineering solutions should continue. Engineers need to work more closely with other interests within a broader management approach whose focus is oriented to long term socio-economic and environmental health. Greater public participation must be part of this approach. In this view, coastal engineering and coastal defense methodology become less a tool for preserving existing economic assets on a short-term basis, and more a means for creatively adapting long-term human needs to the realities of life at the ever-changing meeting of land and sea.

References

Bodovitz, J. E. (1985): The shrinking of San Francisco Bay and How it was Stopped, *California Waterfront Age*, **1**(4), 21–27, Fall, State Coastal Conservancy, Oakland, CA.

Boyd, W. M. (1985): Coastal regulation: Assuring access to California's coast, In: The *Urban Edge: Where the City Meets the Sea* (J. E. Petrillo and P. Grenell, eds.), California State Coastal Conservancy and William Kaufmann, Los Altos, CA.

California Coastal Commission (1985): ReCAP Pilot Project, Findings and Recommendations: Monterey Bay Region, *Regional Cumulative Assessment Project*, San Francisco, CA.

California Coastal Zone Conservation Act of 1972 as amended (Proposition 20) (1972): *Public Resources Code*, Division 18, Section 27000 *et seq.*

California Coastal Zone Conservation Commissions (1975): California Coastal Plan, San Francisco, CA.

California Ocean Protection Council (2011): Resolution of the California Ocean Protection Council on Sea-Level Rise, Oakland, CA.

California State Lands Commission (2009): *A Report on Sea Level Rise Preparedness*: Sacramento, CA.

Carson, R. (1955): *The Edge of the Sea*: Houghton Mifflin Company, Boston, MA.

Child, M. (1995): Taking plans forward through consultation and participation: Are plans sustainable?, *Coastal Management 1995*: *Putting Policy Into Practice*, *Proc. Int. Conf. organized by the Inst. of Civil Engrs.*, Bournemouth, October 1995, London: Thomas Telford.

Cicin-Sain, B. and R. W. Knecht (1998): *Integrated Coastal and Ocean Management: Concepts and Practices*: Island Press, Washington, D. C.

Clark, J. R. (1996): *Coastal Zone Management Handbook*: CRC Press, Lewis Publishers.

Coastal Act of 1976 as amended, *Public Resources Code*, Division 20, Section *et seq.*

Coastal Resources Center (1996): The relevance of the U. S. experience to governance of the world's coasts, *Intercoast Network*, **25**, Winter 1996. Univ. Rhode Island, Narragansett, RI.

Commission on Marine Science, Engineering, and Resources (2004): *Our Nation and the Sea: A Plan for National Action*: U. S. Government Printing Office, Washington, D.C.

Courtney, F. (1991): A regional approach to state coastal management, *Coastal Zone '91, 1, Proc. 7th Symp. on Coastal and Ocean Management*, ASCE, New York, NY.

Ewing, L. and J. Michaels (1991): Effects of sea level rise on the California Coast, *Coastal Zone '91*, 3, *Proc. 7th Symp. on Coastal and Ocean Management*, ASCE, New York, NY.

Faber, P. and J. Liebster (1987): California's fourteen years of coastal zone management, *Coastal Zone '87, 3, Proc. 5th Symp. on Coastal and Ocean Management*, ASCE, New York, NY.

Galanter, R. (1983): California's coastal protection constituency: Conserving a critically valuable resource, *Coastal Zone '83.* ASCE.

Gornitz, V., T. W. White and R. M. Cushman (1991): Vulnerability of the U. S. to future sea level rise, *Coastal Zone '91, 3, Proc. 7th Symp. on Coastal and Ocean Management*, ASCE, New York, NY.

Green, C. and S. Tunstall, E. Penning-Rowsell and A. Coker (1990): The benefits of coast protection: Results from testing the Contingent Valuation Method (CVM) for valuing beach recreation, Presented to the Conference of the River and Coastal Engineers, Loughborough Univ., England, 10 July 1990, Flood Hazard Research Centre, 168. Middlesex Univ. [formerly Middlesex Polytechnic], Enfield, Middlesex.

Grenell, P. (1988a): Coastal conservancy roles in cooperative urban wetland projects, *Urban Wetlands and Riparian Habitat, Proc. Assoc. State Wetland Managers Symp.*, June 26–29, 1988, Oakland, Ca.

Grenell, P. (1988b): The once and future experience of the California Coastal Conservancy, *Coastal Management*, **16**(1), Taylor & Francis, New York, NY.

Grenell, P. (1991): A new framework for integrated coastal zone management. *The California Coastal Zone Experience, Coastal Zone '91*, ASCE, Long Beach, CA.

Grenell, P. (1994): Looking back in light of the future, *California Coast & Ocean*, **10**(1), Winter/Spring, State Coastal Conservancy, Oakland, CA.

Grenell, P. (1995): Resolving coastal conflicts on the California Coast, *Coastal Management 1995: Putting Policy Into Practice, Proc. Int. Conf. Organized by the Inst. Civil Engrs.*, Bournemouth, October 1995, London: Thomas Telford.

Gustaitis, R. (1996): Trying to balance the sand budget, *California Coast & Ocean*, **12**(1), Spring 1996.

Holing, D. *et al.* (1995): State of the Coasts, Coast Alliance, Washington, DC.

Huggett, D. (1995): Strategic planning in coastal defense — a mechanism for sustainable habitat management?, *Coastal Management '95: Putting Policy Into Practice, Proc. Int. Conference organized by the Inst. Civil Engrs.*, Bournemouth, October 1995, London: Thomas Telford.

Intergovernmental Panel on Climate Change (IPCC) (1994): Preparing to Meet the Coastal Challenges of the 21st Century: Conference Report, World Coast Conference 1993. Noordwijk, The Netherlands.

Josselyn, M. and S. Chamberlain, K. Goodnight, H. Hopkins and A. Fiorello (1993): Evaluation of Coastal Conservancy Enhancement Projects: 1978–1992. San Francisco State Univ., Romberg Tiburon Center for Marine Studies, Tiburon, CA.

Kraus, N. C., ed. (1996): *History and Heritage of Coastal Engineering*: ASCE, New York, NY.

Pacific Institute (2009): *Impacts of Sea Level Rise on the California Coast*: Oakland, CA.

Pernetta, J. J. and J. D. Milliman (1995): IGBP Global Change Report No. 33: Land-Ocean Interactions in the Coastal Zone: Implementation Plan, The International Geosphere-Biosphere Programme of the International Council of Scientific Unions, Stockholm, Sweden.

Petrillo, J. E. and P. Grenell (1985) *The Urban Edge: Where the City Meets the Sea*, California State Coastal Conservancy and William Kaufmann, Los Altos, CA.

Pew Oceans Commission (2003): *America's Living Oceans: Charting a Course For Sea Change*: The Pew Charitable Trusts, Washington, D. C.

Richardson, B. D. (1995): Soft Engineering on the Coast — Where to Now?, Coastal Management '95: Putting Policy Into Practice, *Proc. Int. Conf. organized by the Inst. Civil Engrs.*, Bournemouth, October 1995, London: Thomas Telford.

Schmahl, G. P. and E. J. Conklin (1991): Beach Erosion in Florida: A Challenge for Planning and Management, *Coastal Zone '91*, Vol. 1, *Proc. 7th Symp. on Coastal and Ocean Management*, ASCE, New York, NY.

State Coastal Conservancy Act of 1976 as amended (1976): *Public Resources Code*, Division 21, Section 31000 *et seq.*

Stone, K. E. and B. Kaufman (1989): A Legal System to Protect the Shores of the Sea, *California Waterfront Age*, **5**(1), Winter 1989.

Spurgeon, J. and J. Brooke (1995): Use of the contingent valuation method to quantify some aspects of the environmental effects of coastal defense schemes, In: *Coastal Management '95: Putting Policy Into Practice* (C. A. Fleming, ed.), Proc. Int. Conf. organized by the Inst. Civil Engrs., Bournemouth, October 1995, London: Thomas Telford, 1996.

The White House (2010): Executive Order: Stewardship of the Ocean, Our Coasts, and the Great Lakes.

Travis, W. (1987): A comparison of California's coastal programs, *Coastal Zone '87: Proc. of the Fifth Symposium on Coastal and Ocean Management-ASCE*, Vol. 3. Seattle, WA.

U. S. Army Corps of Engineers (1986): An oral history of coastal engineering activities in Southern California, U.S. Army Corps of Engineers, Los Angeles District.

Walker, J. R. (1991): Downdrift effects of navigation structures on the california coast, *Coastal Zone '91*, 3, *Proc. 7th Symp. on Coastal and Ocean Management*, ASCE, New York, NY.

Weber, M. L. (1995): *Healthy Coasts, Healthy Economy: A National Overview of America's Coasts*: Coast Alliance, Washington, D. C.

West Coast Governors' Agreement on Ocean Health-California (2008): Final Action Plan, Oregon, Washington.

West Coast Governors' Alliance on Ocean Health (May 2010): Regional Sediment Management Final Action Plan.

Wiegel, R. L. and T. Saville, Jr. (1996): History of Coastal Engineering in the USA, In: *History and Heritage of Coastal Engineering* (N. C. Kraus, ed.), ASCE, New York, NY.

Chapter 14

Coastal Zone Management (2) — The European Perspective

Hans Kunz

Director & Professor, i.R.

Coastal Research Station, Norderney, Germany

1. Introduction

A large number of social and economic activities are concentrated in the coastal zones of Europe. The present situation is the result of historical developments aimed at optimizing use of the natural system and, if necessary, at protecting life and property against hazards. The coastal communities responded to changes of the natural and of the social surroundings by activities that we nowadays address as Coastal Zone Management (CZM). Traditionally, civil engineering has been applied in support of these activities. The steadily growing industrialization, commercial development, and population pressure have resulted in many coastal areas in over-exploitation of resources, pollution, losses of ecosystems, increase of erosion and flooding. Growing environmental awareness has triggered common agreement on the needs to plan and control coastal development systematically in an integrative and sustainable way. This chapter focuses on selected problems; hence it does not provide a balanced description of the European coastal zones, as a whole, and the respective management approaches. Emphasis is placed on sea defense and coastal protection in the southern North Sea. The German situation is treated as an example, in the context with neighboring countries, to demonstrate the dynamically changing processes of CZM in Europe.

CZM is a term referring to any planning and adjacent development in the coastal area; these applications do not necessarily incorporate common objectives. Integrated CZM (ICZM) means at least that all kinds of influences of certain impacts on each other will be evaluated. In this sense, CZM and ICZM to some extend have traditions in Europe reaching back to several thousand years (i.e. to the Greeks, Romans, Frisians). Since the 1970s, public awareness about environmental degradation and the finiteness of its resources has substantially grown in many parts of Europe. The consequence is that the influence of societal impacts on the coastal environment has been practically and legally incorporated into the management process. Generally, a system perspective and multi-sectoral approach are nowadays the two key characteristics that serve to distinguish ICZM from other types of environmental planning and management programs which occur in coastal areas (Sørensen, 1993).

However, industrialization, commercial, urban and agricultural development, and growing population pressure have resulted in an increase of pollution, over-exploitation of resources, erosion, flooding, losses of dunes, wetlands, habitats, etc., in numerous coastal areas of Europe. As in many areas around the world, it also became obvious in Europe that long-term environmental conservation is essential to preserve the coastal zone as a multi-resource and as a multi-user system for future generations. This has led to the concept of carrying capacity and of sustainable development, which are increasingly accepted.

In Europe, the changing attitude toward human intervention in the coastal zone confronts the multifarious traditions and customs of the coastal community. These developed, in the past, according to the specific societal needs and the related key factors (such as, natural conditions, social surroundings, and options and fixed points) of a certain region. Consequently, conflicts between the various user categories increased, indicating the need to implement common principles by planning and controlling the development of the coastal zones in a systematic way. The CZM objective is to solve present and future problems by finding an acceptable balance between economic welfare and environmental well-being. A representative example is sea defense against flooding of the vulnerable coastal lowlands that border the southern North Sea. Concerning tools and techniques, civil engineering has traditionally entered in protecting life and property against storm surge (a primary need) and in supporting socio-economic activities (economic needs). The damaging effects of societal intervention through engineering on the natural coastal system were not perceived as relevant for the decision-making procedure in the past.

Thus, a technical defense system has evolved within the changing socio-economic system which influences and limits the possibilities for proceeding directly toward ICZM plans, as required to satisfy the recently evolved environmental objectives (ecological needs). ICZM calls for long-term policies, a complex management process driven by the mutual effects between changing needs, natural conditions, and social surroundings. Options for response make it necessary to build the capacity of policy and system — analysis for the specific cultural, social and political situation in each European country. The options for response are to be capable of generating proposals to be used by the decision-makers as well as for public participation. However, in many European countries numerous problems deriving from the demands for intensification of use of the coastal zones are still handled on an ad hoc basis, and a short term policy or solution is frequently preferred by politicians.

It is apparent that coastal engineering in Europe intensively reflects the changing needs and promotes progress towards interdisciplinary co-operation to achieve a sound understanding of the processes. There is now a definite trend toward minimization of irreversible impacts by applying soft techniques and by socio-economic integration of engineering interventions as part of the management process. The literature is large and, therefore, only representative references have been chosen to lead the reader to other works.

2. Coastal Zone Management in Europe

The coastal zone is simultaneously a multi-resource system (natural and societal-modified environment) and a multi-user system (use of natural resources for subsistence, economic activities, recreation etc.) (Hoozemans *et al.*, 1996). Correspondingly, coastal problems are defined and validated from an anthropocentric point of view, being inseparably connected with human expansion in the coastal zone. The complex interactions between resources and uses determine the process of coastal zone development. There is an increasing need to plan and to control this process in a systematic way by management, and CZM has been identified by the international community as an appropriate approach for addressing coastal problems. However, many European countries traditionally coped with problems in their coastal zones vulnerable to hazards by management responses and physical planning. This type of response to hazards has gone on for many centuries (see Sec. 7). Coastal engineering has been increasingly incorporated as an important part of the actions taken (see, e.g. the particular discussions for several countries contained in Kraus, ed., 1996). According to history, special laws, programs, management procedures, institutions, etc., have developed, which are principally suited to plan and to execute in accordance with the targets of ICZM. However, many actions are still based on plans, which predominantly address sectors of special interest. The traditional and cultural perspectives have to be taken into account, as well as the conflicting interests and uses, to understand and to validate the different ways in which CZM is executed in Europe.

The term CZM is nowadays associated with the principle of sustainability, as coined by the Brundtland Commission in 1987 and popularized by the Rio-Earth Summit in 1992 (through Agenda 21). It is rooted in the notion that economic viability, environmental quality, and social justice are related, and a balanced utilization is of advantage both to society and the environment for future generations. Hence, the main objective of sustainable CZM is to allocate scarce resources among competing and often conflicting users with the ultimate goal of optimizing the utilization of these resources for the coastal society, both now and in future. With respect to the integrative character of CZM, as described in Sec. 1, the process is commonly termed ICZM. The World Coast Conference in Noordwijk, Netherlands, 1993 (WCC, 1993) issued a joint statement that calls for progress in developing an operationally oriented CZM concept for vulnerable coastal areas (Coastal Zone Management Plan, CZMP). The CZMP is then to be implemented as a tool to anticipate and respond to the needs of the coastal society. Coastal vulnerability derives from indicators, for example, people and capital at flood risk, expected future damages or losses, costs of countermeasures, and degeneration of ecosystems and habitats (WCC, 1994). A CZMP has to integrate social, political and technical activities. For example, van der Weide *et al.* (1993); Hoozemans *et al.* (1996) describe tools and techniques for the elaboration and implementation of CZMP, especially with respect to coastal engineering and the situation in parts of Europe.

The European Community (EC) has generally agreed upon the need for developing ICZM-programs. Justifying the involvement of the EC-Commission requires, however, that any action taken be in accordance with the principle of subsidiary. This principle is interpreted as aiming to ensure that decisions are taken as close as possible to the citizen and that actions taken at the community level depend on the means available to local, regional or national authorities. The principle of subsidiary has a long tradition in European coastal countries, as well demonstrated for flood defense as described in this paper. This tradition is one reason why progress towards the production of an EC strategy for ICZM is slow. The Fifth EAP Toward Sustainability, published in 1993 (CEC, 1993) is quite general; however, in 1995 a demonstration program was proposed by the Commission, but it is unclear if it will lead to an obligatory operational framework for ICZM (see Chapter 7). The European Environment Agency (EEA) has launched initiatives for ICZM (da Silva, 1996). Also, the council of the Union proposes a Community strategy for the integrated management for the entire Community coastline providing a framework for its conservation and sustainable use. Looking abroad, it is suggested to prove, if an approach like the National Coastal Management Program (NCMP) in the US or the Australian National Coastal Action Program (NCAP) might be suitable for the European situation (Huggett, 1995).

Numerous steps toward integrated management approaches in the sense of Rio Agenda 21 have been taken in Europe during the past two decades. This started in the North Sea and Baltic Sea with conventions and programs mainly addressing sea- and land-based pollution problems (e.g. PARCOM). EC directives toward environmental protection became mandatory (e.g. an Environmental Impact Assessment), and multi-national governmental agreements for special areas were signed (see Chapter 7, Sec. 3). Steps taken toward implementing the objectives of Agenda 21 in the Mediterranean were the Regional Sea Programs, the Mediterranean Action Plan (MAP), and the CAMP-Pilot-programs (Coastal Area Management Programs). As an example, the Mediterranean: ICZM the study CZM for Cyprus (supported through the EC-program MEDSPA, co-financed by Cyprus and carried out by the Public Works Department of Cyprus in joint co-operation with Delft Hydraulics, Netherlands) was completed in 1995. The proposed concept of ICZM was adopted by the Cyprus government, abandoning the sector oriented approach (focusing on fixed targets (i.e. infrastructures for tourism) without integrating concerned interests or drawbacks) which had obviously failed in the past (Loizidou *et al.*, 1997).

The nations of the former Eastern bloc are a special situation, as they have a transition economy and need financial support (e.g. World Bank). An example for integrated planning with international support is the Albanian Coastal Area Management Project (Dobbin *et al.*, 1993). Sørensen *et al.* (1997) point out that economic reality and political necessity for these nations mean that ICZM will have to produce a net socio-economic benefit, a demand, which often is difficult to achieve. However, funding from the EC provides opportunities to reduce the differences in development between the member countries; consequently the ongoing process of enlarging the number of EC members will also promote the planning and implementation of ICZM.

3. Key Factors for CZM

The actual situation of CZM in a country can only be analyzed and validated against the backdrop of its political, social, economical and technical history. To be able to make comparisons of management systems, a converted version of a model of thinking shall be used (Kunz, 1993a). The idea of the model is that four basic forcing factors determine, by their mutual coherence and relation, how management performs. Three factors (natural conditions, social surroundings, needs) are associated with the "why?"-question and one key factor (options and fixed points) is related to the "how?"-question (Fig. 14.2). The primary needs comprise the basic demands of the people (safeguard life and property), the economical needs combine human activities to facilitate and improve life (agriculture, fishing, traffic, industry, tourism, etc.), and the ecological needs are directed towards environmental protection (biotic and abiotic components). Ecological needs have increasingly developed in the course of the last decades, aiming to preserve natural resources for future generations (anthropocentric) and to preserve the natural ecosystems by the concept that nature itself has value (bioethic–Korff *et al.*, 1998). This development was basically a consequence of the extended destruction of the natural environment, deriving from an unhindered growth of human exploitation of natural resources, which in Europe is mainly connected with economical needs.

The natural conditions are influenced by numerous anthropogenic impacts caused by external impact or originating in the coastal zone itself. Examples of external impacts are cutting off sediments by river dams, transport of pollution by air (soil, water) and, changes in climate. Examples of factors along the coast are rising storm flood levels and tidal ranges by dredging a channel into an estuary, lowering of land by drainage or gas extraction, and increased erosion by armoring the coast (cutting off of sediment supply). The importance of technology for addressing the options & fixed points-key factors has grown tremendously during this century. In combination with the available capital in Europe, the engineering services for the activities have had major impacts on the coastal zone (e.g. urban, tourism, industrial usage, sea-defense structures along the shoreline).

The inter-relationships and mutual dependencies between the key factors are complex. For example, the key factors change with time, and the coastal communities respond continually by complicated management and political processes. This process is described with an example in Sec. 7, by developing the coastal defense for the German part of the southern North Sea.

Civil engineering, especially coastal engineering — as a newly defined special branch — was and is highly demanded by the coastal community. The use and exploitation of coastal areas is a basic source for driving and sustaining the economy. In Europe, the engineering service was traditionally asked to respond to special demands, such as: harbor and coastal protection works; to urban, industrial, and tourism infrastructure; and to water supply and sewage treatment (sector oriented approach). Consequently coastal engineering is part of the management traditions and, in agreement with the coastal society, it provided multifarious solutions, which are very successful with respect to the especially addressed tasks (e.g. Kraus, ed., 1996). However, many of the solutions have drawbacks to other sectors, which we nowadays criticize of not being in accordance with the principle of sustainability, as it has developed later along with the politically and legally defined objectives of preservation of nature.

Environmental protection is commonly accepted by the coastal communities of Europe, as a need that has to be integrated into the historically evolved management of the coastal zone and ICZM. It is suited for application in Europe, if it is adequately adapted to the specific boundary conditions of each country or local parts of it, including the politics, socio-economics, organizational and legal structures, and physical infrastructure. The dominating boundary conditions for integrated planning and management differ substantially between the European countries. With respect to the political situation, we have to distinguish between the countries of the European Community, the EC-associate countries, and the countries of the former Soviet Union.

4. European Coastal Zones

Europe is bound by Asia in the East and by Africa in the South, and it occupies about 6.2 million km^2 with 0.63 billion inhabitants, not including the European part of Russia. Nearly all European countries have coastal zones. In Northern Europe, the shorelines run along the open North Atlantic Ocean, the semi-closed North Sea, and the almost-closed Baltic Sea. In Southern Europe, the shorelines extend along the Mediterranean (this coastline, as an example, amounts to approx. 7% of the 500,000 km of the world's coastline (Tabet-Aoul, 1995) and along the Black Sea, length of the shoreline is approx. 4100 km — Kos'yan *et al.*, 1993 — see Fig. 14.1. The length of the shoreline is differently distributed among the European countries, and its relation to the inhabitants and area of the country (a rough estimate of the pressure deriving from the population) varies in a wide range (Table 14.1).

The coastal zones of the individual countries are among the most populated and economically used areas. For example, about 200 million people live in the

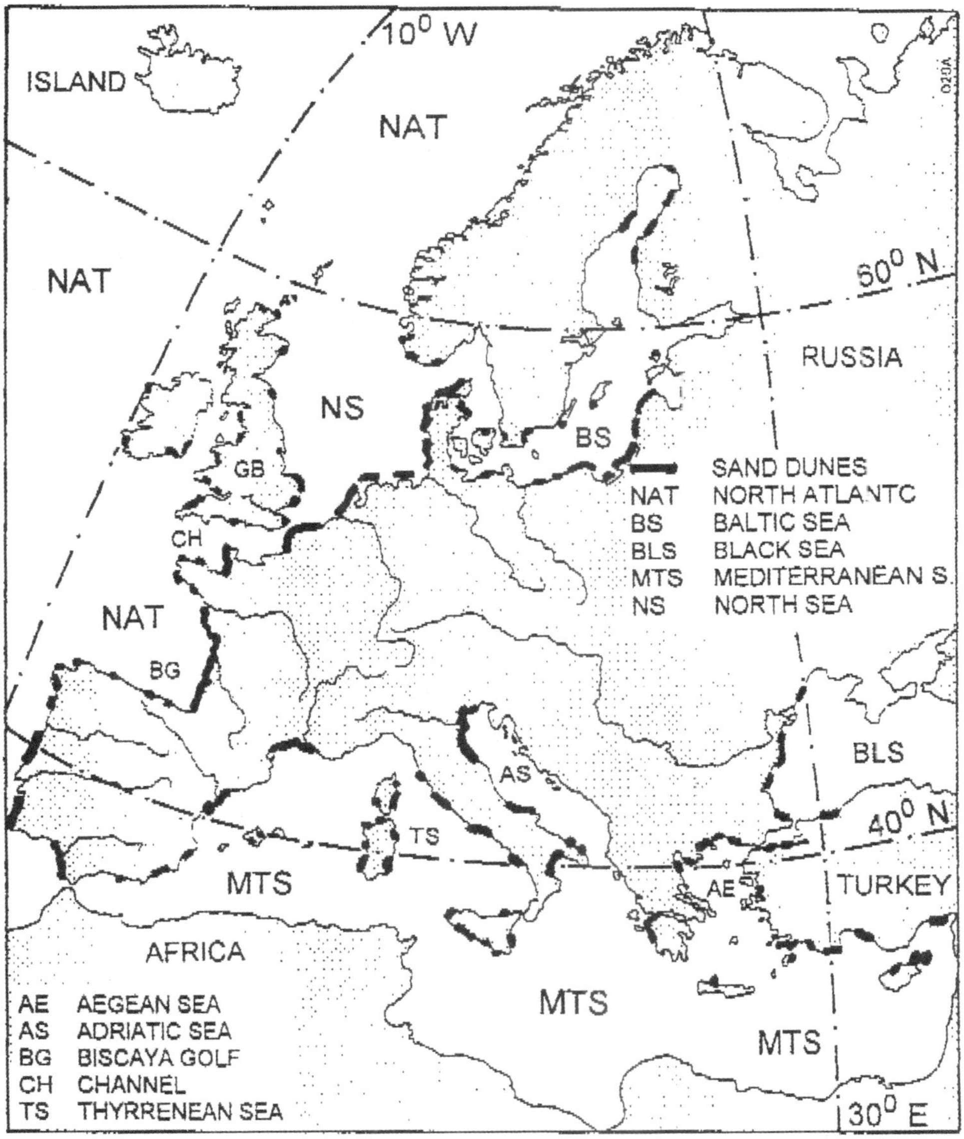

Fig. 14.1 Europe and its coasts with Oceans or Seas and the general distribution of dunes refering to data from Gehu (1985) and Doody (1991).

heavily industrialized North Sea river basins, and they are second only to the North Americans in their use of natural resources and consumer goods. However, the environmental impacts focus on a shallow shelf sea (approx. 575,000 km^2 area, 55,000 km^2 volume, 10 m median depth; see van de Kamer *et al.*, 1993). In the Mediterranean, more than one third of the 370 million inhabitants (1990) reside within the coastal zone, and it has to be considered that tourism additionally

Table 14.1 Length of shorelines (island & estuaries included) in km; shoreline per ihabitant in m/person; shoreline divided by the country area in m/km^2, 'insularity index' (ratio of shoreline length to the circumference of the circle of the equivalent country area), information on the boardering sea (for abbreviations: see Fig. 14.1) data from numerous sources.

Country	Length of the shoreline [km^2]	Shoreline per inhabitant [m/person]	Shoreline per country area [m/km^2]	Insularity index	Boardering sea (see Fig. 14.1)
Denmark	7.300	1,40	168,9	9,9	BS, NS
Greece	13.600	1,31	103,7	10,6	MTS
Cyprus	770	1,05	83,5	2,3	MTS
UK	17.500	0,30	72,3	10,0	NAT, NS
Spain	8.000	0,20	9,8	3,2	NAT
Turkey	8.300	0,14	10,7	2,7	BLS, MTS
Italy	7.500	0,13	24,9	3,9	MTS
France	5.530	0,10	10,2	2,1	NAT, MTS
Portugal	810	0,08	8,8	0,8	NAT
Germany	3.400	0,04	9,5	1,6	BS, NS
The Netherlands	350	0,02	8,4	0,5	NS
Belgium	65	0,006	2,1	0,1	NS

attracts many people. The European coasts are exposed to extremely unequal hydrographic conditions (tides, waves). The semidiurnal tides of the Atlantic Ocean lead to mean tidal ranges in The Channel of up to 12 m (Fig. 14.3). The tidal ranges along the western coast of the Atlantic are still high (e.g. 4–3.5 m on the Spanish and Portuguese continental coasts), and the range is less than 0.5 m in most other areas of the Mediterranean. In the Baltic and the Black Sea, tides are negligible.

The Atlantic coasts are attacked by significant wave heights up to 10 m, and in parts of the Mediterranean they can reach about the same value (e.g. Thyrreanean Sea). On the shore face of the German barrier islands (North Sea), the maximum measured wave height is about 5 m and in front of the mainland it is less than 2 m.

The geological conditions of the European coasts show a high variety, ranging from hard rock cliffs to soft mud flats. The coasts consist of Wadden Sea areas with barrier islands, tidal flats and salt marshes (e.g. southern North Sea — see Chapter VII), as well as of deltas and lagoons (e.g. Ebro-delta — see Chapter VI). Sandy shorelines with dune coasts are distributed all over Europe. There are soft cliffs with sandy or shingle beaches, and Europe has many rocky shorelines (with intercalated pocket-beaches or without any soft parts).

The coastal zones within one country can cover the whole range of morphological features, and most of the unprotected beaches and soft cliffs in Europe experience an increase of structural erosion, both related to natural impacts and to human activities. In France, as an example, the western coasts (approx. 3,850 km along the North Sea, The Channel and the North Atlantic Ocean) are composed of 30% tidal flats and salt marshes, 40% sandy beaches, and 30% rocky shore. The southern coasts of the Mediterranean (approx. 1,700 km) consist of 10% mud flats and salt marshes, 25% sandy beaches, and 60% rocky coast. Almost everywhere

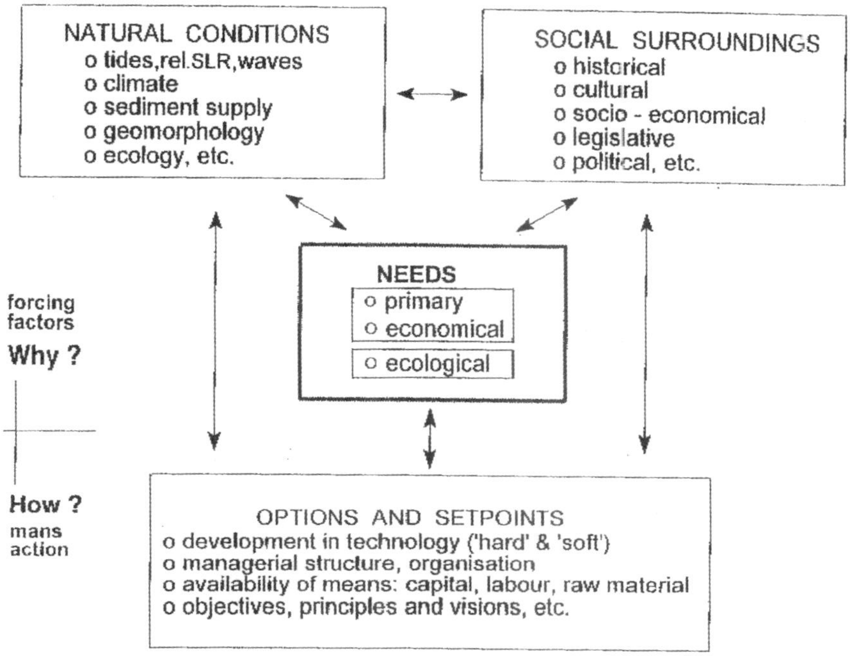

Fig. 14.2 Scheme of combined "key factors" for CZM.

the soft coastline is recently retreating with a rate of more than 1 m/year along approx. 850 km of coastline and 0.5 m/year along another 1,000 km (Hamm, 1996). In Italy, 45% of the coast consists of beaches (55% of the high rock type), and one third of them are suffering erosion problems, although 20% of the soft shoreline are supported by protection works (Franco, 1996). Soft coastlines often are associated with dunes, which are of special interest for CZM. Figure 14.1 shows the distribution of these vulnerable morphological features throughout Europe. According to the International Union for the Conservation of Nature/United Nations Environmental Program (IUCN/UNEP), the total area of wetlands is about 34,000 km^2 in North–West Europe and 6,500 km^2 in the Northern Mediterranean. For both, this is approximately 0.6% of the total country area (van der Weide *et al.*, 1993). Both dunes and wetlands experience losses due to marine dynamics and human impact. For example, during the past 30 yrs almost 75% of Mediterranean coastal dunes have been damaged or destroyed (van der Meulen *et al.*, 1993). Table 14.2 gives data on salt marsh losses in the Wadden Sea (see also Sec. 6).

5. Legislation in Europe for Use and Protection of the Coast

Legislation is an important part of the social surroundings as a key factor. There are laws and regulations on different levels (World, EC, Nation, State) which address the entirety of needs, social surroundings, options, and fixed points. Legislation that has

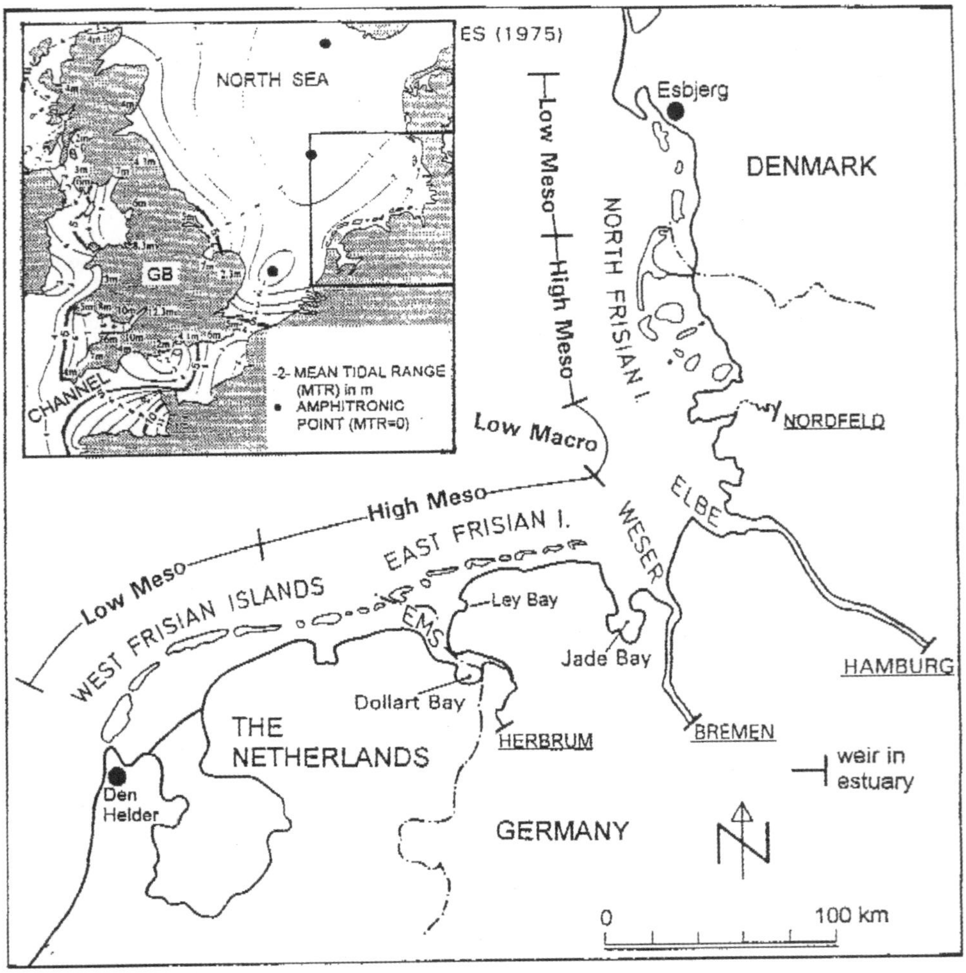

Fig. 14.3 Location map for the southern North Sea and the Channel with information on tidal ranges (small picture) and for the Wadden Sea (Netherlands, Germany, Denmark) with classification applying to Hayes (wave climate dominated) and for German estuaries.

Table 14.2 Surface areas of the Wadden Sea, Salt Marshes and Summer Polder in km^2 and in % of the Wadden Sea area. Data from the 'Report to the Sixth Trilateral Governmental Conference on the Protection of the Wadden Sea, Esbjerg, 13 November 1991 — CWSS *et al.* (1991).

Country	Wadden Sea area [km^2]	Island area [%]	Saltmarsh (incl. Su.Po.) [km^2] | [%]		Summer Polder (Su. Po.) [km^2] | [%]		Embanked Saltmarshes in the past 50 years in [km^2]
The Netherlands	2885	13, 3	85	3, 0	8	0, 28	22
Germany	5035	8, 6	180	3, 6	44	0, 87	127
Denmark	1015	16, 3	81	8, 0	0	0, 00	8

a direct impact on ICZM by regulating the use and the protection of the coast shall be discussed as an example. In the European countries, two types of legislation have developed historically. The first protects the coastal strip, traditionally to ensure free access to the shoreline, but also to keep areas free from infrastructures (set-back lines) and to conserve targets with respect to natural developments (e.g. arable land for agricultural use). The second gives protection to valuable or threatened habitats (EC-directive on Flora, Fauna, Habitat (FFH), National Conservation laws etc.). Additionally, there is special legislation dealing with coastal erosion.

Denmark probably has the most advanced coastal legislation in Europe. The Dune Protection Law, compelled by the migration of dunes, was enacted in 1792. The right of public access to all beaches has been reintroduced by the Nature Preservation Act of 1917. The act was revised in 1937, enforcing a prohibition for erecting buildings in a 100 m wide strip. A Ministry-Directive issued in 1978 has provided a general prohibition for vacation housing and hotels in a 3 km wide zone along open beaches. The Nature Conservation Act of 1993 declares that undeveloped shorelines shall remain to be natural landscape resources. Finally, in 1994 the protection of the nearshore area was extended to ban all constructions, except coastal defense and protection works, in a 300 m wide strip along all Danish coasts (Vaaben, 1993). Additionally, nature conservation objectives are defined by the trilateral (Dutch–German–Danish) Wadden-Sea agreements (see Sec. 7.3).

In most Mediterranean countries, legislation is based on Roman law, which considers the coast as a public domain. In the 1980s and 1990s, coastal legislation generally has been enhanced. In Turkey, the Shore Law of 1990 (amendment 1992) protects a strip of 100 m with special restrictions for arable land (Özhan *et al.*, 1993). In Spain there is a right of way (6 m), a right of protection (100–200 m), and an area of influence (500 m) (Losada *et al.*, 1996). In Portugal, coastal regulations were adopted in 1991 (Nordberg, 1995).

Of the EC-member countries, only Belgium, Finland, Ireland, the Netherlands and UK have no formal coastal legislation. Instead, at least in the Netherlands and in the UK, the role of specific coastal legislation is covered by an effective land use planning system. In the UK, for example, guidelines for coastal planning (structural plans) were adopted in 1992. They declared that undeveloped coasts should not be expected to accommodate development that could be located inland or in existing developed areas (Pethick *et al.*, 1993).

Legislation and regulations must be judged by their results. Just two quotations for Europe: In England and Wales about 44% of the coast has protective designations to landscape quality (National Park, Area of Outstanding Natural Beauty or Heritage Coast). In addition, a substantial length of the coast is protected by statutory designations relating to nature conservation. The National Trust and similar bodies are continuously buying more land to be protected (Nordberg, 1995). Indeed, massive, messy, and unplanned coastal development is a common feature of all countries originating from cultural and economic processes which have been largely similar to those of Italy. Recent and heavy coastal development lines up with no discontinuity for hundreds of miles (Fabbri, 1989). Looking at the reasons for violations of common coastal protection principles, a generic problem in natural resource management becomes obvious as conflicting laws and policies. There exists a large body of legislation and policy which attempts to protect the natural environment. This legislation is positioned against typically more powerful law and policy that exist to promote and facilitate economic development. Often, these two interests (economical and ecological needs) conflict, and economic imperatives typically dominate (Capobianco *et al.*, 1997).

Human development and expansion in the coastal zone create the necessity for coastal engineering works that are inseparably connected with the human needs–key factor. Innumerable outstanding engineering works in the various European coastal zones have been installed in the past as a response to the social demands of the respective time by integrating technical, social and political activities into the process. We nowadays name these works as CZM. These works stand in line with a long tradition in many countries (e.g. Kraus, ed., 1996). For example: in the northern Mediterranean basin, coastal engineering works have taken place for more than 3000 yrs predominantly with the rise of maritime traffic (port structures), and these have always been a key factor in the economic and political growth (Franco, 1996). In the coastal lowlands of the coastal defense structures along the southern North Sea developed over one millennium (Sec. 7.3).

6. European Coastal Zones Under Pressure of Society

The actual situation of the European coastal zones can be globally addressed as a squeeze deriving from steadily increasing attacks from the seaside and growing demographic pressure from the interior. In the following, some more or less arbitrarily selected examples for CZM responses shall be presented. Among the European countries, southern European and particularly the Mediterranean countries experience extreme anthropocentric pressures on their natural and socio-economic environments.

The Mediterranean Sea and its coastline face continuous and intensive pressures from industry, housing expansion, tourism development, aquaculture, and pollution from urban waste and agricultural sources. For instance, it is estimated that 85% of the domestic effluent from about 120 coastal Mediterranean cities is being discharged untreated into the sea, and that more than 1,000,000 tons of crude oil are being released every year in the Aegan and Ionian Seas (Voulgarikis, in Healy *et al.*, eds., 1995). In the Mediterranean deltas — such as the Rhone Delta (France), the Po Delta (Italy), and the Ebro Delta (Spain) — subsidence combined with a reduced sediment input (effects of river regulations) lead to a substantial relative sea level rise. This has serious implications on the natural resources and the related economic activities (Remark: the EC-MEDDELT project (Impact of Climate Change on Northwestern Mediterranean Deltas) will provide basic knowledge on the integral response including socio-economic aspects and feed-backs — see also Sec. 8). The Mediterranean is a major attraction for visitors; with 150 million visits per year, it covers around 30% of the global tourism trade. The tourism industry provides an important contribution to the balance of trade deficit (approx. 20% in Greece and Turkey, and more than 50% in Portugal). Tourism gives employment rates to the populations that are remarkably higher than the worldwide average of about 7%. Although the current rates of growth of tourism are lower than in the past, it is expected that the growth will continue, with around 400 million visitors estimated for the year 2025.

Actually, there is a boom of tourism in many countries. In Turkey, for example, there had been in 1995 almost as many facilities under construction as exist, and the number of beds will almost double to more than half a million. Tourism plans are prepared at the national level within broader objectives and goals. However, implementation planning powers are decentralized, and it is difficult to get integrated plans accepted (Senlier, 1995; Demirayak *et al.*, 1996; Cravidao *et al.*, 1996; Golfi, 1996).

Looking at the results, we must admit that much of the tourism infrastructure (hotels, marinas, artificial beaches and beach walks, traffic ways, supply facilities, etc.) have been built in Europe, changing the coastal zone substantially. Quite frequently violated are the principles fixed in legislation of pollution control and nature conservation (see Sec. 9). However, the willingness to proceed toward sustainable ICZM is declared, and there are many examples showing the importance of coastal engineering to support this process. The Plan de Costas is an example of how Spain responded to its coastal squeeze problems. These problems had been predominantly anthropogenically caused since the 1960s, augmented by newly developed socio-economic demands of an ambitious five-year program (1993–1997) to restore beaches and create new ones with integrated urban and tourism infrastructures. The aim of the Plan "de Costas" is to contribute to the economic revitalization of the coastal zones, but also to include the improvement of its environmental quality, and therefore quality of life (Losada *et al.*, 1996).

In France, coastal defense works are often intended to protect tourism destinations (hotels, villas, etc.) and to prevent shifting or even aggravating erosion. The report "Long-term prospects for the French Coast" draws attention to these problems and has alerted the public and politicians regarding the need for intervention by integrated management means, in order to prevent further deterioration of the coast (Hamm, 1996). In Italy, the earliest regional and inter-regional plans for coastal defense were introduced at the end of the 1970s, predominantly for protection of endangered infrastructures (including roads, railways, paid by state and local authorities). The engineering solutions are typically passive works, also combined with active means (beach nourishment). New policies are called for, which integrate socio-economics (especially related to tourism) and environmental concerns (e.g. Franco, 1996). An ICZM-plan is currently worked out for the Venice lagoon, including the sea defense of Venice as a European and global task especially challenging coastal engineering (see Marchi, 1993).

Pressures caused by the multiple uses with conflicting requirements, by the natural driving forces, by the need for environmentally sound practices, and by financial constraints are concentrated on many of the bigger European Estuaries. Appropriate responses to the pressures frequently cannot avoid hard engineering construction. Arbitrarily chosen examples for estuary zones are: the Storm Surge Barrier in the Thames ship way (UK), which is part of the tidal flood defense system and integrated in a management plan (Kennedy, 1995); the Delta-project (The Netherlands) as an outstanding sea defence system (Bijker, 1996); the Elbe and Weser Waterways to the ports of Hamburg and Bremen with integrated storm flood defense systems (Germany) (Kramer *et al.*, 1992); the Thyboron barrier on the exposed Jutland North Sea coast (Jakobsen *et al.*, 1998); and the stabilization of the tidal entrance and channel at Hornof Jodur (Iceland) (Viggosson *et al.*, 1998). A unique engineering work is the closure of the Ranche bay (The Channel) for the installation of a Hydro-Electric Power Station that uses the extreme high tidal range in this area (Fig. 14.3).

The large variety of coastal engineering works in Europe, which are integrated into CZM as user functions, cannot be addressed in this chapter, even as selected examples. These works cover the complete field of civil engineering with solutions that fit to the specific demands, including management objectives. The development of a coastal protection systems for the low-lying coastal zones of the Southern North Sea are described in the next section as an example of CZM in Europe, emphasizing how specific historical developments impact the ICZM of nowadays.

7. Development of the German Shore-Protection System (Southern North Sea): An Example of the CZM Process

The present shape of the southern North Sea coast is the latest transitional stage of the Holocene warm period. About 15,000 yrs ago, the shoreline was located 250 –300 km further seawards with a sea level (relative sea level) approximately 140 m lower than present and a rising rate up to 2 m per century. The German coastal zones are located in the center of the Wadden Sea which extends from The Netherlands (Den Helder) along the German Bight to Denmark (Esbjerk). This coast has an unique landscape being one of the last large and relatively undisturbed marine ecosystems in North–West Europe (Fig. 14.3). It covers an area of almost 9000 km (56% located in Germany, 33% in the Netherlands, 11% in Denmark) with tidal flats (mudflats, sandbars, channels), islands and salt marshes (Table 14.2).

The mainland of the Wadden Sea coastal zone is low lying marshland extending along the estuaries, far into the mainland (e.g. the Elbe up to Hamburg). It is protected against flooding (storm surges from the seaside and river floods from the land side) by a system of sea dikes along the Wadden Sea coast and open estuaries, supplemented by inland river dikes. For example, more than half of the Netherlands would be flooded without the dikes, and for the German coastal states the percentage is about 15%. Parts of the mainland are sheltered by the Frisian islands (mostly sandy barrier islands). Behind the islands and the open coast spread tidal flats. The flats reach directly the coastal defense construction or they are bordered by saltmarshes. Many of the existing salt marshes have been created by mans support, along with land reclamation and diking which destroyed large parts of the naturally developed salt marshes — Ovesen (ed.) (1990) and Table 14.2.

7.1 *Tidal water levels and storm surges*

The rise of the sea relative to land and extreme storm surges are natural driving forces for decisions on coastal defense and protection (natural condition-key factor — see Sec. 3, Fig. 14.2). From historical research and tide gauge data (since 1700, Amsterdam, The Netherlands), we know that the trend of the relative sea level rise per century has been 25 cm to 30 cm for mean high water and 10 cm to 15 cm for Mean Water. For the water level, as well as for storm flood surges no extraordinary rise can be detected, a result that has been confirmed for several locations of the southern North Sea. The Christmas Storm flood of 1717, for example, is recorded as the most disastrous one in Germany with 11,300 victims. Storm flood disasters with losses of life still occurred in this century. In 1906, about 2000 people drowned in Severnside/England as the sea defenses were overtopped. During the night of 31 January to 1 February, 1953, a severe north-western storm hit the Dutch coast (1835 people drowned and 1,360 km^2 land was inundated) and also the East coast of England (more than 300 deaths). The storm flood on 16–17 February 1962 in Germany caused 340 deaths (315 in the city of Hamburg). The so-called Dutch storm flood of 1953 provoked fundamental changes in the coastal defense management (see Sec. 7.4).

7.2 *Development of sea defense strategies as a socio-economical management process*

The residents of the Wadden Sea lowlands responded to the natural forces deriving from the North Sea with efforts to protect their lives, property, and infrastructures (primary and economical needs). The history of these management reactions is well known and shall be described in the context of the key factors according to Sec. 3. As the rate of the relative sea level rise (RSLR) decreased during the first millenium before Christ, people started to settle permanently in the coastal lowlands. The strategy of society changed from retreat to accommodation by constructing and raising earth hillocks. Around 1,700 years ago they proceeded toward a protection strategy by connecting artificially raised areas with embankments (dikes) creating polders. The generally moderate sea level conditions during the following four centuries allowed successful embanking activities. In Germany, the dikes along the entire North Sea coast were largely completed in the 13th century. Disastrous storm floods hit the coast in the 14th century, which destroyed the defense system and led to enormous loss of life and property in the flooded polders. The risk has increased as population centers and associated infrastructures have accumulated in the polder areas.

The coastal communities were able to repair and enhance the dikes, to reclaim lost land, and to push the defenses towards the sea by combining the protection with a progression strategy. By the 18th century, the available technical tools made it possible to implement solid stone construction and, in the second half of the 19th century, all exposed dikes of the coastal defense line were repaired with bricks or stones. This historically developed defense line is the basis for protection of the German North Sea coast; it became, and still is, a technical key factor for decisions on CZM. The social part of the story is the history of water boards and legislation as a social-economical factor (social surrounding-key factor). Legislation for sea defense and coastal protection always has been linked to drainage of agricultural land and land reclamation. The primary responsibility for dealing with these problems traditionally rests with the land owner, under common law, rather than the state. For more than a 1000 years, sayings such as "dike or depart" combined with "pump or drown" have been main guidelines for the legislation of coastal defense and protection with the related management. Under statutory law, generally a distinction has been made between legislation for coastal (sea) defense (i.e. defense against tidal flooding) and for coastal protection (i.e. protection against coastal erosion). From the 13th century, the democratically organized water boards received the responsibility for the maintenance of dikes and drainage. Citizens had (and have) to pay a contribution corresponding to the value of their protected property. The history of the water boards is closely intertwined with the social surroundings key factor which itself is associated with the incentives for the acceptance of changes, especially those of the physical environment, of commercial activities and of technological skills.

The boards are integrated in a constitutional structure that is adapted to the situation in the respective country. The growing scale of many problems and the need to carry out works for the common good have let to the development of statutory law, vesting powers in specific local boards or authorities or state agencies. Generally there is a national, a provincial and a regional/local level. Wherever possible, the responsibilities are given to the local authorities (principle of subsidiary). Water boards judge the strength of the flood defense system, municipalities carry responsibility for the safety of the citizens. The standards that the flood defenses must satisfy are decided upon by parliament (State in Germany, national government in The Netherlands). Besides serving as a funding source, the role of the government is mainly restricted to supervision and specialist technical support. Based on the cultural and historical development, there is a strong agreement within the coastal communities that the existing coastal defense line must be maintained and strengthened (safety first). Retreat strategies interfere with the traditional defense objectives (priority for the primary and economical needs). Consequently, the increase of driving forces by the ecological needs trigger conflicts (Sec. 7.5).

7.3 *Nature conservation of the Wadden Sea zone*

Nowadays, nature conservation as an ecological need has great influence on society's activities in the coastal zones of the European Wadden Sea. Conservation has to be integrated into the management process, according to numerous laws and regulations (e.g. EC-directive on environmental impacts and related national laws forcing avoidance, mitigation, and compensation). In the following, the development of nature conservation in the Wadden Sea (southern North Sea) is presented as an example for the changing needs, which is a key factor.

Nature conservation in the Dutch–German–Danish Wadden Sea started about one century ago, in particular to protect breeding bird colonies on smaller inhibited islands. Some 25 years ago, it became evident that the traditionally developed terrestrial and species conservation was inadequate to achieve the preservation targets. Advances in coastal protection techniques, including large-scale embankments and harbor and industrial development, as well as a rapid increase in tourism and pollution, resulted in endangering or degrading the more or less existing balance of traditional uses of the Wadden Sea area and the conservation of its ecosystem. Actions were initiated, in particular by environmental groups, and the Wadden Sea was protected, step by step, during the 1980s. Planning started with the Wadden Sea Memorandum in the Netherlands in 1980 (amended in 1993) and establishment of a Nature Reserve in the Danish part in 1982 and National Parks in Germany from 1985 onwards. Nowadays, the Wadden Sea is completely designated as a Wetland of International Importance (Ramsar Site, 1978/1991) and large parts are designated as Special Protection Areas according to the EC-Bird Directive (SPA, 1982/91). Additionally, the Dutch and German parts have been declared as a Man and Biosphere area (MAB) of the United Nations.

Parallel hereto, a trilateral political co-operation has been established. The guiding principle of the trilateral Wadden Sea policy is to achieve, as far as possible, ecosystems in which natural processes proceed in an undisturbed way. Nature protection laws of the countries supported this principle. To narrow the existing differences in legal and administrative systems between the three countries, seven specific common management principles have been adopted for the Wadden Sea. These include the "precautionary and the avoidance principle" against activities which are assumed to have significant damaging impact on the environment, even if there is insufficient scientific evidence to prove a causal link between measures and their impact (Enemark, 1993). With regard to coastal protection, this let to the trilateral agreement prohibiting embankment and minimizing unavoidable loss of biodiversity as an integrated part of a sustainable utilization of the Wadden Sea (CEC, 1993). The ecological needs have been accumulated in a Wadden Sea Plan (1997) which will be part of the trilateral discussions on an mandatory ICZMP.

German policy has curtailed regional and local jurisdiction of the concerned area; a National Park Decree has been fixed (zoning strategy) providing the possibility to the authority to manage and to protect within its limited responsibility. The Dutch policy has brought the Wadden Sea under the jurisdiction of the adjacent provinces and municipalities with early public participation (integration process at the lowest level). The Danish conservation and planning scheme for the Wadden Sea can be addressed as a mixture of the Dutch and the German approaches.

7.4 *Implementation of coastal engineering technology*

The development of technical tools is an important and rapidly changing part of the option and fixed points-key factor. The history of decisions on the technique of coastal defense and protection for the German mainland and the islands (where eroding parts have been stabilized by solid constructions (seawalls, groins) since the middle of the last century), was primarily reactive management until the beginning of this century. It dealt mainly with the impacts of the natural forces, first of all with disastrous storm floods and structural erosion-sedimentation.

The engineering design was basically decided on by experience, trying not to "fight the sea with brute force but with soft persuasion" as quoted from A. Vierlingh, who planned and built in the Netherlands during the 16th century (Bijker, 1996). Dike profiles were optimized with respect to both economical and safety targets, starting as early as A. Brahms, who reconstructed dikes in Germany after its total damage by the storm flood of 1717 (Niemeyer *et al.*, 1996).

The disastrous storm floods that struck the Netherlands and East England in 1953 and Germany in 1962 (see Sec. 7.1) initiated a fundamental change from reaction to planned action. In Germany, the fundamental strengthening of the defense system was based on legal master plans applying semi-empirical design criteria, taking into account the natural future sea level rise (MHW with a rate of approx. 30 cm per century). The defense system consists of dikes, walls (in urban areas, industrial regions, harbors), storm-surge barriers (mainly in tributaries of the estuaries), weir (construction which protects the upper part of the river by cutting off the influence of tides). The system also contains traffic-gates (access), sluices and pumps (drainage, irrigation), and navigation locks (shipping) (Kramer *et al.*, eds., 1992). In the beginning, the management concept itself emphasized combined targets of defense and land reclamation (van der Ven, ed., 1993), aiming at achieving security, basically by shortening the length of the defense line (closure of bays, analogue to the Dutch Delta-Works-concept (Bijker, 1996)).

Since 1962, the length of the defense line along the German North Sea coast has been reduced from 1200 km to 700 km (construction costs of about 5 billion US$ up to now, 70% from Federal and 30% provided by the State). Maintenance funding has been mainly raised by the dike boards. The actual situation of the German North Sea coastline is the result of a long-lasting process of CZM. The coastal defense line on the mainland (Wadden Sea coast and estuaries) runs near the shore and the riverbanks. The islands are partly armored, and the tidal inlets are widely fixed. In extended parts of the coastline there is none or only few salt marshes left (see Table 2, also concerning losses by poldering) which can act as a transition zone between water and land.

7.5 *Conflicts in the Wadden Sea zone and management responses*

The ecological needs, as stated in the principles for nature conservation and fixed in the National Park regulations are in conflict with the primary and the economical needs. Conflict is defined as a disagreement between politically relevant interest groups and may have economic or social consequences. The German National Park Wadden Sea shall develop naturally, an objective which reaches beyond the principle of sustainability as it puts priority on the ecosystem as a value in itself. The term sustainability is quite general, determining a balanced utilization which is of benefit to both people and the environment for future generations. In real life, the main objectives of sustainable development within the framework of operationally oriented CZM-concepts predominantly deal with predicting future demands, planning the supply, controlling the use (Hoozemans *et al.*, 1996).

Referring to the definition of the ecological needs, the traditionally evolved German concept of coastal defense and protection is judged as anthropogenically dominated. It does not appropriately consider the causes and effects (both used in the biological sense (Sidaway, 1995) on the dynamics of natural systems (ability to design themselves — flexibility and adaptability — to be ideally suited to the boundary conditions that are superposed on them). This judgement interferes with the mandatory defense strategy as laid down in the German dike laws and legal plans. Hence conflicts arise, especially in areas with predominantly erosive trends, without large forelands and with high user demands. The conflicts will grow if the natural trend of relative SLR and the frequency of storm floods increase.

In special cases, the addressed contradictory aims can be adapted to each other by compromises on the conceptual design. A German example is the Ley Bay, where closure plans had been given up and a balance had been achieved between the primary and economical needs (Kunz, 1993b). These plans include defense against the sea (dike) and inundation by fresh-water run off (drainage), access to offshore waterways (fishing and pleasure boats). These and the ecological needs, such as salt marsh and tidal flat conservation, were balanced in the process of CZM. A Dutch example is the unique Delta-Project. According to the report of the Delta Commission in 1960, closure dams should be constructed across inlets. In the early 1970s, the realization of ecological needs grew and consequently openings (storm surge barriers) were added to avoid the transformation of the salt-water ecosystem into freshwater lakes. Compromises are also possible by alternative engineering solutions which are validated as soft techniques, such as artificial sand nourishment (foreshore, beach, dune) or groin systems made from brush wood (high tidal flat in front of a foreland), discussed in Sec. 7.6. Compromises may be facilitated by management plans, which deal with special aspects, such as dunes and salt marshes.

7.6 *Coastal management and soft techniques*

Coastal management in northern Europe refers predominantly to coastal defense, including coastal protection means. In Germany, for example, coastal defense is based on the legal strategy (Master Plans) to hold a main defense line. Retreat of the mainland shoreline, created by structural erosion, is counteracted in accordance with the ecological needs, as far as possible. Such means include soft techniques with brush-wood-groin fields and by salt marsh-management to maintain the foreland as a native passive protection means, (e.g. Hofstede, 1996; Boormann *et al.*, 1995; Ovesen, 1990). Erosion of sandy coasts (in Germany only on the islands) is defeated where infrastructure (housing, hotels, drinking water supply, etc.) is endangered.

The first large-scale beach nourishment in Europe was carried out on Norderney-island in 1951–1952 as part of an extended groin-seawall-defence system (Kunz, 1993b). Artifical beach and foreshore nourishment are nowadays increasingly applied in management frameworks. In the Netherlands, a dynamic preservation strategy of the dune coast, based on its presumed 1990 position (RWS, 1990) has been in place since 1990. On average, approximately 7 million m^3/year artificial sand nourishment proved to be an efficient method with a good benefit-to-cost ratio (de Ruig, 1995). The Belgium coastline, only 65 km long, is highly developed, and its maintenance is mandatory. Soft techniques are applied with the aim of stabilizing the existing shoreline (Helewaut *et al.*, 1993). Denmark has long stretches of coastline where retreat is allowed (e.g. soft cliffs). Coastal protection is combined with coastal defense, where needed. Beach and foreshore nourishment is part of the management concept, and accounts for about 80% of all coast protection activities along the Central West Coast (approximately 3 million m^2/year), which changed recession into a net seaward advance of the coastline (Sørensen *et al.*, 1996; Laustrup *et al.*, 1998).

Coastline (Shoreline) management in Europe is increasingly using beach nourishment tools, based on long term coastline management and follow-up strategies (The Netherlands with 110 Million m^3 up to now), on project-level plans (Spain with possibly about 100 Million m^3 placed since 1985), and on ad hoc small scale projects (Italy, excluding the Venice project, and France) — Hamm *et al.* (1998). Shoreline management in Europe incorporates more and more aspects of ICZM, especially concerning the conservation and restoration of dunes (van der Meulen *et al.*, 1993), of saltmarshes (Hilgerloh, ed., 1992), of lagoons (1). The actual discussion on ICZM proceeds towards robust and flexible solutions, according to the concept of resilience. The nourishment concept plays an important role, as it has the feature of dealing flexibly with natural and social dynamics. Therefore, resilient management is about defining strategies which preserve both the capability of the coastal system to cope with major perturbations to its operating environment and to society for a wide range of possible futures (WWF, 1996; Capobianca *et al.*, 1997).

7.7 *Safety concepts and risk management*

Management concepts for coastal defense are basically similar within the European countries around the southern North Sea. The details differ substantially: England promotes Shoreline Management Plans with the aim of providing a framework for the development of sustainable coastal defense policies within a sediment cell or subcell (MAFF, 1995); The Netherlands protect with a management strategy a defined shoreline (RWS, 1990); Germany holds a legally fixed sea defense line on the basis of Masterplans (Kramer *et al.*, 1992). Figure 14.4 shows a step by step approach towards ICZM. A foreland (salt-marsh) management is directed toward coastal protection and nature conservation (Enemark, 1993; Pethik *et al.*, 1993; MAFF, 1995; Hofstede, 1996).

Dune management has to deal with the extraction of drinking water as a traditional use (the Amsterdamer Water Supply Dunes (AWD) is one of many examples). Additionally it has to integrate its function for sea defense, for tourism, as ecological site etc. The exceptional value of these areas and the complexity of the processes involved explain the multifarious European efforts towards sustainable use (Carter *et al.*, eds., 1992; Healy *et al.*, eds., 1995). An area management aims to provide options for flexible sea defense responses. Part of this CZM is the maintenance of old dikes (second dike lines), the promotion of set back lines, the preservation

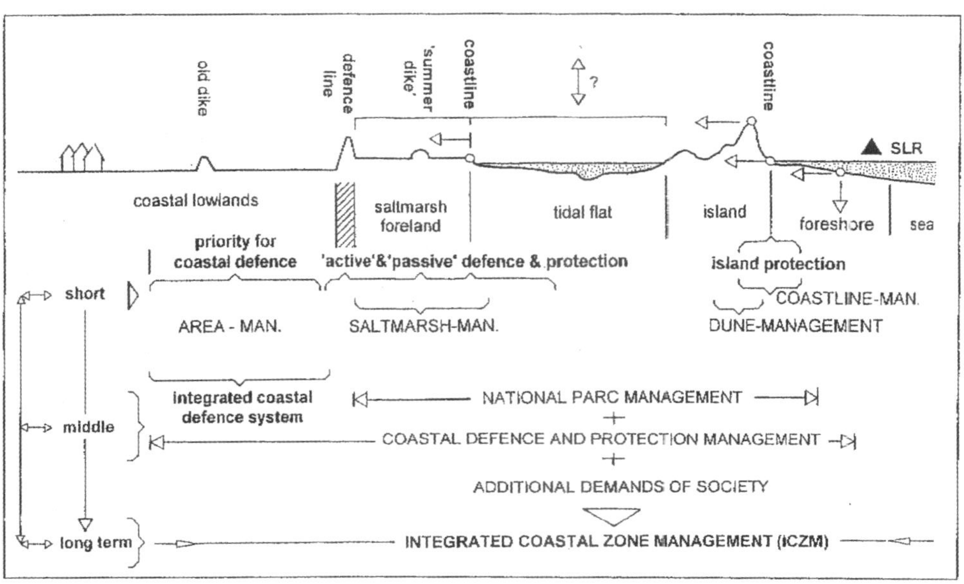

Fig. 14.4 Steps for a development toward ICZM; principal sketch for a situation in Germany, demonstrated by a cross section for the East-Frisian situation (barrier islands).

of special areas (low productivity and very few properties) for coastal protection. Experimental managed retreat and restoration schemes are currently being developed and monitored in some European countries; three sides in Essex (England) are typical examples (Palmer, 1996). Relocating the line of defenses further inland and encouraging the development of salt marsh habitat can also be a preferable alternative from the coastal protection point of view (MAFF, 1995).

Protection against flooding is given by a coastal defense system providing security according to its probability of failure (de Valk *et al.*, 1997). Additionally, the safety of the population and infrastructure against disasters depends on the awareness and the ability of the coastal community to react to hazards and at hazards by adequate management processes (Fig. 14.5). There are uncertainties concerning the design parameters as well as the failure of the construction; hence, we have to deal

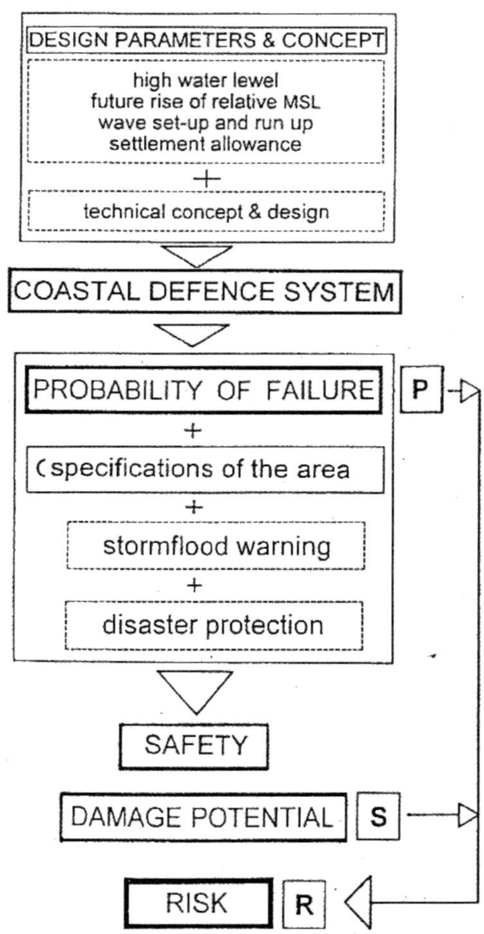

Fig. 14.5 Coastal defence strategy in Germany with related "probability of failure", "damage potential" and "risk".

with risks (probability of failure combined with damage potential). Probabilistic tools have been devised to deal with the problems of uncertainties, both natural and man-made (Peerbolte, 1993; Zhou, 1994). The lowland situation of the Netherlands is extreme. Therefore, there are safety standards expressed as an exceedance frequency of the design water level with return periods of 1:4,000 (tidal areas) and 1:10,000 (risk-tidal areas) (de Ronde, 1996). In the UK, there are generally no statutory levels of protection, and almost all flood defense works are carried out under permission with the aim of choosing safety standards from schemes that achieve the greatest benefit-cost ratio. A recent average exceeds 4:1 (MAFF, 1995). However, indicative standards of protection have been published as a guide (MAFF, 1993b), ranging in tidal areas from 1/5 (low productivity agricultural land with very few properties) to 1/200 (high-density urban area).

Safety standards (level of risk that is deemed acceptable) require a sound calculation of the design storm flood water levels and waves, including aspects of climate change. The legally fixed methods for the calculation differ between the European countries. Numerical models are increasingly applied (Bijl, 1997). Flood warning is an important part of hazard management. Sophisticated structures and means have been built up in the Netherlands and in Germany after the floods of 1953 and 1962 (Sec. 7.1) as well as in other European countries (Utteridge, 1996; Vieira *et al.*, 1995).

The ISOS-study scenarios for greenhouse effects on coastal defenses has shown for the Duch situation that it will not be a national problem to cope with a relative sea level rise of 1 m per century (de Ronde, 1991). In general, the coastal lowlands around the southern North Sea have nowadays achieved a high safety standard against flooding from the sea. The Second Assessment Report of the European Environmental Agency values the risks from fresh water floodings and from other impacts as more important than hazards from the sea (EEA, 1998). Nevertheless, public information is needed about risks as well as about the inevitable conflicts between the primary and economical needs on the one hand and the ecological needs on the other. This information will enable the coastal communities to judge and decide the appropriate coastal defense strategies by ICZM. The Netherlands has intensified its research and discussion towards this subject to such an extent that parts of risk management have been integrated into existing concepts (de Ronde, 1996).

The recent Dutch Flood Protection Act is a transition to a risk-based concept for dike design. This approach discerns a maximum tolerable societal or group risk by taking the consequences of flooding into account (loss of human lives and of material) in setting safety standards. This approach has become common in areas such as planning and transport or health and safety (Jorissen *et al.*, 1998; van der Meer *et al.*, 1998). In Germany, discussion on risk assessment and long-term planning tends towards proposals which maintain the traditional protection strategy, but are looking for more flexibility to achieve less risk and more resilience. This is expected to be achieved by integrating the coastal zones located landward of the existing main dike line into the management-strategy (sea defense system with second dike-lines, set-back regulations, etc.), as discussed by Kunz (1993b) and Klaus *et al.* (1994). A next step could be managed retreat in selected areas, by balancing arguments of risk, B/C-ratio and nature conservation within an ICZM-concept.

England provides possibilities for managed retreat (e.g. experimental schemes in Essex) within the sea defense scope, encouraging the development or restoration of saltmarsh habitats (Palmer, 1996). Creating a resilient coastline by ICZM may also mean that the coastal defense line is pushed forward. As an example, there are plans to extend Europort/Rotterdam harbor, to move Amsterdam-airport into the sea, and to reclaim land along the Holland coast near The Hague (de Ruig, 1995).

8. Database, Research, and Information

Coastal engineering interventions must be based on a sound understanding of the processes active in the coastal zone, and their implementations should be part of CZM. These objectives can only be met if there is a reliable database, research program, and information dissemination and education. The increasing scale of projects and complexity of the management process call for close interdisciplinary cooperation, such as the Coastal Genesis-project in The Netherlands (Louisse *et al.*, 1990), or the European Program on Climatology and Natural Hazards (EPOCH) (Penning-Rowsell *et al.*, 1994) for example. Participation of the decision making bodies as well as the concerned public and interested media are essential. Many European countries are privileged, as they can base plan on comprehensive field data reaching back into historical times and on a long tradition of research. New measuring systems have been set up with the help of international or national funds.

The Commission of the European Communities (EC) has launched 35 demonstration programs (15 million US $) and six special analyses dealing with integrated management of the coastal zones: 32 programs (50% in the Mediterranean and 25% in UK) are related to LIFE (EC-environment commission) and TERRA (EC-fund for structural initiatives). Two programs (EU PHARE and World Bank) are directed to the integration of former Eastern block countries and one deals with the coasts of Norway. There are many links between these demonstration programs and the fourth EU-framework on Science Research and Development, such as Marine Science and Technology (MAST) being of special interest for coastal engineering, European Land Ocean Interaction Studies (ELOISE), Land Cover Changes in Coastal Zones (LACOAST). Results and proposals will be discussed on a broad basis (INTERREG-Conferences), and the final EC-recommendations on ICZM for the European parliament is scheduled for the end of 1999.

There are several informal organizations in Europe dealing with CZM. One of them is the MEDCOAST initiative launched in 1993 with the First International Conference on the Mediterranean Coastal Environment, MEDCOAST 93 held in Antalya, Turkey. This biannual conference series aims to provide a forum for discussion and information exchange among various sectors and nations on issues related to management of the Mediterranean and Black Sea coastal and sea areas. In addition to the conference series, the initiative has two other components. One component is human resources development (training programs, etc.) and collaborated research in various issues of coastal and sea management for producing means to improve our understanding of physical, bio-chemical and ecological processes and their interactions with human activities. The other component is directed towards the utilization of scientific knowledge and modern management tools in achieving integrated solutions by contributing to existing efforts. Notable existing efforts are the Mediterranean Action Plan of the United Nations Environment Program (UNEP) with related Priority Action Programs, the Black Sea Environmental Program, and the Black Sea Strategic Action Plan. Future development plans include setting up of an International MEDCOAST Foundation and the International MEDCOAST Center establishing an international graduate degree program on integrated coastal management. The plans include capacity building for a regional data and information center supporting coastal and marine management in the Mediterranean and the Black Sea countries (Özhan, ed., 1997).

The European Coastal Association for Science and Technology (EUROCOAST) and the series of its international LITTORAL Conferences started in 1990 in Marseille, France. The third conference convened 1996 in Portsmouth, UK, under the heading "Partnership in Coastal Zone Management" (Taussik *et al.*, eds., 1996). Eurocoast is mainly concerned with the physical processes in the coastal zone and related technical aspects, also including interdisciplinary approaches, such as CZM.

The European Union for Coastal Conservation (EUCC) has enlarged its working field from the dunes (former EUDC) to the entire coast. The main objectives are related to environmental protection and/or nature conservation in the coastal zone. Five types of activities can be distinguished: inventories and research; management and policy related activities; information and development of conservation infrastructure. The EUCC runs projects to enhance field research and participation in international cooperation — Healy *et al.*, eds. (1995) and Salman *et al.*, eds. (1996). Such projections include the COMER network supporting ICZM and the Baltic Coastal Corridor Project (BACO) of voluntary conservation organizations in Poland, Lithuania, Latvia. The EUCC provides information by the Coastlines journal and by biannual congresses (Carter *et al.*, eds., 1992), by drawing up special inventories (Coastal Conservation Research Register). The annual "Conferences of River and Coastal Engineers" organized by the Ministry of Agriculture, Fisheries and Food (MAFF), Flood and Coastal Defence Division, London, UK provide broad information on coastal engineering and ICZM, mainly for the United Kingdom.

The communication and collaboration of science and research develops more and more under the umbrella of the European Community (information are available; EC (1997) is an example). The EC promotes multinational and interdisciplinary activities, and many of them address the European coastal zones. There are unions of scientific disciplines like the International Geographical Union (IGU) and related special groups, like the Commission on Coastal Systems (CCS), which cooperate, promote projects, sponsor meetings dealing with ICZM. The North Sea Coastal Management Group shall be mentioned as an informal structure that brings together professional coastal engineers being responsible as managers and advisers to government and public authorities from five countries bordering the Southern North Sea.

The number of publications dealing with CZM in European coastal zones increases continuously. Here, some examples are given: The "Trilateral Governmental Conferences on the Protection of the Wadden Sea" (CWSS *et al.*, 1991) provide comprehensive reports. The "Wadden Sea Newsletter" of the "Common Wadden Sea Secretariat for the Cooperation in the Wadden Sea" (Wilhelmshaven, Germany) deals with actual problems of ICZM, focusing on nature conservation of the southern North Sea (Enemark, 1993). Rijkswaterstaat and the "Coastal Zone Management Center" (RIKZ), The Hague, Netherlands publish comprehensive reports, such as "The riddle of the sands" (Louters and Gerritsen, eds., 1994), informing both specialists and public. The Ministry of Agriculture, Fisheries and Food London, UK publishes reports, guides etc. (MAFF, 1993a,b; 1995 as examples). The special volume "Coastlines of the Southern North Sea" (Hillen *et al.*, eds., 1993) as part of the series "Coastlines of the world" (Magoon, ed.) gives for this European area an overview on coastal engineering and its integration in CZM. The proceedings of the American Society of Civil Engineers "Coastal Zone Conferences" and of the "International Conferences on Coastal Engineering" (ICCE) comprise numerous papers dealing with coastal engineering with respect to ICZM.

Besides compiling the products that have been generated by the efforts related to ICZM, there is a need to evaluate the results. The objectives of numerous European initiatives, supported by the application of Geographical Information Systems (GIS), are directed towards this task. However, many of the existing outputs are buried in the institutions (see also Sørensen, 1993). In many of these institutions there is an increasing shortage of permanent staff, and continuity is not to be guaranteed.

9. Conclusions

The issues at the core of how we respond to the human-induced transformation of our coast are ethical. This aspect accounts especially for coastal conservation. The European coast is our common heritage, and we are responsible for passing it over to the next generations. ICZM has to include this target so that the beauty of the landscape is preserved and the ecosystems with wild flora and fauna can be maintained. Coastal conservation and associated legislation and regulation vary considerably across Europe. The internationally accepted principle of sustainability for ICZM is rooted in the notion that economic viability, social justice, and environmental quality are related, and a balanced utilization is of advantage to both people and the environment for future generations. This objective is not mandatory with respect to the defined ethical precept. In practice, the predominant objectives of sustainable coastal zone management still focus on maximizing the access to resources among competing and often conflicting users with the ultimate goal of optimizing the utilization of these resources for the coastal society, both now and in future. This approach is basically an anthropogenic one, by defining the coastal zone as a multi-resource system and at the same time a multi-user system. By developing and applying soft techniques, coastal engineering can aid positively in future target optimization of societal users that will be more compatible with the ethical precept than is possible nowadays.

There is a growing awareness and basic agreement between coastal engineers in Europe that the preservation of the ecosystems has an intrinsic value and has to have its own agenda. This can only be achieved by understanding coastal zones as being part of a larger system in which damage may move far from the source. Accordingly, an EC-directive on water management is being developed which extends to river basins as a whole. In some coastal areas of Europe there is already an effective integrative management in place, partly including the ethical objectives. At other locations, the works are still carried out on a more or less ad hoc basis with little formal regulation. Only a few countries of the former Soviet Union have been able to achieve significant progress toward sustainable planning or ICZM.

The political and legal framework for ICZM is an integrated coastal zone policy. It represents a conception of long-term thinking in interrelations for the system at a large with all its components. The structure of the system has been be visualized by Price (1991) as a successive support system with coastal engineering as a basis. It refers to the ability of coastal engineers to control the physical system and to bring into effect sustainable development, including terrestrial and water quality aspects. This responsibility, going together with wanted advice, needs sound coastal zone management by thinking in large time and space scales based on a permanently extending knowledge.

References

Andersen, J. O. (1998): Flood protection in the Danish Wadden Sea area, ICCE'98, Copenhagen, Book of abstracts, pp. 624–625. DHI, Denmark.

Bijker, E. W. (1996): History and heritage in coastal engineering in the Netherlands, In: *History and Heritage of Coastal Engineering* (N. C. Kraus, ed.): ASCE, New York, pp. 390–412.

Bijl, W. (1997): Impact of a wind climate change on the surge in the southern North Sea, *Clim. Res.* **8**, 45–59.

Boormann, L. and J. Hazelden (1995): Saltmarsh creation and management for coastal defence, In: *Directions in European Coastal Management* (eds. M. G. Healy and I. P. Doody): Samara Publishing, Cardigan, UK, pp. 175–184.

Capobianco, M. and M. J. F. Stive (1997): Soft protection technologies as a tool for integrated coastal zone management, *Third Int. Conf. on the Mediter. Coastal Environ.*, MEDCOAST 97, Qawra, Malta, (E. Özhan, ed.) **1**, Ankara, Turkey, pp. 469–484.

Carter, R. W. G., T. G. F. Curtis and M. J. Sheehy-Skeffington (eds.) (1992): *Coastal Dunes: Geomorphology, Ecology and Management for Conservation*, EUCC. Publication, A. A. Balkema, Rotterdam, Netherlands.

CEC (1993): Community Programme of Policy and Action in Relation to the Environment and Sustainable Development Towards Sustainability, Resolution 93/C 138/01, 1 February 1993, Council of the European Communities (CEC), No. C 138/1, 17.05.93.

Cravidao, F. and L. Cunha (1996): Tourism and sustainability: The example of coastal Portugal, In: *Partnership in Coastal Zone Management* (I. Taussik and J. Mitchell, eds.): Samara Publication, Cardigan, UK, pp. 235–241.

CWSS and NFNA (1991): The Wadden Sea — Status and development in an international perspective, Report to 6th Trilateral Governmental Conf. on the Protection of the Wadden Sea, Esbjerg, 13 November 1991. The Common Wadden Sea Secretariat (CWSS) and The National Forest and Nature Agency, Ministry of Environ. (NFNA), Denmark.

da Silva, M. C. (1996): Coastal zone management initatives of the european environment agency, In: *Partnership in Coastal Zone Management* (I. Taussik and J. Mitchell, eds.), Samara Publications, Cardigan, UK, pp. 439–448.

de Ronde, J. G. (1991): Rising waters — Impacts of the greenhouse effect for the Netherlands, Publ. Rijkswaterstaat, Tidal Waters Division (L. P. M. de Vrees, ed.), The Hague, Netherlands.

de Ronde, J. G. (1996): Risk assessment and management of potential floodings in The Netherlands, *Proc. Int. Workshop on Floodplain Risk Management*, 77–86. The Hague, Netherlands.

de Ruig, J. H. M. (1995): The Dutch experience: Four years of dynamic preservation of the coastline, In: *Directions in European Coastal Management* (M. G. Healy and I. P. Doody, eds.), Samara Publication, Cardigan, UK, pp. 263–266.

de Valk, C. F., J. G. de Ronde and L. de Haan (1997): Assessing the reliability of coastal structures based on a multivariate description of extreme offshore climate, Rep. RIKZ, 1–25. The Hague, Netherlands.

Demirayak F. and E. Ulas (1996): Man tourism in Turkey and its impact on the Mediterranean coast, In: *Coastal Management and Habitat Conservation* (A. H. P. M. Salman, M. J. Langeveld and M. Bonazountas, eds.), *Proc. 4th EUCC Congr., Marathon, Greece*, pp. 117–123. EUCC, Leiden, Netherlands.

Dobbin, J. A. and I. Trumbic (1993): Planning for the future of the national coast: the Albanian coastal area management project, Second Int. Conf. on the Mediter. Coastal Environ., MEDCOAST95 (E. Özhan, ed.), Tarragona, Spain, **1**, pp. 391–415. Ankara, Turkey.

Doody, P. (ed.) (1991): Sand dune inventory of Europe, Publ. EUCC, Leiden and INCC, Peterborough, UK.

EC (1997): Major European marine research institutes and centres, European Commission, Office for Official Publications, Luxemburg.

EEA (1998): Europe's Environment: The Second Assessment and its Statistical Compendium, Publ. European Environmental Agency, Copenhagen, Denmark.

Enemark, J. A. (1993): The protection of the Wadden Sea in an international perspective — Planing, protection and management of the Wadden Sea, In: Coastlines of the Southern North Sea (R. Hillen and H. J. Verhagen, eds.), Vol. CZ93, New Orleans, pp. 202–213. ASCE, New York.

Fabbri, P. (1989): Italians and their coastal zone: an introduction to some peculiarities, In: *Coastlines of Italy, specifically*, Vol. CZ89, Charleston, pp. V–VIII. ASCE, New York.

Franco, L. (1996): History of coastal engineering in Italy: In: *History and Heritage of Coastal Engrg.* (N, C. Kraus, ed.): ASCE, New York, pp. 275–335.

Gehu, I. M. (1985): European Dune and Shoreline Vegetation, Rep. Council of Europe, Strasbourg.

Golfi, P. (1996): The future of tourism in the Mediterranean, In: *Coastal Management and Habitat Conservation* (A. H. P. M. Salman, M. J. Langeveld and M. Bonazountas, eds.), Marathon, Greece, Proc. 4th EUCC Congr., pp. 133–140. EUCC, Leiden, Netherlands.

Hamm, L. (1996): History of coastal engineering in France, In: *History and Heritage of Coastal Engineering* (N. C. Kraus, ed.): ASCE, New York, pp. 142–168.

Hamm, L., H. Hanson, M. Capobianco, H. Dette, A. Lechuge and R. Spanhoff (1998): Beach fills in Europe — Projects, practices and objectives, ICCE 98, Copenhagen, Book of abstracts, DHI, Denmark, 334–335.

Healy, M. G. and J. P. Doody, eds. (1995): *Directions in European Coastal Management*: Samara Publication, Cardigan, UK.

Helewaut, M. and B. Malherbe (1993): Design and execution of beach nourishments in Belgium, In: *Coastlines of the Southern North Sea* (R. Hillen and H. J. Verhagen, eds.), Vol. CZ93, New Orleans, ASCE, New York, pp. 258–266.

Hilgerloh, G. (ed.) (1992): Dune management in the Wadden Sea area, 3rd Trilateral Working Conf., Publ. National Park Administr. of Lower Saxony, Wilhelmshaven, Germany.

Hillen, R. and H. J. Verhagen, eds. (1993): Coastlines of the Southern North Sea, Spec. Vol. of the series Coastlines of the World (O. T. Magoon, ed.), CZ93, New Orleans, ASCE, New York.

Hofstede, J. (1996): Saltmarsh management for coastal defence in Schleswig-Holstein, Germany, *Proc. 31st MAFF Conf. on River and Coastal Engrs.*, pp. 6.2.1–6.2.9, Ministry of Agriculture, Fisheries and Food, London, UK.

Hoozemans, F. J. M., R. J. T. Klein, A. Kroon and H. J. Verhagen (1996): *The Coast in Conflict — An Interdisciplinary Introduction to Coastal Zone Management*, CZM-Center Publication No. 5, The Hague, Netherlands.

Huggett, D. (1996): Progressing coastal zone management in europe: A case for continental coastal zone planning and management, In: *Partnership in Coastal Zone Management* (J. Taussick and I. Mitchell, eds.), pp. 47–55, Samara Publication, Cardigan, UK.

Jakobsen, P. R., BrØker, I., J. H. Jensen and A. Skou (1998): ThyborØn barriers — A master case of coastal engineering, In: ICCE98, Copenhagen, Book of abstracts, DHI, Denmark, pp. 284–285.

Jorissen, R. E. and J. K. Vrieling (1998): The transition to a risk-based concept for dike design in the Netherlands, In: ICCE98, book of abstracts, Copenhagen, DHI, Denmark, pp. 620–621.

Kennedy, K. H. (1995): Producing management plans for major estuaries — The need for a systematic approach: A case study of the Thames Estuary, In: *Directions in European Coastal Management* (M, G. Healy and I. P. Doody, eds.): Samara Publications, Cardigan, UK, pp. 451–460.

Klaus, J., W. Pflügner, R. Schmidtke, H. Wind and C. Green (1994): Models for flood hazard assessment and management, In: *Floods Across Europe* (E. C. Penning-Rowsell and M. Fordham, eds.), Middlesex University Press, UK.

Korff, W., L. Beck and P. Mikat, eds. (1998): Encyclopaedia of Bioethic (in German), Publ. Gütersloher Verlagshaus, Gütersloh, Germany.

Kos'yan, R. D. and O. T. Magoon (1993): Man on the Black Sea coast, In: *Coastlines of the Black Sea*, Vol. CZ 93, New Orleans, ASCE, New York, pp. 1–13.

Kramer, J. and H. Rohde, eds. (1992): *History of Coastal Defence*, (in German), Konrad Wittwer Publications, Stuttgart, Germany.

Kraus, N. C. (ed.) (1996): *History and Heritage of Coastal Engrg.*, ASCE, New York.

Kunz, H. (1993a): Coastal protection responses to sea level rise and changing social needs — Case study Ley Bay, Southern North Sea, Germany, First Conf. on Mediter. Coastal Environ., MEDCOAST 93 (E. Özhan, ed.), Antalya, Vol. 2, pp. 1059–1073, Ankara, Turkey.

Kunz, H. (1993b): Coastal protection in the past — Coastal zone management in the future? Case study for the Ems-Weser area, Germany, In: Coastlines of the Southern North Sea (R. Hillen and H. J. Verhagen, eds.), spec. Vol. CZ93, New Orleans, ASCE, New York, pp. 314–335.

Laustrup, C., H. Toxvig Madsen (1998): Evaluation of the effect of 20 years of nourishment, In: ICCE'98, book of abstracts. Copenhagen, DHI, Denmark, pp. 360–361.

Loizidou, X. I. and N. G. Iacavou (1997): Coastal zone management for Cyprus: Finalization and implementation, 3rd Int. Conf. on the Mediter. Coastal Environ., MEDCOAST 97 (E. Özhan, ed.), Quawra, Malta, Vol. 1, pp. 533–545. Ankara, Turkey.

Losada, M. A., C. V. Medina and I. J. Losada (1996): History of coastal Engrg. in Spain, In: *History and Heritage of Coastal Engineering* (N. C. Kraus, ed.), ASCE, New York, pp. 465–499.

Louisse, C. J., M. J. F. Stive and H. J. Wiersma, eds. (1990): The Dutch coast — Report of a session on 22nd Int. Conf. on Coastal Engrg., Publ. rijkswaterstaat & delft hydraulics, The Hague, Netherlands.

Louters, T. and F. Gerritzen, eds. (1994): The Riddle of the Sands — A Tidal System's Answer to a Rising Sea Level, Publ. RIKZ, The Hague, Netherlands.

MAFF (1993a): Strategy for Flood and Coastal Defence in England and Wales, Publ. Ministry of Agriculture, Fisheries and Food, London, UK.

MAFF (1993b): Project Appraisal Guidance Notes for Flood and Coastal Defence, Publ. Ministry of Agriculture, Fisheries and Food (MAFF), London, UK.

MAFF (1995): Shoreline Management Plans — A guide for Coastal Defence Authorities, Publ. Ministry of Agriculture, Fisheries and Food, London, UK.

Marchi E. (1993): Coastal engineering in Venice, *Proc. 23rd Int. Conf. on Coastal Engrg.*, ICCE 92, (B. L. Edge, ed.) Venice, Italy, 4–39. ASCE, New York.

Niemeyer, H. D., H. Eiben and H. Rohde (1996): History and heritage of German coastal engineering, In: History and Heritage of Coastal Engrg. (N. C. Kraus, ed.), pp. 169–213. ASCE, New York.

Nordberg, L. (1995): Coastal conservation in selected European States, In: Directions in European Coastal Management (M. G. Healy and J. P. Doody, eds.), Samara Publication, Cardigan, UK.

Ovesen, C. H. (ed.) (1990): Saltmarsh management in the Wadden Sea region, *Proc. 2nd Trilateral Working Conf. National Forest and Nature Agency*, RØmØ, Denmark.

Özhan, E., U. Atila and E. Aktas (1993): Turkish legislation pertinent to coastal zone management, In: *First Int. Conf. on the Mediter. Coastal Environ.*, MEDCOAST 93 (E. Özhan, ed.), Antalya, Turkey, Vol. 1, pp. 333–346. Ankara, Turkey.

Özhan, E. (ed.) (1997): International Conference on the Mediter. Coastal Environ., Proc. 1993, 1995, 1997, MEDCOAST Permanent Secretariat. Civ. Eng. Dep., Middle East Tech. Univ., Ankara, Turkey.

Palmer R. (1996): History of coastal engineering in Great Britain, In: *History and Heritage of Coastal Engrg.* (N. C. Kraus, ed.), pp. 214–274. ASCE, New York.

Peerbolte, E. B. (1993): Sea-level Rise and Safety — A Consideration of Safety Impacts in Low-Lying Coastal Areas with Particular Reference to the Netherlands, Thesis Univ. of Twente. Delft Hydraulics, Marknesse, Netherlands.

Penning-Rowsell, E. C. and M. Fordham, eds. (1994): Floods across Europe, Middlesex Univ. Press, UK.

Pethick, J. and F. Burd (1993): Coastal Defence and the Environment: A Good Practice Guide, Publ. Min. of Agriculture, Fisheries and Food (MAFF), London, UK.

Price, A. (1991): Rethinking our role in coastal zone development, In: *Proc. 22th Int. Conf. on Coastal Engrg.*, ICCE'90 (B. L. Edge, ed.), Delft, Netherlands, Vol. 1, pp. 7–16. ASCE, New York.

RWS (1990): A New Coastal Defence Policy for the Netherlands, Publ. Min. of Transport and Public Works-Rijkswaterstaat. The Hague, Netherlands.

Salman, A. H. P. M., M. J. Langeveld and M. Bonazountas, eds. (1996): Coastal Management and Habitat Conservation, *Proc. 4th EUCC Congress*, Marathon, Greece. Publ. EUCC, Leiden, Netherlands.

Senlier, N. (1995): The effects of tourism on coastal settlements in Turkey from the view point of life quality, In: *2nd Conf. on the Mediter. Coastal Environ.*, MEDCOAST 95 (E. Özhan, ed.), Tarragona, Spain, Vol. 1, pp. 359–371. Ankara, Turkey.

Sidaway, R. (1995): Recreation and Tourism on the Coast: Managing Impacts and Revolving Conflicts, In: Directions in European Coastal Management (M. G. Healy and I. P. Doody, eds.), pp. 71–78. Samara Publication, Cardigan, UK.

Sørensen J., F. Gable, M. Gardner and D. Hinrichsen (1997): The Black Sea: Another environmental tragedy in our times? In: *3rd Intern, Conf. on the Mediter. Coastal Environ.*, MEDCOAST 97 (E. Özhan, ed.), Quawra, Malta, Vol. 2, pp 741–752. Ankara, Turkey.

Sørensen, J. (1993): The international proliferation of integrated coastal zone management efforts, *Ocean and Coastal Management*, arXiv: 0964.5691, 45–80. Elsvier, London, UK.

Sørensen, T., J. Fredsoe and P. Roed Jakobsen, (1996): History of coastal engineering in Denmark, In: *History and Heritage of Coastal Engrg.* (N. C. Kraus, ed.), pp. 103–141. ASCE, New York.

Taussik, T. and J. Mitchell, eds. (1996): Partnership in Coastal Zone Management, Samara Publ., Portsmouth, UK.

Utteridge, B. (1996): Flood warning into the next decade, In.: *Proc. 31st MAFF Conf. on river and coastal Engrs.*, 4.1.1–4.1.12. Ministry of Agriculture, Fisheries and Food (MAFF), London, UK.

Vaaben, J. (1993): Coastal planning for recreation in Denmark along the North Sea coast, In: Coastlines of the Southern North Sea (R. Hillen and H. J. Verhagen, eds.), spec. Vol. CZ93, New Orleans, pp. 227–232. ASCE, New York.

van de Kamer, I. P. G., K. J. Wulfraat, A. Cramer and M. J. P. H. Waltmans (1993): Water quality management in the Southern North Sea, In: *Coastlines of the Southern North Sea* (R. Hillen and H. J. Verhagen, eds.), spec. Vol. CZ93, New Orleans, pp. 145–161. ASCE, New York.

van der Meer, J. W., H. de Loof and P. C. G. Glas (1998): Integrated approach on the safety of dikes along the great Dutch lakes, In: ICCE98, Copenhagen, book of abstracts, pp. 80–81. DHI, Denmark.

van der Meulen, F. and A. H. P. M. Salman (1993): Management of Mediterranean coastal dunes, In: *First Int. Conf. on the Mediter. Coastal Environ.*, MEDCOAST93 (E. Özhan, ed.), Antalya, Turkey, pp. 167–183. Ankara, Turkey.

van der Ven, G. P. (ed.) (1993): Man-Made Lowlands — History of Water Management and Land Reclamations in The Netherlands, Stichting Matrijs, Utrecht, Netherlands.

van der Weide, I. and F. M. J. Hoozemans (1993): Engineering tools and techniques for coastal zone management, In: *Coastlines of the Southern North Sea* (R. Hillen and H. J. Verhagen, eds.), spec. Vol. CZ93, New Orleans, pp. 202–213. ASCE, New York.

Vieira, J. R., J. Føns and H. Kofoed-Hansen (1995): Venice flood protection-implementation of an operational flood warning system, In: *2nd Int. Conf. on the Mediter. Coastal Environ.*, MEDCOAST 95 (E. Özhan, ed.), Tarragona, Spain, Vol. 3, pp. 1645–1659. Ankara, Turkey.

Viggosson, G. and S. Sigurdarson (1998): Stabilization of the tidal entrance at Hornaf Jordur, Iceland, In: ICCE98, Copenhagen, book of abstracts, pp. 136–137. DHI, Denmark.

WCC (1994): Preparing to Meet the Coastal Challenges of the 21st Century, Rep. World Coast Conf. 1993, Min. of Transport, Public Works and Water Management, National Institute for Coastal and Marine Management/RIKZ, The Hague, Netherlands.

WWF (1996): Growing with the Sea — Creating a Resilient Coastline, World Wide Fund for Nature (Netherlands), Zeist, Netherlands.

Zhou, H.-M. (1994): Towards an Operational Risk Assessment in Flood Alleviation, Thesis Tech. Univ. Delft. Delft Univ. Press, Delft, Netherlands.

Chapter 15

Coastal Zone Management (3) Sediment Movement and Beach Transformation Due to Port Construction in Japan

Kazumasa Katoh

Environmental Consultants for Ocean and Human, Japan

1. Critical Water Depth of Sand Movement

With a rapid economic revival in Japan after the World War II, full scale industrial port plans were established and constructions of ports began in 1950s. When large industrial ports began to be planned in a stream of sandy beaches, the grasp of littoral drift characteristic and its mechanism in each beach became the essential and urgent technology in harbor engineering. It may be said that construction of many ports on the sandy beaches which are unique in the world is one of the factors for pushing the research on littoral drift forward in this country. In those days, fundamental knowledge about various hydraulic phenomena in the nearshore zone was not sufficient, and, moreover, the information about characteristics of waves, currents, and littoral drift in areas along the coast of this country lacked extremely. Accordingly, in every area where construction of a port was planned, a pre-investigation had to be done on the natural conditions. On littoral drift, it was the major technological subject to prevent shoaling in an anchorage and a channel. In other words, the subject was to understand the critical water depth of sand movement, which is related to which water depth a breakwater should be extended.

From engineering standpoint, it is important to estimate the critical water depth of sand movement due to waves, in which offshore bottom sediments are not moved. Sato *et al.* (1963) started laboratory study. The data obtained in experiments are

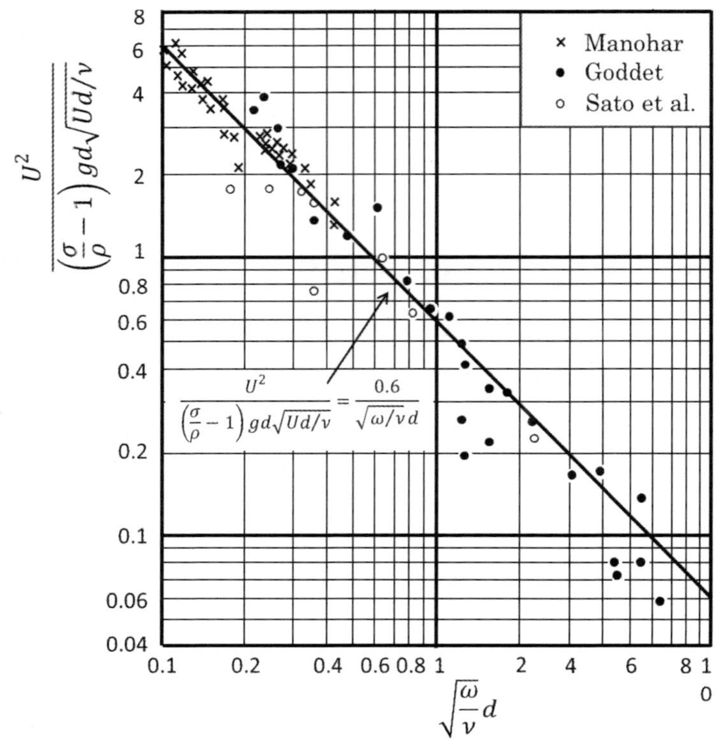

Fig. 15.1 The beginning of general movement by laboratory test.

plotted in Fig. 15.1 with the data by Manohar (1955) and Goddet (1960). From this figure, the following relation for the initiation of general sand movement is obtained,

$$\frac{U^2}{\left(\frac{\sigma}{\rho}-1\right)gd\sqrt{Ud/v}} = \frac{0.6}{\sqrt{\omega/vd}}, \tag{1}$$

where U is the maximum velocity due to waves at the bottom, σ is the density of sand grains, ρ is the density of fluid, g is the gravitational acceleration, d is the diameter of sand grains, ν is the kinematic viscosity, and ω is the angular frequency. By substituting the following relations into Eq. (1);

$$U = \frac{\pi H}{T}\cdot\frac{1}{\sinh(\frac{2\pi h}{L})}, \quad \omega = \frac{2\pi}{T}, \quad T = \sqrt{\frac{2\pi L_o}{g}}$$

and taking the representative values for σ, ρ and ν, we have

$$\frac{H_0}{L_0} = 0.565\left(\frac{d}{L_0}^{1/3}\cdot\frac{H_0}{H}\cdot\sinh\frac{2\pi h}{L}\right), \tag{2}$$

where H_0 and L_0 are the offshore wave height and wavelength respectively, H and L are the wave height and wavelength at the water depth, h, which is a critical water depth of sand movement.

Equation (2) was obtained as a result in the laboratory. In order to examine whether this formula could be applied in the actual field, Sato *et al.* (1963) carried out field experiments by utilizing a newly developed radioactive tracer. The radioactive glass sand had mainly been made in a nuclear reactor, by irradiating glass sand which contained cobalt or scandium. The specific gravity of glass sand was adjusted to be 2.65, and its grain distribution was also adjusted to be the required one. Some amount of radioactive glass sand was injected on the sea bottom. After an event of large waves, the radioactivity of sea bottom around the injection point was traced by means of a watertight detection unit connected with a cable to a ratemeter on a survey boat. The detection unit contained the Geiger–Muller tube or the scintillation tube. The count per second measured by the detection unit, substracting the natural count, were plotted and equi-count lines were drawn. This measurement had been done repeatedly at many coasts around Japan, which were facing to the Pacific Ocean, the Japan Sea, an inland sea, and a sea between the ocean and an inland sea, during a period from 1953 to 1962.

Sato *et al.* (1963) have classified the distribution patterns of radioactive glass sand injected in the offshore zone into the following four groups which are schematically shown in Fig. 15.2.

(1) The first group: This is a case that the point of the maximum count moves in the direction of the wave propagation and all equi-count lines also extended in the same direction.
(2) The second group: This is the case that the point of the maximum count does not move, but all equi-count lines extend in the same direction of the wave propagation.
(3) The third group: This is the case that the point of the maximum count does not move and only a part of equi-count lines extend in the direction of wave propagation. This group also contains the case that the count decreases markedly in total, despite the distribution of equi-count lines scarcely changes.
(4) The fourth group: This is a case that the distribution of equi-count lines remains unchanged and the count also does not decrease significantly. The cases that all equi-count lines extend in the same direction as the tidal current or in the direction opposite to the wave propagation are also included in this group.

Then, classifying the field data into these four groups, the result that they have been related with the wave characteristics is shown in Fig. 15.3. In the figure, the wave height H and wave period T are those of the maximum significant wave at the injection point of glass sand during the period when the change of count lines was monitored, and depth h is the mean water depth at the injection point. The symbols, H_0 and L_0, are the height and length of the deep water wave for H and T, respectively, and d is the median diameter of the glass sand which is the same as the bottom sand in the vicinity. In this figure, the transition from the second group

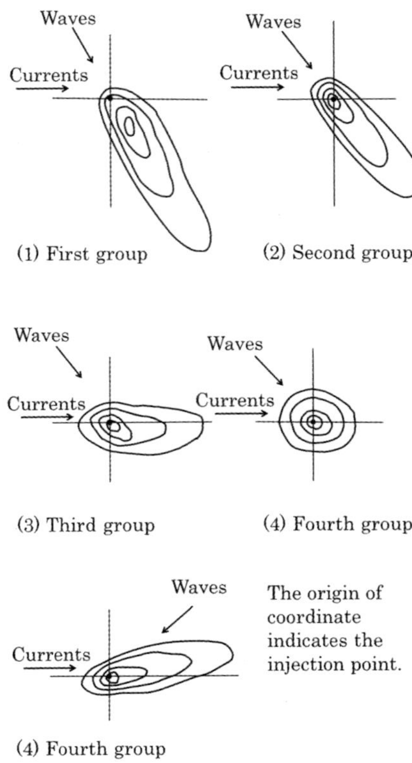

Fig. 15.2 Classification of the distribution of radioactive glass sand.

to the third group is shown by the following equation;

$$\frac{H_0}{L_0} = 1.35 \left(\frac{d}{L_0}\right)^{1/3} \cdot \frac{H_0}{H} \cdot \sinh\left(\frac{2\pi h}{L}\right), \tag{3}$$

and the critical condition on which points enter into the first group is shown by the following equation;

$$\frac{H_0}{L_0} = 2.40 \left(\frac{d}{L_0}\right)^{1/3} \cdot \frac{H_0}{H} \cdot \sinh\left(\frac{2\pi h}{L}\right), \tag{4}$$

That is, Eq. (3) corresponds to the state where all sand particles on the surface layer of the sea bed move collectively in the direction of wave propagation and Eq. (4) corresponds to such great movement of the bottom sand as to produce the change of water depth. The former is called the surface layer movement and the latter is called the completely active movement.

These results become essential to give a clear index to decide an arrangement of breakwater, a water depth of its tip, and a location of mouth of port when a port is built in sandy coast. In most of the big harbors in Japan, for example, in order to prevent the sand from entering into an access channel, the breakwater are extended up to a location where the water depth is deeper than the value given by Eq. (4), in which the design wave conditions are adopted.

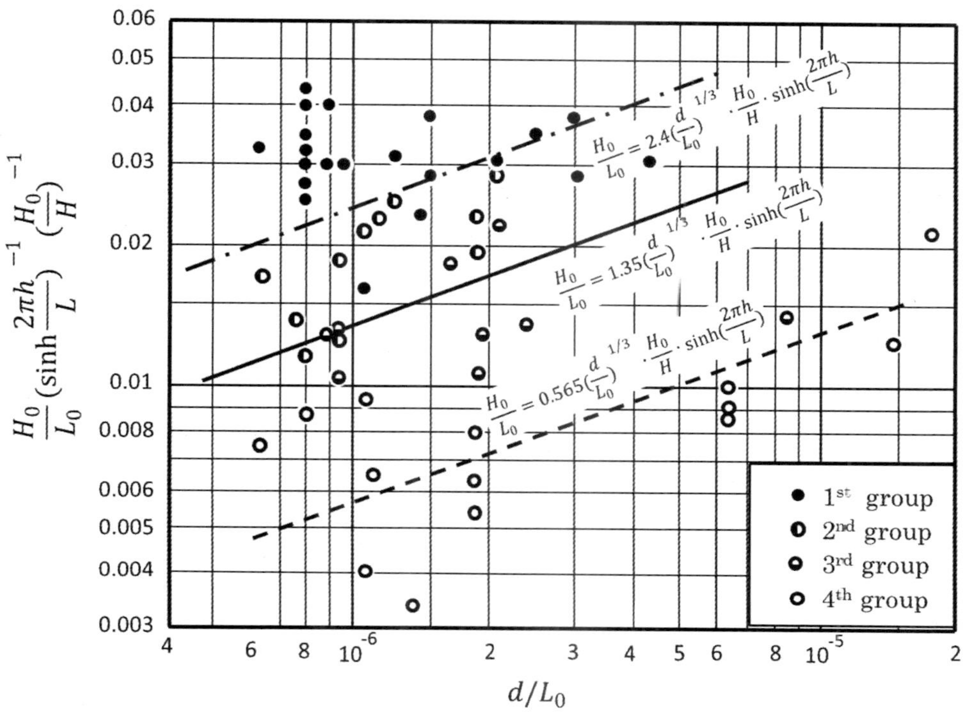

Fig. 15.3 Relation between the wave characteristics and the medium diameter of sand for each distribution of radioactive glass sand.

Figure 15.4 is the calculation diagram of Eqs. (3) and (4), where a is a constant;

$$a = \begin{cases} 1.35 & \text{for Eq. (3)} \\ 2.40 & \text{for Eq. (4).} \end{cases}$$

By utilizing Fig. 15.4, the critical water depths can be easily estimated with the offshore wave conditions and the median diameter of sand.

Since the critical depth of sand movement was formulated, it has become possible to build a port without shoaling at the sandy coast, where a port construction was difficult till then. However, the port constructed made a new problem to cause deformations of its peripheral beaches.

2. Patterns of Beach Change Around Structures

There are certain patterns of beach changes caused by construction of man-made structures along the coastline. Tanaka (1983) made a detailed study of morphological changes of beach shape based on comparison of aerial photographs over many years. From the analysis of some 120 examples, he classified the beach changes, in accordance with the following criteria, into 11 groups or types as shown in Fig. 15.5.

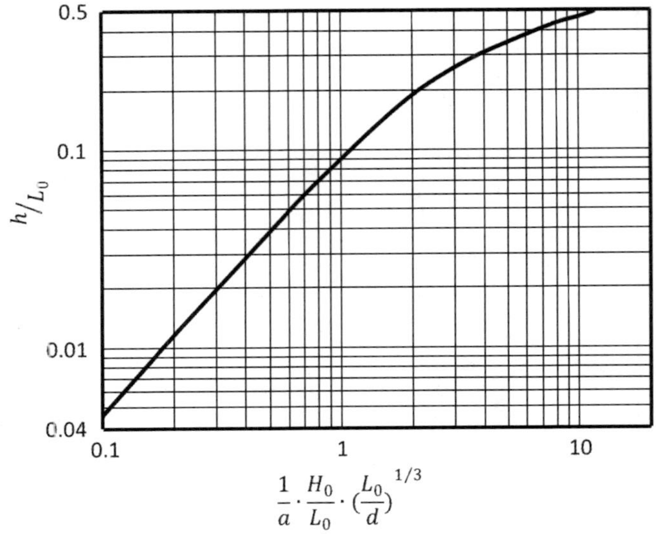

Fig. 15.4 Diagram for estimating the critical water depth of sand movement.

(1) similarity in morphological features;
(2) similarity in locality (e.g., open sea, bay, etc.), topography (e.g., convex, concave, etc.), and layout of the structures;
(3) similarity in the cause and mechanism of the morphological changes; and
(4) frequency of occurrences.

The characteristics of 10 groups or types are explained in the following:

Type l: Longshore drift, checked; updrift side accretion, downdrift side erosion.

This is a typical case in which a jetty blocks the longshore transport. Sediment accumulation takes place on the updraft side and erosion takes place on the downdrift side. The shoreline change of this type is found clearly on coasts of following conditions. The first is a coast on the side of deep gulf, where waves approach with a large angle of incidence, then the predominant direction of longshore drift is distinct, and waves become comparatively smaller due to the effect of wave refraction, then the width of surf zone is relatively narrower. The second is a coast that the beach slope is steep and grain size is coarse, where the surf zone width is narrow, and an active domain of littoral drift is limited to narrow zone near to the shoreline. On these beaches, the capture efficiency of longshore drift by jetty is high.

(a) Depending on the length of jetty.

When a jetty is short, there is the tendency that the shoreline and the jetty make an acute angle at the deposition side. There are many cases that a submerged bank is formed in a downdrift side of jetty tip. This is because the longshore currents that occurred at the updrift side spread to the offshore from the tip of jetty and slow down, which promotes the accumulation of sand there. When the jetty extension is

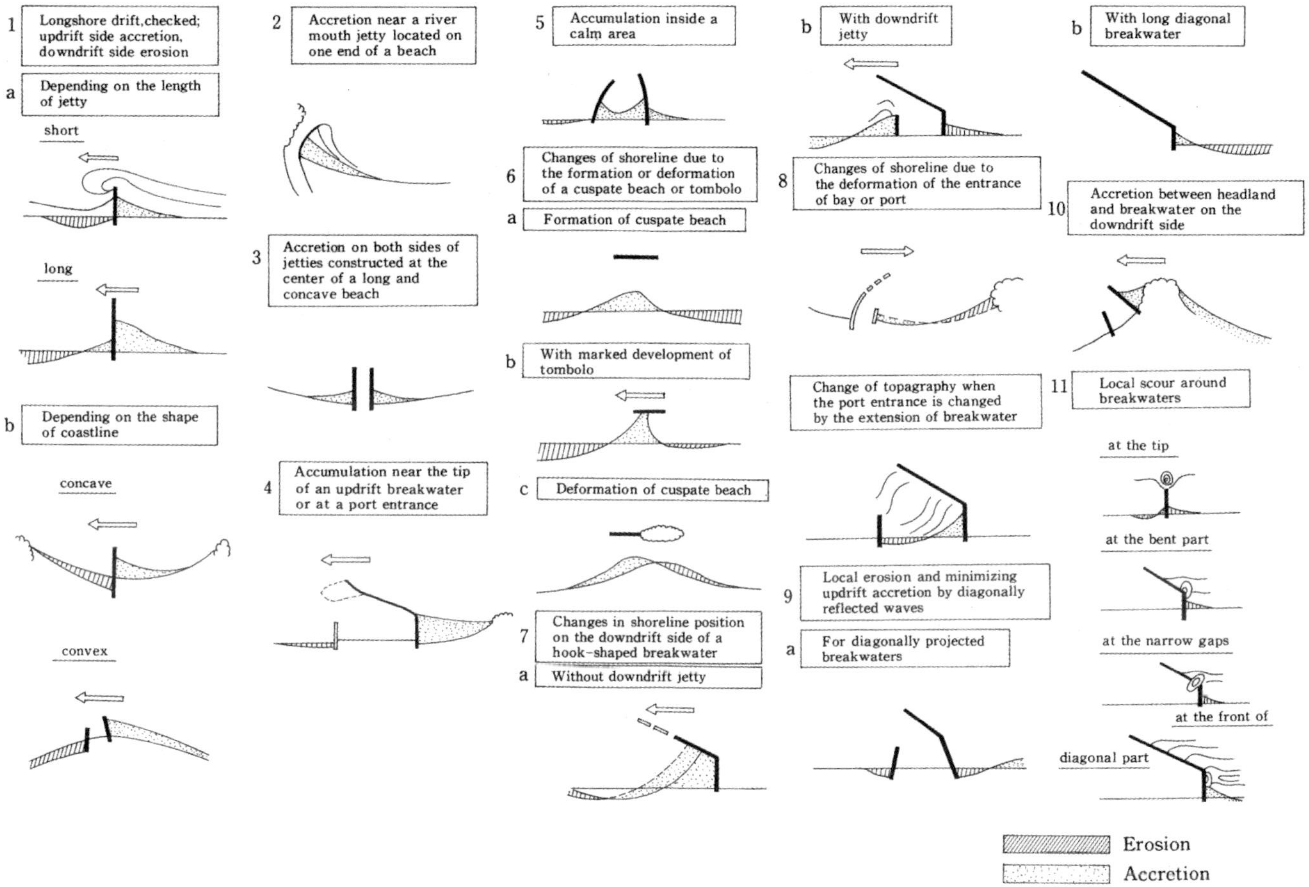

Fig. 15.5 Classification of morphological changes due to the presence of structures by Tanaka (1983). The arrow indicates the direction of net sediment transport. This figure is reproduced from Horikawa (1988).

long, the shoreline is at nearly right angle to the jetty due to the effect of incident waves along the jetty.

The extent of erosion depends on the length of the jetty; the longer the jetty is, the wider the erosion area is. Even on the downdrift side, however, there are many cases rather to show the accumulation near the long jetty. This is because the downdrift side of long jetty is sheltered from waves, where the sand gather due to the similar mechanism as explained later in Type-7.

(b) Depending on the shape of coastline.

If a structure is constructed in the center of a coast, where the coastline ranging from several km to 10 km in length is concave in plan view, the shoreline advances on the updrift side and recedes on the downdrift side so that those shorelines maintain approximately the same direction to each other. In this case, the accumulation area on the updrift side is narrow and the erosion on the downdrift side is apt to extend over the wide area. If the shoreline is convex, on the other hand, the shoreline on the updrift side advances almost to the tip of the jetty and erosion extends very far along the downdrift coast.

Type 2: Accretion near a river mouth jetty located on one end of a beach.

There are many cases that a river flows out into the sea along a hard topography such as a headland located at one end of a beach. In particular, an example of these cases can be easily found on the sandy beach, less than several km in extension. In this situation, the longshore sand transport toward the river mouth is prevailing. Sand transported toward the river mouth is carried sequentially to the offshore by river flows and deposited there, and the one part of deposit sand is transported in the onshore direction and cast up again on the beach by waves. There exists this circulation, which is shown diagrammatically in Fig. 15.6(a).

If a training jetty is constructed at the mouth of a river as shown in Fig. 15.6(b), it checks sand transport toward the river mouth, which interrupts the contact of longshore sand and river flows, and prevents sand from being transported to the offshore. As a result, a flat wide beach appears near the jetty and submerged topography is shallow to a great distance from the shore.

Type 3: Accretion on both sides of training jetties constructed at the center of a long and concave beach.

If long jetties are placed at the center of a long and concave beach, distinct accumulation develops on both sides of the jetties. This type involves a larger area of accumulation and little erosion, as compared to Type 1.

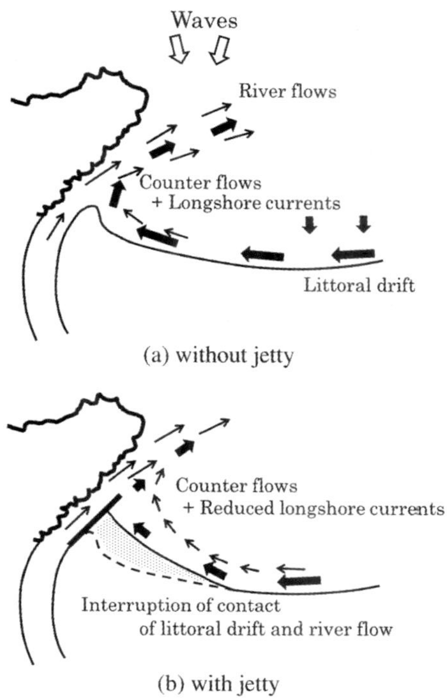

Fig. 15.6 Mechanism of sand accumulation near a river mouth jetty located at the end of beach.

As the cause of the accumulation on both sides, following three items are pointed out.

(1) At the center part of the long and concave beach, waves approach the shoreline almost in a right angle to it. Then, a predominant longshore direction of littoral drift is non-clear.
(2) When a river flows out to the offshore, counter currents from both sides to the river mouth are generated, which transport the sand to both sides of jetties.
(3) Since generally the training jetties are long, clam areas sheltered from waves are formed on the both sides, where sand is apt to come to accumulate.

Type 4: Accumulation near the tip of an updrift breakwater or at a port entrance.

Accumulation, a submerged sand-spit, takes place immediately downdrift of the tip of a breakwater, caused by the same mechanism as that for the accumulation at the downdrift of a jetty, as in Type 1. However, this case is classified as an independent type because of its implications to shoaling in harbor basins.

Although, conditions of accumulation differ slightly from port to port in actual examples, there are three common characteristics as follows:

(1) The accumulation of sand on the updrift side is remarkable. Accordingly, the predominant direction of littoral drift is considerably clear.

(2) For the shoreline of beach formed well by the accumulation of sand, a breakwater is extended by diagonal to an extreme, and the direction of shoreline and that of breakwater are nearly parallel to each other. On this account the sand from updrift side is easily transported up to the mouth of port, where it accumulates, forming submerged sand-spit.
(3) A headland exists on the updrift side beach in a position comparatively near to the port, where a pocket beach bounded by them is established. Because the capacity of updrift beach is small, the shoreline advances remarkably by the accumulation of a little volume of sand.

Type 5: Accumulation inside a calm area.

Accumulation takes place only because the interior of the basin is calm. In this meaning, this phenomenon is the same as that related to the formation of a tombolo in Types 6 and 7. In the formation of the tombolo, however, the flows which come from the outside to the calm area sheltered from waves is generated, by which the sand is transported into the calm area, as explained later, while in this type the formation of such flows is not found.

There are many conditions for the topographic change of this type, but the following ones are factors common to every port.

(1) Because the water depth at the entrance of a port is shallow, the whole port is inside the surf zone in the stormy weather.
(2) Bottom materials around the port is fine sand.
(3) Topographic changes of this type do not occur simultaneously with Type 1, namely the remarkable accumulation on the updrift side and erosion on the downdrift side do not occur.
(4) The sand deposits inside the port when the steepness of incident waves is small, while rather erosion occurs inside the port when it is large.

Type 6: Changes of shoreline due to the formation or deformation of a salient or tombolo

A detached breakwater creates a shadow zone to waves incident on the shoreline behind it. Wave diffraction always occurs in this shadow area, but waves frequently undergo refraction as well as diffraction while propagating in the lee toward the shore. Such combined diffraction and refraction processes give rise to an alongshore difference in wave height in the lee of the detached breakwater, inducing converging longshore currents from both ends of the shadow area. The processes lead to the formation of salients, a cuspate spit behind the detached breakwater. At both sides of shadow area, it is general that an erosion of shoreline occurs at the same time (Type 6(a)). Even if there is not a sandy beach, a cuspate spit may be formed behind a detached breakwater if there is sand on the offshore bottom. This is because that the sand in the offshore is suspended by waves and transported to the shadow area, where it deposits.

When the accumulation of sand is remarkable, the shoreline advances to a position of the detached breakwater. A salient that connects a detached breakwater with the beach is called a tombolo. In this case, the tombolo functions as a jetty, and magnifies the erosion on the downdrift coast (Type 6(b)). In the field of coastal engineering, the salient and the tombolo are often called the tombolo collectively.

If the shadow area is expanded by the extension of detached breakwater or by construction of a new harbor on one side of an island, a cuspate spit already in existence shifts its position toward the new center of the sheltered area (Type 6(c)).

Figure 15.7 shows some geometrical relationship of depositional landforms developing behind an island, based on data collected from Japanese open coast (Sunamura and Mizuno, 1987). Each physical symbol is defined in Fig. 15.7(a), and the curves in Fig. 15.7(b) and (c) indicate general trends of the data. For example, Fig. 15.7(b) suggests that a tombolo forms if $J/I \leq 1.5$, a salient develops if $1.5 \leq J/I \leq 3.5$, and no island influence appears if $J/I \geq 3.5$, which can be applied for the empirical prediction of shoreline changes behind a detached breakwater on the sandy beach.

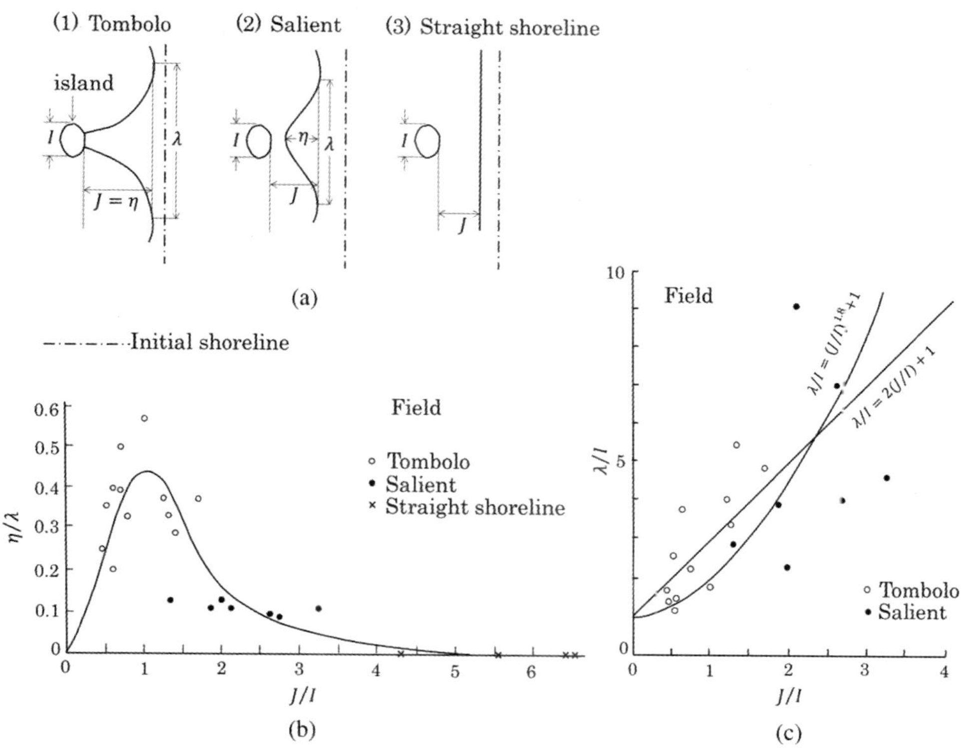

Fig. 15.7 Geometrical relationship of depositional landforms behind island (Sunamura and Mizuno, 1987).

Name of Port	α_1	α_2	Definition of degrees
Kohchi	42	22.5	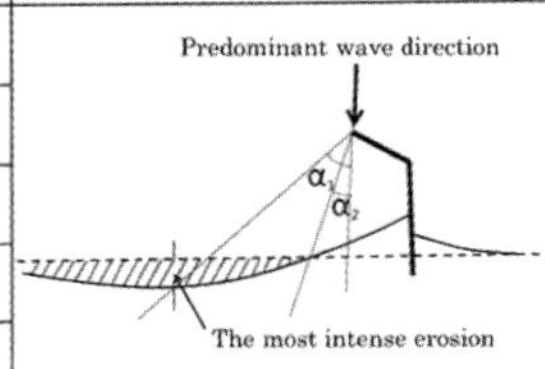
Kashiwazaki	44	22.5	
Hitachi	51	26	
Kashima	45	30	
Ooarai	49 ~ 55	27 ~ 31	

Fig. 15.8 Locations of the most intense erosion and the border point.

Type 7: Changes in shoreline position on the downdrift side of a hook-shaped breakwater

(a): Without downdrift jetty.

Accretion takes place in the sheltered area landward of a hook-shaped breakwater and erosion takes place on the downdrift side of the sheltered area. Mechanism of topographic change in this type is certainly the same as that of tombolo in Type 6. A point of the most intense erosion and a border point of the accumulation area and the erosion area are dependent on the position of a tip of breakwater and the characteristics of waves such as a predominant direction, a period, and an angular spreading. Figure 15.8 shows the locations of the most eroded point, α_1, and the border point, α_2, in relation with wave direction for several ports in Japan. According to this figure, $\alpha_1 = 40 \sim 50$ degrees° and $\alpha_2 = 20 \sim 30°$, respectively. For example, by assuming $\alpha_1 = 45°$ and using the physical symbols for the salient defined in Fig. 15.7(a), we have

$$\frac{\lambda}{I} = 2\left(\frac{J}{I}\right) + 1, \tag{5}$$

which is well applicable to the field data related to the configuration of tombolo behind islands (see Fig. 15.7(c)). Since the location of the most eroded point is strongly dependent on the location of the tip of breakwater, the accumulation area spreads out with extension of breakwater and the erosion area shifts away from the port progressively.

(b): With downdrift jetty

Installation of a jetty on the downdrift side of a hook-shaped breakwater would check the development of accretion inside a harbor basin. If the hook-shaped breakwater is extended markedly, however, notable accretion takes place on the left side of the jetty, frequently resulting in shoaling of the navigation channel. There are many cases of this type of shoaling in Japan, because many ports in Japan are usually under expansion and the main breakwater is extended far beyond the jetty to secure a enough stopping distance of shipping.

Type 8(a): Changes of shoreline due to the deformation of the mouth of small bay, or prolongation of breakwater

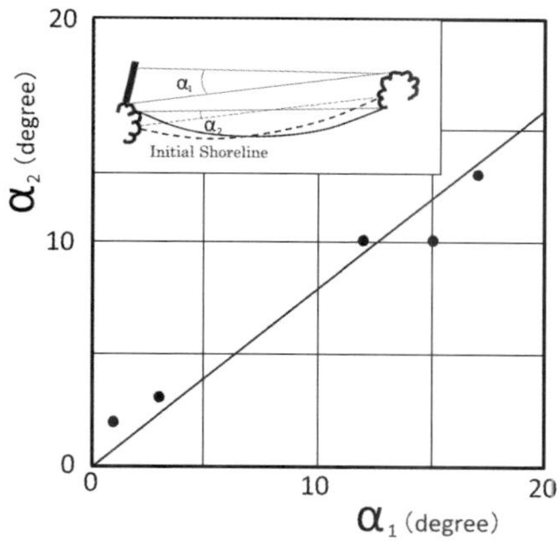

Fig. 15.9 Changes of shoreline orientation due to the deformation at mouth of small bay.

The stable configuration of small scale sandy beach, of which extension is ranged from several hundred meters to 1 km, is similar to a circular arc. Prolongation of a jetty or a breakwater would result in a newly stabilized concave shoreline on the small scale beach. If the width or position of the entrance of the port or bay is changed due to the extension of breakwaters or jetties, the shoreline changes its orientation and curvature to reach a new stabilized form. According to the actual data, there is a following relation between the angular changes of the bay mouth and the shoreline in their orientation (see Fig. 15.9).

$$\alpha_2 = 0.8\,\alpha_1. \tag{6}$$

If the jetty for preventing the sand transport into the port basin is constructed far away from the breakwater and there exists the beach between the jetty and the breakwater, the shoreline inside the port changes its orientation in the same manner as explained.

Type 9: Local erosion and minimizing updrift accretion by diagonally reflected waves

(a) For diagonally projected breakwaters

It has been mentioned that a local scouring occurs at the toe of the jetty due to the reflection of waves when the extension of jetty gets longer in Type 1. When the jetty is extended with a large angle more than 90° to the shoreline, the effect of reflected waves is strengthened more. Then if breakwaters of wall type are extended diagonally from the beach, local erosion takes place at the toe of the breakwaters due to the reflection of waves.

(b) With long diagonal breakwater

Similar erosion will take place if the shore normal part of the breakwater is short and the diagonal part of the breakwater is long. However, the erosion does not occur when the angle of the diagonal breakwater to the shoreline becomes large furthermore. Figure 15.10 shows directions of waves reflected by two kinds of breakwaters

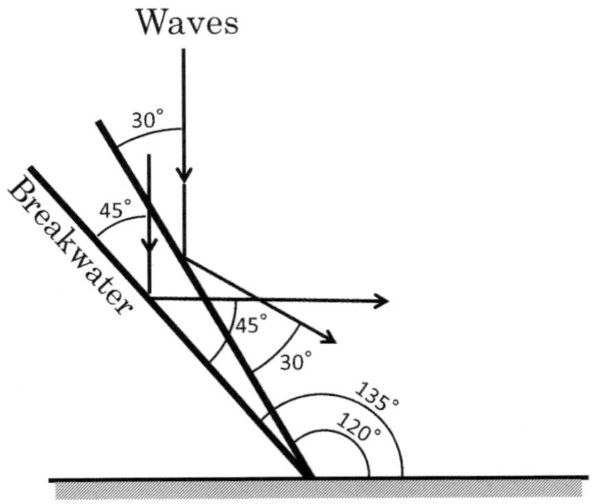

Fig. 15.10 Directions of reflected waves.

which constructed diagonally with the angles of 120° and 135°, respectively, to the shoreline. When the waves approach normal to the shoreline, the direction of reflected waves by the breakwater of 135° is parallel to the shoreline, which can hardly reach the beach. Examinations of 27 ports in Japan reveal that the erosion take place at the toe of the breakwater in the cases that the breakwaters are extended diagonally with the angles of 110° to 120° to the shoreline, while there are no erosion in the cases that these angles are more than 130°.

Type 10: Accretion between headland and breakwater on the downdrift side

If a port is constructed at the downdrift side of a projected coast such as a headland, accumulation rapidly occurs between the breakwater and the headland. There are numerous examples of shoreline changes of this type. Particle size of sand accumulated is apt to be coarse, because the fine sand is washed away by waves on the way of making a detour around the headland and the coarser sand is transported along the coast to the beach.

In planning of any coastal project, we should always expect certain changes will occur along the shoreline: some accretion and some erosion. Before finalization of the planning, we must predict what type of beach change will occur and must answer what will be the best solution to mitigate a possible damage. Examples of Fig. 15.5 will hopefully serve for the qualitative prediction of morphological changes around the coastal structures.

References

Goddet, J. (1960): Etude du debut d'entrainement des materiaux mobiles sous l'action de la houle, *La Houille Blanche*, **15**(2).

Horikawa, K. (1988): *Nearshore Dynamics and Coastal Processes*: Univ. of Tokyo Press, 522 pp.

Manohar, M. (1955): Mechanics of Sand Movement due to Wave Action, Tech. Memo. No.75, Beach Erosion Board, US Army Corps of Engineers.

Sato, S., T. Ijima and N. Tanaka (1963): A study of critical depth and mode of sand movement using radioactive glass sand, *Proc. 8th Int. Conf. on Coastal Engrg.*, 04–323.

Sunamura, T. and O. Mizuno (1987): A study on depositional shoreline forms behind an island, Ann. Rep., Inst. Geosci., Univ. Tsukuba, No. 13, 71–73.

Tanaka, N. (1983): A study on characteristics of littoral drift along the coast of Japan and topographic change resulted from construction of harbor on sandy beach, Technical Note of PHRI, No. 453, 148 pp. (in Japanese).

Chapter 16

Coastal Zone Management (4) — Downcoast Erosion Triggered by Exhaustion of Sand Supply from Sea Cliffs with Northern Kujukuri Beach in Japan Taken as Example

Takaaki Uda

Executive Director, Public Works Research Center, Japan
Visiting Professor, Nihon University, Japan

1. Introduction

The relationship between the protection of sea cliffs from which a large amount of sand was supplied to the downcoast and the resultant downcoast erosion was investigated, taking northern Kujukuri Beach in Japan as an example. A seawall has been constructed along with the installation of concrete blocks to protect the downcoast locally, resulting in shoreline recession on further downcoast. These actions have repeatedly occurred on this coast, and a long stretch of the coastline has been covered with concrete blocks, resulting in the loss of the natural sandy beach. On the basis of the obtained field data, future measures are considered.

Beach erosion caused by anthropogenic factors has been severe at many of Japan's coasts. Uda (2010) has classified the causes of beach erosion into seven categories. One of them is related to the selection of the countermeasure itself; the protection of sea cliffs from which a large amount of sand was supplied to the downcoast, resulting in a marked decrease in sand supply to the downcoast and subsequent beach erosion. When downcoast erosion occurs, a seawall is constructed along with the installation of concrete blocks to protect the site locally, resulting in shoreline recession further downcoast. These actions repeatedly occur, causing a long stretch of the coastline to be covered with concrete blocks and the loss of the natural sandy beach. We investigate this issue, taking Kujukuri Beach, as shown in Fig. 16.1, as an example.

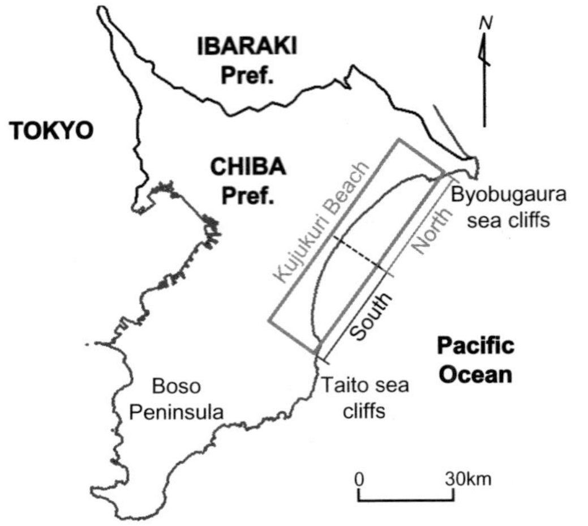

Fig. 16.1 Location of Kujukuri Beach in Chiba prefecture.

This beach is a 60 km long sandy beach facing the Pacific Ocean and extending between Byobugaura and Taito sea cliffs located at the north and south ends, respectively, and has been formed by the deposition of sand, supplied from sea cliffs mainly composed of unconsolidated mudstone and sandstone (Horikawa and Sunamura, 1971). Byobugaura sea cliffs, which constitute the main source of sand, have been retreated at a rate of 0.7 m/yr and a large amount of sediment has been supplied to the coast (Horikawa and Sunamura, 1969; 1970). However, sand supply to Kujukuri Beach has markedly decreased mainly because of the installation of concrete blocks along the sea cliffs as a measure against erosion since the 1960s. Longshore sand transport has also been obstructed by the extension of Iioka and Taito fishing ports located at the north and south ends of the beach, respectively. Moreover, the volume of littoral sand was reduced by the offshore disposal of materials dredged at Iioka, Katagai and Taito fishing ports, where Katagai fishing port is located at the center of Kujukuri Beach. All these activities resulted in the expansion of the eroded zone from the vicinity of both ends of the pocket beach to the central part. Thus, this beach is in a critical condition at present in terms of the preservation of the natural sandy beach. The investigation of the background of this problem reveals that the beach erosion of Kujukuri Beach arises from structural problems related to the Japanese management system of the coastal zone, and the present situation is close to an ultimate condition in which any type of effort will be ineffective for fundamentally resolving the issue.

2. Shoreline Changes

Figure 16.2 shows an aerial photograph of northern Kujukuri Beach in 2010 extending north of Katagai fishing port, as shown in Fig. 16.1. A gradually curved

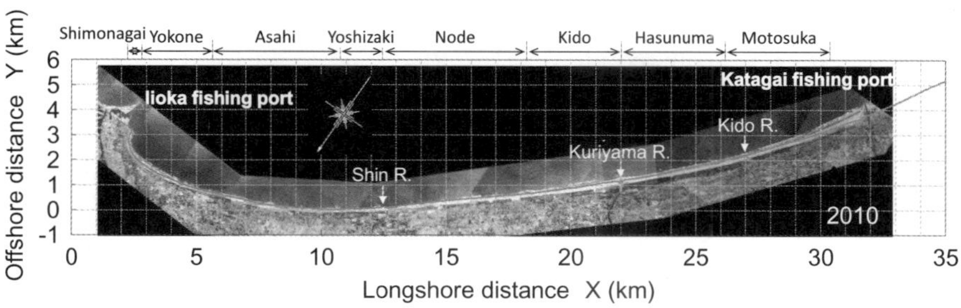

Fig. 16.2 Aerial photograph of northern Kujukuri Beach between Iioka and Katagai fishing ports in 2010.

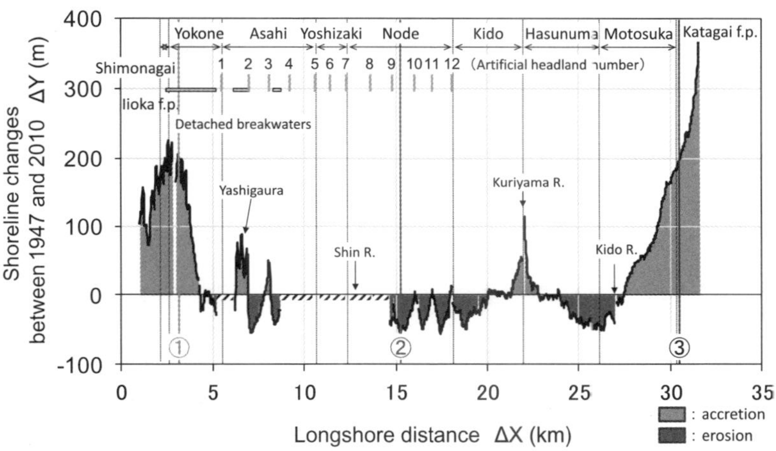

Fig. 16.3 Shoreline changes between 1947 and 2010.

embayment shoreline extends from Iioka fishing port located at the north end of the study area to Katagai fishing port in the northeast–southwest direction. The shoreline changes in the entire area shown in Fig. 16.2 were investigated using aerial photographs. Figure 16.3 shows the shoreline changes between 1947 and 2010, where the shoreline changes are measured in the direction normal to the shoreline in 2010 between Iioka fishing port and the Shimonagai coast because the shoreline has a large curvature, as shown in Fig. 16.2.

Figure 16.3 shows the marked contrast between the area with a wide foreshore and that without a sandy beach, and the shoreline advance has been concentrated on the Shimonagai coast and in the area northeast of Katagai fishing port with maximum shoreline advances of 226 m and 385 m, respectively. Except for the local shoreline advance at the Yashigaura coast ($X = 6.5$ km) and the Kuriyama River mouth ($X = 22$ km), almost the entire area between the two sand accumulation zones is erosive, the sandy beach has completely disappeared between $X = 4.3$ and 14.8 km, and the seawall is exposed to waves. In the area where a sandy beach is left, erosion on the Node coast has been the most severe and a maximum shoreline recession of 56 m has occurred at $X = 15.2$ km.

In 1947, Kujukuri Beach was a natural sandy beach in which sand supplied from the Byobugaura sea cliffs northeast of the beach was transported without obstruction by artificial structures toward the central part of Kujukuri Beach by longshore sand transport. By 2010, the Iioka fishing port breakwater had been extended, as shown in Fig. 16.4, and the south breakwater of the fishing port had

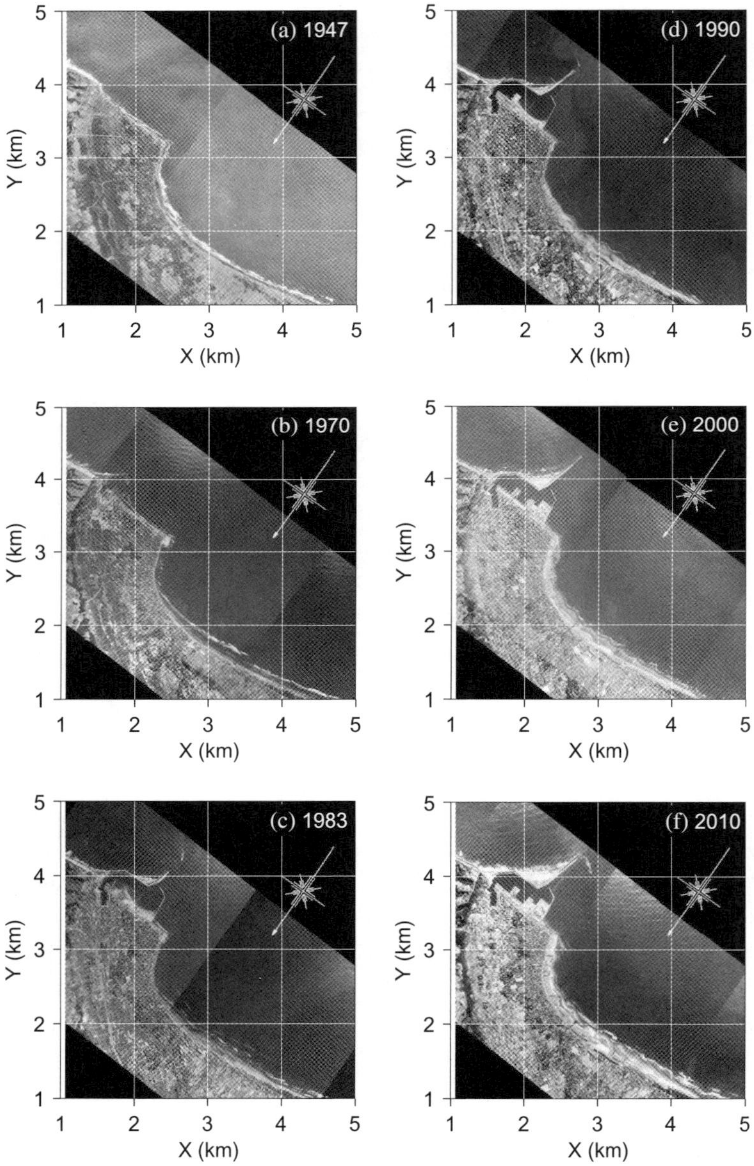

Fig. 16.4 Aerial photographs of Iioka fishing port and nearby coast.

been extended 1 km offshore from the previous coastline. Due to the extension of the south breakwater, westward longshore sand transport has been obstructed, and a large wave-shelter zone has formed west of the south breakwater, causing sand deposition. In contrast, severe erosion has occurred between the Yokone and Hasunuma coasts further southwest of the area because of the marked decrease in sand supply from the upcoast. In contrast, a large amount of sand has been deposited with the shoreline advance between the Motosuka coast and Katagai fishing port in the central part of Kujukuri Beach because of the obstruction of longshore sand transport by the breakwater of Katagai fishing port, which has been extended over time, as shown in Fig. 16.5.

Figure 16.6 shows the changes in shoreline position between 1947 and 2010 at points 1, 2 and 3 on the Yokone, Node and Motosuka coasts, respectively, where

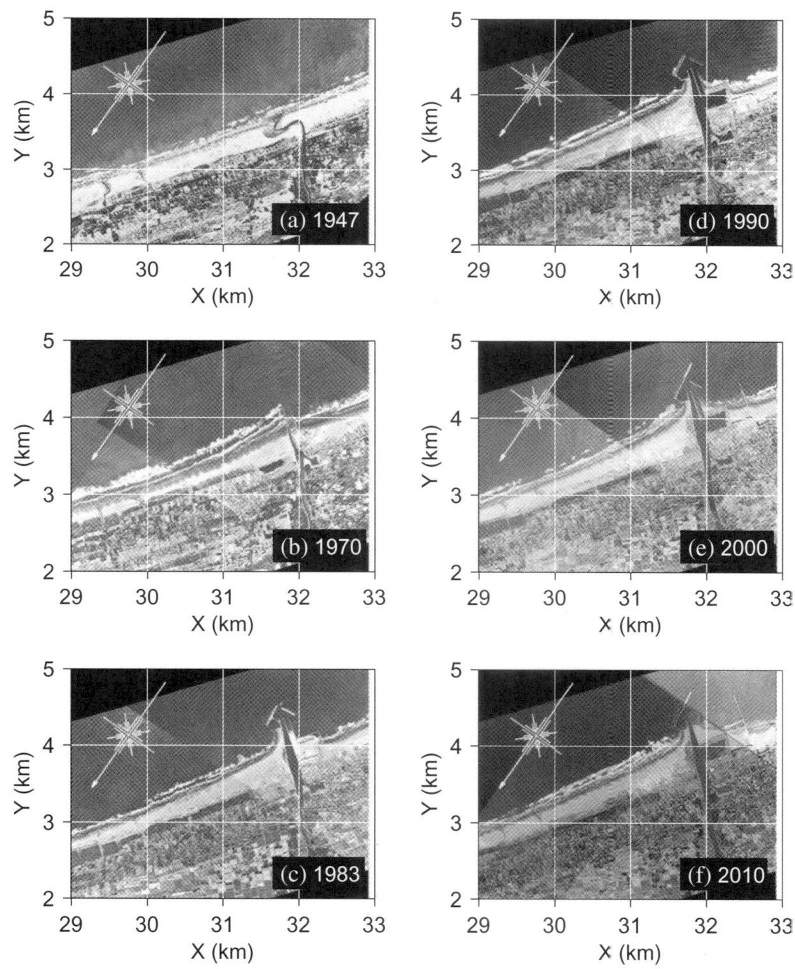

Fig. 16.5 Aerial photographs of Katagai fishing port and nearby coast.

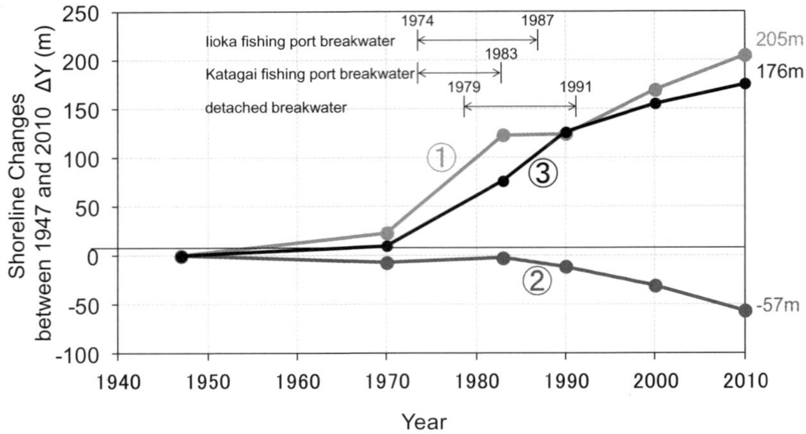

Fig. 16.6 Shoreline changes with time at location (1) on Iioka coast, location (2) on Node coast and location (3) on Motosuka coast.

marked shoreline changes were observed in Fig. 16.3. Although no shoreline changes were observed before 1970 when the extent of artificial alteration was low, after 1970, the shoreline at point 1 advanced at a rate of 6.8 m/yr with the construction of the Iioka fishing port breakwater and 15 detached breakwaters on the Iioka coast, resulting in a total shoreline advance of 205 m until 2010. In contrast, the shoreline position at point 2 retreated at a rate of 1.3 m/yr with a 13 year time lag compared with that at point 1 with a total shoreline recession of 57 m until 2010. In addition, the shoreline at point 3 advanced at a rate of 5.8 m/yr with a total shoreline advance of 176 m until 2010. At all three points, the shoreline is still continuing to change at present.

The sand supplied from Byobugaura sea cliffs had been transported westward under natural conditions before the construction of Iioka fishing port, and the longshore sand transport was estimated to be $4.0^{*}10^{5}\,m^{3}/yr$ on the basis of the shoreline changes from 3500 year ago (Uda *et al.*, 2000). Then, it was reduced to $3.0^{*}10^{5}\,m^{3}/yr$ in the period between 1960 and 1967 because of the installation of concrete blocks along the sea cliffs since 1966 and the construction of the Iioka fishing port breakwater from 1953 (Uda, 1997). Moreover, the breakwater of Iioka fishing port was extended, resulting in the reduction in sand supply to the west of the fishing port, and sand supply from the sea cliffs was assumed to have been reduced to $9^{*}10^{4}\,m^{3}/yr$ since 1967 (Uda, 1997). Recently, Uda *et al.* (2010) have estimated that the rate of longshore sand transport obstructed by the south breakwater was $1.4^{*}10^{4}\,m^{3}/yr$ on average between 1985 and 2005 after the extension of the south breakwater, and the same amount of longshore sand transport has been discharged downcoast, while turning around the tip of the breakwater, since 2005. Note that the rate of longshore sand transport is one order of magnitude lower than that previously observed under natural conditions.

At Katagai fishing port, a 400 m long jetty was extended in 1970 at the mouth of the Sakuta River, and then in 1983, two oblique breakwaters were constructed at the tip of the jetty to prevent sand deposition, as shown in Fig. 16.5. Thereafter, the breakwater of Katagai fishing port was further extended until 2000, resulting in further shoreline advance northeast of the fishing port because of the blockage of longshore sand transport. Regarding the beach changes around Katagai fishing port, San-nami *et al.* (2009) estimated the amount of longshore sand transport turning around the tip of the breakwater to be $7^*10^4\,m^3/yr$ on the basis of the results of numerical simulation using the contour-line-change model. Thus, to summarize the sand budget in the study area, sand is supplied from upcoast (Iioka fishing port) at a rate of $1.4^*10^4\,m^3/yr$, whereas sand is transported away from the downcoast boundary (Katagai fishing port) at a rate of $7^*10^4\,m^3/yr$ without balance in the sand transport, resulting in erosion becoming severe with time.

3. Changes in Beach Width

On Kujukuri Beach, the seaward advance of the marginal line of the coastal forest significantly affected the decrease in beach width (Hoshigami *et al.*, 2005). Therefore, the shoreline position and the location of the seaward marginal line were read from the six sets of aerial photographs obtained between 1947 and 2010, and the changes in beach width were calculated, as shown in Fig. 16.7. The vertical axis shows the offshore distance from the seawall to the shoreline and the marginal line of the coastal forest in 2010. From the definition of variables, the distance between the shoreline and the marginal line of the coastal forest corresponds to the beach width.

Although the beach was as narrow as 28 m in the vicinity of Iioka fishing port in 1947, when the natural sandy beach extended, it gradually expanded southwestward with a maximum width of 175 m on the Motosuka coast at $X = 26.4$ km. By 1970, the breakwaters of Iioka fishing port and the jetty of the Sakuta River (breakwater of Katagai fishing port) had been constructed, and the shoreline started to advance northeast of the jetty of the Sakuta River. However, as a result of the seaward advance of the marginal line of the coastal forest, which occurred simultaneously with the construction of the breakwaters, the beach was narrowed.

By 1983, the breakwater of Iioka fishing port had been extended along with the construction of seven detached breakwaters on the Yokone coast as a measure against beach erosion, and then the beach behind these detached breakwaters extended to a maximum width of 192 m at $X = 2.8$ km. In the area between the Yokone and Asahi coasts ($X = 4.5$ and 6.4 km) further downcoast of the detached breakwaters, the sandy beach was eroded away and the seawall was exposed to waves. Also, the Katagai fishing port breakwater was further extended, resulting in further sand deposition and shoreline advance upcoast of the fishing port owing to the obstruction of longshore sand transport.

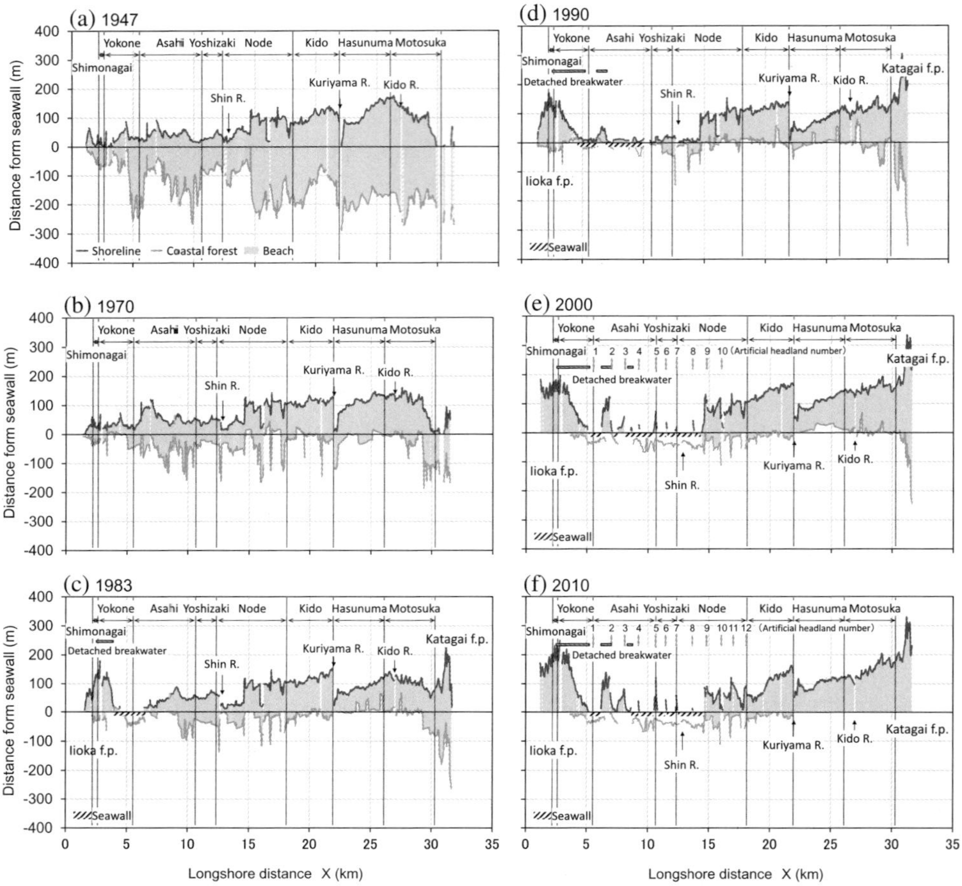

Fig. 16.7 Offshore distance to shoreline and marginal line of coastal forest from seawall.

By 1990 and 2000, the south breakwater of Iioka fishing port had been further extended along with the construction of 12 detached breakwaters and 10 artificial headlands. The beach was further widened between the Shimonagai and Yokone coasts ($X = 2.2$ and 5.3 km), whereas on the Asahi and Node coasts, the sandy beach disappeared. The area where the sandy beach disappeared was between $X = 4.7$ and 10.6 km (extension: 5.9 km) in 1990, it expanded to the area between $X = 5.5$ and 14.3 km (extension: 8.8 km) until 2000, resulting in the southwestward expansion of the area with the exposed seawall. Furthermore, the additional extension of the Katagai fishing port breakwater caused the shoreline advance on the northeast side of the fishing port.

By 2010, the number of detached breakwaters had increased to 15 between the Shimonagai and Asahi coasts along with the construction of 12 artificial headlands further downcoast. Between the Shimonagai and Yokone coasts ($X = 2.2 - 5.3$ km), the sandy beach further extended, and simultaneously, the sandy beach between the Asahi and Node coasts ($X = 5.5 - 14.6\,\text{km}$) completely disappeared, resulting in the exposure of the seawall along a length of 9.1 km, which is equal to one-third of northern Kujukuri Beach. The length at which the seawall was exposed to waves gradually increased downcoast between 1983 and 2010 at a mean velocity of 270 m/yr.

The reduction in the beach width between the shoreline and the marginal line of the coastal forest was triggered by both the shoreline recession (landward movement) and the advance of the marginal line of the coastal forest (seaward movement). Regarding these two factors related to the reduction in beach width, the coastal forest had extensively been planted after World War II to stabilize the devastated coastal land and to reduce the amount of windblown sand. However, the excess advance of the coastal forest zone up to the vicinity of the shoreline became one of the causes of the reduction in sandy beach width as a buffer zone against beach changes due to waves (Uda, 2010).

4. Field Observation

To investigate the present condition of northern Kujukuri Beach, site observation was carried out on 18–19 November 2011. Figure 16.8 shows a photograph of Byobugaura sea cliffs, which are the main source of sand supplied to Kujukuri Beach, obtained from the shoreline east of Iioka fishing port. Although Byobugaura sea cliffs have a height of 30–50 m and are composed of unconsolidated layers of sandstone and mudstone, many concrete blocks have been constructed continuously

Fig. 16.8 Photograph of Byobugaura sea cliffs obtained from shoreline east of Iioka fishing port.

Fig. 16.9 Long breakwater of Iioka fishing port, facing west from Gyobu Point located at southwest end of Byobugaura.

along the sea cliffs to prevent erosion. This induced a marked reduction in sand supply downcoast, resulting in beach erosion.

Figure 16.9 shows a photograph of the long south breakwater of Iioka fishing port obtained from Gyobu Point located at the southwest end of Byobugaura. Previously, without the breakwaters of the fishing port, westward longshore sand transport freely passed this location, but now it is obstructed by the breakwater. The formation of the sandy beach on the left (east) side of the breakwater evidences the blocking of smooth longshore sand transport. Furthermore, because the predominant wave direction is from the southeast at this location, an extensive wave-shelter zone has formed west of the south breakwater, inducing beach changes downcoast.

Although the foreshore has completely disappeared between $X = 8.6$ and 14.6 km, as shown in Fig. 16.3, and the seawall is now completely exposed to waves, the seawall has been undermined at $X = 9$ km owing to the discharge of the filler material behind the seawall and has partly collapsed, as shown in Fig. 16.10. The concrete blocks in front of the seawall have also subsided and their wave dissipation effect has been lost. Similarly, at $X = 13$ km on the Node coast, where the beach is severely eroded, many large concrete blocks have been piled to the level at which the ocean view from the shoreline has been totally lost, as shown in Fig. 16.11. Further south, the seawall and foot protection extend in a straight line for a long distance (Fig. 16.12).

The sandy beach reappears southwest of artificial headland No. 9 on the Node coast, but erosion is under way at that site. Immediately south of artificial headland No. 9, a scarp of 1.5 m height has formed, as shown in Fig. 16.13, and downcoast of this area, isolated dune vegetation consisting of an agave, which originally grew

Fig. 16.10 Seawall undermined by discharge of filler material behind seawall at $X = 9$ km on Asahi coast.

Fig. 16.11 Seawall and many concrete blocks ($X = 13$ km, Node coast, facing north).

on the sand dune, has been left on the backshore owing to erosion, as shown in Fig. 16.14.

At a location 600–850 m downcoast of artificial headland No. 11 on the Node coast, there originally was a wide sandy beach, and a playground was constructed very close to the shoreline along with a seawall to protect the facility. However, this

Fig. 16.12 Seawall and subsided concrete blocks ($X = 13$ km, Node coast, facing south).

Fig. 16.13 Scarp of 1.5 m height formed downcoast of artificial headland No. 9 on Node coast.

seawall was exposed to waves by successive beach erosion, which in turn caused the obstruction of southward longshore sand transport. As a result, a 1.8 m high scarp was formed immediately downcoast of the seawall (Fig. 16.15).

Figure 16.16 shows the damaged gently sloping revetment built at a location 19 km from the Kido coast. Almost all the concrete blocks on the slope of the revetment have subsided and it is severely damaged. Because the original elevation

Fig. 16.14 Isolated dune vegetation consisting of agave left on backshore owing to erosion.

Fig. 16.15 Beach erosion in front of playground constructed south of artificial headland No. 11.

of the crown concrete blocks was 0.2 m below the top of the revetment, where a white concrete wall can be seen in Fig. 16.16, it is clear that the concrete blocks have subsided by 0.7 m. The collapse of this gently sloping revetment is mainly due to the subsidence of the foot of the revetment associated with the erosion and the resultant discharge of the filler material under the revetment (Uda *et al.*, 2005). Thus, it is clear that the gently sloping revetment was easily damaged when the foot of the revetment was exposed to waves after the shoreline recession.

Fig. 16.16 Collapsed gently sloping revetment ($X = 19$ km, Kido coast).

Fig. 16.17 Natural sandy beach with very gentle slope of 1/50 remaining south of Kido River mouth.

At present, the Kido coast is in the state that any type of coastal structure will lose its effectiveness owing to the successive shoreline recession caused by the exhaustion of sand supply from the upcoast. However, a sandy beach with a very gentle foreshore slope of 1/50, with the appearance of the past Kujukuri Beach, remains south of the Kido River mouth, located at $X = 27$ km, in contrast to the severe erosion observed north of the Kido River mouth, as shown in Fig. 16.17.

5. Discussion

Northern Kujukuri Beach has developed over several 1000 years under the condition that a large amount of sand was continuously supplied from the sea cliffs of Byobugaura. This situation completely changed upon the construction of the breakwater along the sea cliffs to protect the tableland behind the sea cliffs in the 1970s. In addition, the breakwater of Iioka fishing port had been extended at the north end of Kujukuri Beach since 1974, obstructing longshore sand transport, which should supply sand to Kujukuri Beach. As a result, the northern beach became erosive.

The south breakwater of Iioka fishing port had been extended up to a point of 1 km offshore of the original coastline by 1987, and therefore a large wave-shelter zone was formed on the west side of the fishing port against the waves predominantly incident from the southeast, inducing longshore sand transport from the outside to the inside of the wave-shelter zone and the resultant sand deposition on the Iioka coast.

In contrast to the sand deposition on the Iioka coast, the beach on the Asahi, Yoshizaki and Node coasts, located south of the Iioka coast, eroded owing to the marked decrease in longshore sand supply from the upcoast, and the foreshore completely disappeared with the shoreline recession, exposing the seawall to waves. Because of the countermeasures to protect the shoreline against recession by using the seawall, beach erosion became severe further downcoast and the natural sandy beach rapidly disappeared.

Although various measures including the construction of artificial headlands have been carried out, it has become difficult to control beach erosion. Because Kujukuri Beach is composed of fine sand with a grain size ranging between 0.1 and 0.2 mm, and the equilibrium slope of fine sand is as gentle as 1/40, even though the artificial headland has been extended, longshore sand transport easily turns around the tip of the headland and proceeds downcoast. Thus, the construction of artificial headlands is ineffective for controlling beach erosion. On northern Kujukuri Beach, the seawall was exposed to waves along the coastline of 9.1 km length in 2010, i.e., one third of the entire length of the sandy beach. The area with the exposed seawall is still expanding southwestward. Furthermore, the gently sloping revetments already constructed have often been damaged by subsidence, and each time, recovery works to regain their original form have been carried out to maintain the present coastline. Thus, Kujukuri Beach with its natural scenic beauty has been rapidly lost.

Although the reconstruction of the damaged gently sloping revetments to recover their original shape has been carried out, the fundamental cause of beach erosion cannot be removed by such works, and a chain reaction between the damage of the revetments and the recovery recurrently occurs with the downcoast expansion of the erosion zone (Uda *et al.*, 2006). To solve the erosion problem fundamentally, the increase in sand supply is necessary. However, since the land behind the sea cliffs is also used extensively, it is impossible for the recession of Byobugaura sea cliffs to occur again. Also, it is difficult to procure a large amount of sand from other areas cost-efficiently. Because the volume of sand deposited upcoast of Iioka fishing port is small, sand bypassing using the deposited sand will be ineffective and sand back passing using the deposited sand at Katagai fishing port will have a limited effect in recovering the sandy beach because of the limited volume of sand. Thus, the natural environment of Kujukuri Beach has been rapidly lost and the artificial coast covered with the seawall and concrete blocks has been formed.

The most fundamental cause of the erosion of northern Kujukuri Beach is the disruption of the continuous sand transport, which had continued for several 1000 year, at the upcoast end of a littoral cell. When the continuous movement of longshore sand transport is disrupted, resulting in downcoast erosion, measures involving such as the construction of a seawall or detached breakwaters at the most severely eroded sites have been widely adopted in Japan. This allows further downcoast erosion and the extension of the coastline protected by hard structures. In northern Kujukuri Beach taken as an example, the erosion zone has expanded downcoast at a rate of 270 m/yr. The setting back of the coastline as a measure against beach erosion is possible in principle. However, the coastal zone is subject to the Coast Law, and that landward of the coastal protection zone is under the jurisdiction of the Forest Law, and the adjustment of the activities by two independent authorities is normally difficult and time-consuming. In fact, such a measure is difficult to adopt.

6. Conclusion

In northern Kujukuri Beach, taken as an example, shore protection has been carried out only in a narrow band along the coastline using hard structures. Also, fishing ports have been constructed independently of shore protection works without the comprehensive management of continuous sand transport in a widespread area of the littoral cell. When many people realized the difficulty in solving the erosion problem, it was too late to fundamentally resolve the issue. Thus, we cannot foresee a successful solution regarding the erosion of northern Kujukuri Beach at present. The only acceptable way considered is that we concentrate on preventing the erosion zone from expanding further downcoast, with the most severely eroded sites covered with concrete blocks left as they are.

References

Horikawa, K. and T. Sunamura (1969): Erosion of Byobugaura Sea Cliffs in Chiba Prefecture — Investigation of recession of sea cliffs using aerial photographs (2), *J. Coastal Eng.* **16**, 137–145 (in Japanese).

Horikawa, K. and T. Sunamura (1970): Erosion of Byobugaura Sea Cliffs in Chiba Prefecture — investigation of recession of sea cliffs using aerial photographs (3), *J. Coastal Eng.* **17**, 289–296 (in Japanese).

Horikawa, K. and T. Sunamura (1971): Study on direction of predominant sand transport on Kujukuri Beach in Chiba Prefecture, *J. Coastal Eng.* **18**, 417–422 (in Japanese).

Hoshigami, Y., A. Kobayashi, T. Uda and T. Kumada (2005): Recent formation and deformation of Kujukuri Beach, *J. Coastal Zone Studies* **17**(3), 47–56 (in Japanese).

San-nami, T., T. Uda, K. Furuike and Y. Hoshigami (2009): Beach erosion of south Kujukuri Beach triggered by compound anthropogenic factors, *Asian and Pacific Coasts* 2009, *Proc. 5th Int. Conf.* **4**, 252–258.

Uda, T. (1997): *Beach Erosion in Japan*: Sankaido Press, 442 p. (in Japanese).

Uda, T., O. Takada, Y. Hoshigami, M. Serizawa, T. San-nami and K. Furuike (2000): Estimation of longshore sand transport of Kujukuri Beach in geological time scale, *J. Coastal Eng.* **47**, 686–690 (in Japanese).

Uda, T., M. Serizawa, S. Seino, Y. Hoshigami, T. San-nami and K. Furuike (2005): Summary of gently sloping revetment in Japan, *Proc. Int. Conf. on Coastlines, Structures and Breakwaters*, ICE, 1–10.

Uda, T., T. San-nami and T. Ishikawa (2006): Improvement of recovery system of damaged coast necessary for preventing rapid expansion of artificial coast, *J. Coastal Eng.* **53**, 1321–1325 (in Japanese).

Uda, T., K. Sakai, S. Seino and Y. Hoshigami (2010): Formation of sandy beach and environmental change on upcoast of breakwater associated with extension of south breakwater of Iioka fishing port, *Proc. Civil Eng. in the Ocean, JSCE*, **26**, 1071–1076 (in Japanese).

Uda, T. (2010): *Japan's Beach Erosion — Reality and Future Measures*: World Scientific, 418 pp.

Chapter 17

View on Future Prospects of Coastal Engineering

Kiyoshi Horikawa

Member, The Japan Academy
Professor Emeritus, The University of Tokyo
Professor Emeritus, Saitama University
Professor Emeritus, Tokyo City University
(formerly Musashi Institute of Technology), Japan

In this chapter, the author's view on future prospect of coastal engineering discipline is described on the basis of his experience during the past 60 years. The interest of coastal engineers has dramatically changed in response to the social need. That is to say, prevention of coastal disaster has been one of the most important subjects from the beginning of coastal engineering activity to the present. The succeeding subjects were coastal development, coastal pollution, coastal environment in the order corresponding with the social demands. Nowadays, global environment such as sea level rise is seriously concerned to be solved internationally. Considering the dramatic change of social needs, the author extends his discussion on the following subjects as future prospect. These are (1) necessity of further development of coastal engineering discipline, (2) establishment of coastal environment engineering, and (3) international exchange and corporation.

1. Introduction

1.1 *Role of coastal engineering*

During the past 60 years, the art of coastal engineering has been expanding very much to perform its important role in preserving and developing coastal region for human activities. Establishment of the coastal engineering discipline was declared by O'Brien (1950) at the First Conference on Coastal Engineering which was held at Long Beach, California in the United States of America. His primary intention was to establish a new engineering field related coastal treatments with a combined engineering/science approaches on the favorable occasion of secret release of the ocean wave forecasting technique developed by Sverdrup *et al.* (1942).

On the other hand, Civil Engineers in Japan encountered the severe storm surge damage that never happened in the past on the coast of Ise Bay by the attack of Typhoon No. 13 in 1953. This unfortunate accident became a strong motivation to initiate coastal protection activities in Japan. Hence the initial concerns of coastal engineers in Japan were: How to protect coastal region against natural disaster, with the purpose of guarding human lives first and then of preserving property?

Going with advancement of coastal utility, the importance of disaster prevention has highly been recognized by the people, thus a lot of public works have continuously been accomplished all over the world. In order to establish these works successfully, countless coastal engineering research works have actively been carried out in various countries. Therefore, we have exchanged valuable information on the achievement of these investigations through domestic as well as international conferences and numerous publications. As a result, the discipline of coastal engineering has remarkably been developed during the past 60 years as engineering science and technology. It is really true that such advancement in this field was based on not only the world wide social needs but also the rapid expansion of human knowledge.

1.2 *Development of coastal engineering discipline*

In parallel to the social and economical development of the nations, the needs of people diversify from time to time, hence research subjects to be treated have been expanding so widely that the interest of coastal engineers has a lot of variety at present. In addition to the above fact, technological advancement in various fields, such as measuring techniques and computer technology, makes the quality of research output extremely high and reliable. In spite of such advancement, our knowledge on natural phenomena is still inadequate to solve numerous problems appearing in our practical cases. Hence we have to devote continuously our effort to clarify the mechanism of coastal phenomena.

Considering the building process of coastal engineering as a new engineering discipline, we can realize that close cooperation with numerous specialists in different form but related fields to coastal engineering, should contribute very much for the advancement of coastal engineering by ways of exchanging views and seeking together the most favorable solution for our difficulties. Coastal engineers therefore accept, without prejudice, the comments raised by the specialists in the different disciplines, such as oceanography, particularly nearshore oceanography, meteorology, seismology, geology, geography, geomorphology, geotechnical engineering, sanitary engineering and earthquake engineering. However, the stated effort is still inadequate to match with the actual social needs.

1.3 *Necessity of global consideration*

Since 1970s, we have truly realized that environmental aspect became a crucial and essential factor for coastal development. Due to the lack of prudence, coastal waters were once seriously contaminated mainly by untreated waste water discharged from industrial plants and household. Since then, we have deeply realized that the seas have a limited capacity for purifying waste water, and we are at present quite sensitive to the influence of pollutant on not only marine growth but also wild life in general. Coastal engineers have therefore come to take into account biological or ecological aspect as an important element to design and proceed toward future coastal works in order to maintain favorable or acceptable environment. Importance of mitigation concept has gradually been recognized in Japan as well. Nowadays, cooperation between coastal engineers and fishery scientists or ecologists is hence a common and indispensable feature in coastal works.

Oil pollution occurs occasionally causing a serious effect on fishery, tourism, and other coastal activities in a wide area. From this fact it is awfully needed to extend our environment concept or concerns from the local scale to global one.

1.4 *Overview and scope*

In this chapter, the following subjects will be discussed as essential parameters for seeking future prospect of coastal engineering. These are (a) necessity for further development of coastal engineering discipline, (b) establishments of coastal environment engineering, and (c) international exchange and cooperation.

2. Necessity for Further Development of Coastal Engineering Discipline

2.1 *Insight into natural phenomena*

Recalling my real situation in the olden days when I just started my research activities in the field of coastal engineering at the University of Tokyo in 1954, I was suffering from shortage of research fund, reference materials, laboratory facilities, measuring instruments and others as usually at the most of universities in Japan. Starting from such unfavorable circumstances, I devoted my great effort with my colleagues to understand the fundamental mechanism of natural phenomena in the nearshore area. Based on the result of frequent exchange of knowledge through domestic as well as international conferences, we have formulated the scientific and technological system of coastal engineering as an independent engineering discipline.

It is really true that remarkable progress has been accomplished in coastal engineering during the past 60 years by the joint effort of researchers and practice engineers. The results of technological advancement in various fields have been positively introduced in numerous phases of field investigations and research activities. Introduction of computer technology is just one example. Our knowledge on the subject becomes, as a result of stated effort, extremely abundant and fairly qualified to understand the mechanism of nearshore phenomena. With the increase of our knowledge, the range of subjects which should be clarified is extending so rapidly that additional information on the related subjects is needed. In order to get a clear picture of our insight for producing favorable coastal environment, it is essentially requested to simulate real coastal phenomena as much as possible in the sense of analytical and experimental treatments. In this case, numerical simulation technique should be a powerful means or in many cases a unique procedure to find out our feasible solution.

2.2 *Comprehensive field observations*

In the period of the mid-1970s to the mid-1980s, several attempts were made in various countries to measure numerous natural physical parameters involved in the coastal phenomena, such as wave characteristics, nearshore currents, coastal sediment transport, and nearshore bottom configuration. Typical examples were the NSTS (Nearshore Sediment Transport Study) program directed by Seymour in the U.S.A. and the NERC (Nearshore Environment Research Center) program directed by Horikawa in Japan.

The former NSTS program was officially accepted by the National Science Foundation as a Sea Grant Program in 1977 and was continued until 1982. This was to be a field-oriented program, and the results of this program were compiled in the book of "Nearshore Sediment Transport" edited by Seymour (1989).

On the other hand, the latter NERC program took place in the period of 1978 to 1984. The final target of this program was to develop feasible predictive models of beach evolution in order to respond to the urgent need from the electric power industry in Japan. The important outcome was compiled first in the Japanese edition entitled "Coastal Environment Engineering " under Horikawa's editorship and published in 1985 through the University of Tokyo Press. Later on, the book of "Nearshore Dynamics and Coastal Processes — Theory, Measurement, and Predictive Models" was edited by Horikawa (1998) as an entirely revised English version of the former Japanese edition.

In addition to these two representative programs, two other similar programs were, as far as I know, organized almost in the same period. One was conducted in Canada (Willis, 1987) and the other in the US (Mason *et al.*, 1987). It is a truly interesting fact that number of programs were organized independently in different countries almost in the same period and that in every programs field observations were set up as an essential part of these investigations. Tremendous amount of field data were accumulated at different locations under varied physical conditions by each group, and these data played an important role in grasping the nature of nearshore phenomena. At least in the NERC program, the obtained field data were extremely valuable in setting up and verifying the proposed simulation models of beach evolution.

Even if the above data were truly valuable for clarifying the characteristics of natural phenomena in the neareshore area, our understanding is still incomplete and far from our ideal. Therefore, it would be the time to organize new comprehensive and cooperative programs at various locations or in various countries with the aim to learn much more from nature. The reason is that we are able, at present, to apply newly advanced technology for field observations. It is true that the work to organize, establish and accomplish such a program might be really time consuming and laborious, however the reward gained from such effort should be splendid for educating young scholars and extending their abilities. Difficulties to be overcome are how to search out a competent project leader and how to get research fund for the intended program. Nowadays younger scholars are likely fond of doing research works in office by using an electronic computer rather than in laboratory and/or in field. However, they should always recall that the modeling of nature will not be possible without any related data of natural phenomena and sure insight gotten out through observations.

2.3 *Necessity of cooperation among researchers in related but different fields*

Coastal engineering itself is a kind of synthetic engineering science from the beginning of its history. This discipline has been established in combining numerous science and engineering fields which are closely related to coastal features. In recent years, we have realized again the importance of cooperation with other fields such as earthquake engineering, geotechnical engineering, sanitary engineering, and landscape planning. It is needless to say that each researcher has his/her own specialty. In order to get his/her cooperation, strong motivation should be offered beforehand to him/her. Therefore, generally speaking, it is a difficult task to establish real cooperation among researchers in related but different fields. Even if it is so, such effort might be effective to break through barriers existing at present in practical problems.

The following is a typical case of cooperative investigation work in Japan. As you know, Japan is one of the island countries, where tectonic movement is extremely active and a number of earthquakes occur annually. Tsunamis are hence generated rather frequently by submarine earthquakes and attack the coastal region in producing tremendous disasters on human lives and properties.

Nowadays Japanese Government has a strong intention to provide a useful guideline for assisting local governments to establish their own plans of preventing tsunami disasters. Thus Japanese Government organized an investigation committee in the period of 1994 to 1997. The importance of local as well as national governments' functions was further recognized after the 2011 Great Tohoku Earthquake and tsunami in Japan. The committee members consisted of scientists and engineers with different academic background. Their specialties are coastal engineering including tsunami engineering, seismology and regional planning.

Recent advancement in seismology is so remarkable that generation mechanism of submarine earthquake has been clarified fairly well. One of the achievements in this field is the fault model of submarine earthquakes. Various physical quantities of the proposed fault models are closely related to the magnitude of earthquake. Therefore if the expected magnitude of future earthquake were determined in some ways, the physical characteristics of corresponding fault model will be determined rationally.

Based on the stated fault model, we can continue with numerical calculation to get the temporal and spatial distribution of water surface elevation of generated tsunamis. The reliability of numerical calculation is largely dependent on the applied governing equations, linear or nonlinear, the mesh size of numerical calculation, and the boundary conditions. Therefore, it is quite necessary to evaluate seriously the limitation of the applied computation technique for various cases beforehand. In order to do so, verification analysis should be carried out by using the record of the past known submarine earthquakes. The computed results of tsunami height at grid points, mostly adjacent to shoreline, are used to be compared with the corresponding tsunami run-up height traced after the questioned tsunami attack. The overall agreement of these stated values seems to be good in general, hence the application of the numerical model technique can offer very valuable information for establishing tsunami disaster prevention plan.

Comparison between the computed tsunami height and the crown height of existing coastal structures such as seawalls and revetments is helpful to evaluate the function or effectiveness of present structures for tsunami protection. Additional calculation can be made by using nonlinear governing equation and finer meshes to estimate the tsunami run-up height more precisely. Such results can be used to establish evacuation plan for dwellers and to make future prevention plan against tsunamis.

The above is only one example of cooperation works, although it is a remaining task as well as others to establish the fault models which produce giant tsunamis such as the 2004 Indian Ocean Tsunami and the 2011 Tohoku Earthquake Tsunami. Cooperation between coastal engineering scientists and researchers in the various related fields should be promoted more actively to solve our numerous practical problems.

3. Establishment of Coastal Environment Engineering

3.1 *Improvement of coastal engineering scope*

In the word of "environment", an individual keeps his/her own image which may have broad variety. However at the present time, everybody agrees that environmental situation should be kept to be natural as much as possible for not only human being but also wildlife. Before reaching the above widely accepted concept, we had spent once a long and painful period of nearshore pollution caused by thoughtless disposal of sewerage from industrial plants and domestic houses.

In addition to the above pollution problems, we have induced a great environmental impact on coast, among which beach erosion must be the most serious problem during the past 60 years. Particularly after the Second World War, we have earned tremendous amount of construction power in engineering practices. Keeping step with the advancement of high technology, a lot of construction works were conducted in Japan with the purpose of achieving economical and social development.

These works include the projects of dam construction for electric power generation as well as water resources development. The purposes of these projects were fulfilled successfully, but unexpected negative effect appeared gradually on river structures such as embankments and intake structures. The main reason is as follows. Most of the sediment transported through the upstream of the river deposited at the man-made reservoir, hence the sediment flow through the river is cut down at the damsite. Finally the sediment supply to the coast decreased very much causing severe beach erosion. In addition to the above fact, construction of coastal structures such as breakwaters, seawalls, and jetties is used to distract the balance of sediment budget along the coast. Beach erosion appears hence in some places and beach deposition happens in other places.

In most cases, beach erosion is irreversible. Various measures have thus been proposed frequently to recover its former natural environment as much as possible. However, this task, in general, is very difficult to be accomplished.

It has been recognized in these days that the subjects in the sense of coastal environment should cover much more broader area beyond the problems of pollution and beach erosion. A lot of people are likely to enjoy relaxed time by observing the behavior by feeding the wildlife such as migratory birds at tidal wetland. Earlier shallow water areas were considered as favorable sites for land reclamation, their tidal wetland decreased its area very rapidly during the past 40 years in Japan. The concept of mitigation has widely been accepted first in the US and then in Japan, rather very recently. The principal idea of this concept is as follows. When a governmental agency or a developer has an intention to proceed with a particular project for development purposes, the organization has to propose before-hand an additional program at another site in order to replace the natural conditions which may be lost by the stated development project. We really need to maintain the concept hereafter, otherwise we may not be able to slow-down the destruction rate of favorable environment for not only the wildlife, but also human being.

3.2 *Induction of new advancement of related sciences*

Our daily life receives a lot of favor from modern science and technology. One example in our field is the meteorological satellite through which we can observe with strong interest numerous meteorological phenomena, such as migration of a typhoon cell, in a global scale. Hence we could open our eyes to look at and think about the environment surrounding us. In the same ways, coastal engineers are stretching their arms to maintain closer cooperation with specialists in the related fields of science as mentioned previously in the subject of the future tsunami prediction.

3.3 *Coastal environment engineering*

In recent years in Japan, young scholars in the field of coastal engineering have initiated their trial approach to establish a new discipline for saving coastal environment engineering, covering and extending the present coastal engineering. As far as I know, there were two attempts on this subject. The first one was conducted by the group presided by Isobe (1994). Twenty-two members came together from various organizations such as universities, government research institutes, and private sectors. The result of their discussion was compiled in a comprehensive monograph entitled "Creation of Coastal Environment — An Introduction to Waterfront Learning." The other attempt was made by the subcommittee members in the Committee on Coastal Engineering, Japan Society of Civil Engineers. Their discussion covers the following three themes: (1) wave dynamics in coastal region; (2) effect of global warming on coastal environment; and (3) coastal sediment environment. The results of the discussion were printed as a style of committee report from JSCE (1994a; 1994b; 1998). It is unfortunate that these publications were written in Japanese only. This kind of effort should be continued to facilitate the establishment of a new and attractive discipline such as coastal environment engineering.

In Japan the Sea Coast Act was issued in 1956 as a result of strong impact of the storm surge disaster caused by the Typhoon No. 13 in 1953 for the purpose of coastal management, particularly preservation of coastal region. The Act has fulfilled its aims effectively for the subsequent years. However, the region treated in the Act is quite limited in the coastal zone with the width of several hundred meters depending on the slope of beach face. At present, the UN Law of the Sea is now in effect by the nations in the world and the territorial seas have extended to 12 nautical miles from the coastline. Considering the stated worldwide tendency we should cover hence the coastal region for management much more broadly than the present one, as similar to the Coastal Zone Management Act in the US.

4. International Exchange and Cooperation

Natural phenomena in the nearshore area are truly complex and are normally far beyond the scope of human knowledge. Hence we need a lot of reliable data and related information in order to clarify those phenomena more precisely and to solve our practical problems. From such a view point, international as well as domestic conferences have taken an important role in exchanging data and idea.

It is well-known that O'Brien initiated Coastal Engineering Conference in 1950 with Johnson having foreseeing intelligence. From inauguration of this engineering discipline, this is a kind of interdisciplinary field covering not only engineering sciences but also related natural and social sciences. The above conference was at the initial stage an entirely domestic one in the US, but became an international one since the 5th Conference in 1955 which was held in Grenoble, France. Then the stated international conference has been organized biennially in a different country selected in considering the geographical distribution of the site in the world. The conference gave a strong influence on strengthening the motivation of coastal engineering activities in the host country like Japan in 1966.

On the other hand, serious storm surge disaster happened along Ise Bay coast as the effect of Typhoon No. 13 in 1953 as mentioned previously. This was the moment when coastal engineering in Japan started its activity. As a result of long and heated discussion in the National Diet, the Sea Coast Act was finally established in 1956. The aim of the stated act was primarily concerned with preservation of the nation land against future coastal disasters caused by strong waves, storm surges, tsunamis and beach erosion. In parallel to the nation action, Committee on Coastal Engineering was organized in 1954 by Hom-ma in the Japan Society of Civil Engineers as a technical committee to intensify the research ability of Japanese scientists and engineers. Therefore, the main activity of this committee has been to organize the annual conference on coastal engineering. The domestic conferences have taken a very important role to cultivate men and women of ability, rather in a short period.

The subjects treated by coastal engineers are normally restricted to those in local area, however the results of these investigations supply valuable information to us as case studies. The interest of coastal engineering scientists is expanding very rapidly corresponding to the social needs. During the past 60 years, science and technology in various fields have advanced in an enormous scale. We have to thank the glorious development of space technology to realize clearly that the earth is finite in the sense of physical scale. It is our common understanding that social activity in one country gives regretful influence on environment in the neighboring countries in numerous ways. Therefore, cooperation and assistance among all countries in the world, whether developed or developing, are of essential importance to let the people on the globe enjoy their sustainable development in their social life. A typical example of our common interest is now the mean sea level rise which may be caused by increase of the amount of green house effect gases diffused from the earth as a result of human activities. In order to detect actual sea level rise and to reduce the emission of greenhouse effect gases, international cooperation must be the remaining unique way. Hence, we have to devote much more effort to make an international forum in order to disseminate the achievements made by coastal engineering scientists toward the people in the world.

Acknowledgments

The author would like to express his sincere appreciation to Dr. Masahiko Isobe, Professor at the University of Tokyo, for kind and effective assistance in preparing the manuscript. The author is also indebted to Ms. Chieko Okano for her great effort in typing the manuscript, repeatedly.

References

Committee on Coastal Engineering (1994a): Nearshore Wave Dynamics, JSCE (in Japanese).

Committee on Coastal Engineering (1994b): Effect of Global Warming on Coastal Environment, JSCE (in Japanese).

Committee on Coastal Engineering (1998): Creation of Coastal Sediment Environment, JSCE (in Japanese).

Horikawa, K., ed. (1988): *Nearshore Dynamics and Coastal Processes — Theory, Measurement, and Predicted Models*: University of Tokyo Press.

Isobe, M., ed. (1994): *Creation of Coastal Environment — An Introduction of Waterfront Learning*: Asakura Publishing, Tokyo (in Japanese).

Mason, C., W. A. Birkemeir and P. A. Howd (1987): Overview of DUCK85 nearshore processes experiment, *Coastal Sediment* '87, ASCE, 818–833.

O'Brien, M. P. (1950): *Preface, Proceedings of the first Conference on Coastal Engineering*: University of California.

Seymour, R. J., ed. (1989): *Nearshore Sediment Transport*: Plenum Press, New York.

Sverdrup, H. U. and W. H. Munk (1947): Wind, sea and swell: Theory of relations for forecasting, U. S. Nary Hydrographic Office, H. O. Pub. 601.

Willis, D. H. (1987): The Canadian coastal sediment study: An overview, *Coastal Sediment* '87, ASCE, 682–693.